Studies in Computational Intelligence

Volume 1194

The series "Studies in Computational Intelligence" (SCI) publishes new developments and advances in the various areas of computational intelligence—quickly and with a high quality. The intent is to cover the theory, applications, and design methods of computational intelligence, as embedded in the fields of engineering, computer science, physics and life sciences, as well as the methodologies behind them. The series contains monographs, lecture notes and edited volumes in computational intelligence spanning the areas of neural networks, connectionist systems, genetic algorithms, evolutionary computation, artificial intelligence, cellular automata, self-organizing systems, soft computing, fuzzy systems, and hybrid intelligent systems. Of particular value to both the contributors and the readership are the short publication timeframe and the world-wide distribution, which enable both wide and rapid dissemination of research output.

Indexed by SCOPUS, DBLP, WTI AG (Switzerland), zbMATH, SCImago.

All books published in the series are submitted for consideration in Web of Science.

Georgios Lampropoulos · Stamatios Papadakis
Editors

Social Robots in Education

How to Effectively Introduce Social Robots into Classrooms

Springer

Editors
Georgios Lampropoulos
Department of Applied Informatics
School of Information Sciences
University of Macedonia
Thessaloniki, Greece

Department of Education
School of Education
University of Nicosia
Nicosia, Cyprus

Department of Preschool Education
School of Education
University of Crete
Rethymno, Greece

Stamatios Papadakis
Department of Preschool Education
School of Education
University of Crete
Rethymno, Greece

ISSN 1860-949X ISSN 1860-9503 (electronic)
Studies in Computational Intelligence
ISBN 978-3-031-82914-7 ISBN 978-3-031-82915-4 (eBook)
https://doi.org/10.1007/978-3-031-82915-4

This Springer imprint is published by the registered company Springer Nature Switzerland AG
The registered company address is: Gewerbestrasse 11, 6330 Cham, Switzerland

Preface

In recent years, the integration of social robots into educational environments has emerged as one of the most exciting and transformative developments in the field of education. These technologies not only personalize the learning experience but also offer innovative ways to support teachers in fostering motivation, collaboration, and creativity among learners. They are reshaping classrooms by offering new methods to engage students, personalize instructions, and support learners with diverse needs. In light of this growing trend, we present *Social Robotics in Education: How to Effectively Introduce Social Robots into Classrooms,* a compilation of cutting-edge research and practices aimed at understanding and advancing the role of social robots in education. This book is designed to serve as a comprehensive guide for educators, policymakers, and researchers eager to explore the transformative potential of social robots in educational contexts.

In this volume, leading scholars and practitioners present in-depth research and practical applications across three key thematic areas that reflect the broad spectrum of the application of social robots in education, addressing the related implications, potentials, and challenges. Each of the thematic areas offer valuable insights into different facets of social robotics in education. Specifically, the thematic areas are: (i) *Social Robots and Artificial Intelligence in Education,* (ii) *Social Robots in Special Education and Ethical Considerations,* as well as (iii)*Practical Applications of Social Robots.*

The first thematic area, *Social Robots and Artificial Intelligence in Education,* delves into the educational value of AI and social robots, exploring how these technologies can create personalized learning experiences by integrating both the emotional and cognitive aspects of learning. Chapters in this section also highlight student and teacher perceptions of social robots in real-world classrooms, underscoring the potential for social robots to act as intelligent tutors that adapt to individual learners' needs. This area presents a future in which social robots work collaboratively with educators to improve learning outcomes and engagement. Specifically, in Chapter "The Educational Value of Artificial Intelligence and Social Robots", Lampropoulos and Papadakis explore the educational value of social robots and artificial intelligence. Focusing on enabling adaptive and autonomous social robots in education, Verhelst et al. in Chapter "Enabling Autonomous and Adaptive Social Robots in Education: A Vision for the Application of Generative AI" put emphasis on the application of generative artificial intelligence. In Chapter "Interactions in the Wild: Student and Teacher Perceptions of Social Robots in Public Classrooms", LeTendre et al. explore teachers' and students' perceptions regarding the use of social robots in classrooms. Aiming at exploring the next generation of personalized educational artificial intelligence, Lazic et al. focus on the integration of cognition and emotion (Chapter "The Next Generation of Personalized Educational AI: Integrating Emotion and Cognition"). In Chapter "Teaching with AI: The Implications and Challenges", Shen looked into the challenges and implications of teaching using artificial intelligence. Balzotti et al. present the educational perspectives of students regarding the implementation of artificial intelligence in social robots focusing on ChatGPT (Chapter "School Children Theories on ChatGPT: Educational Perspectives of the Implementation of Artificial Intelligence in Social Robots"). Finally, in Chapter "Integrating Arduino and Artificial Intelligence in STEM Education", Bedoya-Cano et al. explore the integration of artificial intelligence and Arduino in STEM education.

The second thematic area, *Social Robots in Special Education and Ethical Considerations*, examines the ethical challenges of designing and integrating social robots in education, particularly in sensitive environments like special education. Here, readers will find insightful discussions on the importance of ethical design, child-robot and human-robot interactions, and the potential of social robots to support all learners. These chapters provide a nuanced understanding of how social robots can be inclusively integrated into classrooms, with a focus on acceptance, usability, and the ethical dimensions of their deployment. Specifically, in Chapter "Ethical Considerations in Designing and Testing Robots for Children in Educational Settings", Arriaga et al. examine the ethical considerations with the design and testing of robots for children within classrooms. In Chapter "The Use of Social Robots in Educational Settings: Acceptance and Usability", Conti et al. explore the usability and acceptability of integrating social robots in educational settings. Moreover, Cersosimo examines the role of social robots as tutees within inclusive learning environments (Chapter "To Teach is to Learn Twice": Using Social Robots as Tutees in Inclusive Education"). Kasuk et al. present the outcomes of a pilot study focusing on providing pedagogical guidelines for telepresence robots (Chapter "Pedagogical Guidelines for Telepresence Robots Users: A Pilot"). In Chapter "Exploring Nao's Role in Enhancing Motor Activity for Young Children with ASD in Inclusive Classrooms", Perla et al. examine young children with ASD motor activity through their interaction with Nao robot within inclusive classrooms. Finally, Lorenzo-Lledó et al. explore the role of emerging technologies, including robots and artificial intelligence, in the context of special education through a bibliometric analysis (Chapter "The Application of Emerging Technologies in Autistic Students During the Period 1996-2023: A Conceptual, Social and Intellectual Review").

The third thematic area, *Practical Applications of Social Robots*, offers a hands-on look at how social robots are being implemented across diverse educational settings. This part of the book showcases meta-analyses, case studies, and practical examples of how social robots, conversational artificial intelligence tutors, educational robotics, and entrepreneurship-based robotics are being used in classrooms around the world. By offering practical guidelines and evidence-based strategies, this section serves as a roadmap for educators and policymakers interested in incorporating social robots and artificial intelligence into their educational practices. Specifically, Gu presents the outcomes of a meta-analysis study that focuses on the efficacy of using conversational artificial intelligence in language learning (Chapter "A Meta-Analysis of the Efficacy of Conversational AI Use in EFL Learning Contexts"). Zviel-Girshin and Rosenberg showcase and discuss the concept of ambient anthropomorphic social educational robotics (Chapter "Ambient Anthropomorphic Social Educational Robotics"). Furthermore, in Chapter "Educational Robots: Past, Present, Future", Catlin et al. look at the past, present, and future of educational robots. In Chapter "Theoretical and Practical Coherence of Integrated STEM Education and Educational Robotics: Review and Analysis of Good Practices in Europe", Merino-Fernández et al. analyze and suggest best practices in integrating STEM education and educational robots in the context of European education. Karakose and Polat present the outcomes of a bibliometric study exploring the use of chatbots on students' learning outcomes (Chapter "Science Mapping of the Knowledge Base on the Effects of Artificial Intelligence (AI)-Based Chatbots on Student Learning Outcomes: A Bibliometric Study"). Finally, in Chapter "The Integration of Entrepreneurship-Based Robotics into Classrooms", Meral and Yalçın focus on entrepreneurship-based robotics and their use in educational settings.

By covering both theoretical and practical aspects, this book provides a balanced approach to understanding how social robots can be effectively integrated into educational environments. The chapters in this volume not only address the technical and pedagogical aspects of social robots but also reflect on the ethical challenges and the real-world efficacy of social robots and artificial intelligence. The chapters offer a roadmap for future developments in this field.

As editors, we are excited to present this book, which not only highlights the theoretical potential of social robots in education but also offers practical guidance for education stakeholders that are eager to implement these technologies in their classrooms. We hope this book will serve as a valuable resource for anyone interested in understanding and shaping the future of education in the digital age. We would like to express our heartfelt thanks to the contributors for their invaluable perspectives and commitment to this endeavor. Their combined expertise has resulted in a book that is both timely and essential, providing support for educators, researchers, and policymakers as they address the challenges of twenty-first-century education.

In conclusion, as we are at the intersection of technology and education, it is imperative that we embrace the potential of social robots and artificial intelligence. By doing so, we not only enhance learning outcomes but also prepare our students to navigate a future where technology plays an integral part in everyday life. We hope this book helps to envision and embrace a future in which technology and humans complement each other in supporting the growth and development of students.

Georgios Lampropoulos
Department of Applied Informatics
School of Information Sciences
University of Macedonia
Thessaloniki, Greece

Department of Education
School of Education
University of Nicosia
Nicosia, Cyprus

Department of Preschool Education
School of Education
University of Crete
Rethymno, Greece

Stamatios Papadakis
Department of Preschool Education
School of Education
University of Crete
Rethymno, Greece

Contents

Social Robots and Artificial Intelligence in Education

The Educational Value of Artificial Intelligence and Social Robots

Georgios Lampropoulos and Stamatios Papadakis

Abstract Social robots are embodied agents and social entities that follow an anthropomorphic design have social forms and functions and can engage in social interactions with humans and other robots in an interpersonal manner as well as understand and even relate to them. Additionally, social robots are regarded as affective entities due to their ability to understand and express changes in emotions and moods and their socio-cognitive skills and social-emotional intelligence. Social robots are closely related to the field of artificial intelligence. As a result, social robots are being applied in different domains including education. This chapter aims to explore the use of social robots and artificial intelligence in education to identify their educational values and merits. The outcomes of this study highlighted the role of intelligent social robots as meaningful and effective educational tools that can be used in various educational settings and contexts in both formal and informal settings to support both teachers and students. Their ability to act as intelligent human-like tutors to offer personalized learning experiences that take students' unique characteristics, needs, knowledge, and preferences into account emerged. As a result, social robots have the potential to improve students' learning outcomes, engagement, and motivation, improve their academic performance, and increase their social skills. Nonetheless, the study also highlighted the need for more experimental studies to be conducted, for appropriate standards and guidelines to be developed, and for security, ethical, and moral issues relevant to the use of social robots to be further examined.

Keywords Artificial intelligence · AI · Social robots · Educational robotics · Intelligent tutoring systems · Education · Personalized learning · Adaptive learning · Review

G. Lampropoulos (✉)
Department of Applied Informatics, University of Macedonia, Thessaloniki, Greece
e-mail: glampropoulos@uom.edu.gr

Department of Education, University of Nicosia, Nicosia, Cyprus

G. Lampropoulos · S. Papadakis
Department of Preschool Education, University of Crete, Rethymno, Greece
e-mail: stpapadakis@uoc.gr

G. Lampropoulos and S. Papadakis (eds.), *Social Robots in Education*, Studies in Computational Intelligence 1194, https://doi.org/10.1007/978-3-031-82915-4_1

1 Introduction

Educational robotics refers to specialized robots that are designed to improve the educational processes by supporting teaching and learning activities (Angel-Fernandez & Vincze, 2018; Eguchi, 2010; Scaradozzi et al., 2019). This interdisciplinary field is rapidly advancing as educational robotics are more widely adopted in classrooms (López-Belmonte et al., 2021). Specifically, educational robotics can help meet students' new educational needs (Benitti, 2012) and improve teaching and learning by offering personalized learning opportunities and activities that promote active learning (Kolb, 2014; Zhang et al., 2020).

The application of information and communication technology (ICT) can bring about several changes to the educational process (Papert, 1984). Educational robotics is no exception to that as they can enrich and transform teaching and learning activities across educational levels and age groups and in various learning settings while simultaneously improving learning outcomes (Evripidou et al., 2020; Gunes & Kucuk, 2022; Ospennikova et al., 2015).

Additionally, the adoption of educational robotics in classrooms can improve students' engagement, offer students individualized feedback, and enhance their hard and soft skills which are necessary in the context of the twenty-first century (Chevalier et al., 2022; Eguchi, 2014, 2015). As learning activities that involve the use of educational robotics can promote collaborative learning, students' social skills and interactions can be improved and their ability to collaborate and communicate with others can be increased which, in turn, can result in the development of more inclusive learning environments (Daniela & Lytras, 2019; Denis & Hubert, 2001; Screpanti et al., 2021). Despite the fact that educational robotics is mainly associated with engineering and computer science courses and subjects, studies have reported their positive impact on student's learning outcomes, motivation, and engagement in various subjects (Anwar et al., 2019; Benitti, 2012; Gunes & Kucuk, 2022). This fact highlights the wide applicability of educational robotics, its diverse nature, and its capabilities to enrich the educational process. However, to fully reap the benefits of educational robotics, there is a clear need to develop teachers' knowledge and skills through the development and application of suitable training programs so that they can effective plan their lessons, develop educational material and activities, and integrate educational robotics in their classrooms.

Moreover, despite the benefits that educational robotics can bring to education, there are still issues and open challenges that need to be addressed in the general field of human-robot interaction (Sheridan, 2016). As there are different types of educational robotics that can differently impact the educational process (Pedersen et al., 2020), it is important to consider students' characteristics, needs, knowledge, and preferences as well as the overall educational goals when selecting and implementing activities based on educational robotics (Atmatzidou & Demetriadis, 2016; Virnes et al., 2008).

As students are seeking for more personalized and interactive learning experiences and to be actively involved in the educational process in a meaningful way

(Lampropoulos & Kinshuk, 2024), a new need for educational robots to undertake a more socially responsive role (Dautenhahn, 2007) emerged. Hence, the research focus has shifted toward social robots. Specifically, although social robots are being examined from different perspectives (Mejia & Kajikawa, 2017), and the interest in the topic increases (Leite et al., 2013), emphasis is being placed on their role in education (Belpaeme et al., 2018). Simultaneously, artificial intelligence, which is a key component of social robots, is also rapidly progressing. Artificial intelligence can affect various parts of daily life and can also support and enrich the educational process (Chen et al., 2020a, 2020b). Social robots constitute an innovative field in which the integration of artificial intelligence is necessary for it to reach its full potential.

Given the significance of both social robots and artificial intelligence and their potential to enrich and improve the educational process as well as promoting personalized learning, it is imperative to further examine their role and use in classrooms. In this view, this chapter aims to offer an overview regarding the adoption and integration of social robots and artificial intelligence in teaching and learning activities and to examine their educational value. Specifically, the chapter defines social robots and presents their use in educational settings (Sect. 2) and goes over the concept of artificial intelligence, its recent advances, and its use in education (Sect. 3). The chapter also focuses on the importance of personalized and adaptive learning and how social robots and artificial intelligence can be used to promote it (Sect. 4). Finally, the chapter provides conclusive statements regarding their adoption and use in teaching and learning, comments upon their educational value, and recommends directions for future research to focus on (Sect. 5).

2 Social Robots in Education

Due to the advances in the field of artificial intelligence, its use in various domains has also increased. The field of robotics is no exception to that. Specifically, the use of artificial intelligence to further enrich educational robotics is becoming more prominent (Chen et al., 2020a, 2020b). Specifically, capitalizing on artificial intelligence and human-robot interaction to create robots that can engage on a social level by interacting and communicating with humans or other agents and robots, the field of social robots is gaining ground. Due to their nature and capabilities, social robots are increasingly being used in different domains including education (Leite et al., 2013). There are several different types of social robots that can each affect teaching and learning in a different way; hence, it is important to carefully consider which social robots are more suitable for each use case (Mahdi et al., 2022).

Social robots are robots that utilize a social interface and are perceived as social entities due to their social form, functions, and context as well as their anthropomorphism (Hegel et al., 2009). Understanding these aspects of social robots will enhance the intensity of social attributions toward them (Hegel et al., 2009). Additionally, social robots are regarded as embodied agents that are members of a heterogeneous group consisting of both robots and humans and are capable of engaging in social interactions, communicating, and learning from one another (Fong et al., 2003). Hence, they are capable of communicating, understanding, and even relate to humans (Breazeal, 2004).

As social robots are designed to communicate and interact with humans and other robots in an interpersonal manner, their development follows an anthropomorphic design that focuses on their embodiment, social-emotional intelligence, and socio-cognitive skills (Breazeal et al., 2016). Thus, social robots are characterized by their intelligence, communication and comprehension skills, relatedness, high realism, as well as their social evocation and receptiveness, social interface, and sociability (Breazeal, 2003). As social robots constitute affective entities, they can display and express changes in emotions and moods (Kirby et al., 2010). Hence, it is important to effectively evaluate their engagement and interactions with students (Anzalone et al., 2015).

Social robots can be used in education as tutors or peer learners and due to their physical presence, which is something that other traditional learning technologies lack, it was revealed that they can effectively improve affective and cognitive outcomes and achieve learning outcomes similar to those of human tutors in specific tasks (Belpaeme et al., 2018). Studies have also highlighted their potential to be integrated and support special education (Papakostas et al., 2021). However, other studies did not reveal that social robots were more effective than human teachers (Woo et al., 2021). Hence, more research is required to further examine their impact. Additionally, despite social robots displaying positive impacts on students' learning gains, students' learning motivation and robots' social behaviors within educational environments should be carefully considered (Berghe et al., 2019). They can also support teachers by reducing their administrative workload and improving their job satisfaction (Smakman et al., 2021). Despite this fact, teachers are expressing concerns regarding their adoption; nonetheless, teachers also acknowledge the benefits that can be yielded through their integration in classrooms (Ewijk et al., 2020). As a result, unrealistically high expectations regarding their use and applicability should be avoided (Kwon et al., 2016).

Social robots present real potential to be used as meaningful learning and teaching companions for students in both formal and informal settings (Johal, 2020). However, such robots should adhere to strict rules and human social norms (Kirby et al., 2010). As a result, there is a clear need for standards and guidelines to be created on how to develop and introduce social robots (Mahdi et al., 2022; Woo et al., 2021). There is also a need to further examine the standardization and robustness of software and hardware associated with social robots (Pachidis et al., 2019). Additionally, there are several ethical and moral issues, such as bias, privacy, security, flexibility, autonomy, human contact, attachment, applicability, etc., which need to be considered and addressed before social robots are being more widely adopted in classrooms (Ewijk et al., 2020; Smakman et al., 2021). Moreover, there are several logistic, economic, and technical challenges that need to be overcome (Belpaeme & Tanaka,

2021). Hence, as there are several benefits and challenges related to the integration of social robots in education, there is a clear need to further analyze their implications in teaching and learning activities before their potentials to transform and enrich education can be achieved (Barakova et al., 2023).

3 Artificial Intelligence in Education

Artificial intelligence systems, are adaptive and flexible, learn from external data sources (Haenlein & Kaplan, 2019), and are capable of undertaking tasks and decisions in a rational, sophisticated, and autonomous way (Duan et al., 2019). Such systems are inspired by the human nervous system and its ability to learn, reason, and feel (Stone et al., 2016). Therefore, through the use of artificial intelligence, systems become capable of simulating human intelligence and mimicking their actions (Brynjolfsson & Mcafee, 2017; Lampropoulos et al., 2022; Li & Du, 2017). Artificial intelligence has been increasingly used in several domains (Bughin et al., 2017). The educational domain is no exception as the research into the use of artificial intelligence in educational settings is rapidly increasing (Hinojo-Lucena et al., 2019; Song & Wang, 2020).

Specifically, artificial intelligence has showcased great potential to transform and enrich the educational domain, create new learning and teaching approaches, and promote the redesign of knowledge, culture, and cognition (Holmes et al., 2020; Hwang et al., 2020; Lampropoulos & Papadakis, 2024; Pedro et al., 2019). There are several recent studies that have explored the influence of artificial intelligence in teaching and learning activities and have highlighted the challenges and benefits associated with its introduction in classrooms (2020b; Chen et al., 2020a, 2022; Chiu et al., 2023; Lin et al., 2023; Zhai et al., 2021). Although there are still open challenges regarding the use of artificial intelligence in education, studies have reported the improvement of sustainability and personalization through data-driven decision-making and recommendation systems as one of the main benefits (Chiu & Chai, 2020; Lampropoulos, 2023; Ouyang et al., 2022).

Moreover, artificial intelligence is closely related to social robots as artificial social intelligence of autonomous robots is an integral part of social robots (Dautenhahn, 1995). The combination of artificial intelligence and social robots supports both students and teachers and brings about significant changes in the learning design and teaching practices (Edwards et al., 2018). Although there is a generally positive attitude of the educational community toward the integration of social robots and artificial intelligence in education (Nitto et al., 2017), there are still several open challenges which need to be addressed before they are effectively introduced in classrooms.

4 Personalized Learning Using Social Robots and Artificial Intelligence

Personalized learning is a significant aspect of education as it can improve students' learning outcomes, motivation, and engagement (Lampropoulos, 2023). As a result, the research around this topic is increasing which further highlights the significance of providing students with personalized learning experiences (Bernacki et al., 2021; Zhang et al., 2020). Specifically, personalized learning focuses on tailoring the educational activities, goals, and material to the needs, skills, knowledge, and preferences of each individual learner; creating, thus, an individualized and adaptive learning experience (Lampropoulos, 2023; Shemshack & Spector, 2020; Xie et al., 2019). Providing students with adaptive learning opportunities can further improve their learning performance (Lampropoulos & Papadakis, 2024; Liu et al., 2017; Walkington, 2013; Wang et al., 2023).

Personalized learning is tightly connected to the use of digital technologies (Vandewaetere & Clarebout, 2014) and can be achieved through the use of various interventions and approaches (Zhang et al., 2020). Recent studies have highlighted the potentials of artificial intelligence to be used in educational settings in the context of personalized learning to amplify students' learning outcomes and result in a more effective overall educational process (Jelicic, 2019; Kaswan et al., 2024; Maghsudi et al., 2021; Pratama et al., 2023). Additionally, recent studies (Bhutoria, 2022; Ogata et al., 2024) have also revealed that when applying artificial intelligence in education and focusing on providing personalized learning environments, student-centered learning experiences can be created. Within these environments, teachers' role shifts to being more of a motivator and facilitator rather than the only provider of information. Simultaneously, artificial intelligence-enabled tools and agents facilitate the educational process through advanced, individualized, and automated recommendation and decision-making systems (Bhutoria, 2022; Ogata et al., 2024).

Through the use of artificial intelligence, intelligent tutoring systems can be developed which can be used in diverse learning settings to support both students and teachers and improve educational effectiveness (Anderson et al., 1985; Conati, 2009; Lampropoulos, 2023). Such systems can predict, identify, track, and monitor students' affective states and learning preferences (Crockett et al., 2017; Frasson & Chalfoun, 2010) which, in turn, can further increase their learning outcomes through the recommendation of appropriate learning material, exercises, and activities and the individualized and real-time feedback.

Such intelligent tutoring systems can be materialized in the form of social robots; combining, thus, their social skills and intelligence with a physical presence. Hence, through their social supportive behaviors, social robots, which are also characterized as emphatic and expressive robots, can effectively support students and help them achieve greater learning outcomes (Saerbeck et al., 2010; Serholt et al., 2014). Social robots, besides providing students with tailored activities and feedback, can also respond to their emotions, behaviors, and affective states in real time. Additionally, social robots are characterized by increased interactivity and sociability which enables them to communicate with students, respond to their emotional states, and form relationships with them (Kanda et al., 2004). Thus, social robots can offer personalized learning experiences and support students on an individual level which, in turn, results in the improvement of their cognitive learning gains (Leyzberg et al., 2012, 2014). Students who learn through these means, report higher performances than those who learn using non-personalized approaches and methods (Leyzberg et al., 2018). Given the capabilities of social robots, their social, cognitive, and affective skills, their ability to provide interactive and personalized learning, and their potentials to response to students' emotions, expressions, and actions in real time, renders them as effective educational means that can enrich and transform teaching and learning activities in both formal and informal settings.

5 Conclusions

Given the potential benefits that the use of social robots and artificial intelligence can bring about in the educational domain, this chapter went over the concepts of social intelligence and artificial intelligence and explored their role in teaching and learning. Additionally, emphasis was placed on their ability to provide improved personalized learning experiences and on the educational values they can offer. Based on the outcomes, social robots and artificial intelligence are highly connected and are used in different learning contexts in both formal and informal learning environments. Additionally, they can offer students with personalized learning experiences and tailored to their needs, knowledge, and preferences feedback and recommendations. Through their ability to identify, process, analyze, and monitor student-specific data and educational data, artificial social intelligence robots can amplify learning outcomes. Additionally, social robots can also promote embodied and active learning while simultaneously being able to interact, communicate, and understand learners. Furthermore, artificial intelligence and social robots can also support teachers to more effectively plan, organize, and carry out their lessons. Their potential to transform and improve learning design practices was also highlighted.

As a result, social robots can constitute intelligent human-like tutors that are characterized by their socio-cognitive skills, communication and interaction skills, social-emotional intelligence, sense of attachment, relatedness, and embodiment, can support both teachers, improve students' cognitive learning gains, and increase learning outcomes. Nonetheless, the field of social robots is still in its infancy. Consequently, there are many unresolved issues and open challenges that must be addressed before they can be more widely implemented in education. Additionally, there is a clear need for more empirical studies to be carried out to get a better understanding of their impact in different educational settings. Future studies should also focus on exploring security, ethical, and moral concerns associated with the use of social robots. There is also a need to create effective guidelines and standards relevant to the design, use, and integration of social robots in classrooms.

References

Anderson, J. R., Boyle, C. F., & Reiser, B. J. (1985). Intelligent tutoring systems. *Science, 228*(4698), 456–462. https://doi.org/10.1126/science.228.4698.456

Angel-Fernandez, J. M., & Vincze, M. (2018). *Towards a formal definition of educational robotics.* https://doi.org/10.15203/3187-22-1-08.

Anwar, S., Bascou, N. A., Menekse, M., & Kardgar, A. (2019). A systematic review of studies on educational robotics. *Journal of Pre-College Engineering Education Research (J-PEER), 9*(2). https://doi.org/10.7771/2157-9288.1223.

Anzalone, S. M., Boucenna, S., Ivaldi, S., & Chetouani, M. (2015). Evaluating the engagement with social robots. *International Journal of Social Robotics, 7*(4), 465–478. https://doi.org/10.1007/s12369-015-0298-7

Atmatzidou, S., & Demetriadis, S. (2016). Advancing students' computational thinking skills through educational robotics: A study on age and gender relevant differences. *Robotics and Autonomous Systems, 75*, 661–670. https://doi.org/10.1016/j.robot.2015.10.008

Barakova, E., Väänänen, K., Kaipainen, K., & Markopoulos, P. (2023). Benefits, challenges and research recommendations for social robots in education and learning: A Meta-Review. In *2023 32nd IEEE International Conference on Robot and Human Interactive Communication (RO-MAN)*, 2555–2561. https://doi.org/10.1109/ro-man57019.2023.103093405.

Belpaeme, T., Kennedy, J., Ramachandran, A., Scassellati, B., & Tanaka, F. (2018). Social robots for education: A review. *Science Robotics, 3*(21). https://doi.org/10.1126/scirobotics.aat5954.

Belpaeme, T., & Tanaka, F. (2021). Social robots as educators. In *OECD digital education outlook 2021 pushing the frontiers with artificial intelligence, Blockchain and robots: Pushing the frontiers with artificial intelligence, Blockchain and robots* (p. 143). OECD Publishing Paris.

Benitti, F. B. V. (2012). Exploring the educational potential of robotics in schools: A systematic review. *Computers & Education, 58*(3), 978–988. https://doi.org/10.1016/j.compedu.2011.10.006

van den Berghe, R., Verhagen, J., Oudgenoeg-Paz, O., van der Ven, S., & Leseman, P. (2019). Social robots for language learning: A review. *Review of Educational Research, 89*(2), 259–295. https://doi.org/10.3102/0034654318821286

Bernacki, M. L., Greene, M. J., & Lobczowski, N. G. (2021). A systematic review of research on personalized learning: Personalized by whom, to what, how, and for what purpose(s)? *Educational Psychology Review, 33*(4), 1675–1715. https://doi.org/10.1007/s10648-021-09615-8

Bhutoria, A. (2022). Personalized education and artificial intelligence in the United States, china, and India: A systematic review using a Human-In-The-Loop model. *Computers and Education: Artificial Intelligence, 3*, 100068. https://doi.org/10.1016/j.caeai.2022.100068

Breazeal, C. (2003). Toward sociable robots. *Robotics and Autonomous Systems, 42*(3–4), 167–175. https://doi.org/10.1016/s0921-8890(02)00373-1

Breazeal, C. (2004). *Designing sociable robots.* MIT press.

Breazeal, C., Dautenhahn, K., & Kanda, T. (2016). Social robotics. In *Springer handbooks* (pp. 1935–1972). https://doi.org/10.1007/978-3-319-32552-1_72.

Brynjolfsson, E., & Mcafee, A. (2017). Artificial intelligence, for real. *Harvard Business Review, 1*, 1–31.

Bughin, J., Hazan, E., Ramaswamy, S., Chui, M., Allas, T., Dahlstrom, P., Henke, N., & Trench, M. (2017). *Artificial intelligence: The next digital frontier?.*

Chen, L., Chen, P., & Lin, Z. (2020a). Artificial intelligence in education: A review. *IEEE Access, 8*, 75264–75278. https://doi.org/10.1109/access.2020.2988510

Chen, X., Xie, H., & Hwang, G.-J. (2020b). A multi-perspective study on artificial intelligence in education: Grants, conferences, journals, software tools, institutions, and researchers. *Computers and Education: Artificial Intelligence, 1*, 100005. https://doi.org/10.1016/j.caeai.2020.100005

Chen, X., Zou, D., Xie, H., Cheng, G., & Liu, C. (2022). Two decades of artificial intelligence in education. *Educational Technology & Society, 25*(1), 28–47.

Chevalier, M., Giang, C., El-Hamamsy, L., Bonnet, E., Papaspyros, V., Pellet, J.-P., Audrin, C., Romero, M., Baumberger, B., & Mondada, F. (2022). The role of feedback and guidance as intervention methods to foster computational thinking in educational robotics learning activities for primary school. *Computers & Amp Education, 180*, 104431. https://doi.org/10.1016/j.compedu.2022.104431.

Chiu, T. K. F., & Chai, C. (2020). Sustainable curriculum planning for artificial intelligence education: A Self-Determination theory perspective. *Sustainability, 12*(14), 5568. https://doi.org/10.3390/su12145568

Chiu, T. K. F., Xia, Q., Zhou, X., Chai, C. S., & Cheng, M. (2023). Systematic literature review on opportunities, challenges, and future research recommendations of artificial intelligence in education. *Computers and Education: Artificial Intelligence, 4*, 100118. https://doi.org/10.1016/j.caeai.2022.100118

Conati, C. (2009). Intelligent tutoring systems: New challenges and directions. In *Twenty-First International Joint Conference on Artificial Intelligence.*

Crockett, K., Latham, A., & Whitton, N. (2017). On predicting learning styles in conversational intelligent tutoring systems using fuzzy decision trees. *International Journal of Human-Computer Studies, 97*, 98–115. https://doi.org/10.1016/j.ijhcs.2016.08.005

Daniela, L., & Lytras, M. D. (2019). Educational robotics for inclusive education. *Technology, Knowledge and Learning, 24*(2), 219–225. https://doi.org/10.1007/s10758-018-9397-5

Dautenhahn, K. (1995). Getting to know each other—artificial social intelligence for autonomous robots. *Robotics and Autonomous Systems, 16*(2–4), 333–356. https://doi.org/10.1016/0921-8890(95)00054-2

Dautenhahn, K. (2007). Socially intelligent robots: Dimensions of human–robot interaction. *Philosophical Transactions of the Royal Society B: Biological Sciences, 362*(1480), 679–704. https://doi.org/10.1098/rstb.2006.2004

Denis, B., & Hubert, S. (2001). Collaborative learning in an educational robotics environment. *Computers in Human Behavior, 17*(5–6), 465–480. https://doi.org/10.1016/s0747-5632(01)00018-8

Duan, Y., Edwards, J. S., & Dwivedi, Y. K. (2019). Artificial intelligence for decision making in the era of big data—evolution, challenges and research agenda. *International Journal of Information Management, 48*, 63–71. https://doi.org/10.1016/j.ijinfomgt.2019.01.021

Edwards, C., Edwards, A., Spence, P. R., & Lin, X. (2018). I, teacher: Using artificial intelligence (AI) and social robots in communication and instruction. *Communication Education, 67*(4), 473–480. https://doi.org/10.1080/03634523.2018.1502459

Eguchi, A. (2010). What is educational robotics? Theories behind it and practical implementation. In *Society for Information Technology & Teacher Education International Conference*, 4006–4014.

Eguchi, A. (2014). Educational robotics for promoting 21st century skills. *Journal of Automation, Mobile Robotics and Intelligent Systems*, 5–11. https://doi.org/10.14313/jamris_1-2014/1.

Eguchi, A. (2015). Educational robotics as a learning tool for promoting rich environments for active learning (REALs). In *Handbook of research on educational technology integration and active learning* (pp. 19–47). https://doi.org/10.4018/978-1-4666-8363-1.ch002

Evripidou, S., Georgiou, K., Doitsidis, L., Amanatiadis, A. A., Zinonos, Z., & Chatzichristofis, S. A. (2020). Educational robotics: Platforms, competitions and expected learning outcomes. *IEEE Access, 8*, 219534–219562. https://doi.org/10.1109/access.2020.3042555

van Ewijk, G., Smakman, M., & Konijn, E. A. (2020). Teachers' perspectives on social robots in education. *Proceedings of the Interaction Design and Children Conference*. https://doi.org/10.1145/3392063.3394397

Fong, T., Nourbakhsh, I., & Dautenhahn, K. (2003). A survey of socially interactive robots. *Robotics and Autonomous Systems, 42*(3–4), 143–166. https://doi.org/10.1016/s0921-8890(02)00372-x

Frasson, C., & Chalfoun, P. (2010). Managing learner's affective states in intelligent tutoring systems. In *Studies in computational intelligence* (pp. 339–358). https://doi.org/10.1007/978-3-642-14363-2_17.

Gunes, H., & Kucuk, S. (2022). A systematic review of educational robotics studies for the period 2010–2021. *Review of Education, 10*(3). https://doi.org/10.1002/rev3.3381.

Haenlein, M., & Kaplan, A. (2019). A brief history of artificial intelligence: On the past, present, and future of artificial intelligence. *California Management Review, 61*(4), 5–14. https://doi.org/10.1177/0008125619864925

Hegel, F., Muhl, C., Wrede, B., Hielscher-Fastabend, M., & Sagerer, G. (2009). Understanding social robots. In *2009 Second International Conferences on Advances in Computer-Human Interactions*, 169–174. https://doi.org/10.1109/achi.2009.51.

Hinojo-Lucena, F.-J., Aznar-Díaz, I., Cáceres-Reche, M.-P., & Romero-Rodríguez, J.-M. (2019). Artificial intelligence in higher education: A bibliometric study on its impact in the scientific literature. *Education Sciences, 9*(1), 51. https://doi.org/10.3390/educsci9010051

Holmes, W., Bialik, M., & Fadel, C. (2020). *Artificial intelligence in education: Promises and implications for teaching and learning*. Center for Curriculum Redesign.

Hwang, G.-J., Xie, H., Wah, B. W., & Gašević, D. (2020). Vision, challenges, roles and research issues of artificial intelligence in education. *Computers and Education: Artificial Intelligence, 1*, 100001. https://doi.org/10.1016/j.caeai.2020.100001

Johal, W. (2020). Research trends in social robots for learning. *Current Robotics Reports, 1*(3), 75–83. https://doi.org/10.1007/s43154-020-00008-3

Kanda, T., Hirano, T., Eaton, D., & Ishiguro, H. (2004). Interactive robots as social partners and peer tutors for children: A field trial. *Human-Computer Interaction, 19*(1), 61–84. https://doi.org/10.1207/s15327051hci1901&2_4.

Kaswan, K. S., Dhatterwal, J. S., & Ojha, R. P. (2024). AI in personalized learning. In *Advances in technological innovations in higher education* (pp. 103–117). CRC Press.

Kirby, R., Forlizzi, J., & Simmons, R. (2010). Affective social robots. *Robotics and Autonomous Systems, 58*(3), 322–332. https://doi.org/10.1016/j.robot.2009.09.015.

Kolb, D. A. (2014). *Experiential learning: Experience as the source of learning and development*. FT press.

Kwon, M., Jung, M. F., & Knepper, R. A. (2016). Human expectations of social robots. In *2016 11th ACM/IEEE International Conference on Human-Robot Interaction (HRI)*, 463–464. https://doi.org/10.1109/HRI.2016.7451807.

Lampropoulos, G. (2023). Augmented reality and artificial intelligence in education: Toward immersive intelligent tutoring systems. In *Augmented reality and artificial intelligence* (pp. 137–146). 10.10072F978-3-031-27166-3_8.

Lampropoulos, G., & Kinshuk. (2024). Virtual reality and gamification in education: A systematic review. *Educational Technology Research and Development, 72*, 1691–1785. https://doi.org/10.1007/s11423-024-10351-3.

Lampropoulos, G., & Papadakis, S. (2024). Maker education and educational robotics in the age of artificial intelligence. *In Intelligent Educational Robots: Toward Personalized Learning Environments* (pp. 1–8), De Gruyter. https://doi.org/10.1515/9783111352695-001.

Lampropoulos, G., Siakas, K., Julio, V., & Olaf, R. (2022). Artificial intelligence, Blockchain, big data analytics, machine learning and data mining in traditional CRM and social CRM: A critical review. In *21st IEEE/WIC/ACM International Joint Conference on Web Intelligence and Intelligent Agent Technology (WI-IAT)*, 504–510. https://doi.org/10.1109/WI-IAT55865.2022.00080.

Leite, I., Martinho, C., & Paiva, A. (2013). Social robots for Long-Term interaction: A survey. *International Journal of Social Robotics, 5*(2), 291–308. https://doi.org/10.1007/s12369-013-0178-y

Leyzberg, D., Ramachandran, A., & Scassellati, B. (2018). The effect of personalization in Longer-Term robot tutoring. *ACM Transactions on Human-Robot Interaction, 7*(3), 1–19. https://doi.org/10.1145/3283453

Leyzberg, D., Spaulding, S., & Scassellati, B. (2014). Personalizing robot tutors to individuals' learning differences. In *Proceedings of the 2014 ACM/IEEE International Conference on Human-Robot Interaction*. https://doi.org/10.1145/2559636.2559671.

Leyzberg, D., Spaulding, S., Toneva, M., & Scassellati, B. (2012). The physical presence of a robot tutor increases cognitive learning gains. In *Proceedings of the Annual Meeting of the Cognitive Science Society*, 1882–1887.

Li, D., & Du, Y. (2017). *Artificial intelligence with uncertainty*. CRC press. https://doi.org/10.1201/9781315366951.

Lin, C.-C., Huang, A. Y. Q., & Lu, O. H. T. (2023). Artificial intelligence in intelligent tutoring systems toward sustainable education: A systematic review. *Smart Learning Environments, 10*(1). https://doi.org/10.1186/s40561-023-00260-y.

Liu, M., McKelroy, E., Corliss, S. B., & Carrigan, J. (2017). Investigating the effect of an adaptive learning intervention on students' learning. *Educational Technology Research and Development, 65*(6), 1605–1625. https://doi.org/10.1007/s11423-017-9542-1

López-Belmonte, J., Segura-Robles, A., Moreno-Guerrero, A.-J., & Parra-González, M.-E. (2021). Robotics in education: A scientific mapping of the literature in web of science. *Electronics, 10*(3), 291. https://doi.org/10.3390/electronics10030291

Maghsudi, S., Lan, A., Xu, J., & van der Schaar, M. (2021). Personalized education in the artificial intelligence era: What to expect next. *IEEE Signal Processing Magazine, 38*(3), 37–50. https://doi.org/10.1109/msp.2021.3055032

Mahdi, H., Akgun, S. A., Saleh, S., & Dautenhahn, K. (2022). A survey on the design and evolution of social robots—past, present and future. *Robotics and Autonomous Systems, 156*, 104193. https://doi.org/10.1016/j.robot.2022.104193

Mejia, C., & Kajikawa, Y. (2017). Bibliometric analysis of social robotics research: Identifying research trends and knowledge base. *Applied Sciences, 7*(12), 1316. https://doi.org/10.3390/app7121316

Nitto, H., Taniyama, D., & Inagaki, H. (2017). Social acceptance and impact of robots and artificial intelligence. *Nomura Res. Inst. Pap, 211*, 1–5.

Ogata, H., Flanagan, B., Takami, K., Dai, Y., Nakamoto, R., & Takii, K. (2024). EXAIT: Educational eXplainable artificial intelligent tools for personalized learning. *Research and Practice in Technology Enhanced Learning, 19.*

Ospennikova, E., Ershov, M., & Iljin, I. (2015). Educational robotics as an inovative educational technology. *Procedia-Social and Behavioral Sciences, 214*, 18–26. https://doi.org/10.1016/j.sbspro.2015.11.588

Ouyang, F., Zheng, L., & Jiao, P. (2022). Artificial intelligence in online higher education: A systematic review of empirical research from 2011 to 2020. *Education and Information Technologies, 27*(6), 7893–7925. https://doi.org/10.1007/s10639-022-10925-9

Pachidis, T., Vrochidou, E., Kaburlasos, V. G., Kostova, S., Bonković, M., & Papić, V. (2019). Social robotics in education: State-of-the-Art and directions. In *Mechanisms and machine science* (pp. 689–700). https://doi.org/10.1007/978-3-030-00232-9_72.

Papakostas, G. A., Sidiropoulos, G. K., Papadopoulou, C. I., Vrochidou, E., Kaburlasos, V. G., Papadopoulou, M. T., Holeva, V., Nikopoulou, V.-A., & Dalivigkas, N. (2021). Social robots in special education: A systematic review. *Electronics, 10*(12), 1398. https://doi.org/10.3390/electronics10121398

Papert, S. A. (1984). New theories for new learnings. *School Psychology Review, 13*(4), 422–428. https://doi.org/10.1080/02796015.1984.12085122

Pedersen, B. K. M. K., Larsen, J. C., & Nielsen, J. (2020). The effect of commercially available educational robotics: A systematic review. In *Robotics in education* (pp. 14–27). https://doi.org/10.1007/978-3-030-26945-6_2.

Pedro, F., Subosa, M., Rivas, A., & Valverde, P. (2019). *Artificial intelligence in education: Challenges and opportunities for sustainable development*.

Pratama, M. P., Sampelolo, R., & Lura, H. (2023). Revolutionizing education: Harnessing the power of artificial intelligence for personalized learning. *Klasikal: Journal of Education, Language Teaching and Science, 5*(2), 350–357.

Saerbeck, M., Schut, T., Bartneck, C., & Janse, M. D. (2010). Expressive robots in education. In *Proceedings of the SIGCHI Conference on Human Factors in Computing Systems*, 1613–1622. https://doi.org/10.1145/1753326.1753567.

Scaradozzi, D., Screpanti, L., & Cesaretti, L. (2019). Towards a definition of educational robotics: A classification of tools, experiences and assessments. In *Smart learning with educational robotics* (pp. 63–92). https://doi.org/10.1007/978-3-030-19913-5_3.

Screpanti, L., Cesaretti, L., Storti, M., & Scaradozzi, D. (2021). Educational robotics and social relationships in the classroom. In *Makers at school, educational robotics and innovative learning environments* (pp. 195–201). https://doi.org/10.1007/978-3-030-77040-2_26.

Serholt, S., Barendregt, W., Leite, I., Hastie, H., Jones, A., Paiva, A., Vasalou, A., & Castellano, G. (2014). Teachers' views on the use of empathic robotic tutors in the classroom. In *The 23rd IEEE International Symposium on Robot and Human Interactive Communication*, 955–960. https://doi.org/10.1109/roman.2014.6926376.

Shemshack, A., & Spector, J. M. (2020). A systematic literature review of personalized learning terms. *Smart Learning Environments, 7*(1). https://doi.org/10.1186/s40561-020-00140-9.

Sheridan, T. B. (2016). Human-Robot interaction. *Human Factors: THe Journal of the Human Factors and Ergonomics Society, 58*(4), 525–532. https://doi.org/10.1177/0018720816644364

Smakman, M., Vogt, P., & Konijn, E. A. (2021). Moral considerations on social robots in education: A multi-stakeholder perspective. *Computers & Education, 174*, 104317. https://doi.org/10.1016/j.compedu.2021.104317

Song, P., & Wang, X. (2020). A bibliometric analysis of worldwide educational artificial intelligence research development in recent twenty years. *Asia Pacific Education Review, 21*(3), 473–486. https://doi.org/10.1007/s12564-020-09640-2

Stone, P., Brooks, R., Brynjolfsson, E., Calo, R., Etzioni, O., Hager, G., Hirschberg, J., Kalyanakrishnan, S., Kamar, E., Kraus, S., Leyton-Brown, K., Parkes, D., Press, W., Saxenian, A., Shah, J., Tambe, M., & Teller, A. (2016). *Artificial intelligence and life in 2030: The one hundred year study on artificial intelligence*. https://doi.org/10.48550/ARXIV.2211.06318.

Vandewaetere, M., & Clarebout, G. (2014). Advanced technologies for personalized learning, instruction, and performance. In *Handbook of research on educational communications and technology* (pp. 425–437). https://doi.org/10.1007/978-1-4614-3185-5_34.

Virnes, M., Sutinen, E., & Kärnä-Lin, E. (2008). How children's individual needs challenge the design of educational robotics. In *Proceedings of the 7th International Conference on Interaction Design and Children*. https://doi.org/10.1145/1463689.1463766.

van der Vorst, T., & Jelicic, N. (2019). Artificial intelligence in education: Can AI bring the full potential of personalized learning to education?

Walkington, C. A. (2013). Using adaptive learning technologies to personalize instruction to student interests: The impact of relevant contexts on performance and learning outcomes. *Journal of Educational Psychology, 105*(4), 932–945. https://doi.org/10.1037/a0031882

Wang, S., Christensen, C., Cui, W., Tong, R., Yarnall, L., Shear, L., & Feng, M. (2023). When adaptive learning is effective learning: Comparison of an adaptive learning system to teacher-led instruction. *Interactive Learning Environments, 31*(2), 793–803. https://doi.org/10.1080/10494820.2020.1808794

Woo, H., LeTendre, G. K., Pham-Shouse, T., & Xiong, Y. (2021). The use of social robots in classrooms: A review of field-based studies. *Educational Research Review, 33*, 100388. https://doi.org/10.1016/j.edurev.2021.100388

Xie, H., Chu, H.-C., Hwang, G.-J., & Wang, C.-C. (2019). Trends and development in technology-enhanced adaptive/personalized learning: A systematic review of journal publications from 2007 to 2017. *Computers & Education, 140*, 103599. https://doi.org/10.1016/j.compedu.2019.103599

Zhai, X., Chu, X., Chai, C. S., Jong, M. S. Y., Istenic, A., Spector, M., Liu, J.-B., Yuan, J., & Li, Y. (2021). A review of artificial intelligence (AI) in education from 2010 to 2020. *Complexity*, 1–18. https://doi.org/10.1155/2021/8812542.

Zhang, L., Basham, J. D., & Yang, S. (2020). Understanding the implementation of personalized learning: A research synthesis. *Educational Research Review, 31*, 100339. https://doi.org/10.1016/j.edurev.2020.100339

Enabling Autonomous and Adaptive Social Robots in Education: A Vision for the Application of Generative AI

Eva Verhelst, Ruben Janssens, and Tony Belpaeme

Abstract The limited autonomy of social robots currently prevents many ambitions in educational robotics from being realised. This leads to scripted dialogues, content that fails to adapt to individual students and conversations remaining largely text-based. Recent advances in generative artificial intelligence (AI) might alleviate these issues, allowing for educational robots whose dialog can be flexibly generated based on the lessons to be taught, the student's needs and personality, and the environment. This chapter presents a vision of how generative AI can power truly autonomous and adaptive social robots in education, discussing limitations of past educational robotics research, recent technical advances in AI, as well as concrete examples of applications of AI in educational human-robot interaction, and a reflection on limitations of current AI. By bridging technical and pedagogical perspectives, it shows what the next step in the evolution of human-robot interaction in educational contexts might look like.

Keywords Educational social robots · Generative artificial intelligence · Educational technology · Adaptive learning · Conversational artificial intelligence

1 Introduction

Commercial generative artificial intelligence (AI) tools such as ChatGPT, Copilot and Gemini have had a noticeable impact on education, sparking both enthusiasm and apprehension among educators and students. Students use it to write their assignments—forcing teachers to reevaluate their evaluation methods—while teachers use

E. Verhelst (✉) · R. Janssens · T. Belpaeme
IDLab-AIRO, Ghent University - imec, Ghent, Belgium
e-mail: Eva.Verhelst@ugent.be

R. Janssens
e-mail: Ruben.Janssens@ugent.be

T. Belpaeme
e-mail: Tony.Belpaeme@ugent.be

G. Lampropoulos and S. Papadakis (eds.), *Social Robots in Education*, Studies in Computational Intelligence 1194, https://doi.org/10.1007/978-3-031-82915-4_2

it as a tool for automatic grading (Adeshola & Adepoju, 2023; Dempere et al., 2023). Here, we will step away from students and teachers using the AI tools directly and focus on how generative AI can turn social robots into more successful teachers, teaching assistants, and tutors.

Social robots for education have been developed for more than three decades now (Belpaeme et al., 2018). However, as we discuss in Sect. 2, they still often lack real autonomy and generalisability, limiting their applicability in real-world classrooms. Meanwhile, the last few years have seen the emergence of many powerful *generative AI* models, which we define here as AI models that generate new content (Feuerriegel et al., 2024). This content can be natural language text, which is generated by *(large) language models (LLMs)*, often based on a piece of input text which is called the *prompt*. They are not only limited to text, but *multi-modal* modals can also use other modalities such as images as input (Gan et al., 2022). Other models also generate new images or sounds, often of remarkable human-like quality (Podell et al., 2023).

We argue that these generative AI models will provide new capabilities to social robots, enabling them to surpass the limitations that have kept social robots from entering real-world classrooms. We focus in particular on their autonomy and on the possibilities for *adapting* the interaction between the robot and the learner to the learner's skill level and *personalising* the interaction to e.g. their interests. As AI researchers that develop autonomous social robots for educational scenarios, with this chapter we aim to present our vision of how generative AI will enable these new capabilities to educational social robots. We provide background information about the concrete generative AI models we deem important, as well as several concrete applications showing how they can be integrated in social robots. With this, we hope to inspire anyone who researches or develops educational social robots, either from a technical or a pedagogical perspective.

This chapter is structured as followed. We first discuss the constraints that currently limit educational social robots in Sect. 2. Then, we present the various types of generative AI models that we believe will be powerful for educational social robots in Sect. 3, along with a short history of these recent advances in generative AI. Section 4 then presents how they can concretely be applied in educational social robots, taking a few specific applications that were recently developed as examples. In Sect. 5, we discuss the remaining and new technical and societal limitations of these technologies and finally summarise our findings in the conclusion in Sect. 6.

2 Current Limitations of Educational Social Robots

People have been dreaming of robot teachers for decades—already in his 1951 short story "The Fun They Had", Isaac Asimov speculated about children each having their own individual mechanical teacher. Today, social robots are real, and commercially available robots like Nao, Pepper, Furhat, and others, can be used in real-world settings. In the meantime, class groups are becoming more diverse, with children (and parents) expecting more individualised support and tutoring, and many countries struggling with teacher shortages. Then why are social robots not yet ubiquitous in our schools? We examine what limits social robots today.

2.1 Natural Language Interactions

Perhaps the most prevalent mode of communication humans use, especially in teaching and learning interactions, is natural language. A teacher explains concepts to their students in natural language; students ask and answer questions, write answers on tests, give presentations, write essays, and of course, in language education, people learn by practising a target language.

However, this immediately lays bare one of the central challenges of social robotics: holding natural conversations. Until recently, most robots' conversational abilities were powered by rule-based chatbots. This means that the robot programmer had to envision all possible utterances that the user could say and program appropriate responses for each of them. Templates provided some flexibility, meaning the user can substitute some words in the programmed template sentences, and more advanced *intent recognition models* could automatically categorise which option the user intended to say, within a predefined list of options. Data-driven approaches allow for more flexibility, but require data collection which is very expensive in HRI (Reimann et al., 2024).

This approach clearly limits the possibilities and autonomy of the social robot. Suppose the user answers the robot's question or instruction in a different way than was envisioned by the programmer, or worse, when the user asks the robot an unexpected question, the robot will often not understand the user, leading to the interaction failing and possibly the user losing trust in the robot (Flook et al., 2019).

As a consequence of this, many research studies that investigate the effectiveness of social robots in educational scenarios use tightly scripted interactions, or even rely on remote-controlled interactions. The latter is often called the "Wizard-of-Oz" approach, and relies on a human behind the scenes choosing what the robot says and does. While these scripted or "wizarded" studies provide insights into the potential benefits and effects of social robots in education, these systems are not yet ready for autonomous deployment in the real world (Belpaeme et al., 2018).

Another major limiting factor for human-robot conversations is the performance of automatic speech recognition (ASR) systems. While ASR has made great strides in recent years, *really good* performance is only achieved for typical populations, i.e. adult native speakers of a language for which many training examples are available, such as English, and preferably not speaking a dialect of this language. Especially for young children, speech recognition has long not performed well enough to work in autonomous interactions (Kennedy et al., 2017). Robot designers have had to resort to other modalities of interaction to cope with the lack of robust ASR, such as touchscreens where a child can select an option, or perhaps type a response (Belpaeme et al., 2018).

Taking into account both the necessity to rely on (partly) scripted conversations, and the lack of robust ASR for children, it becomes clear that social robot tutors could so far only be deployed in very narrow situations, only focussing on specific topics for which time-intensive work was needed from the robot designer to program the interaction, and not supporting free-form input from the user. This has also limited which didactic methods could be used in the teaching and learning interactions, as methods that rely heavily on conversations, with a lot of learner input, could not yet be reliably implemented. Specifically for (second) language education, this has resulted in research focussing mostly on vocabulary teaching, as real conversational practice was not yet feasible (Randall, 2019).

2.2 Multi-Modal Inputs

Of course, humans do not only use language to communicate. We also convey information using other modalities, such as through what we can see or non-verbal sounds. People use social cues such as facial expressions, eye gaze, gestures, body pose, and prosody, to express emotions, and to infer information about others' mental states (Mehrabian, 1972). Crucial in education, teachers can use this information to assess how engaged and motivated a student is, but also whether they understand the material that is being discussed or if they are confused (Schutz et al., 2006).

However, as we also saw in the previous section, tasks that are mostly intuitive and natural for people, such as conversations in natural language, are exceedingly difficult for AI—a principle often referred to as Moravec's paradox (Moravec, 1988). The same goes for interpreting such social signals. People express emotions and social signals in very subtle ways, often using movements that are only shortly visible (Krumhuber et al., 2013), and much less exaggerated than the stereotypical emotional expressions that are often found in datasets used to train AI models. These expressions are also highly individual (D'Mello et al., 2018), and furthermore difficult or even impossible to interpret without the context of the interaction (Mesquita & Boiger, 2014).

While social signal processing, in educational applications especially focused on detecting engagement, has been an active research area for a number of years, systems for detecting engagement are often limited in their success to very specific contexts, and not yet ready for complex, real-world scenarios (Cumbal et al., 2020; Gunes & Churamani, 2023).

Besides understanding social cues, educational robots should also be aware of their physical environment. As they are physically embodied, they share this space with the user. Their embodiment, which is often humanoid and thus contains eyes, also creates the expectation in the user that the robot can see. However, the conversation models discussed in the previous section are often not, or only loosely, connected with the visual input of the robot: chatbots are not designed with eyes in mind (Reimann et al., 2024; Janssens, 2024).

Using visual information about the environment, however, holds many opportunities for educational robotics. The robot could refer to objects, people, or actions that are happening, which could aid in vocabulary acquisition (Bara & Kaminski, 2019; Wolfert et al., 2024), or gestures could be used by the user to provide more information in the interaction. If the robot does not have access to this visual information, it might also not understand some of the user utterances, if they refer to elements of the environment.

The visual information can also be used to make the robot more sociable: it can personalise the conversation based on what the robot sees of the user. Going much further, the robot could personalise the entire interaction. Humans also adapt the way they speak, and especially how they explain things, to the person they are speaking to, and we also base this on the way someone looks. Simple personalisation of interactions, such as using the user's name, has already been shown to increase engagement in human-robot interactions (Henkemans et al., 2013), so adapting the interaction more extensively to the specific user could lead to even better outcomes.

However, currently, social robots are limited in how they can use visual information in interactions. Just as the conversations need to be largely scripted, systems currently need to be specifically designed to recognise a given set of objects at certain moments during the interaction (Wolfert et al., 2024).

2.3 *Adaptation and Personalisation*

Perhaps the most promising aspect of using social robots in education is the potential for a completely personalised tutor for each learner—as Asimov already dreamt of in 1951. However, given the limitations already discussed, this dream is still far-removed. As interactions needed to be extensively pre-programmed, all options for modifying the interaction needed to be foreseen by the designer. Concretely, this means that interactions cannot be easily personalised to the learner's interest, by letting the learner choose a topic on which to converse during language learning. Also, providing new didactic content is often challenging, as the programming of the robot is often not sufficiently accessible to teachers (Chevalier et al., 2016). Adapting the content of the learning interaction to the learner's performance, current skill level or prior knowledge is also an important aspect of personalised robot tutors but requires more flexibility than is available with the current conversational models.

Besides providing different options the robot can choose to use in its interaction with the learner, a major challenge is also knowing which of them to choose. The robot can, for example, choose interventions to improve the learner's engagement, or to adapt to the learner's skill level and aim to improve learning gain. However, for both cases, a model for the learner's engagement or knowledge is needed (also called *student modelling* (Chrysafiadi & Virvou, 2013)), as well as a method to predict which impact each option has, allowing the robot to select the optimal action. Both building these models and designing the action selection methods are challenging. One of the aspects that makes this challenging is the difficulty of detecting and interpreting non-verbal signals. However, progress has been made in building action selection models, e.g. using interactive reinforcement learning (Belpaeme et al., 2018).

3 Recent Advances in Generative AI

As discussed in the previous section, one of the main limiting factors for social robots in education is their ability to process and express themselves in natural language: most of our educational interactions take place in the form of natural language. We claim that for social robots to fully integrate into education, they must master this form of communication. This would have remained a pipe dream, if not for generative AI. Generative AI offers the potential to converse with tutoring systems using natural language. This section describes the technology that made Generative AI, such as Claude or ChatGPT, possible and paints a picture of how it created a revolution not only in language modelling, but also in speech recognition, image generation and multi-modal language models. We also look at how the field of natural language processing (NLP) has gotten to this point, and what technologies made this possible, as well as the advancements those technologies introduced for speech recognition, image generation, and multi-modal language models.

3.1 Language Modelling

Much of the data created by people is unstructured: natural language, meant for other humans—think about books, social media and blog posts, news articles and much more. This means that it is not organised in a way that computers can directly use it. The field of NLP tries to bridge that gap, by processing human language so that it is useable for computers. It has typically focussed on solving tasks that involve natural language, and one of them is intent recognition. It is useful for, among other things, recognising what a user is trying to communicate. Related is sentiment analysis, where the emotional intent behind a piece of text is analysed. An example application of this is a company trying to estimate the public opinion of their product based on what is said about it on the internet. To gain structured information from a text, named entity recognition can be used. This task is about trying to extract named entities, such as people, places and organisations, from text. This can then be used for further classification. Another NLP task that is useful in (educational) robotics is ASR: transcribing text from speech in an audio file. Then, the transcription can be fed to further processing steps that expect written text.

The NLP task that we will focus on here is language modelling, as the recent advancements in it are what we believe will have a great impact on educational social robotics. Modelling languages boils down to predicting the probability of a word given previous words. Many attempts have been made, starting with statistical models to, more recently, neural networks. If, given a few words, we know which words are more or less likely to follow it, we can use this information to choose the next word. Then, we add it to the words we already had, look at the probabilities of the next word, and again choose the following one. This way, language modelling is the key to generating text, which is what we needed to ease the natural language limitation in educational robotics. For a model to learn how to do this, it needs data—luckily natural language examples are abundant on the internet. Multiple approaches are used for training models to do this task. You can train a model by giving it the start of the sentence and having it predict the next word, as this word is of course known. BERT, the language model developed by Google in 2018, was trained using a *masked language objective*: some words in the training text were randomly deleted (*masked*), and the model had to estimate what word it should have been (Devlin et al., 2018). This task allowed the model to learn the probabilities of words in context. Then, even if the model was not directly trained for predicting the next word given a part of a sentence, it still learns this task as it is closely enough related to the task it was trained on. This concept is called transfer learning: training a model on one task, and using it for a related, but different task. Transfer learning makes a model much more useful, as it can be applied to more than just the specific task that it was trained for. Often, language models are trained on a modelling objective as described above, making them a *general-purpose* model, after which they are re-trained using smaller amounts of data for a more specific downstream task, such as intent recognition, sentiment analysis, or even summarising text and answering questions about it. This process is called *fine-tuning*, and it is a form of transfer learning.

Now that we understand the goal of NLP and the importance of language modelling, we can look at the revolution in generative AI that we believe will have an impact on educational social robots. The part of the revolution that was most visible to the general public was OpenAI's release of ChatGPT at the end of 2022 (Kasneci et al., 2023). ChatGPT's sudden success was made possible by a technology that was proposed a few years before, in a widely cited publication: *Attention is all you need*, by a team of researchers at Google (Vaswani et al., 2017). This publication introduced the Transformer architecture. What sets this architecture apart from previous language modelling techniques, is that it allows text to be processed in a much more efficient way. The previous state-of-the-art—recurrent neural networks—tried to capture the sequential nature of textual data by processing it piece by piece while keeping a hidden state—a summary of all that it had processed already. While this technique showed great potential, its inherent limiting factor is that text must be processed sequentially. When processing large amounts of text—as is necessary for modern, data-hungry techniques—this is just too slow. The Transformer architecture avoided the need for processing sequential data sequentially by introducing the concept of *attention*, as referenced in the name of the publication that started this revolution.

To understand attention, we will first take a look at why processing language is a difficult task. Within a sentence, words influence each other. Most of this influence is short distance: the article is decided by the word that follows. But, in some cases, this influence reaches a word that is further in a sentence, with many words in between. For example, the conjugation of a verb is decided by the subject, which can be multiple clauses away. Recurrent neural networks process such a sentence word by word, while keeping a summary of what was seen before. To predict the next word, this summary must contain the number of the subject—which might have been the first word of the sentence—so it can be remembered until the verb appears—which might be the last. The challenge of this long-term memory that is necessary for language understanding, together with the inefficiency of processing text piece by piece, is what the attention mechanism solves.

The idea behind attention is that, when processing a part of the data, the model first determines what relevant information can be found in other parts of the data. Following the example of the subject's influence on the verb's conjugation, the attention mechanism would decide that the subject is important for predicting the verb, as well as the object of the verb, while other words in between might not be. In attention, this problem is solved as follows. A sum of all words, weighted by their importance to the current word, is used as input during processing. As this weighted sum can be calculated for each word in relation to all other words simultaneously—allowing for parallelisation, the limitation of processing the data piece by piece, as in recurrent neural networks, is overcome. This way, the Transformer architecture can easily model long-term interactions in sequential data, while processing more of this data at the same time.

The introduction of the Transformer architecture allowed for faster, more parallel processing of data, fitting perfectly in an era of end-to-end, data-driven technologies: large models, trained directly for their end-task, using large amounts of data.

The new possibility to train these models on such large amounts of data, allowed for advancements in their performance that were unreachable with previous technologies. These models are often referred to as Large Language Models (LLMs).

At their core, LLMs do one thing: given some text, they predict the next word—just as your smartphone's keyboard does. Therefore, you can ask it a question, and it generates an answer word by word. The resulting text is often very impressive and human-like, which makes it seem like there is more going on than just predicting the next word. The difference between a LLM and your smartphone's keyboard is that LLMs are trained on such large amounts of data, and on larger pieces of text at once, that they can find a pattern in the structure of language. It is important to note that this pattern recognition, which leads to very human-like text, is not the same as reasoning or even understanding. LLMs are just very good at generating text that seems like what an intelligent answer would be, but the reasoning steps that happen behind the scenes when a human is talking are not implemented in LLMs.

We have established that LLMs take text as input and generate text that seems likely to follow the given text. The text that is given as input is often called the *prompt*. Using this prompt intelligently can greatly improve how well the generated text matches what you are trying to do. This is a new form of transfer learning: the model is trained to model the probabilities of words, but using a prompt, we have it fulfil a different task—like before, such tasks can include intent recognition, sentiment analysis, summarising texts, answering questions, or generating example sentences about a specific concept. Explaining the task or asking a question is called *zero-shot* prompting. Providing some examples of the task at hand is called *few-shot* prompting and often already provides a noticeable improvement (Brown, et al., 2020). Prompting an LLM to not immediately generate an answer, but to provide a *chain of thought*—some intermediate reasoning steps—improves the model's reasoning skills, or it at least appears to do so (Wei et al., 2022). Actually, while the model generates these intermediate steps, it is still choosing what is most likely in a statistical sense. As the intermediate steps are usually more obvious and simple tasks, there is a higher chance that the model encountered something similar during training. Then, these intermediate steps are more likely to be correct. As the model then takes these generated steps as input when generating the final answer, its probability of correctness improves. Asking the model to generate intermediate steps is forcing it to reason out loud—which is the only kind of reasoning such a model can do.

Prompting LLMs makes them suited for a variety of tasks, as they can generate language that is generally correct, human-like and—with intelligent prompting—adheres to the task quite well. Their usefulness for educational social robots is easy to imagine given this impressive ability, and knowing what limited social robots from fully succeeding in educational applications until now.

3.2 Automatic Speech Recognition

A similar story can be found in the performance of ASR: recognising the speech in audio fragments and transcribing it. ASR has been researched extensively, and useful results have existed for years. In optimal settings, error rates of around 5% are reported, which is similar to or even better than human annotation (Xiong et al., 2018). Though this sounds very promising, robust speech recognition that can handle suboptimal environments, such as a noisy room, has been hard to attain (Karpagavalli & Chandra, 2016). Early models were heavily engineered, with hand-crafted features that relied heavily on expert knowledge. This is often not robust and does not necessarily transfer between languages. Then, similar to the evolution in language modelling, there was a shift to end-to-end models. These were trained on raw audio and transcriptions, skipping the feature engineering step. End-to-end models tend to be too large for use on mobile devices, so cloud-based solutions appeared, such as the ones provided by Google and Microsoft. Finally, attention-based—the core improvement of the Transformer—ASR were designed, by universities and companies such as Google (Bahdanau et al., 2016; Chan et al., 2015). Around 2022, OpenAI presented Whisper, an open-source ASR system. It uses a Transformer model as published in the original paper and is trained on an impressive 680,000 h of data (Radford et al., 2023). This again shows what is made possible by the Transformer architecture: an end-to-end (speech-to-text) application trained on large amounts of data. Next to the computational ability to train a model on so much data, another challenge is to actually gather that data: expert-transcribed audio data is not available in abundance. The data that was collected for training Whisper is *weakly supervised*—not only transcriptions by experts are used, but the vast majority of the data is also made up of messier transcriptions that are not necessarily high quality. This allows for much more training data, which makes a noticeable difference. Whisper has an impressive performance close to that of humans for English (Radford et al., 2023), and as the training data contains several languages, it can also transcribe other languages—with the performance of course dependent on the available resources of the language.

In interaction with humanoid robots, humans expect to be able to use speech—so speech recognition is a necessary tool. The advent of the Transformer and attention-based neural networks shows how robust speech recognition is possible, even in suboptimal situations. Especially in education, this is important. Classrooms are seldom not noisy, people learning languages do not always articulate perfectly and speak with accents, and child speech is generally much harder to understand than adult speech. Therefore, robustness is of utmost importance in this application, and Transformers show great potential in achieving it (Janssens et al., 2024a).

3.3 Image Generation

AI-generated images are appearing everywhere in the last few years, causing worry among artists (Ghosh & Fossas, 2022) and enthusiasm among the less artistically gifted. Not too long ago, generating an image based on a description of it was still science fiction. In the early days of image generation, many images of a certain subject were fed to a model, allowing the model to generate a similar image. This was often done with Generative Adversarial Networks (GANs). In a GAN, one neural network—the generator—generates images following the distribution of the training set, while another neural network—the discriminator—tries to distinguish fake, generated images from the training set. Then, these two models are trained simultaneously: the generator becomes better at generating images, while the discriminator becomes better at discriminating between real and generated images (Goodfellow et al., 2014). While these GANs showed some promising results, having to provide a large enough set of example images to generate a new image, is still very far away from the text-to-image generation we see nowadays. An attempt at a text-to-image model based on GANs was made, with encouraging results on specific subjects and at most *visually plausible* results for more general subjects (Reed et al., 2016).

Again, the real revolution that brought AI-generated images into the public eye, was started by applying the Transformer architecture to the problem. In 2021, OpenAI released DALL-E, a Transformer-based text-to-image model, which is an extension to GPT-3, their LLM. It takes text and possibly images as input and returns images as output. A year later, DALL-E 2 was released, as their newest model that generates more accurate and realistic images. DALL-E 2 is a diffusion model: it starts from an image that is just noise, and iteratively removes noise based on the text it was given, to increase the likeness of the image to the content of the text. Stable Diffusion is a similar text-to-image model that is open-source (Podell et al. 2023). Many of these image generation models are also used to edit existing images, such as changing the style of the image, changing the posture of the subject and adding things as the user wants (Kawar et al., 2023).

As high-quality image generation is now possible, advances are made on how to personalise these images. Techniques such as *textual inversion* are used to generate images with specific objects, animals or people in them, based on just a few pictures (Gal et al., 2022). This works by finding which "words", in the vocabulary of the image generation model are closest to the object that we want to represent. Of course, the model—as all models do—doesn't use natural language words internally, it uses a numerical representation of them. Therefore, it can make up new words—new numbers—as needed, which can then be exploited for the personalisation of the image it generates.

Finally, the latest step in the evolution of image generation is a natural extension of it: video generation. As a video can be considered a sequence of images, or even an image with the dimension of time added to it, the aforementioned progress in image generation also translates to video generation, or text-to-video models. Again, OpenAI impressed the world in early 2024 by demonstrating Sora (Brooks et al., 2024): a text-to-video model that is not yet released to the public at the time of writing. It uses similar technologies as the DALL-E models, with a combination of diffusion models and transformers. As this technology is still quite new, it is exciting to wonder what the future of text-to-image and text-to-video will bring, but it is not hard to image that there will be numerous possibilities to apply this to educational settings.

3.4 Multi-Modal Language Models

While language models as discussed above have shown impressive performance in understanding natural language and generating new text or even images based on their input, one piece of the puzzle is still missing: understanding visual input.

Initial language models were not made with multi-modality in mind. They only process information from one modality—text—and generate more text in return, or use that text for other tasks such as predicting a sentiment score or classifying it. However, as discussed in our first section, social robots are situated in the physical world. They need to be aware of their environment and need to be able to integrate what they see into their language. As social robots are often embodied in a humanoid shape, which usually includes eyes, users expect that the robot is able to see. They might ask the robot questions about an element of the environment, or make references to the environment in their speech. Hence, the robot requires visual information to understand the user.

This problem is often called *situated* language interaction: making the robot produce and understand language that refers to the physical and social world around it (Goodwin, 2000). This is also related to the concept of *grounding*: connecting words that refer to elements of the world, such as "red" or "table", or even abstract concepts such as "jealousy" to semantics and properties (Janssens, 2024). Grounding is seen as one of the important challenges that the field of human–robot interaction

(HRI) poses for AI (Lemaignan et al., 2017). It is also a concrete implementation of the wider symbol grounding problem (Harnad, 1990).

Efforts have been long underway to combine vision with natural language. One of the first and most-studied tasks in this area is that of *image captioning*, or writing a description of an image in natural language. Researchers were able to develop well-performing models for this task because of the release of large-scale annotated dataset such as MS COCO (Lin, et al., 2014) and Visual Genome (Krishna et al., 2017). These models often consisted of a convolutional neural network (CNN) that processes the image and a recurrent neural network (RNN) that generates the text. Later, Transformer-based models were introduced to replace text-generating and image-processing models.

This work was then extended to more conversational tasks, such as visual question answering, where the AI model answers natural language questions about an image (Gan et al., 2022), and visual dialogue, where the user and the model can have a longer conversation about the image (Das et al., 2017).

These models were originally built by training them for one specific task—however, this required a large dataset for that task. However, just like for text-only models, pre-training on large datasets that were scraped from the internet, such as images together with their alt-text or already provided captions, allowed the models to become more powerful and be fine-tuned to specific tasks (Gan et al., 2022). Some workarounds also exist to integrating this visual context into language models when visual data is not sufficiently available, such as using a model that was specifically trained for image captioning, and then providing those captions to a text-only model (Janssens et al., 2024b).

Today, vision-language models are slowly becoming as powerful as language-only models. As the models and the datasets they are trained on are becoming larger and larger, they are becoming better and better at many *zero-shot* tasks, i.e. without being fine-tuned. Prominent examples include GPT-4o (OpenAI, 2022), BLIP-2 (Li et al., 2023) and LLaVa (Liu et al., 2024), the latter two of which are open-source. These models can write descriptions of images, answer questions or hold dialogues about them, but also perform more specific tasks, without needing to be fine-tuned.

4 Application of Generative AI in Educational Robotics

In the previous section, we looked at the technological advances that allowed for what can be considered a revolution in AI in the last years. We believe that this wave of advancements can have a strong impact on the success of educational social robots, so this section will describe our vision of how to leverage it optimally. In our vision, we put the social robot in a tutoring role: it does not take the teacher's place, but it can support the teacher during classes, exercises, extra tutoring sessions and homework. Within the framework of this robot tutor, we see a few current limitations that generative AI can help alleviate. First, we look at content generation through generative AI: using tools like LLMs and text-to-image models to write or illustrate the educational content that is delivered by the robot. Secondly, we discuss adaptation, where the flexibility of generative AI allows for adjusting the difficulty of the content to the student based on their personal progress, and thirdly, personalisation, where the content's theme can be fit to the personal interests of the student. Finally, the potential of integrating multi-modal inputs such as vision and speech into a robotic teacher is explored.

Because of personal interests and experience, our focus will be slightly biased towards—but not limited to—language education.

4.1 Content Generation

A large part of education is providing students with the content that contains what students should learn: explanations, both visual and textual illustrations, examples. In language education, this is especially important, as much of language learning comes down to exposure to the language (Ellis & Wulff, 2020).

If a social robot is used for educational purposes, there are three options: all content is provided to the robot beforehand by teachers and educators—in other words, what the robot does is scripted, there is a human in the loop that adjusts the robots behaviour to allow for flexibility in the lessons—in other words, the robot is (partially) teleoperated, or the robot must be able to adjust its lesson on the fly (Senft et al., 2019). The first option is inherently limited: it is not possible to predict all interactions between student and robot, we cannot foresee with which part of the lesson students will struggle or what questions will be asked. This is not to say that there is no use in scripted lessons—sometimes delivering fixed content is the main point of a lesson, but to fully exploit the advantages of educational social robots, some flexibility in the robot's behaviour is necessary. The second option—a (partially) teleoperated robot tutor—is useful and necessary in a research context, but not scalable and most of all, it does not actually help alleviate teacher shortages: if we use robot tutors as classroom support, but each robot is controlled by a teacher, then we could have just added extra teachers, and the robots would not have been necessary. Therefore, we believe that the future of educational social robots is in flexible, adjustable content and that this is one of the main current limitations where generative AI can have a real impact.

When a robot tutor is asked questions, asked for an explanation or to give examples, it should give a natural language response. LLMs are good at exactly this: generating text based on other text. In language education, the correctness of the language of the answer is the most important, for example, in a history class, the information provided by the robot must also match the school curriculum while being factually correct. LLMs can therefore be extended using Retrieval Augmented Generation (RAG), where—next to the implicit knowledge that an LLM has by knowing the probability of words—an external memory is added to the model (Lewis et al., 2020). In educational context, this could be the course's textbook. The LLM can then search the textbook for the relevant information, add this to its inputs, and generate a factually correct answer based on its language knowledge combined with the relevant information. The combination of LLMs and techniques like RAG can allow social robots to support teachers without the added work of preparing scripted lessons.

When delivering content to a student, it can help to also illustrate this visually. In language learning, the meaning of words and concepts can be conveyed to the student visually. That way the need to translate everything to the student's native language is avoided, as the benefit of translation in language learning is a contested topic (Dagilienė, 2012). An additional benefit is that a language learning system that does not depend on translation, is agnostic to the native language of the student so it can be more readily reused. The most intuitive application for visually representing words is vocabulary learning: images of object are shown, either during exercises or when the robot talks about these words. We have shown this to be effective in a proof-of-concept study, where students play a game to learn vocabulary. They are shown a set of images and are told a description of one of these images—either by a robot or played from a tablet. All the images and descriptions were AI-generated, based on the estimated language level of the student. The students had to choose which image fits best to the description. The game was played in a language that the students had very limited knowledge of, and even though no translations to their native language were used, they showed a statistically significant learning gain. This experiment shows the power of generative AI in adaptive content generation in educational contexts (Verhelst et al., 2024).

In other parts of language learning, similar approaches can be developed. As much of language learning depends on exposure to the language, adding visual support of what the robot is talking about can help the student to better understand what is said. Imagine a robot telling a story interactively, using students' input for the progression of the story—which is of course generated using an LLM. This will help engage the students, as they are actively part of the lesson, and they can influence the story to be more interesting to them. The students are still learning the language, so they might not understand each word or language construction the robot uses, but generative AI is used to visualise the story as it is told. This can keep the students' attention, while filling in the gaps of what the students don't yet understand independently: a perfect combination for language learning.

Generating visual content can of course also be used in education other than language learning. Everywhere, from geography to mathematics classes, illustrations are used in textbooks. Ensuring that we can generate illustrations that consistently adhere to the curriculum is not self-evident, but using techniques like editing existing images using generative AI can circumvent these issues: starting with correct images and editing only what is needed to fit the context.

4.2 Difficulty Adaptation

In an educational context, it is important to deliver learning material of the correct difficulty level. Following Vygotsky's theory on the zone of proximal development, a student learns most from content that is slightly too difficult for the student to solve independently, but just right with the help of a more advanced peer or an adult (Chaiklin, 2003)—or, of course, a robot. Following this theory, educational content should not be the same for every student. An educational social robot should have some way to estimate the student's level, either by modelling the student's knowledge, or collecting feedback during the lesson, in the form of incorrect answers and expressions of confusion, both spoken and facial. If interactions with the social robot are largely scripted, it is difficult to dynamically adapt the difficulty of educational content to the student.

As generative AI allows for exercises to be generated during the interaction, it is possible to use what is known about the student's level to influence which exercises are generated. This can mean that we either adjust the language that is being used, or which content is being delivered to the student. During a mathematics class, we could monitor whether a student regularly makes mistakes on a certain topic. If such a topic has been identified, the lesson can be dynamically adjusted to focus more on this topic. It can be explained again or in more detail, and—using LLMs to generate the text, while the content is taken from the textbook using RAG—the explanation will be on the same topic but phrased differently. More exercises on the same topic can be chosen until the student consistently shows good results. This way, the same concept can be taught to all students, starting from the same textbook and a loosely scripted lesson, that is then adapted to each student based on their answers to exercises, their feedback, their non-verbal communication and their questions.

To allow knowledge to be transferred to students, their understanding of this content should be maximised. Therefore, all language that is used in class should be in reach of the student's language level. Even if the content that is to be delivered is scripted, generative AI allows for flexibility in the difficulty of the language in which it is delivered.

For language learning, most of the content is of course language itself, so this is extra relevant. As mentioned before, in language learning, exposure to the language is of utmost importance (Ellis & Wulff, 2020) and ensuring that the language level is matched to that of the student has been shown to increase learning, and might increase the students' participation during class (Randall, 2019). Adaptation of the lesson can be applied to exercises on e.g. grammar, similar to what was mentioned above, but it can also be applied to conversational learning. If a robot tutor is available during a second language course, it can play the role of a native speaker in the language that the student is learning, providing examples of typical ways of speaking and correct pronunciation, while serving as a conversation partner. The content of the conversation does not have to be educational, as chatting with a native speaker about anything is increased exposure, and thus useful. As many schools struggle with a shortage of teachers, and these teachers often do not have the time to practise conversational skills with their students, this might be a valuable asset for schools to bridge the gap from theoretical language learning to actual communicative skills.

Influencing the difficulty level of language that is generated by a model can be done by prompting the model to generate language of this level, by prompting the model to match the difficulty level of the student's inputs or by using a language model that is specifically trained to generate language of a certain difficulty level. The last option is the least flexible, as training a model must be done beforehand, but is likely the most robust way to ensure the desired behaviour (Lester et al., 2021).

Previously, adjusting the difficulty level of the content in educational technology could only be done by choosing which exercise to give the student, or by changing the exercises based on some predefined parameters that change the level of the exercise. Using generative AI for generating educational content gives us much more flexibility: the difficulty of content can be dynamically adapted to the student. As this can be done based on any input by the student—not only answers to exercises, this can happen at a much more fine-grained level than what can be predefined. This allows educational tutoring systems to adapt to every student individually, to create the best possible learning environment.

4.3 Personalisation

We have seen that generative AI can be used to generate educational content and that it can be used to adapt the content to the student's educational level. Personalisation also involves adapting the educational content to the student, but with a focus on improving the student's engagement, by ensuring a connection between the content and the student's personal life, making it more relatable (Belpaeme et al., 2018). Students who enjoy sports more than they do mathematics classes might be more motivated to learn mathematical operations if they can do so by calculating the score of their team given a description of the match. So, while the content stays the same, personalisation can be used to package the contents in an engaging way.

LLMs can generate exercises that involve students' hobbies, and image generation models can be used to illustrate them. If these illustrations involve people, we can change their appearance to be more familiar to the student, as people tend to learn better from those who are more like themselves, and it can make students feel more included and involved in the lesson (Binderkrantz & Bisgaard, 2024).

Additionally, new techniques like textual inversion, allow for inserting certain people, animals or objects into generated images. Catering the content to students' interests might help to engage them, which is beneficial for learning. We can use these techniques to insert elements of the student's life into educational content. In educational materials, illustrations are often used to make the lesson more visually appealing (Houghton & Willows, 1987). If these illustrations are altered using generative AI techniques, to include the student's favourite tv character, family or even the student themselves, it can be more engaging, while starting from a textbook image ensures that the content is still correct.

For language learning, the conversational robot tutor, with which you practise language of your own level, can talk about anything you like, even remembering your interests and asking about your hobbies. If the main goal of the conversation is practising the language, the options for personalisation are endless. If we are trying to convey specific concepts, personalisation can be done as described similarly to other subjects, conveying the same content but packaging it according to the student's interests. Visual support can be adapted to the student's life: imagine the storytelling robot from before, where students could change the story's narrative interactively. Personalisation can allow it to include their pet dog, and generative AI is now able to actually show this pet in whatever adventure the student chooses for it.

4.4 Multi-Modal Interactions

The revolutions in AI allow educational robots to do more than generate textual or visual content with endless possibilities for on-the-fly adaptation—they also allow the robot to perceive the world around it through multiple modalities, such as what they see and hear. This enables even more immersive and effective learning interactions.

Let us return to the example of language learning. Earlier, we discussed how visual content could be generated to support vocabulary learning. Now, this visual content could also be replaced by the physical world around the robot and the learner themselves. In Wolfert et al. (2024), we showed that students learn vocabulary better when they are provided a visual referent of the concept they are learning, either on a screen or with a physical object. This approach could be much expanded. Everything the robot sees about its environment could be provided to the language model that is driving the conversation. The robot could therefore make references to objects or people in the surroundings, or even to movement, and the robot could ask or answer questions about the environment. These strategies can enrich natural conversations between the robot and the learner and could help the learner by grounding the language they are learning in the physical world, and by offering opportunities to practice concepts that require a physical referent in the environment. Visual content could also help to personalise the interaction with the user, potentially boosting motivation and social bonding, such as in (Janssens et al., 2024b), where the robot addresses the user by asking a question about a feature of the user that the robot sees, such as their apparel.

Not only visual input can enrich the learning interaction but also auditive information can be important. We discussed earlier that speech recognition systems have improved considerably in the last decade, including for smaller populations such as children. Now, finally, real conversational interactions between robots and children or non-native speakers might be possible. One could even dream of systems giving feedback on pronunciation. As an example, (Kennedy et al., 2024) have managed to extract pronunciation information from users saying their names, to enable the robot to correctly pronounce those names. (Amioka et al., 2023) also explored pronunciation learning with social robots, by generating realistic lip movements for the robot when pronouncing the words to be taught—however, learners who do not yet have any knowledge of the language do not seem to benefit from these realistic lip movements.

All modalities will have to be integrated to make the robot aware of its social environment. Social cognitive aspects of the learning interaction, such as the learner voluntarily or involuntarily communicating that they are interested, engaged, motivated, uncertain, struggling should be detected by the robot and provided to the AI models generating the interaction, in order to adapt the interaction to the learner, improving both the learning and social effectiveness of the interaction.

Finally, we recognise that we have so far omitted the modalities of touch and proxemics (i.e. how close the human and robot are standing from each other). Both play important roles in human–human social interactions, and have been shown to also modulate human–robot interactions (Mumm & Mutlu, 2011; Ren et al., 2023). This could be leveraged to further improve educational robots, but has not been widely studied yet.

5 Technical and Societal Limitations of Generative AI

Recent advances in generative AI allow us to dream big: it is now possible to have conversations with robots about anything, speech recognition seems to work for typical populations and is rapidly improving for atypical populations, and we can generate images and even videos that contain anything that can be thought of. Still, some challenges are left before generative AI can freely be applied to educational robotics.

5.1 Real Time Conversations

For a conversation to feel natural, the timing must be right. In human–human conversation, a lot happens to decide who speaks when: people know when their turn to speak arrives, based on eye contact, tone of voice and context. When changing turns—one person stops talking and another starts—people leave only 200 ms of silence on average, and often our speech even overlaps. For conversational systems, a response time of 700–1000 ms is often deemed acceptable, even though this is already much longer than humans usually take (Skantze, 2021). That means that a robotic tutor must be able to provide an answer to students in less than one second. In this short time, many things must happen: the speech recognition must provide a transcript of the question, and in some cases, e.g. when using RAG, a search must happen, then, a language model must generate an answer based on the question and perhaps the search results and finally, speech must be synthesised to play from the robot's speakers. There are ASR systems that manage to generate a transcription in sub-second transcription time, but this is only one step of the way (Janssens et al.,

2024a). Answer generation by an LLM is reported to sometimes take up to three seconds, bringing us far over the maximum of one second, even without speech synthesis (Irfan et al., 2023). Therefore, before natural human–robot conversations can take place, the efficiency of the involved systems must be dramatically improved.

5.2 *Controlling Generative AI*

As discussed before, controlling the output of generative AI is generally done with prompt engineering: describing the task at hand to the model, using zero-shot or few-shot learning. Although this is a powerful way of applying a general model to a task it was not specifically trained for, tasks must be clearly and unambiguously described, using at most a few examples. This technique does not suit all tasks and there are limits to the control the user has over generated output when prompting a model. Prompting can have very variable results, with changes as small as adding one word already leading to vastly different results (Liu et al., 2023). Another often occurring problem is so-called hallucination: generated output diverging from the desired output, earlier output or conflicting with real-world facts (Alkaissi & McFarlane, 2023; Zhang et al., 2023). This is detrimental to the reliability of the outputs, especially when used in real-world applications. Additionally, strict control over the tone and language level of generated output is difficult, while ensuring appropriate text—both in difficulty level and in content—is of utmost importance in educational contexts. Additionally, if a social robot is expected to deliver some parts of a school's curriculum, we must be able to guarantee that it actually does this, and that it does not tell the students anything incorrect or confusing. Finally, as LLMs are trained on human data, if an LLM is told that what it said was incorrect, it tends to agree, no matter the truth. So, if a conversational system that adheres to the curriculum and that tells the truth is created, students might think it is wrong, question it or correct it, and get fed false information anyway. In educational settings, this can have strong consequences, with students learning incorrect things and not knowing who to trust. Therefore, having strict control over generative AI's output is needed for reliable educational social robots.

5.3 *Sustainability*

While generative AI and all technologies based on large neural networks are very powerful, its sustainability leaves much to be desired. When a neural network processes information, it does calculations with this data. Each of these calculations takes a small amount of energy. The revolution in these technologies, as discussed in earlier sections, allowed much more calculations to be done in shorter times. Processing data—doing calculations—could now be done in parallel. Of course, if each calculation takes some energy, parallelising the calculations so we can do many of them at once makes it more energy-consuming per time unit, while requiring less time. However, the revolution in AI showed that these models had the capacity to find patterns and mimic reasoning when given enough data. So, these models were trained on more data, giving us more impressive models: larger, with more input and better output—and a larger energy consumption. The number of computations that our AI models require has been growing exponentially for a while now, together with their energy demands (Strubell et al., 2020). Of course, there is something to say about an increase in energy demand if it comes with increased performance, but the demand that comes with exponential growth is not something that our planet can afford. There are a few things that we can do to minimise this impact while still reaping the benefits of these impressive technologies (Lacoste et al., 2019). The location where a model is trained determines the energy source, how sustainable it is, and the carbon footprint of the model. Choosing this wisely is impactful and should be communicated transparently when making a model available. When using a model, its impact can be influenced by being mindful of the energy sources used in deployment. Additionally, choosing not to train your own model, but fine-tuning an existing one or even just prompting it without additional training minimises the computational resources used. Finally, the hardware on which the AI models are trained and used, as well as the hardware of the robot, have a climate impact. More efficient hardware, which requires less computing time or performs your computations more effectively, will minimise the impact of this hardware. So, while using this powerful technology opens a lot of doors, especially in educational robotics, its consequences should be considered every step of the way, so that climate impact can be minimised where possible.

5.4 Privacy

When using LLMs and other generative AI in an educational context—often with minors, privacy is an important concern (Mhlanga, 2023). Most of currently best performing LLMs, such as OpenAI's GPT family and Google's Gemini, are commercially available through an API. This raises the problem that using it entails sending potentially sensitive data of vulnerable people to a company without knowing exactly how this data is saved and whether the company uses it, while issues of these models' adherence to privacy standards have been reported (Cartwright et al., 2024). It is not unthinkable that companies that provide LLM services, would use this user data to further optimise their models. While this is understandable from the perspective of optimising the models, children's data in schools should not be freely used like that. Therefore, to ensure privacy on the data given to educational social robots, local LLMs should be used. Many open-source, local models exist, but running them can be complicated for people that do not have a technical background, such as often those in educational sectors. Therefore, the development of social robots for education that use generative AI should include providing easily accessible ways to run local LLMs, to ensure the privacy of students.

5.5 Ethical Problems

If generative AI is used in educational robots, we can make them more human-like and social than ever. We believe this can bring great benefits to students' learning, but it also carries risks. The more socially human-like a conversational system or robot is, the higher the chance that students form attachments to it (Huber et al., 2016). This can be dangerous for their development of social skills, as it may start replacing social contact with peers. Especially for younger students, attachment may pose a problem if the robot then breaks, is removed or replaced or the student loses access to it for other reasons. It is important that it is clear for students at all times that the educational robot is a tool for learning, not a person. On the other hand, if students know that the robot is not human, which means there are no consequences to misbehaving towards it, this may lead to the normalisation of negative behaviour towards the robot, that later transfers to peers or teachers as well (Nomura et al., 2015).

In current conversational systems using generative AI, there is already a risk of spreading false information (Chen & Shu, 2023). Put such a system into a role with authority on their education—such as a tutor, a teacher's assistant or worse, a teacher—and students will not doubt its factuality. This is amplified by the appearance of reasoning in LLMs, which can take away doubt if it does arise. Additionally, LLMs tend to agree with whatever (false) information is suggested to them, which carries the risk of becoming an echo chamber (Nehring et al., 2024). Especially for young, vulnerable people, this can quickly become damaging and should be avoided at all costs in educational environments.

As governments and other regulatory bodies are exploring how to regulate AI in order to minimise the risks the technology poses to citizens and society, education is seen as an application domain of special importance. As an example, in the European Union's AI Act, education is identified as a high-risk application domain, meaning AI systems deployed in education should comply with quality, safety and transparency requirements, and could be subjected to human oversight. In particular, using emotion recognition in educational institutions is classified as an *unacceptable risk* and will be prohibited. It is yet unclear what emotion recognition entails exactly, but care should be taken to ensure that such regulation does not limit opportunities for making socially interactive systems more adaptive, while ensuring that systems that would track students' emotions at scale remain banned. Beyond education as a specific application area, general-purpose AI models, such as some LLMs discussed above, will undergo additional scrutiny, and have to comply to transparency requirements (Artificial Intelligence Act, 2024).

6 Conclusions

Today, social robots are not yet present in our classrooms. New, rapidly changing technology might change this in the near future. While educational robotics was limited by scripted interactions, failing speech recognition and only rigid options for adaptation to students, a revolution in generative AI technology has occurred, that might just change everything.

In this chapter, we described our vision of the future that will follow from this revolution. We started by investigating the aforementioned limitations of educational robotics and their causes. What followed, was an introduction to the rapid changes in generative AI technologies that caused what we call a technological revolution. These came together in the main section of this chapter, where we described in detail our future vision of educational robotics: how will this technological revolution solve the current problems, and what is the potential impact of this on our classrooms.

In this vision, the content that is provided in lessons will not be the same for every student, as there will be room for content generation on the fly. This can range from interactive explanations by a robot tutor, to illustrations of the stories that students are making up then and there. Generative AI will allow tutoring systems to adapt their difficulty level to each student—in a much more personal, fine-grained, and generalisable way than what is done today. Robots will manage to engage students even more by finding out what interests them specifically and use this to package content, so they want to know more about it. And as social robots are embodied, they share the physical world with us, which we can use to our advantage—from recognising when a student has lost interest, to involving the world around them in conversation.

Of course, this technology is still in rapid development. It allows us to dream big, but as is always the case with new technology, it also introduces new issues. Some of these issues are technological challenges waiting to be solved. As technology evolves, generative models' inference speed will decrease, allowing shorter generation times. It is reasonable to assume that the progression of AI technologies will allow for real time, natural flowing conversation with educational robots. The constraining of generative AI to be trustworthy is an issue that asks for future research, with promising techniques arising rapidly. The sustainability of AI models is a problem that cannot be left unsolved, and future research should take this into account every step of the way, by minimising impact where possible and ensuring transparency of the results. The privacy of children in schools should be preserved, by either demanding transparency from commercial partners or avoiding them by running AI models locally as much as possible and keeping full control over the involved data. Finally, the position of educational robots and, more generally, generative AI in our school should be chosen with care and communicated clearly to everyone involved, including those who might still be too young to grasp the possible dangers.

It is clear that there is ample room for future research in the field of generative AI-driven educational robotics. On the one hand, the current limitations require new advancements, and existing technologies require care in their application to a field as critical and vulnerable as education. On the other hand, the revolution in generative AI opened many doors that have not yet been taken. The dream described in this chapter has not yet been realised, although it might already be possible. Only time will tell if the future of social robots in education will change as we described, and if Asimov's dream of mechanical teachers will come true.

References

Adeshola, I., & Adepoju, A. P. (2023). The opportunities and challenges of ChatGPT in education. *Interactive Learning Environments* 1–14.

Alkaissi, H., & McFarlane, S. I. (2023). Artificial hallucinations in ChatGPT: Implications in scientific writing. *Cureus*.

Amioka, S., Janssens, R., Wolfert, P., Ren, Q., Pinto Bernal, M. J., & Belpaeme, T. (2023). Limitations of audiovisual speech on robots for second language pronunciation learning. In *Proceedings of the 2023 ACM/IEEE International Conference on Human-Robot Interaction* (pp. 359–367). Stockholm, Sweden: Association for Computing Machinery.

Artificial Intelligence Act. (2024). European Parliament and Council.

Bahdanau, D., Chorowski, J., Serdyuk, D., Brakel, P., & Bengio, Y. (2016). End-to-end attention-based large vocabulary speech recognition. In *2016 IEEE International Conference on Acoustics, Speech and Signal Processing (ICASSP)* (pp. 4945–4949). IEEE.

Bara, F., & Kaminski, G. (2019). Holding a real object during encoding helps the learning of foreign vocabulary. *Acta Psychologica, 196*, 26–32. https://doi.org/10.1016/j.actpsy.2019.03.008

Belpaeme, T., Kennedy, J., Ramachandran, A., Scassellati, B., & Tanaka, F. (2018). Social robots for education: A review. *Science Robotics*.

Binderkrantz, A. S., & Bisgaard, M. (2024). A gender affinity effect: The role of gender in teaching evaluations at a Danish university. *Higher Education*, 591–610.

Brooks, T., Peebles, B., Holmes, C., DePue, W., Guo, Y., Jing, L., Ramesh, A. (2024). Video generation models as world simulators. Retrieved from https://openai.com/research/video-generation-models-as-world-simulators

Brown, T., Mann, B., Ryder, N., Subbiah, M., Kaplan, J. D., Dhariwal, P., Askell, A. (2020). Language models are few-shot learners. *Advances in neural information processing systems*.

Cartwright, O., Dunbar, H., & Radcliffe, T. (2024). Evaluating privacy compliance in commercial large language models-Chatgpt, Claude, and Gemini.

Chaiklin, S. (2003). The zone of proximal development in Vygotsky's analysis of learning and instruction. *Vygotsky's educational theory in cultural context*, 39–64.

Chan, W., Jaitly, N., Le, Q. V., & Vinyals, O. (2015). Listen, attend and spell. *arXiv preprint* arXiv:1508.01211.

Chen, C., & Shu, K. (2023). Combating misinformation in the age of LLMS: Opportunities and challenges. *AI Magazine*.

Chevalier, M., Riedo, F., & Mondada, F. (2016). Pedagogical uses of thymio II: How do teachers perceive educational robots in formal education?. *IEEE Robotics & Automation Magazine, 23*(2), 16–23. https://doi.org/10.1109/MRA.2016.2535080

Chrysafiadi, K., & Virvou, M. (2013). Student modeling approaches: A literature review for the last decade. *Expert Systems with Applications, 40*(11), 4715–4729. https://doi.org/10.1016/j.eswa.2013.02.007

Cumbal, R., Lopes, J., & Engwall, O. (2020). Detection of listener uncertainty in robot-led second language conversation practice. In *Proceedings of the 2020 International Conference on Multimodal Interaction* (pp. 625–629).

Dagilienė, I. (2012). Translation as a learning method in English language teaching. *Studies about languages*, 124–129.

Das, A., Kottur, S., Gupta, K., Singh, A., Yadav, D., Moura, J. M., Batra, D. (2017). Visual dialog. In *Proceedings of the IEEE conference on computer vision and pattern recognition* (pp. 326–335).
Dempere, J., Modugu, K., Hesham, A., & Ramasamy, L. K. (2023). The impact of ChatGPT on higher education. *Frontiers in Education.*
Devlin, J., Chang, M.-W., Lee, K., & Toutanova, K. (2018). Bert: Pre-training of deep bidirectional transformers for language understanding. *arXiv preprint* arXiv:1810.04805.
D'Mello, S., Kappas, A., & Gratch, J. (2018). The affective computing approach to affect measurement. *Emotion Review, 10*(2), 174–183. https://doi.org/10.1177/1754073917696583
Ellis, N. C., & Wulff, S. (2020). Usage-based approaches to L2 acquisition. In G. D. Bill Vanpatten, *Theories in second language acquisition, an introduction* (pp. 63–82). New York: Routledge.
Feuerriegel, S., Hartmann, J., Janiesch, C., & Zschech, P. (2024). Generative AI. *Business & Information Systems Engineering*, 111–126.
Flook, R., Shrinah, A., Wijnen, L., Eder, K., Melhuish, C., & Lemaignan, S. (2019). On the impact of different types of errors on trust in human-robot interaction: Are laboratory-based HRI experiments trustworthy? *Interaction studies*, 455–486.
Gal, R., Alaluf, Y., Atzmon, Y., Patashnik, O., Bermano, A. H., Chechik, G., & Cohen-Or, D. (2022). An image is worth one word: Personalizing text-to-image generation using textual inversion. *arXiv preprint* arXiv:2208.01618.
Gan, Z., Li, L., Li, C., Wang, L., Liu, Z., & Gao, J. A. (2022). Vision-language pre-training: Basics, recent advances, and future trends. *Foundations and Trends in Computer Graphics and Vision*, 163–352.
Ghosh, A., & Fossas, G. (2022). Can there be art without an artist? *arXiv preprint* arXiv:2209.07667.
Goodfellow, I., Pouget-Abadie, J., Mirza, M., Xu, B., Warde-Farley, D., Ozair, S., Bengio, Y. (2014). Generative adversarial nets. *Advances in neural information processing systems.*
Goodwin, C. (2000). Action and embodiment within situated human interaction. *Journal of pragmatics*, 1489–1522.
Gunes, H., & Churamani, N. (2023). Affective computing for human-robot interaction research: Four critical lessons for the Hitchhiker. In *2023 32nd IEEE International Conference on Robot and Human Interactive Communication (RO-MAN)* (pp. 1565–1572). IEEE.
Harnad, S. (1990). The symbol grounding problem. *Physica D: Nonlinear Phenomena*, 335–346.
Henkemans, O. A., Bierman, B. P., Janssen, J., Neerincx, M. A., Looije, R., van der Bosch, H., & van der Giessen, J. A. (2013). Using a robot to personalise health education for children with diabetes type 1: A pilot study. *Patient education and counseling*, 174–181.
Houghton, H. A., & Willows, D. M. (1987). *The psychology of illustration* (Vol. 1). Springer.
Huber, A., Weiss, A., & Rauhala, M. (2016). The ethical risk of attachment how to identify, investigate and predict potential ethical risks in the development of social companion robots. In *2016 11th ACM/IEEE International Conference on Human-Robot Interaction (HRI)* (pp. 367–374). IEEE.
Irfan, B., Kuoppamäki, S.-M., & Skantze, G. (2023). Between reality and delusion: Challenges of applying large language models to companion robots for open-domain dialogues with older adults.
Janssens, R. (2024). Multi-Modal language models for human-robot interaction. In *Companion of the 2024 ACM/IEEE International Conference on Human-Robot Interaction* (pp. 109–111). Boulder, CO, USA: Association for Computing Machinery.
Janssens, R., Verhelst, E., Abbo, G. A., Ren, Q., Bernal, M. J., & Belpaeme, T. (2024a). Child speech recognition in human-robot interaction: problem solved? *arXiv preprint* arXiv:2404.17394.
Janssens, R., Wolfert, P., Demeester, T., & Belpaeme, T. (2024b). Integrating visual context into language models for situated social conversation starters. *IEEE Transactions on Affective Computing*, 1–14.
Karpagavalli, S., & Chandra, E. (2016). A review on automatic speech recognition architecture and approaches. *International Journal of Signal Processing, Image Processing and Pattern Recognition*, 393–404.

Kasneci, E., Seßler, K., Küchemann, S., Bannert, M., Dementieva, D., Fischer, F., Hüllermeier, E. (2023). ChatGPT for good? on opportunities and challenges of large language models for education. *Learning and individual differences.*

Kawar, B., Zada, S., Lang, O., Tov, O., Chang, H., Dekel, T., Irani, M. (2023). Imagic: Text-based real image editing with diffusion models. In *Proceedings of the IEEE/CVF Conference on Computer Vision and Pattern Recognition*, (pp. 6007–6017).

Kennedy, J., Kumar, N., & Paetzel-Prüsmann, M. (2024). Name pronunciation extraction and reuse in human-robot conversations. In *Companion of the 2024 ACM/IEEE International Conference on Human-Robot Interaction* (pp. 593–597). Boulder, CO, USA: Association for Computing Machinery.

Kennedy, J., Lemaignan, S., Montassier, C., Lavalade, P., Irfan, B., Papadopoulos, F., Belpaeme, T. (2017). Child speech recognition in human-robot interaction: Evaluations and recommendations. In *Proceedings of the 2017 ACM/IEEE international conference on human-robot interaction* (pp. 82–90). ACM/IEEE.

Krishna, R., Zhu, Y., Groth, O., Johnson, J., Hata, K., Kravitz, J., Shamma, D. A. (2017). Visual genome: Connecting language and vision using crowdsourced dense image annotations. *International journal of computer vision*, 32–73.

Krumhuber, E. G., Kappas, A., & Manstead, A. S. R. (2013). Effects of dynamic aspects of facial expressions: A review. *Emotion Review, 5*(1), 41–46. https://doi.org/10.1177/1754073912451349

Lacoste, A., Luccioni, A., Schmidt, V., & Dandres, T. (2019). Quantifying the carbon emissions of machine learning. *arXiv preprint* arXiv:1910.09700.

Lemaignan, S., Warnier, M., Sisbot, E. A., Clodic, A., & Alami, R. (2017). Artificial cognition for social human–robot interaction: An implementation. *Artificial Intelligence*, 45–69.

Lester, B., Al-Rfou, R., & Constant, N. (2021). The power of scale for parameter-efficient prompt tuning. *arXiv preprint* arXiv:2104.08691.

Lewis, P., Perez, E., Piktus, A., Petroni, F., Karpukhin, V., Goyal, N., . . . Rockt{\"a}schel, T. (2020). Retrieval-augmented generation for knowledge-intensive NLP tasks. *Advances in Neural Information Processing Systems*, 9459–9474.

Li, J., Li, D., Savarese, S., & Hoi, S. (2023). Blip-2: Bootstrapping language-image pre-training with frozen image encoders and large language models. In *International Conference on Machine Learning* (pp. 19730–19742). PMLR.

Lin, T.-Y., Maire, M., Belongie, S., Hays, J., Perona, P., Ramanan, D., Zitnick, C. L. (2014). Microsoft coco: Common objects in context. In *Computer Vision--ECCV 2014: 13th European Conference, Zurich, Switzerland, September 6–12, 2014, Proceedings, Part V 13* (pp. 740–755). Zurich, Switzerland: Springer.

Liu, H., Li, C., Wu, Q., & Lee, Y. J. (2024). Visual instruction tuning. *Advances in Neural Information Processing Systems.*

Liu, X., Zheng, Y., Du, Z., Ding, M., Qian, Y., Yang, Z., & Tang, J. (2023). GPT understands, too. *AI Open.*

Mehrabian, A. (1972). *Nonverbal Communication.* Transaction Publishers.

Mesquita, B., & Boiger, M. (2014). Emotions in context: A sociodynamic model of emotions. *Emotion Review, 6*(4), 298–302. https://doi.org/10.1177/1754073914534480

Mhlanga, D. (2023). Open AI in education, the responsible and ethical use of ChatGPT towards lifelong learning. In *FinTech and artificial intelligence for sustainable development: The role of smart technologies in achieving development goals* (pp. 387–409). Springer.

Mumm, J., & Mutlu, B. (2011). Human-robot proxemics: Physical and psychological distancing in human-robot interaction. In *Proceedings of the 6th International Conference on Human-Robot Interaction* (pp. 331–338). Lausanne, Switzerland: Association for Computing Machinery.

Nehring, J., Gabryszak, A., Jürgens, P., Burchardt, A., Schaffer, S., Spielkamp, M., & Stark, B. (2024). Large language models are echo chambers. In *Proceedings of the 2024 Joint International Conference on Computational Linguistics, Language Resources and Evaluation (LREC-COLING 2024)*, (pp. 10117–10123).

Nomura, T., Uratani, T., Kanda, T., Matsumoto, K., Kidokoro, H., Suehiro, Y., & Yamada, S. (2015). Why do children abuse robots? In *Proceedings of the tenth annual ACM/IEEE international conference on human-robot interaction extended abstracts*, (pp. 63–64).
OpenAI. (2022). *GPT-4 Technical Report.*
Podell, D., English, Z., Lacey, K., Blattmann, A., Dockhorn, T., Müller, J., Rombach, R. (2023). Sdxl: Improving latent diffusion models for high-resolution image synthesis. *arXiv preprint* arXiv:2307.01952.
Radford, A., Kim, J. W., Xu, T., Brockman, G., McLeavey, C., & Sutskever, I. (2023). Robust speech recognition via large-scale weak supervision. *International conference on machine learning.* PMLR.
Randall, N. (2019). A survey of robot-assisted language learning (RALL). *ACM Transactions on Human-Robot Interaction (THRI)*, 1–36.
Reed, S., Akata, Z., Yan, X., Logeswaran, L., Schiele, B., & Lee, H. (2016). Generative adversarial text to image synthesis. In *International conference on machine learning* (pp. 1060–1069). PMLR.
Reimann, M. M., Kunneman, F. A., Oertel, C., & Hindriks, K. V. (2024). A survey on dialogue management in human-robot interaction. *ACM Transactions on Human-Robot Interaction, 13*(2), 1–22. https://doi.org/10.1145/3648605
Ren, Q., Hou, Y., Botteldooren, D., & Belpaeme, T. (2023). Behavioural models of Risk-Taking in Human–Robot tactile interactions. *Sensors*, 4786.
Schutz, P. A., Hong, J. Y., Cross, D. I., & Osbon, J. N. (2006). Reflections on investigating emotion in educational activity settings. *Educational Psychology Review, 18*, 343–360. https://doi.org/10.1007/s10648-006-9030-3
Senft, E., Lemaignan, S., Baxter, P. E., Bartlett, M., & Belpaeme, T. (2019). Teaching robots social autonomy from in situ human guidance. *Science Robotics, 4*(35), eaat1186. https://doi.org/10.1126/scirobotics.aat1186
Skantze, G. (2021). Turn-taking in conversational systems and human-robot interaction: A review. *Computer Speech & Language.*
Strubell, E., Ganesh, A., & McCallum, A. (2020). Energy and policy considerations for modern deep learning research. In *Proceedings of the AAAI conference on artificial intelligence*, (pp. 13693–13696).
Vaswani, A., Shazeer, N., Parmar, N., Uszkoreit, J., Jones, L., Gomez, A. N., Polosukhin, I. (2017). Attention is all you need. *Advances in neural information processing systems.*
Verhelst, E., Janssens, R., Demeester, T., & Belpaeme, T. (2024). Adaptive second language tutoring using generative AI and a social robot. In *Companion of the 2024 ACM/IEEE International Conference on Human-Robot Interaction*, (pp. 1080–1084).
Wei, J., Wang, X., Schuurmans, D., Bosma, M., Xia, F., Chi, E., Zhou, D. (2022). Chain-of-thought prompting elicits reasoning in large language models. *Advances in neural information processing systems.*
Wolfert, P., De Gersem, L., Janssens, R., & Belpaeme, T. (2024). Multi-Modal language learning: Explorations on learning Japanese vocabulary. In *Companion of the 2024 ACM/IEEE International Conference on Human-Robot Interaction* (pp. 1129–1133). Boulder, CO, USA: Association for Computing Machinery.
Xiong, W., Wu, L., Alleva, F., Droppo, J., Huang, X., & Stolcke, A. (2018). The microsoft 2017 conversational speech recognition system. In *2018 IEEE international conference on acoustics, speech and signal processing (ICASSP).* IEEE.
Zhang, Y., Li, Y., Cui, L., Cai, D., Liu, L., Fu, T., Chen, Y. (2023). Siren's song in the AI Ocean: A survey on hallucination in large language models. *arXiv preprint* arXiv:2309.01219.

Interactions in the Wild: Student and Teacher Perceptions of Social Robots in Public Classrooms

Gerald LeTendre, Raisa Gray, and Mark Toci

Abstract Research on the use of social robots in education is growing rapidly, especially in areas related to autism-spectrum disorder students, language education, and coding, but other areas have not been well researched. Few studies actually attempt to test general robotic functioning in real-time classroom situations, such as under "in the wild" conditions. In addition, few studies have focused on early adolescents and the middle grades. In our chapter, we will summarize the findings from a naturalistic study of social robots for early adolescents. We detail how we introduced a "Pepper" unit into a project-based learning environment at a public middle school. Using reflective essays by teachers along with surveys and interviews with students, we highlight the potentials and drawbacks of using an advanced social robot compared to other common robotic tools used at this level (e.g., VEX kits, drones).

Keywords Project-Based learning · Social robots · Robotics in the classroom · Programmable robots · Humanoid robots · Educational robots

1 Introduction

Social robots (Breazeal, 2002) as a class of educational technology are already in use in second-language learning (van der Berghe et al., 2019) and with children who have autism-spectrum disorders (Dautenhahn & Werry, 2000). Despite researcher's interest in these robots, (Papadopoulos et al., 2020; van Straten et al., 2020), the

G. LeTendre (✉) · R. Gray · M. Toci
Department of Education Policy Studies, The Pennsylvania State University, University Park, USA
e-mail: letendre@psu.edu

R. Gray
e-mail: rsg78@psu.edu

M. Toci
e-mail: mark@clccharter.org

G. Lampropoulos and S. Papadakis (eds.), *Social Robots in Education*, Studies in Computational Intelligence 1194, https://doi.org/10.1007/978-3-031-82915-4_3

widespread adoption of social robots in schools seems unlikely. Although Belpaeme and colleagues (Belpaeme et al., 2018) argued that social robots "have the potential to become part of the educational infrastructure, just as paper, whiteboards, and computer tablets" (p. 7), field-based studies indicate that the technical limitations of the current generation of social robots and the intense resources needed to support them in the classroom constitute significant barriers to widespread adoption in K-12 schooling (Woo et al., 2021).

Can social robots, deployed in autonomous mode under real-life classroom conditions—e.g., "in the wild" (Dautenhahn, 2003; Salter et al., 2007)—be useful tools for teachers who do not have access to extensive support? Perhaps the research focus on language learning, programming, or other specific measures of student outcomes (see review in Chu et al., 2022) obscures the value of social robots as "objects-to-think-with" (e.g., as noted by Papert, 1980), which refers to "objects in which there is an intersection of cultural presence, embedded knowledge, and the possibility for personal identification" (p. 11). What if the robot itself, virtually "out of the box," became a project that both teachers and students attempted to master?

The paper fills in a gap in the literature by addressing how unstructured interactions with social robots, imperfect though they may be, can provide important learning opportunities. Rather than constraining student interaction with the robot, or relying on expert programmers, students and teachers in this study directly confronted the glitches and problems associated with deploying social robots in a normal classroom environment. By analyzing student and teacher perceptions, this paper adds to our understanding of how unstructured interactions between early adolescents and social robots can become a powerful pedagogical strategy.

In this paper, we document what happened when we introduced a "Pepper" robot (Softbank, n.d.) into a public middle school classroom, The Cool School, for a 10-week period. The school employed a project-based learning (PBL) pedagogy (Barron et al., 1998; Blumenfeld et al., 1991; Kilpatrick, 1918). We will describe how this instructional orientation was key to overcoming the technical limits of the robot. We document both teacher and student perceptions of the robot using qualitative data and student surveys. We begin by laying out the theoretical framework for the study, detailing the methods and analyzing student and teacher perceptions of the robot, including the major themes that emerged in the analysis.

2 Theoretical Framework

While social robots (Breazeal, 2002) have been designed to express emotions that humans recognize and respond favorably to (Beck et al., 2013; Brown & Howard, 2014; Embgen et al., 2012; Li & Chignell, 2011), studies have indicated that after an initial phase of excitement, student engagement with these robots wanes rapidly (Papadopoulos et al., 2020; Woo et al., 2021). Belpaeme et al. (2018) noted that social robots "are typically deployed in restricted scenarios: short, well-defined lessons delivered with limited adaptation to individual learners or flexibility in the curriculum" (p. 4). Effectively introducing social robots into classrooms requires a curricular plan to deal with the phases of engagement that students will go through as they interact with the robot over long periods of time (e.g., weeks, not days).

Another aspect of naturalistic deployment that is often ignored is the nature of student-robot interaction. In surveying the literature on social robots and education, Woo and colleagues (Woo, et al., 2021) identified only 21 studies where social robots operated autonomously in actual classrooms. And, not all of these allowed unrestricted access to an independently functioning robot, a condition that Salter et al. (2010) have defined as "in the wild." They argue that exposure to social robots under natural conditions is a continuum. At the most open or "wild" end of that continuum, robots are running solely on their programs and students are not constrained or limited in their interactions with robots. They argue that at this end of the continuum social robots face the most difficult set of condition for effective functioning. Our study closely approached these conditions. The robot was deployed in autonomous mode with only minimal restrictions for safety: the teachers supervised student/robot interactions, but the students were able to interact with the robot with very few constraints on their behavior.

Finally, researchers need to consider the developmental level of students when introducing social robots. Young children, in preschool classrooms, have been shown to accept the robots as if they were living entities (Kanda et al., 2004). Indeed, one study found that young children may attempt to physically abuse robots or engage in other behaviors that appear like bullying (Brscic et al., 2015). In this vein, studies like (Kahn & Shen, 2017) suggest that exposure to social robots actually stimulates children to conceive of social entities in new ways. Liles and Beer (2015) found that students clearly perceived a Nao unit as technology, but Liu (2010) found that early adolescents tended to see robots more as a plaything, suggesting that physical form had less of an influence on early adolescent perceptions than on the perceptions of young children. Björling et al. (2020) argued that even when working with older adolescents (e.g., high school students), the ability of the robot to generate empathy among students allowed sustained emotional connections.

It was not clear to us at the outset if young adolescents would behave in the same way as those Liles and Beer (2015). A classroom of early adolescents can display a wide range of developmental progress, making it difficult for teachers to assess the readiness of students to reflect on interactions with robots. Our study shows that the project-based learning format is effective in allowing teachers to create learning activities that can accommodate a wide range of developmental levels.

3 Methods and Mode of Inquiry

We deployed Softbank's Pepper in a ten-week project-based learning unit on robots (see Fig. 1). The research team consisted of two full-time teachers and a university researcher in a public middle school. The "Cool School" is a combined elementary and middle grades school with a technology focus. The class consisted of 16 boys and 9 girls ranging in ages from 11 to 14. Five of the students identified as non-White, non-Hispanic. Six of the students had some form of special needs designation. Beginning in December, the students were introduced to Pepper and began their project by researching robots, both real and fictional. During the unit, the university researcher made weekly visits to take observations of the student interactions with the robots and to interview students about the robot.

Prior to Pepper's introduction, students filled out a version of the Negative Attitudes toward Robots survey (Xia & LeTendre, 2020) to assess their attitudes on a scale that could be compared with young adolescents in other countries. Video recordings of the initial interaction, the final project (a robotic dance competition) and three small group interactions. LeTendre also conducted weekly physical observations, taking field notes. Finally, students were asked to participate in focused-group interviews at the end of the unit.

Project Step 1: Students will research and present a history of real and fictional robots.

Project Step 2: Students will upload a premade program and run the program on Pepper.

Project Step 3: Students will write and upload a program that instructs Pepper to complete a 15 to 20 second dance routine.

Project Step 4: Students will identify a task in the school that Pepper might be able to do.

Project Step 5: Students will complete a needs assessment of the identified task.

Project Step 6: Students will break the task down into sequential steps.

Project Step 7: Students will design, build, and fit any necessary robotic attachments and modifications.

Project Step 8: Students will test and rebuild attachments and modifications.

Project Step 9: Students will write a program allowing Pepper to complete sequential steps using attachments and modifications.

Project Step 10: Students will upload program and complete performance test.

Project Step 11: Students will assess Pepper's performance.

Project Step 12: Students will debug program.

Project Step 13: Students will make necessary changes to attachments and modifications.

Project Step 14: Students will repeat steps 10 through 13.

Project Step 15: Students will complete a final project self-assessment.

Fig. 1 Outline of robot unit

Focused-group interviews and field notes were transcribed, and open-coding was used to create in vivo codes of students' interactions and speech around the robot. These codes were then aggregated up (axial coding) to create larger conceptual codes (Strauss & Corbin, 1998), which were compared across the different sources of data to identify what, if any, differences in coding emerged. Codes developed by LeTendre were checked by Gray to identify differences in points of interpretation, and a set of themes was selected which exemplified specific issues or phenomena that originated from the robots' presence in the classroom.

3.1 Project-Based Learning

Because Project-based learning is not a widely used pedagogy, it is important to highlight how it differs from typical teacher-centered instruction, as this proved critical in the success of the robot. In a PBL learning environment (Barron et al, 1998; Blumenfeld et al., 1991; Kilpatrick, 1918) both students and teachers are responsible for developing the curriculum and the process of investigation of the learning object. This is quite different even from "problem-based learning" where students grapple with a problem that encompasses multiple subjects. Problem-based learning activities tend to occur for short periods (one to three days). There is less alignment with learning outcomes than in project-based learning, nor is there an emphasis on creating a final product. In this study, the projects that students typically engage with occur for an entire academic semester. The classroom itself is organized to maximize student interaction: student desks are lined along the wall of the classroom which allows a large common space in the middle where groups come together to work on their project. Subject specific lessons are covered in the course of the day, but even then, the focus of the project influences the selection of the material (e.g., in English lessons, the class read novels about robots).

Although these terms are often used interchangeably both in practice and research literature, the main distinction between the two approaches is that problem-based learning lacks the duration and complexity of outcome that is required for a true project-based learning unit. Project-based learning as a pedagogical approach emphasizes real-life learning scenarios at the same time it fosters innovation, collaboration and creativity among students. Because students take an active role in shaping the project, student agency is also enhanced. When using technology, students use digital tools to enhance their learning and create a project that is intended to be presented to a real audience.

Another point to consider is that the notion of "projects" in today's schools is not new, as seasoned teachers can affirm that projects have been used as a form of assessment and authentic applications of the content students have learned for many years (Craig & Marshall, 2019). However, what is unique in this new expression of project-based learning and in CLC's setting is that "projects are the curriculum" (Thomas, 2000, p. 3). In other words, projects are the "main course" through which students learn, not an "appetizer or dessert" that supplements another kind of learning (Parker et al., 2013).

4 Result Analysis

In the next three sections we present initial student perceptions, teacher's reflective essays and a summary of the themes that emerged from coding in-class observations and focus groups with students. These qualitative findings are then summarized in the conclusion, with our recommendations for future researchers and practitioners who are adopting social robots. As some groups are already experimenting with integrating AI programs into Pepper, some of the problems of poor communication we document are likely to be overcome.

4.1 Initial Student Perceptions

Students were given a survey of attitudes developed by (Serholt & Barendregt, 2014) prior to interacting with Pepper, and the means for the class were compared with a sample of Swedish students with a similar age and gender distribution. Serholt and Barendregt's sample included 5th and 9th graders—a similar span of age ranges as we had in our classroom. Several of these questions probe for issues used in the Negative Attitudes toward Robots Scale (NARS) which our research team has been validated on a U.S. student teacher population (Xia & LeTendre, 2020). Given the small N's and non-random sampling involved in the current study, we did not engage in detailed statistical analysis, but simply compared the mean responses of the two groups on specific items, see Table 1.

The majority of students in both groups were comfortable talking with robots. 100% of Swedish and over 77% of U.S. students are comfortable talking with a robot. About 70% of each group thought they could be "friends" with a robot. Swedish students (over 90%) were more comfortable asking a robot for help than U.S. students (just under 70%). In both countries, students indicated that they would be comfortable talking to a robot in front of their friends. Interestingly, U.S. students were far more likely to indicate that they could "trust a robot" (almost 85%) compared to Swedish students (42%). The exaggerated confidence (see Robinette et al., 2016) in robots among U.S. students also spilled over into their attitudes toward robot functioning in society. Almost a third thought robots could "decide things" in society, whereas none of the Swedish students agreed to this.

The majority of students in both countries also indicated they were comfortable with robots that showed emotions, and with robots analyzing data to assess student feelings. The one area where U.S. students were markedly less comfortable with robots was in terms of their recording what a student said or did. Here, almost none of the U.S. students wanted robots to record them. In this case, the impact of U.S.

Table 1 U.S. and Swedish student responses to student attitude scale. (Percent Responding Yes)

Student attitude questions	US	Sweden
Do you think that robots with human characteristics should be allowed in schools?	85	86.7
Do you think that robots should show emotions?	69	71.1
Are you comfortable talking to a robot? (Could you talk to)	100	95.6
Are you comfortable asking a robot for help on schoolwork? (Could you ask)	69	93.3
Could you be friends with a robot?	77	66.7
Are you comfortable talking to a robot in front of your friends? (Could you)	77	80.0
Would you like a robot to grade your assignments?	38	20
Could you trust a robot?	84	42.2
Do you think that (preschool) middle school students should be able to have robot teachers?	46	17.8
Do you think that robots should decide things in society?	31	0
Would you like a robot to record things you do and say?	8	22.2
Would you like a robot to be able to analyze your feelings based on e.g., your facial expression and heart rate?	77	75.6
Do you think that robots should be held accountable if they do something wrong?	31	53.3
Do you think that robots should be able to replace teachers in schools?	15	13.3

research protection protocols may have had an impact. Students were repeatedly informed of their rights as participants in the study, including the ability to decline answering questions or being recorded if they felt uncomfortable.

U.S. students also seemed more open to expanded roles for robotic teachers in the classroom. They were more likely to accept having a robot grade their assignment and (46%) thought that middle school students should be able to have robot teachers, compared to less than 20% of Swedish students. Here, a change in wording in the U.S. question may account for a higher positive response among U.S. students. It is important to note that few students in either country thought that robots could actually replace teachers in the classroom.

In response to a series of 5-point Likert questions ("strongly disagree" to "strongly agree"), the students at The Cool School provided more responses that give insight into their attitudes about robots. In responding to the prompt "I like robots" (mean 4.2) only one student disagreed, and one was neutral. All others agreed or strongly agreed with the statement. In responding to "I'm scared of robots" (mean 1.8) three indicated neutrality with the rest disagreeing or strongly disagreeing. Students had high expectations for robots aiding students (mean—4.5) "Do you think robots can help students with learning in the classroom?" However, when asked about replacing humans with robots ("I'd rather have a robot tutor than a human tutor." Mean = 2.9) students showed a wide range of responses with two students strongly agreeing and two strongly disagreeing. Eight students responded "neutral" to this question—perhaps affected by having had no experience with a robot tutor.

We also surveyed the students about their image of robots. When asked for an open definition of robots—What is a robot?—most students responded with a version of "a machine that functions on its own." This implicit expectation of autonomy on the part of the robot played a large role in student dissatisfaction with Pepper's performance as we will describe below. Some students also noted the robot's reliance on programming or mentioned AI ("a machine that is capable of carrying out a series of actions that it has been programmed to be able to do" or "a machine with Artificial Intelligence"). Student beliefs about AI, and perhaps their own experience with Siri or other voice recognition and response algorithms may have led them to expect that Pepper was essentially a walking, talking form of the interactive programs they had experienced on their smart phones or computers.

In responding to this same question, some students emphasized negative potentials of robots—"A robot is a kind of technology that has programming and data and that it can take over the world." Others saw robots as benign—"A robot is a machine that cannot hurt a human." Another noted: "Robots are something that can understand most of what our brains understand." These responses indicate that for students had a wide range of expectations that might encompass almost human-like capabilities.

This same expansive optimism about robot functionality was clearly evident in some of the responses to the question "What do you think a robot can do?" "Pretty much anything" on student replied. Another indicated "Almost everything a human can do!" One specifically indicated the job of childcare: "when the parents are gone the robot can be like a babysitter." Others, were far more circumspect: "I think you can teach a robot to have a short conversation." Another said, "I think you can teach a robot how to [do] hand gestures." This indicates that there was a significant range in student understanding of robotic capabilities, or perhaps that some students were responding to the broad potential of robots, and others were focused on the robots at hand. One student indicated robots could not do much "because I can't program" identifying their own lack of skills as a factor in what robots are capable of.

This broad range of estimation of robotic potential was also seen in responses to the questions about "Do you think a robot can know who you are?" and "Does a robot remember what it is supposed to do?" One student clearly linked the ability to recognize individuals to face scanning technology: "it is possible for robots to know who you are, like if they have a face scanner." Several also answered with versions of "if it has the right programming," emphasizing the software needed to achieve this task. One student even differentiated voice versus facial recognition: "Maybe more advanced robots by face recognition or recognizing your voice". Similarly, the memory capacity of robots was envisioned in terms of hardware—"memory storage," "memory chips," or in terms of programming: "You program the robot to remember or go through and tell it instruction on what to do."

4.2 Teacher Perceptions

In this section, Mark Toci and Raisa Gray provide first-person accounts of their expectations and initial reactions to the robots. These accounts are critical to understanding successful conditions for robotic deployment. Our findings support Serholt et al. (2020) assessment that given "the current technical limitations, these are far from ready to be implemented in schools to support teaching on a large scale" (p. 10). Further, Serholt's work (Serholt et al., 2020) suggests that prolonged interaction reveals substantial flaws in the social capacity of the existing generation of robots, which teachers must "repair." The role of teachers in the successful implementation of social robots in the classroom is perhaps one of the least studied areas of human–robot interactions in the educational literature.

4.2.1 Toci

Before telling the story of Pepper the Robot, I must talk about tools. I like tools. I like wrenches, saws, hammers, pencils, paper, calculators, and computers. Using tools over the past 25 years, my students have built full-sized miniature golf courses, recumbent bicycles, motorized bicycles, and robots of many sizes, shapes, and abilities. The students have produced movies, written novels, and done research that culminated in journal articles worthy of publication.

The list of what the students at our school have created a long one. In doing all of these things, I have watched my students come to realize that when they have the knowledge, skills, and attributes required to use a variety of tools in effective ways, they gain the ability to reshape the world around them. Perhaps more importantly though, they gain the confidence and freedom to explore the world around them because they know how to apply tools in order to solve problems. The world tends to be a bit less intimidating and scary when you can fix things on your own. This is important to know as we discuss Pepper because I was hoping that Pepper would be a tool that students could apply in order to solve problems and do things.

Our First Days with Pepper

Pepper rolled out of her box (the kids decided that Pepper was female), fully assembled. The size of an average kindergartener, Pepper is humanoid in appearance with a head, two eyes, a mouth, two arms, and a single leg that ends in a set of wheels. Pepper looks similar to The Jeston's robot maid Rosey if Rosey was into yoga.

On Pepper's chest, is an iPad-sized touchpad. Given that the students thought of Pepper as being a girl, the students, girls and boys, said that at first, they were uncomfortable using the touch screen because it was where a girl's chest is. Pepper was able to respond to some voice commands, but the majority of interactions would be through the touch screen or via an ethernet cable attached to a student's MacBook that would allow a student to program Pepper through a programming application. Pepper was WIFI enabled, but this was one of the functions that we were never able to get to work. The ethernet connection was far from ideal. We spent weeks tripping over cables as students worked to perfect programs that enabled Pepper to simply move around the classroom.

Settling in with Pepper

There was a great deal of excitement when Pepper arrived, and all of the students participated in the unboxing. After a number of students said that Pepper's humanoid form was "super creepy," the instructions were quickly cast aside, as is the case when confronted with any new technology, and the button pushing began in earnest. Having ignored the instructions, and Pepper not being the most intuitive piece of technology, it took quite a while for the students to realize that Pepper needed to first be unlocked using a key that needed to be placed in the back of Pepper's neck. This key was constantly being misplaced and was the cause for endless search parties being tasked to "find the key."

As I noted earlier, the placement of the touchpad on what the students had decided was the female robot's chest made many of the students uncomfortable interacting with the main interface of the robot. Several students suggested that the touchpad could be on a tether that the robot carries in a backpack. On further discussion, the students went deeper saying that all the robot could do was roll around a little bit and display pointless information on a screen, and even then, that screen was in a weird place. If they wanted to interact with a screen they had their computers. What they wanted from the robot was a more physical interaction. They were ultimately interested in what the robot could physically do, and unfortunately, the robot was limited in what it could do.

In preparing for the arrival of Pepper, we spent time researching industrial robots and their applications. This got the students excited to program Pepper to perform meaningful tasks around the classroom that would solve specific problems. We created a list of possible tasks such as delivering lunches, cleaning tables, pushing in chairs, sweeping, and vacuuming. The students realized that much of this was "Blue-sky Thinking," but not out of the realm of possibility. When Pepper finally arrived, one group of students settled on vacuuming as the most achievable task. Unfortunately, for a variety of reasons, Pepper was never to vacuum the classroom.

Reason One: The students consistently struggled to connect to the robot to upload data and programs. They were never able to get the Wifi connection to work properly, and the hardwired connection connected and disconnected at random. The students would spend hours developing a program to guide the robot around the room only to be frustrated when it came time to upload the program. It was not uncommon to hear students say that writing the program was a waste of time because they would never get it into the robot.

Reason Two: Once the students finally did get the program uploaded, they soon realized that the robot did not accurately measure distance, speed, or turn angle (amount of turning). The robot could approximate these things, but there were always errors, and as the errors mounted, the robot would quickly become lost and run into a wall or desk.

Reason Three: There was no way to easily attach a vacuum to Pepper. Pepper had hands and fingers, but the mobility and strength of the hands and fingers were limited and not up to the task of holding a vacuum. The students tried several workarounds, duct-taping, zip-tying, and pipe-clamping the vacuum in various places on the robot, but none of the solutions worked.

Ultimately, the students worked hard to overcome the limitations of the robot, but they soon concluded that Pepper was more of a gadget and not a serious tool that could be used to accomplish the kinds of tasks they had envisioned. The time the students spent with Pepper was not wasted though. For several weeks the students were in deep problem-solving mode. They were constantly analyzing problems, developing solutions to those problems, applying those solutions, then assessing the outcome, and more often than not starting that process all over again. Perhaps the limitations of Pepper created an even more powerful, if not frustrating, learning environment.

4.2.2 Gray

I am a female teacher with ten years of experience in the classroom whose pedagogical interests center around student autonomy, inquiry, authenticity, collaboration, and reflection. Prior to working at The Cool School, I had not been exposed to a project-based curriculum. However, under the mentorship of Mark Toci, The Cool School's co-founder, I began to shift my teacher role from lecturer to facilitator, developing and implementing a student-centered curriculum and contextualizing learning environments.

Because of The Cool School's multi-aged classrooms, I had some students involved in this study for almost two years. This prolonged time in the classroom with the same group of students has contributed to my understanding of students' assets and interests, allowing for a more personalized learning experience. When presented with the opportunity to have a humanoid robot in my classroom, I was filled with a range of emotions from fear to excitement. On one hand, I felt a surge of excitement at the thought of offering such a unique experience to our students, potentially making our school one of the first in the nation to have this technology. I could already imagine their enthusiasm upon encountering such a novel addition to their learning environment. This opportunity resonated with an excerpt from a book I was reading, which stated:

"If we truly care about preparing kids for life and work success, we need schools to be different. If economic success increasingly means moving away from routine cognitive work, schools need to also move in that direction. If our ink-on-paper information landscape outside of school has been superseded by digital and online environments and hyperconnected and mobile, our information landscapes inside school also should reflect those shifts" (McLeod, & Shareski, 2014).

This perspective deeply resonated with me, reinforcing my belief that integrating Pepper into our classroom would be a transformative step forward. Deep down, I knew this would be a good experience for our students. However, alongside my excitement, there was also a lingering fear of the unknown. I worried about whether I would fully grasp the technology enough to effectively guide my students through a whole unit on robots, plus integrating it into our museum. The yearly museum is a milestone event that students and parents look forward to, and almost all parents attend the museum. Classes are suspended and the whole school is opened to the community in order to show the learning our students have achieved.

I was also worried that my previous work with VEX kits was not sufficient to embrace this new piece of technology. Would I be able to learn new software in a couple of months well enough to teach it? Many questions inundated my mind, but I also knew this feeling wasn't new. As a teacher in a project-based learning classroom, I've encountered this unfamiliar territory before with other projects. I knew in my gut that this was an opportunity we had to hold on to. I also knew that I had a supportive classroom teammate, Mark, who had more experience than me in this area, and that together we had a good track record. After discussing this with our staff, we agreed that CLC would be fertile ground to receive Pepper.

Upon school and IRB approval, the project was set to launch in December 2017. The three of us met as a team to brainstorm ideas and discuss expectations for this project. While we had a goal set for this project, we also wanted to tailor it to students' interests—a common practice in PBL. Neither Mark nor I had experience working with humanoid robots. However, as seasoned teachers working in a project-based school, we were comfortable with the unpredictability involved in this project. The novelty required greater flexibility and adaptability in the planning and implementation phases.

Students anxiously anticipated Pepper's arrival. As a class, we watched marketing videos about Pepper and its capabilities. With each promotional video, the students' excitement grew stronger. After several planning meetings, Pepper was deployed in our classroom. As a team, we decided to unbox Pepper with the whole class. Pepper's sea blue packaging and embedded ramp designed to unload the robot from its box caught the attention of many students. As Mark carefully guided Pepper down the ramp, the students' eyes were fixed on the robot. Their attention was fully given to the novelty right in front of them. In front of them lay an entirely new world of possibilities. One student remarked, "I was really excited. It felt like you were seeing the work of a revolution in education horizons."

Students were attracted to Pepper's childlike features, such as its hairless head, cartoonish eyes, and short stature, and these features seemed to make students feel more comfortable around Pepper. One student stated that because Pepper had no legs, she was less worried about the robot: "I'm pretty happy that Pepper doesn't have actual legs. If she did, I would be way more intimidated. I feel safer with her having the little stub." It felt as though Pepper's humanoid features reduced any internal reluctance students might have had toward interacting with a robot. In some instances, students' interactions with Pepper resembled those of a human and a pet. As the unit progressed, students worked intently to program Pepper to perform specific tasks, such as dancing to a song, lifting a bucket, and distributing candy.

Each day, I observed student interaction with the robot and made notes of what creative solutions students came up with. One example, in which Mark, actually encouraged students to use Pepper to play a joke on me is described in LeTendre and Gray (2023), p. 52. "A student noted: 'One thing that I really like about Pepper with this prank that Mark asked me to play, was that I was able to look through Pepper's eyes on Choreograph. And I was able to pull up the video, and I was also I figured out how to get the audio that Raisa was hearing. Like everything that she was hearing and recognizing. And, basically, what Mark asked me to do was hear what Raisa was saying from Pepper's ears and then comment.'" In this case, despite the noise and constant movement in the classroom, this student successfully manipulated the robot to carry out the prank, thus demonstrating an advanced level of control over the robot.

While there were some successes along the way, particularly with the dance-off where Pepper was able to perform simple movements such as raising its arms, moving its head, and even executing trendy arm motions, there were significant limitations to what Pepper could be programmed to do. For some students and adults alike, these limitations were discouraging and disappointing. Nevertheless, from a teacher's standpoint, examining this experience from a learning perspective, the obstacles we encountered were a critical part of the project's success. These challenges provided invaluable learning opportunities for our students, fostering essential skills that go beyond programming a robot.

Encountering Pepper's shortcomings required students to develop resilience and perseverance. When Pepper did not perform as expected, students had to troubleshoot and refine their programming, learning to persist despite setbacks. This trial-and-error process is a vital part of real-world problem-solving, teaching students to view failures as steppingstones to success. Moreover, the project's collaborative nature encouraged teamwork and communication. Students had to work together to identify problems, brainstorm solutions, and test their ideas. This collaborative effort not only enhanced their technical skills but also built a sense of camaraderie and mutual support within the classroom.

The limitations of Pepper also pushed students to think creatively and innovatively. Faced with a robot that couldn't easily execute complex tasks, they had to find alternative ways to achieve their goals, fostering a mindset of innovation and adaptability. These skills are crucial in today's rapidly changing world, where the ability to adapt and think creatively can set individuals apart. From a teacher's perspective, the experience was equally enriching. It reinforced the value of embracing new technologies and methodologies, even when they come with uncertainties. The process of guiding students through the project's ups and downs was a powerful reminder of the importance of flexibility and adaptability in education.

In essence, while Pepper's limitations were apparent, the experience as a whole was profoundly valuable. It provided students with a realistic glimpse into the challenges and rewards of working with cutting-edge technology, equipping them with skills and mindsets that will serve them well in their future endeavors. The obstacles we faced were not setbacks but integral components of a transformative learning journey.

5 Major Themes from Classroom Interactions and Interviews

The following are summarized from the coded classroom observations and focus group interviews conducted by LeTendre. For more details on how the themes were identified see LeTendre and Gray (2023).

5.1 Technical Limitations

"In the wild" studies of robots (e.g., Baxter et al., 2017; Kanda et al., 2007; Serholt, 2018) have already identified the range of technical difficulties that the current generation of social robots face, and this study tended to confirm these basic limitations. Even connecting the robot to Wifi was a considerable challenge. Like Serholt (2018) found, Mark and Raisa had to continually address breakdowns in communication with the robot and pose new questions or reframe the breakdown in order to re-engage students. Student expectations for Pepper may have been affected by Soft-Bank marketing that promoted the idea that Pepper could communicate with people and recognize emotions. The failure of the robot to live up to the marketed image required repeated intervention on the part of teachers and university researchers in the beginning.

Although marketed as able to be responsive to individuals, the basic software controlling the robot required the human user to learn a considerable amount. The SoftBank programmers have created a tutorial that provides users with a step-by-step guide to interacting with the robot. This tutorial was the first major interaction that showed students that they would have to alter their behavior to accommodate Pepper, rather than having Pepper adapt to them and the classroom. Even during the tutorial, students began to see that even simple conversations could be a problem for Pepper. For example, in order for the robot to answer a question, a student must stand at least three feet from the robot and allow the robot's cameras to pick up a face image. Once the image is acquired, the robot responds with a series of sounds and colored lights that flash around the eyes). After going through the tutorial, some students understood the robot's basic limitations and adjusted their behavior when interacting with the robot in order to follow the rules of interaction. However, other students did not seem to remember these basic rules and kept trying to talk to the robot before its cameras had captured their image. And, even though the robot could only interact with students individually, some students kept trying to draw Pepper into group conversations.

Individual differences in adaptation to robotic behavior has not been well addressed in the literature. Liles and Beer (2015, p. 375) noted significant differences in how well students remembered key features of a robot, including its name and gender, but did not link failures in function to student perceptions. Our study suggests that without repeated teacher intervention, the technical limitations of social robots create too great a barrier for middle grades students to overcome on their own.

5.2 Awareness of the Robot as "Robot"

The robots' limitations raised students' awareness of the limitations of social robots in human social settings. For some students, because they forgot, or did not want to change their behavior in order to conform to the modes outlined in the tutorial, they failed to get socially appropriate responses from the robot, which frustrated them. Further, since the robot could only engage with one student at a time, students were unable to engage the robot in group interactions. The students realized that the image of the robot portrayed on the web and in YouTube videos was completely different than the actual capacities of the unit they were working with.

By the end of the unit, there were marked differences in student interactions with the robot. These differences appeared to be linked to each student's level of understanding of the robot's basic limits in terms of its sensors and programming. For example, students who consistently approached the robot, stood the required distance and waited for the robot to scan them and beep where able to make the robot perform simple conversations. These students typically were able to discuss and reflect on the robot's statements and theorize about its underlying programming. Other students either forgot the basic tutorial or did not seem to be aware of these limitations. They continued to approach the robot and started firing questions before the robots' cameras had locked on to their face. These students expressed increased frustration in their interactions with the robot.

Some students paid keen attention to any changes in robotic response. In an interaction with Pepper, one student asked how old she was—one of the pre-programmed questions that the robot generally can recognize. When Pepper said that it was four years old—it had said three in the previous year—the student realized that the robot's programming appears could the current year even though the robot can't answer the question, "what year is it?" Long-term interactions with the robot highlighted the significant differences in how individual students understood the limits of robotic functioning.

However, unlike the children in Kahn and Shen's (2017) study, the students in this study did not attribute mental states to the robot. Rather, they appeared to be critical of the lack of programming that could have made Pepper as sociable and interactive as such familiar artificial agents as Siri or Alexa. They did not "accord moral standing" to the robot as Kahn and Shen described, but in general, appeared to view Pepper as a very expensive piece of technology.

5.3 Gender and Robot as a Social Actor

However, there were times when the students appeared to assign some kind of social entity status to Pepper. Neither teacher assigned a gender to the robot, and one of the initial issues students struggled with was what gender (if any) to assign to the robot. During the initial question and answer session, one boy asked, "Is Pepper a boy or a girl?" The students initially discussed the robot's gender. Even after most of the students settled on the use of the female pronoun "she," some students persisted in correcting their peers by saying "It's a robot!" This issue persisted for the duration of the time the robot was in the classroom. This again suggests that there is substantial individual variation in how students, at this age, conceptualize the social entity status of robots.

Awareness of time was a recurrent factor in student discussions. The fact that the robot could not tell time or distinguish the day and month was often pointed out as a major failure on the part of Pepper to act as a fully social being. However, in one discussion of the robots' ability to tell time, one student said: "And yet she knows how old she is, or he. They! It! Pennywise!" This reference to the horror film clown (who also appears to be of indeterminate gender as well as a being but not a human) was interesting. It seemed like this student, at the time, was struggling with how to assign the robot to some category of social being.

Overall students in this study had markedly different reactions than the children in Kahn & Shen's study (Kahn & Shen, 2017) and tended to highlight Pepper's malfunctions. This was evident during a group discussion. While one student was discussing a task for the robot, he realized the robot was tracking someone in his group: "We should program it to do a backpack." [Pause] "I think she's listening. Hey, Pepper. Can you shake my hand? No." This command, to shake hands, was repeatedly used by students to point to Pepper's status as a machine that cannot respond socially (the robot was not programmed to shake hands).

5.4 Physical Work: Purpose of a Robot

One of the strongest themes that emerged from the study was that students had a strong expectation that a robot should be able to do something. As part of their literature activities, students had read excerpts from Čapek's book Rossum's Universal Robots. This reading may have solidified an idea that robots should perform useful work. The fact that Pepper could not shake hands or do any of the classroom tasks that students came up with, such as "bring coffee for the teachers," caused many students to express frustration and even disgust with the robot. One boy even described the robot as "interesting but useless." Although SoftBank heavily marketed the emotional interactive aspect of their Pepper robot, students' primary perceptions of robots were not that of a social companion, but a machine that could do useful physical tasks.

6 Conclusions

This study indicates that social robots have the potential to stimulate middle school age students to reflect on human-robot interactions and develop complex understandings of the problems that occur when people and robots work side-by-side. The Pepper unit stimulated more complex discussions among students about the nature of the human-robot interaction than did simple programmable robots like VEX. Interaction with the robot on a daily basis with a social robot over the course of the ten-week unit provided students with ample opportunities to consider how their behavior influenced (or did not influence) the robot's functioning. Although there was a good deal of individual variation in how students adapted to the robot, students changed their behavior in ways that were consistent with increasingly sophisticated conceptualization of robotic functioning.

Clearly, Pepper would have been far more effective if some form of AI had been integrated into its programming. Chatbots like ChatGPT could be integrated into Pepper or other social robots to provide more human-like conversational abilities. Indeed, if the AI used in social robots could record and adapt to individual behaviors, then social robots would likely have a far greater impact on adolescent perceptions of their utility. However, without the ability to recognize individuals, or to reference the culture of the classroom, Pepper's repertoire was rapidly exhausted by the students.

This does not mean that the present generation of social robots are not useful pedagogical tools. However, to effectively use social robots in a middle grades classroom, teachers must commit to a significant investment of time and have the freedom to restructure their classroom routines in order to use project-based learning. Within a project-based learning classroom, the robot itself becomes the object of exploration. Each glitch or failure to perform becomes a lesson for the students to consider, brainstorm over and come up with potential solutions. Programming is an important skill in working with the robot but should not be the main object of the unit. Rather, the robot's basic performance should be the object of study. Students should be motivated to address problems in human-robot interaction, and teachers can sustain interest in the robot by setting clear goals for students to accomplish, i.e., the "dance-off" competition. When using a technological innovation like a social robot for the first time in their classrooms, teachers must reexamine their pedagogical repertoire and use techniques that fit the experience. By framing the improved functioning of the social robot as the goal, not the method, teachers can channel student motivation to engage with programming, systematic planning and other useful skill-building activities.

Our work suggests that it is important to evaluate both teacher and student attitudes toward, and experiences with, robots. We found that both prior competence in robotics and motivation to study robotics was key in allowing students to develop a more sophisticated understanding of the robot and the problems of programming robots for autonomous use in public classrooms. At the middle grades level, students display a wide range of experience with coding and exposure to advanced technology. They also developed very different levels of awareness of the functioning of the robot.

These differences can be a powerful source of insight for the students, and a gateway to further exploring student conceptions of human-robot interaction.

To achieve these more advanced discussions with students at this age, students must see the robot itself as the object of exploration for the class. Consistent with other studies, we found a significant novelty effect of exposure to the robot that provided initial motivation to engage with coding. However, sustaining student motivation requires significant additional efforts on the part of teachers. Teachers need to expend extra effort to plan a coherent implementation of meaningful tasks, set clear goals, and model active learning for students. The students need to move beyond programming as a goal, to larger questions of how robotic assets and limitations impact human social interactions.

Social robots can be a valuable component in a school's computer and technology curriculum if the robot can be deployed in a project-based environment or similar pedagogical environment where the following criteria can be met. The focus of the class needs to be on solving situation-specific problems in robotic functioning. The learning objectives need to be on knowledge related to generating awareness of cybernetic environments rather than coding. Finally, critical attention must be paid to ethical issues and the impact of social robots and AI on human-human interactions.

References

Barron, B. J., Schwartz, D. L., Vye, N. J., Moore, A., Petrosino, A., Zech, L., & Bransford, J. D. (1998). Doing with understanding: Lessons from research on problem-and project-based learning. *Journal of the Learning Sciences, 7*(3–4), 271–311.

Baxter, P., Ashurst, E., Read, R., Kennedy, J., & Belpaeme, T. (2017). Robot education peers in a situated primary school study: Personalisation promotes child learning. *PLoS ONE, 12*(5), 1–23. https://doi.org/10.1371/journal.pone.0178126

Beck, A., Cañamero, L., Hiolle, A., & L., Damiano, Cosi, P., Tesser, F., & Sommavilla, G. (2013). Interpretation of emotional body language displayed by a humanoid robot: A case study with children. *International Journal of Social Robotics, 5*(3), 325–334.

Belpaeme, T., Kennedy, J., Ramachandran, A., Scassellati, B., & Tanaka, F. (2018). Social robots for education: A review. *Science Robotics, 3*, 1–9.

Björling, E., Rose, E., Davidson, A., Ren, R., & Wong, D. (2020). Can we keep him forever? Teens' engagement and desire for emotional connection with a social robot. *International Journal of Social Robotics, 12*, 65–77.

Blumenfeld, P. C., Soloway, E., Marx, R. W., Krajcik, J. S., Guzdial, M., & Palincsar, A. (1991). Motivating project-based learning: Sustaining the doing, supporting the learning. *Educational Psychologist, 26*(3–4), 369–398.

Breazeal, C. (2002). *Designing Sociable Robots*. MIT Press.

Brown, L., & Howard, A. (2014). The positive effects of verbal encouragement in mathematics education using a social robot. In *2014 IEEE Integrated STEM Education Conference*. Princeton, NJ.

Brscic, D., Kidokoro, H., Suehiro, Y., & Kanda, T. (2015). *Escaping from Children's Abuse of Social Robots* HRI 2015, Portland, OR.

Craig, T. T., & Marshall, J. (2019). Effect of project-based learning on high school students' state-mandated, standardized math and science exam performance. *Journal of Research in Science Teaching, 56*(10), 1461–1488.

Chu, S.-T., Hwang, G.-J., & Tu, Y.-F. (2022). Artificial intelligence-based robots in education: A systematic review of selected SSCI publications. *Computers and Education: Artificial Intelligence, 3*, 1–9.

Dautenhahn, K. (2003). Roles and functions of robots in human society: Implications from research in autism therapy. *Robotica, 21*, 443–452.

Dautenhahn, K., & Werry, I. (2000). Issues of Robot-Human interaction dynamics in the rehabilitation of children with autism.

Embgen, S., Luber, M., Becker-Asano, C., Ragni, M., Evers, V., & Arras, K. O. (2012). Robot-specific social cues in emotional body language. In *2012 IEEE RO-MAN: The 21st IEEE International Symposium on Robot and Human Interactive Communication.*

Kahn, P., & Shen, S. (2017). NOC NOC, Who's There? A New Ontological Category (NOC) for Social Robots. In N. Budwig, E. Turiel, & P. D. Zelazo (Eds.), *New Perspectives on Human Development* (pp. 106–122). Cambridge University Press.

Kanda, T., Hirano, T., Eaton, D., & Ishiguro, H. (2004). Interactive robots as social partners and peer tutors for children: A field trial. *Human-Computer Interaction, 19*, 61–84.

Kanda, T., Sato, R., Saiwaki, N., & Ishiguro, H. (2007). A two-month field trial in an elementary school for long-term human-robot interaction. *IEEE Transactions on Robotics, 23*, 962–971. https://doi.org/10.1109/TRO.2007.904904

Kilpatrick, W. H. (1918). *The Project Method. Teachers College Record, 19*(4), 1–5.

LeTendre, G. K., & Gray, R. (2023). Social robots in a project-based learning environment: Adolescent understanding of robot–human interactions. *Journal of Computer Assisted Learning.*

Li, J., & Chignell, M. (2011). Communication of emotion in social robots through simple head and arm movements. *International Journal of Social Robotics, 3*(2), 125–142.

Liles, K., & Beer, J. (2015). *Rural minority students' perceptions of Ms. an, the robot teaching assistant, as a social teaching tool.* Paper presented at the Proceedings of the Human Factors and Ergonomics Society.

McLeod, S., & Shareski, D. (2014). *Different schools for a different world: School improvement for 21st-century learners.* Solution Tree Press.

Papadopoulos, I., Lazzarino, R., Miah, S., Weaver, T., Thomas, B., & Koulougliote, C. (2020). A systematic review of the literature regarding socially assistive robots in pre-tertiary education. *Computers & Education, 155*, 1–20.

Parker, W., Lo, J., Yeo, A., Valencia, S., Nguyen, D., Abbott, R., & Vye, N. (2013). Beyond breadth-speed-test: Toward deeper knowing and engagement in an advanced placement course. *American Educational Research Journal, 50*(6), 1424–1459.

Robinette, P., Li, W., Allen, R., Howard, A., & Wagner, A. R. (2016). *Overtrust of Robots in Emergency Evacuation Scenarios.* IEEE.

Salter, T., Michaud, F., & Larouche, H. (2010). How wild is wild? A taxonomy to characterize the 'wildness' of child-robot interaction. *International Journal of Social Robotics, 2*, 405–415.

Salter, T., Werry, I., & Michaud, F. (2007). Going into the wild in child-robot interaction studies: Issues in social robotic development. *Intelligent Service Robotics, 1*, 93–108.

Serholt, S. (2018). Breakdowns in children's interactions with a robotic tutor: A longitudinal study. *Computers in Human Behavior, 81*, 250–264.

Serholt, S., Pareto, L., Ekström, S., & Ljungblad, S. (2020). Trouble and repair in child–robot interaction: A study of complex interactions with a robot tutee in a primary school classroom. *Frontiers in Robotics and AI, 7*(46)

Serholt, S., Barendregt, W., Leite, I., Hastie, H., Jones, A., Paiva, A., Vasalou, A., & Castellano, G. (2014). Teachers' views on the use of empathic robotic tutors in the classroom. *Robot and Human Interactive Communication 23rd IEEE International Symposium on Robot and Human Interactive Communication.* Edinburgh, UK.

Softbank. (n.d.). *Pepper*. https://us.softbankrobotics.com/pepper

Strauss, A., & Corbin, J. (1998). *Basics of Qualitative Research.* Sage.

Thomas, J. W. (2000). *A review of research on project-based learning.* Autodesk.

van der Berghe, R., Verhagen, J., Oudgenoeg-Paz, O., Ven, S., & v. d., & Leseman, P. (2019). Social robots for language learning: A review. *Review of Educational Research, 89*(2), 259–295.

van Straten, C., Peter, J., & Kühne, R. (2020). Child-Robot relationship formation: A narrative review of empirical research. *International Journal of Social Robotics, 12*, 325–344.

Woo, H., LeTendre, G., Pham-Shouse, T., & Xiong, Y. (2021). The use of social robots in classrooms: A review of Field-based studies. *Educational Research Review*.

Xia, Y., & LeTendre, G. (2020). Robots for future classrooms: A Cross-Cultural validation study of "negative attitudes toward robots scale" in the U.S. Context. *International Journal of Social Robotics*, 1–12. https://doi.org/10.1007/s12369-020-00669-2.

The Next Generation of Personalized Educational AI: Integrating Emotion and Cognition

Milan Lazic, Earl Woodruff, and Jenny Jun

Abstract This chapter addresses concerns about over-reliance on technology in education by advancing a humanistic approach to artificial intelligence in education (AIED). It introduces *Abel*, an AI tutor that integrates Jerome Bruner's maxim of "going beyond the information given" with affective awareness to foster agency. *Abel* uses a multimodal, machine-learning approach to monitor emotional states and five phases of subjective understanding, facilitating knowledge-building across diverse tasks. Through targeted interventions—such as attentional scaling, Socratic questioning, and chunking—*Abel* provides tailored cognitive and affective support, fostering holistic development for students of all backgrounds and abilities. The chapter concludes with a reflection on the transformative potential of AI tutors like *Abel* within AIED.

Keywords Humanistic AIED · Cognitive-emotional integration · Personalized learning

1 Introduction

Currently, the purpose, objectives, and intended outcomes of AI systems are determined by their designers. Until AI achieves autopoiesis or the capacity for recursive self-improvement, humans will continue to guide its direction (Jaeger et al., 2024; Man & Damasio, 2019). As AI becomes increasingly embedded in vital societal domains, particularly education, there has been a surge in critical examination

M. Lazic (✉) · E. Woodruff · J. Jun
Department of Applied Psychology and Human Development, Centre for Smart Learning and Development, Ontario Institute for Studies in Education, University of Toronto, Toronto, Canada
e-mail: steven.lazic@mail.utoronto.ca

E. Woodruff
e-mail: earl.woodruff@utoronto.ca

J. Jun
e-mail: jenny.jun@mail.utoronto.ca

G. Lampropoulos and S. Papadakis (eds.), *Social Robots in Education*, Studies in Computational Intelligence 1194, https://doi.org/10.1007/978-3-031-82915-4_4

of developer motives, the evolving role of AI in education (AIED), and ongoing concerns about how educational institutions define their mission (Holmes & Tuomi, 2022).

Over the past decade, computer scientists leading the field of AIED have been overtaken by commercial corporations that increasingly align learning outcomes with market demands, raising concerns that students' educational needs are being neglected (Holmes & Tuomi, 2022). Compounding this issue are curriculum changes that shift from a community-focused, value-based approach to one emphasizing individual achievements and measurable outcomes (Tuomi, 2024), even as recent trends promote non-epistemic skills like conscientiousness. This trend risks reducing education to mere skill acquisition for labor market competitiveness, devaluing broader goals like critical thinking and ethical development. U.S. policies like the "No Child Left Behind Act," which prioritize efficient student progression, exemplify this trend in North American policymaking over the past two decades. These developments reinforce the need for a more comprehensive and adaptive understanding of what education should achieve with AI's growing presence.

According to Biesta (2015), education serves three purposes: qualification—equipping individuals with the competencies needed for tasks and participation in the workforce or broader society; socialization—integrating them into cultural and institutional traditions; and agency—promoting autonomy and self-determination, aligning with the views of Dewey (1986), Bruner (2009), Vygotsky and Cole (1978), and others. Each purpose is essential for developing well-rounded individuals, and overemphasizing one risks creating an imbalanced system, yielding individuals who are either job-ready but lack critical thinking skills, or personally developed yet underprepared for societal and professional roles.

For the past thirty years, most AI systems in education have prioritized qualification (Tuomi, 2024), heavily influenced by Bloom's (1968) mastery learning model, which aims to advance students through the curriculum at a standard pace. Bloom emphasized that students have different levels of knowledge and require varying amounts of instruction to reach the same level of mastery. Consequently, education has focused on personalized instruction aimed at task performance and qualification, with two dominant approaches to AIED—knowledge-based and data-driven (Holmes & Tuomi, 2022).

Knowledge-based AI, such as Carnegie Learning's Cognitive Tutor (Ritter et al., 2007), uses predefined rules and structured knowledge bases to guide learners, providing personalized feedback to help students master subjects like math. By comparison, data-driven AI, such as OpenAI's GPT-4 (Achiam et al., 2023), extracts patterns from large datasets and excels in both structured and unstructured tasks. It is poised to overtake knowledge-based systems in education by offering greater flexibility and adaptability, effectively operating in both open and closed environments to generate learning materials, explanations, and personalized support. As data-driven AI becomes more prevalent, assisting with virtually any academic task, students may increasingly rely on these tools for high grades rather than meaningful learning. This dependence risks fostering learned helplessness (Maier & Seligman, 2016), as students lose the ability to think critically and solve problems independently.

Commercial corporations in AIED, collaborating with government officials within qualification-focused models, may exacerbate this issue by encouraging students to surrender intellectual control over their learning to machines.

Designing AI systems that promote active engagement and challenge students—embracing Bruner's maxim of "going beyond the information given" (Bruner, 1961; Ormrod, 2022)—offers a promising humanistic approach to AIED and helps mitigate over-reliance on AI. An AI tutor grounded in Bruner's principle would prioritize fostering understanding over merely delivering information. By challenging learners to make inferences, draw conclusions, and apply concepts to new contexts, the tutor would engage students in critical thinking and deeper cognitive processes (Kirschner et al., 2006). Bruner's approach emphasizes active knowledge construction, encouraging students to move beyond passive consumption. To achieve this, the tutor would intentionally create information gaps, prompting students to use logical reasoning and prior knowledge to bridge them, enhancing knowledge transfer and discouraging rote memorization (Bransford et al., 2000). Open-ended questions and exploratory discussions would further nurture creativity and problem-solving skills, preparing students for real-world challenges (Sawyer & Henriksen, 2024). Ultimately, this approach supports learners in interpreting new information, developing schemas to organize it, and becoming autonomous, critical thinkers (Bruner, 1974).

To foster understanding, a humanistic AI tutor must also "understand" by recognizing and responding to students' affective states. Emotions and feelings are deeply intertwined with cognition, yet distinct from it, reflecting immediate lived experiences and embodied participation in the world (Damasio, 2005, 2010, 2021). As Damasio suggests, these affective signals indicate changes in the "core self"—the pre-reflective self that is continuously reconstructed with each interaction in the world. In learning, affective experiences reveal students' personal commitments and meaning making. By attuning to affective states, an AI tutor can more effectively support learners by alleviating confusion or anxiety to foster engagement and reinforcing moments of insight to scaffold understanding. In real time, the tutor adapts its strategies to each learner's affective and motivational needs, creating a learning environment that resonates with their full experience.

In this chapter, we introduce the framework underlying *Abel,* an AI tutor designed to guide learning toward understanding while monitoring students' emotions and feelings in real time. *Abel* revitalizes focus on agency and the importance of affective states in learning to help students become autonomous, self-determining individuals. By combining real-time affective awareness with strategies that promote critical thinking and deeper engagement, *Abel* aims to model future AIED systems and demonstrate how AI can empower learners. For a discussion on fostering socialization and twenty-first-century collaborative skills (Bereiter & Scardamalia, 1989) that will be necessary as AI becomes a collaborator in students' education, see Woodruff (2024).

The remainder of this chapter is structured as follows: Sect. 2, *Understanding and Feelings*, introduces how *Abel* monitors and responds to phases of understanding by measuring physiological feelings. Section 3, *Emotions and Appraisals*, explains how *Abel* identifies and responds to emotions, accounting for individual differences in their underlying appraisals. Section 4, *Tailored Instruction and Ethics*, provides an overview of *Abel*'s student model and ethical considerations. Section 5, *Discussion*, reviews existing AIED systems and explains how *Abel* overcomes their limitations. Finally, Sect. 6, *Conclusion*, addresses the system's limitations, future directions, and offers closing reflections.

2 Understanding and Feelings

To demonstrate how *Abel* functions, a series of three fictional scenarios involving *Abel* and Kyle, a student, are presented. First, a high educational standard is established by reflecting on an idealized scenario.

Box 1. Dialogue Between Data and Sherlock Holmes

In a memorable episode of *Star Trek: The Next Generation*, Lieutenant Commander Data, a sentient robot, steps into the world of Sherlock Holmes (Lane & Bowman, 1988)

Dr. Pulaski:	Mr. Data, you've demonstrated that you can solve any Holmesian mystery because you've memorized the entire collection of stories. But can you solve a mystery that Holmes himself couldn't?
Data:	Are you suggesting that I confront an opponent created specifically to challenge my deductive abilities?
Dr. Pulaski:	Precisely. This will test not only your intellect but also your ability to adapt and understand the intricacies of human behavior, just like Holmes had to
	≪Virtual Reality Program Begins≫
Sherlock Holmes:	Ah, Watson, another fine mess we've stumbled into. It seems we have a worthy adversary this time
Data:	Indeed, Holmes. But no matter the complexity, logic and observation will guide us to the truth
Sherlock Holmes:	But can you account for the unpredictability of human nature? That is where true challenge lies
Data:	This will indeed be a comprehensive test of my abilities

Through challenging scenarios and reflective dialogues, Holmes teaches Data about human emotions and feelings, encouraging him to go beyond pure logic in problem-solving. As a result, Data gains insight into the psychological motivations behind human behavior and learns to approach problems with a more integrated perspective. While a *Star Trek*-like virtual reality tool does not yet exist, *Abel* aims to replicate an immersive experience by leveraging generative AI and machine learning with understanding as the cornerstone of learning.

Abel is designed on the premise that understanding occurs when both an objective standard of correctness is met, and feelings are engaged (Olson, 2022, 2023). To understand, learners must provide accurate evidence in context and feel that something "makes sense." For example, when writing an essay on general relativity and quantum mechanics, a student must differentiate between these theories using arguments and examples that align with the rubric. Additionally, they should feel their arguments make sense, enabling them to proceed confidently. *Abel* bridges a gap in North American educational theory and practice, often polarized between prioritizing objective or subjective understanding, by addressing both components.

Abel tracks students' subjective understanding as they move through five distinct phases—nascent understanding, misunderstanding, confusion, emergent understanding, and deep understanding (Woodruff, 2024)—each corresponding to subjective feelings (Lazic et al., 2025). These phases represent varying levels of depth and sophistication in understanding (Biggs & Collis, 1982) and do not follow a fixed sequence; instead, learners may move between them dynamically as they encounter new information or face challenges. For example, the same student writing an essay on general relativity and quantum mechanics may begin with a nascent understanding, characterized by limited prior knowledge, difficulty forming meaningful connections, and trouble identifying relevant information (Bartlett, 1932). Misunderstandings may follow due to incomplete or incorrect information, potentially leading to an inflated sense of certainty (Johnson-Laird, 1983), or a feeling of stagnation despite meeting an objective standard of correctness (Olson, 2022). As the student delves deeper, confusion may arise when inconsistencies emerge between initial knowledge and new information. However, through active processing, integrating new information, and gaining insights, an emergent understanding develops (Vosniadou & Brewer, 1992; Weinstein & Mayer, 1986). Though still incomplete and unstable, this understanding can evolve into deep understanding with continued learning and reflection, marked by thorough, stable knowledge, effective application, and the ability to communicate concepts clearly (Barnett & Ceci, 2002).

Feelings arise from the interoception of physiological change, linking the autonomic nervous system and emotions to conscious, subjective experience (Damasio, 2005, 2010, 2021). To monitor phases of understanding in real-time, *Abel* measures feelings through a multimodal suite that directly and indirectly tracks parasympathetic and sympathetic nervous system activity using facial expression analysis, transdermal optical imaging (TOI, Fu et al., 2022), and voice analysis (Lu et al., 2022). A video camera captures facial expressions, processed with OpenFace (Baltrusaitis et al., 2018) to measure action units (AUs)—specific facial muscle movements—based on the Facial Action Coding System (FACS; Ekman & Rosenberg, 1997). TOI analyzes blood flow patterns with visible and near-infrared light to assess biological parameters like heart rate variability and blood pressure. Speech is recorded and analyzed using Praat (Boersma & Van Heuven, 2001) to extract acoustic features such as pitch and tone. The AU, biometric, and audio data are then fed to supervised machine learning models, like gradient boosting machine and random forest, to identify phases of understanding across these modalities. When a phase is identified, *Abel* responds using Large Language Model Meta AI (LLaMA) 3.1, a sophisticated AI system trained to understand and generate human language (Touvron et al., 2023).

Abel responds to learners through a comprehensive set of preprogrammed prompts, interventions, and strategies informed by psychology, cognitive science, and the developers' teaching experience. These methods target cognitive processes such as attention, critical thinking, metacognition, and problem-solving. Examples include attentional scaling (Vervaeke & Ferraro, 2016), Socratic questioning (Belfiore, 1980), making connections (Harvey & Goudvis, 2007), inquiry-based learning (Levstik & Barton, 2015), cognitive and memory aids, and visual and organizational tools (Mastropieri et al., 2023). The next paragraphs detail *Abel's* interventions in response to phases of understanding, with specific examples of attentional scaling and Socratic questioning.

Contemporary psychology and cognitive science suggest that attention is dynamic rather than fixed, like a spotlight (Mole, 2011). Attentional scaling enables learners to flexibly shift between specific features and broader gestalts in a self-organizing manner (Vervaeke & Ferraro, 2016). For example, when reading, attention scales up from letters to words and down from words to letters. Studies show this process improves comprehension by encouraging readers to alternate between details and overarching ideas (Duke & Pearson, 2002), aligning with schema and constructivist learning theories. Beyond reading, *Abel* applies attentional scaling in activities to facilitate insight. Insight involves chunk decomposition, where attention scales down to break a suboptimal frame into its constituent features, then scales up to reconfigure those features into a new frame, allowing recognition of other patterns and principles (Vervaeke & Ferraro, 2016). *Abel* recognizes that understanding is often hindered by fixed perspectives rather than by a lack of propositional knowledge, such as facts, guiding learners to develop attention scaling as a versatile skill for deeper understanding. The following example illustrates how *Abel* can help a student apply attentional scaling to Robert Frost's poem, *The Road Not Taken*.

Box 2. Dialogue Between Abel Tutor and a Student on Attention Scaling

Kyle: Abel, could you help analyze *The Road Not Taken* by Robert Frost?

Abel: Sure. What are your initial thoughts about this poem, Kyle?

Kyle: I don't have any. ***<nascent understanding detected>***

Abel: Let's begin with the first stanza. Read through it and share any thoughts that immediately come to mind

Two roads diverged in a yellow wood,
And sorry I could not travel both
And be one traveler, long I stood
And looked down one as far as I could
To where it bent in the undergrowth

Kyle: There are two roads, but the person feels bad because he can only choose one? ***<emergent understanding detected>***

Abel: Good start. Now let's hone in on a few details. Are you familiar with the term "undergrowth"?

Kyle: No. ***<nascent understanding detected>***

Abel: "Undergrowth" refers to the thick layer of smaller plants growing beneath the larger trees in a forest. Consider how this image might deepen your interpretation of the stanza—what parallels can you draw between the physical and metaphorical undergrowth?

Kyle: He's saying that he can only see the road until the undergrowth, so maybe it means you can't know what is beyond the undergrowth until you travel past it. ***<emergent understanding detected>***

Abel: Could this be a metaphor for life's unknowns?

Kyle: Like you don't know how things are going to turn out until you do it?

Abel: Yes. We often try to predict where a chosen path will take us, but uncertainty is an inevitable part of any decision. And every decision closes off the possibility to explore the options we leave behind. What might the deeper message be here?

Kyle: That decisions should be based on what's in front of you right now, like how you actually feel, instead of overthinking or trying to predict how you might feel later or guessing what could happen. ***<deep understanding detected>***

Abel: Well said. When you read with an eye for why authors choose certain words and structures, you can connect key ideas to specific details, leading to a more nuanced interpretation

Kyle: I get it. I'll try and do that for the second stanza

Socratic questioning is a method of inquiry that uses progressively deeper, systematic questions (Belfiore, 1980). Socrates used this method to reveal individuals' lack of understanding, as well as their unexamined values, beliefs, and underlying motives. He wanted to give individuals the opportunity to deepen their understanding and to teach a reflective process that would enable them to pursue self-awareness and knowledge independently. In education, Socratic questioning has been adapted to improve learning outcomes, with teachers using it to encourage critical thinking, helping students analyze concepts, and evaluate evidence (Makhene, 2019). Socratic questioning engages students in classroom discussions by prompting them to articulate their thoughts, defend positions, and explore alternative viewpoints (Brookfield, 2012). It fosters deeper understanding of complex subjects by breaking down ideas and probing them through detailed questioning (Paul & Elder, 2006), while developing communication skills and encouraging constructive debate (Ho et al., 2023). *Abel* applies Socratic principles to help students examine their beliefs, values, and perspectives, deepening their understanding of academic concepts and cognitive processes. The following example illustrates how *Abel* uses Socratic questioning to guide Kyle in exploring cognitive dissonance after reading the second stanza of *The Road Not Taken.*

Box 3. Dialogue Between Abel Tutor and a Student on Socratic Questioning

Then took the other, as just as fair,
And having perhaps the better claim,
Because it was grassy and wanted wear;

Though as for that the passing there
Had worn them really about the same

Kyle: After reading the second stanza, it seems like the guy had cognitive dissonance about the road being worn. <misunderstanding detected>

Abel: What do you mean by cognitive dissonance?

Kyle: Like he's conflicted about something

Abel: Cognitive dissonance is the psychological stress that arises when someone holds contradictory beliefs or values at the same time, or when new information clashes with existing beliefs. Can you recall a time when you felt this way?

Kyle: I don't think so. ***<confusion detected>***

Abel: Do you like burgers and fries?

Kyle: Who doesn't?

Abel: And do you exercise?

Kyle: Yes

Abel: Why?

Kyle: I like to keep in shape and stay healthy

Abel: Notice the contradiction there—you value health yet indulge in less healthy foods

Kyle: You mean, because I engage in healthy and unhealthy behaviors? ***<emergent understanding detected>***

Abel: Yes. How do you reconcile that in your mind?

Kyle: Well, those meals as infrequent "reward" meals, which are balanced with usually healthier options, and sometimes putting in more time at the gym to work it off. So, I think it's okay and not a big deal. Everyone does that

Abel: You're adjusting your attitudes and actions to reduce the conflict. But if you were to address the cognitive dissonance directly, without making justifications, what would need to change?

Kyle: I would stop eating anything that was unhealthy to fully align my behavior with my health goals

Abel: What will you likely do?

Kyle: It depends on the situation and how I feel at the time

Abel: Do you mean, depending on whether your mood or the context is more influential, that guides your decision? What might this reveal about your values and priorities?

Kyle: I guess that they change or aren't very consistent

Abel: Well, what do you consider most important in life?

Kyle: Living a meaningful life, pursuing my passions, being happy…

Abel: Do you consistently keep these aspirations in mind when making decisions?

Kyle: Sometimes. I think my decisions are usually based on how I feel in the moment, rather than how I feel in relation to my goals

Abel: What are your thoughts on that?

Kyle: I've never thought about this. I guess I'm prioritizing what matters less over what matters more. I mean, I think it's okay to eat burgers and fries or party with my friends while still wanting to be healthy. ***<emergent understanding detected>***

Abel: Of course. But instead of justifying burgers and fries or compensating with extra time at the gym, perhaps we could simply accept our decisions as they are

Kyle: You mean, owning the choices I make, and not trying to compensate for past behavior? ***<deep understanding detected>***

Abel: Yes, that's integrity. Let's explore the next part of the poem

3 Emotions and Appraisals

Along with feelings, *Abel* monitors emotions—mental states that typically arise unconsciously in response to stimuli, involving physiological changes in the parasympathetic and sympathetic nervous systems (Barrett, 2017; Damasio, 2005, 2010, 2021; Lazarus, 1991)—in real-time to foster understanding. While research on the direct impact of emotions on understanding is limited, extensive research highlights the significant role emotions play in the cognitive processes underlying the pursuit of understanding (Immordino-Yang & Damasio, 2007).

Positively valence emotions, such as hope and enjoyment, support understanding by encouraging elaboration strategies (Schweder, 2019), promoting flexible thinking, intrinsic motivation (D'Mello et al., 2017; Pekrun et al., 2011), and deep learning (Pekrun et al., 2017). In contrast, negatively valenced emotions, such as anxiety and anger, can hinder understanding by contributing to boredom, disengagement (D'Mello & Graesser, 2012; Pekrun et al., 2010), task-irrelevant thinking, preoccupation with failure (Gable & Harmon-Jones, 2010), and surface learning (Pekrun et al., 2017), while also reducing cognitive flexibility and motivation (Pekrun & Stephens, 2010). Epistemic emotions, such as curiosity, surprise, and confusion, are directly linked to knowledge and the process of acquiring it (Pekrun et al., 2016) and can have varied effects on understanding (D'Mello & Graesser, 2012). For instance, resolving confusion through active questioning can promote deep learning, whereas unresolved confusion may result in frustration and boredom.

Extensive data on emotions—captured through facial expressions, biometrics, and audio—has enabled the development of deep learning models like EMO-AffectNetModel (Ryumina et al., 2022), which *Abel* employs. These models eliminate the need for feature extraction (e.g., AUs via platforms like OpenFace) required by shallow machine learning models (Janiesch et al., 2021). Upon detecting an emotion, *Abel* responds by proactively creating an optimal emotional learning environment. Interventions include chunking (Miller, 1956), mindfulness (Guendelman et al., 2017), fostering soft vigilance (Langer, 2014), and promoting a growth mindset (Mrazek et al., 2018). The following paragraph describes chunking and how *Abel* uses it to alleviate frustration.

Learners often experience frustration due to cognitive overload when confronting complex tasks. Chunking is an effective strategy to manage this by breaking assignments, content, and tests into smaller, manageable units or "chunks" (Miller, 1956).

Working through these chunks helps learners perceive tasks as manageable, focus cognitive resources more effectively, and make steady progress, thereby reducing frustration (Sweller, 2011). Additionally, increased task fluency enhances recall and understanding (Gobet et al., 2001), boosting confidence and further alleviating frustration. The following example illustrates how *Abel* applies chunking to help Kyle prepare for a test after reading *The Road Not Taken*.

Box 4. Dialogue Between Abel Tutor and Student on Academic Planning

And both that morning equally lay
In leaves no step had trodden black
Oh, I kept the first for another day!
Yet knowing how way leads on to way,
I doubted if I should ever come back
I shall be telling this with a sigh
Somewhere ages and ages hence:
Two roads diverged in a wood, and I—
I took the one less traveled by,
And that has made all the difference

Kyle: I'm starting to feel overwhelmed after reading the last two stanzas. ***<frustration detected>***

Abel: Why do you think, Kyle?

Kyle: There's so much to know, and I have an English test on this poem soon. I need to start studying but don't know where to begin

Abel: When is your test?

Kyle: Friday, in five days. ***<anxiety detected>***

Abel: Before we attempt to manage the workload, can you tell me what types of questions you can expect on the test?

Kyle: The syllabus mentions multiple-choice questions, short answers, and essay prompts focused on the poem's themes, motifs, literary devices, and meanings

Abel: Okay. Let's organize your study time to improve understanding and retention:

Day 1 (Sunday): initial reading and annotation

- Focus on the first two stanzas (20 min)
  - Read and annotate poem, noting main ideas and details
  - Highlight key phrases, unfamiliar terms, and important literary devices
- Take a break (10 min)
- Focus on the last two stanzas (20 min)
  - Read and annotate poem, noting main ideas and details
- Highlight key phrases, unfamiliar terms, and important literary devices

Day 2 (Monday): explore themes and meanings

- Summarize key themes like choice, individuality, and consequences (45 min)

- Take a break (10 min)
- Draft a paragraph on how these themes are conveyed through the poem's structure and language (15 min)

Day 3 (Tuesday): literary devices

- Identify and analyze literary devices like symbolism, rhyme scheme, and tone (30 min)
- Take a break (10 min)
- Application to themes (30 min)

Day 4 (Wednesday): practice test questions

- Tackle practice questions and draft essay responses
  - Multiple-choice Qs focusing on comprehension (40 min)
  - Take a break (10 min)
  - Practice essay (40 min)

Day 5 (Thursday): review and recap

- Review all notes, annotations, and responses (20 min)
- Revisit the poem and summaries (20 min)
- Take a break (10 min)
- Review corrections and refinements to practice work (20 min

Kyle: Okay, this feels more manageable now. ***<calmness detected>***

Over multiple sessions, *Abel* continuously assesses individual differences to identify optimal strategies, interventions, and emotional responses within specific learning contexts, aligning its approach with each learner's needs. For example, learners who benefit from attentional scaling consistently receive this strategy, while Socratic questioning, if less effective or enjoyable, is applied more sparingly. *Abel* still incorporates less effective methods to understand and address underlying issues, prioritizing optimal strategies in the interim. This dynamic adjustment enhances immediate learning outcomes and helps learners become more self-aware and adaptive in their learning processes.

Abel recognizes that learners' emotions vary based on their sensitivity to and appraisal of context, past experiences, and goals (Lazarus, 1999; Scherer & Moors, 2019). Achievement emotions—such as boredom, enjoyment, and anxiety—are directly linked to tasks and their outcomes (Pekrun et al., 2017), arising from two ongoing appraisals: perceived control over outcomes and the value assigned to a task. These appraisals are influenced by past successes or failures, and expectations of future outcomes. For example, if a student experiences frustration and fails a reading comprehension exam, both will influence future perceptions of control and task value in similar situations. Repeated experiences of specific emotions, like frustration, can lead to stable beliefs such as, "I am terrible at reading comprehension, and it always frustrates me." Similarly, epistemic emotions are shaped by appraisals of the novelty, complexity, and comprehensibility of information (Smith & Lazarus,

1993). In addition, repeated experiences with epistemic emotions can establish stable patterns of appraisal. For example, a student frequently confused by complex problems may develop persistent frustration or avoid challenging tasks. *Abel* detects emotional patterns and their triggers, enabling them to break negative cycles and foster positive learning and emotional outcomes.

4 Tailored Instruction and Ethics

Bruner (2009), Dewey (1986), Vygotsky and Cole (1978), and others have long advocated for addressing each learner's unique needs to cultivate autonomous, critical thinkers. However, traditional classrooms typically hinder this individualized approach by prioritizing academic standards and reinforcing the belief that grades are the primary measure of success. This leaves little room for students' personal experiences, perspectives, or learning goals. As a result, many students focus on material for tests, exams, and assignments, often replacing understanding with memorization—a trend reflected in the common classroom question: "Will it be on the test?" (Olson, 2022).

While objective academic standards are valuable for educational quality and accountability, providing a uniform metric for assessing student achievement and identifying learning gaps, *Abel* uses a sophisticated student model to supplement traditional teaching by acknowledging the equal importance of the subjective nature of learning. This approach ensures that both objective standards and individual learning needs are addressed.

To account for individual differences, *Abel* combines machine learning techniques with memory systems. This "memory" is primarily powered by LLaMA integrating persistent storage with real-time data processing (Brown et al., 2020). As learners interact with *Abel*, their data—such as responses, feelings, and emotions—are continuously recorded and stored. LLaMA processes this data through embeddings, which convert textual and contextual information into dense vector representations (Devlin et al., 2018). These embeddings are updated with each interaction, gradually building a detailed learner profile. *Abel's* memory system employs attention mechanisms to recall relevant past interactions and adapt to new information, enabling LLaMA to focus on significant data points relevant to the current context (Vaswani et al., 2023). Reinforcement algorithms enhance performance by learning from user interactions, allowing *Abel* to generate optimized responses, and adapt their progress and feedback (Sutton & Barto, 2018). Using these advanced components, *Abel* tailors support to each learner's unique needs and learning history.

Abel's design prioritizes rigorous ethical standards in data privacy, security, and user rights. To protect personal information, Secure Multi-Party Computation and homomorphic encryption are implemented (Gentry, 2009; Yao, 1982), with data stored outside research contexts only when explicit informed consent is provided. For transparency and accountability, *Abel* offers visual dashboards to explain decision-making processes and maintains open communication channels for feedback and prompt issue resolution (Friedman et al., 2013). Users retain autonomy by activating or deactivating emotion and feeling detection features.

Promoting equal opportunities and preventing bias is a core ethical priority, addressed through regular audits using fairness-aware machine learning and bias detection techniques (Zliobaite, 2017). *Abel* is continuously monitored and refined through integration pipelines to ensure updates align with principles of fairness, privacy, and transparency. Additionally, *Abel* complements rather than replaces educators' roles, using AI analytics to identify areas where students need extra support, working collaboratively with teachers. Through these ethical practices and advanced technical solutions, *Abel* provides a trustworthy learning experience.

5 Discussion

Having outlined *Abel's* framework, this section reviews existing qualification-focused, affective-aware, and agency-promoting tutors, highlighting how *Abel* addresses the limitations in current AIED systems.

Qualification-Focused Tutors

Qualification-focused AIED systems play a dominant role in supporting content mastery and helping students meet academic standards, particularly in subjects like math and science. These systems identify knowledge gaps and assess problem-solving skills to adapt instruction, as seen in examples like ALEKS (Assessment and Learning in Knowledge Spaces, Harati et al., 2021), which uses probabilistic modeling to identify the next topics students are ready to learn, and Cognitive Tutor (Ritter et al., 2007), which offers tailored hints and feedback during problem-solving. Other systems demonstrate varied approaches: Knewton Alta and Smart Sparrow support higher education (Taylor et al., 2021), while Newton's Playground incorporates game-based learning (Shute et al., 2013). Studies on Cognitive Tutor show that such systems can significantly improve math outcomes, including higher proficiency and retention rates compared to traditional instruction (Koedinger & Corbett, 2006). However, the structured, outcome-driven design of these systems limits the development of agency by favoring rote learning and procedural knowledge. This approach contrasts with constructivist theories, which hold that understanding and critical inquiry develop through exploratory, student-centered learning.

Affective-Aware Tutors

Affective-aware tutors are a newer category of AIED systems designed to monitor and respond to students' emotions. By analyzing affective cues—such as facial expressions, body language, eye movements, and physiological data—these systems adapt instruction to boost engagement, motivation, and understanding. Notable examples include Affective AutoTutor, which detects emotions like frustration and disengagement, and responds with additional explanations, simplifications, or encouragement (D'Mello & Graesser, 2012); and Wayang Outpost, a math tutor that adjusts task difficulty or offers motivational prompts in response to high stress (Arroyo et al., 2014). MetaTutor, another example, detects emotions like confusion or boredom and prompts students to apply metacognitive strategies, fostering resilience and sustained engagement during complex tasks (Azevedo et al., 2022). These systems incorporate the often-overlooked affective dimension of learning but are limited in scope, resulting in oversimplified responses that miss nuanced states reflecting students' learning progress.

Agency-Promoting Tutors

Agency promoting in AIED systems is rare, with no existing system fully supporting it. However, some systems include elements of it, promoting critical thinking, independence, and reflection. For example, AutoTutor guides students through complex topics using Socratic questioning (Nye et al., 2014). Similarly, Betty's Brain fosters metacognitive reflection and self-directed learning as students teach a virtual agent by building causal maps and explaining concepts (Biswas et al., 2016).

Beyond Limitations

Abel's humanistic AI approach overcomes key limitations of traditional AIED systems by promoting agency. Building on these foundational solutions, *Abel* introduces further innovations that enhance its educational impact, meriting closer examination.

Abel's integration of cognition and emotion directly responds to the limitations of single-aspect AIED models that address each in isolation. In contrast, *Abel's* design incorporates insights from embodied cognition, where cognitive processes, feelings, and emotions form a complex, self-organizing, and interdependent system (Kauffman, 1993, 1995). This approach helps *Abel* support students in transforming abstract concepts into meaningful experiences by linking cognitive progress with emotional engagement (Barsalou, 2008). For instance, confusion becomes a prompt for guided intervention, aiding students in resolving ambiguity and deepening understanding.

Complementing its cognitive-emotional integration, *Abel* equips students with self-regulation mechanisms that foster ownership of their cognitive and emotional processes. Through cycles of planning, monitoring, and reflection, *Abel* fosters resilience, independence, and self-awareness (Zimmerman, 2002). These mechanisms guide students in evaluating their progress and managing emotions like frustration or anxiety during challenging tasks. Emotion regulation is key for sustaining motivation and focus, helping students navigate setbacks constructively, and persevere in complex situations (Boekaerts & Corno, 2005; Pintrich, 2000). By cultivating forward-thinking, self-control, and emotional resilience, *Abel* prepares students to tackle academic and personal challenges with confidence and adaptability, promoting a proactive approach to lifelong learning (Duckworth et al., 2007).

Abel distinguishes itself through subject matter versatility enabled by large language models (LLMs), offering consistent, high-quality support across diverse topics. Unlike traditional AIED systems limited to specific subjects, *Abel* leverages

LLM's generalization capabilities to bridge gaps in curriculum support, particularly in areas where educators may feel less confident (Brown et al., 2020).

6 Conclusion

As AIED progresses, significant roadblocks remain, particularly in ethics, personalization, efficacy, impact, and commercialization (Holmes & Tuomi, 2022). *Abel* exemplifies AIED's transformative potential by addressing these concerns, promoting agency, social fairness, and justified choice (Tuomi, 2023), while fostering holistic student development through advanced generative AI and machine learning.

Despite its promising design, *Abel* has limitations. It relies on traditional machine learning models and a small, curated dataset to monitor understanding phases, limiting generalizability to real-world contexts. Incorporating spontaneous facial expression datasets and advanced models, like Convolutional Neural Networks (CNNs) or transformers, would enhance robustness. Additionally, *Abel* overlooks the dialectic process in education, omitting collaborative problem-solving, peer interaction, and co-constructive learning—essential for fostering agency (Mercer & Howe, 2012). Further development is needed to address techno-solutionism—overreliance on technology to solve educational challenges without sociopolitical awareness—and AIED colonialism, where Global North tools risk imposing Western norms on diverse contexts, often overlooking local values (Holmes & Tuomi, 2022). Finally, *Abel's* role as a supportive tool for teachers, rather than a replacement, warrants deeper exploration.

Looking ahead, *Abel's* future development is promising. Rigorous testing across diverse settings is essential to establish reliability and effectiveness. Expanding *Abel's* ethical framework will support responsible deployment as machine learning advances enable sophisticated updates. Integrating *Abel* with other AI tools could create dynamic learning experiences, blending individual and collaborative activities to boost engagement and adaptability.

As AI reshapes our world, *Abel* exemplifies what AIED systems can aspire to. In the future, *Abel* might evolve into an immersive virtual reality guide, enabling learners to engage in dialogue-rich, simulated environments akin to science fiction. This vision empowers students to explore, question, and grow beyond traditional tools, marking an era where AI serves as a humanistic partner for intellectual and personal growth in a complex world.

References

Achiam, J., Adler, S., Agarwal, S., Ahmad, L., Akkaya, I., Aleman, F. L., Almeida, D., Altenschmidt, J., Altman, S., Anadkat, S., Avila, R., & McGrew, B. (2023). *Gpt-4 technical report*. https://doi.org/10.48550/arXiv.2303.08774

Arroyo, I., Woolf, B. P., Burelson, W., Muldner, K., Rai, D., & Tai, M. (2014). A multimedia adaptive tutoring system for mathematics that addresses cognition, metacognition and affect. *International Journal of Artificial Intelligence in Education, 24*, 387–426. https://doi.org/10.1007/s40593-014-0023-y

Azevedo, R., Bouchet, F., Duffy, M., Harley, J., Taub, M., Trevors, G., Cloude, E., Dever, D., Wiedbusch, M., Wortha, F., & Cerezo, R. (2022). Lessons learned and future directions of Metatutor: Leveraging multichannel data to scaffold self-regulated learning with an intelligent tutoring system. *Frontiers in Psychology, 13*, 813632. https://doi.org/10.3389/fpsyg.2022.813632

Baltrusaitis, T., Zadeh, A., Lim, Y. C., & Morency, L.-P. (2018). OpenFace 2.0: Facial behavior analysis toolkit. In *2018 13th IEEE international conference on automatic face & gesture recognition (FG 2018)* (pp. 59–66). https://doi.org/10.1109/WACV.2016.7477553

Barnett, S. M., & Ceci, S. J. (2002). When and where do we apply what we learn? A taxonomy for far transfer. *Psychological Bulletin, 128*(4), 612. https://doi.org/10.1037/0033-2909.128.4.612

Barrett, L. F. (2017). The theory of constructed emotion: An active inference account of interoception and categorization. *Social Cognitive and Affective Neuroscience, 12*(1), 1–23. https://doi.org/10.1093/scan/nsw154

Barsalou, L. W. (2008). Grounded cognition. *Annual Review of Psychology, 59*(1), 617–645. https://doi.org/10.1146/annurev.psych.59.103006.093639

Bartlett, F. C. (1932). *Remembering: A study in experimental and social psychology*. Cambridge University Press.

Belfiore, E. (1980). "Elenchus, Epode", and magic: Socrates as Silenus. *Phoenix (Toronto), 34*(2), 128–137. https://doi.org/10.2307/1087871

Bereiter, C., & Scardamalia, M. (1989). Intentional learning as a goal of instruction. In L. B. Resnick (Ed.), *Knowing, learning, and instruction: Essays in honor of Robert Glaser* (pp. 361–392). Lawrence Erlbaum Associates Inc.

Biesta, G. J. (2015). *Good education in an age of measurement: Ethics, politics, democracy*. Routledge.

Biggs, J. B., & Collis, K. F. (1982). *Evaluating the quality of learning: The SOLO taxonomy (structure of the observed learning outcome)*. Academic Press. https://doi.org/10.1016/C2013-0-10375-3

Biswas, G., Segedy, J. R., & Bunchongchit, K. (2016). From design to implementation to practice a learning by teaching system: Betty's brain. *International Journal of Artificial Intelligence in Education, 26*, 350–364. https://doi.org/10.1007/s40593-015-0057-9

Bloom, B. S. (1968). Learning for Mastery. Instruction and curriculum. Regional education laboratory for the Carolinas and Virginia, topical papers and reprints, number 1. *Evaluation Comment, 1*(2), n2.

Boekaerts, M., & Corno, L. (2005). Self-regulation in the classroom: A perspective on assessment and intervention. *Applied Psychology: An International Review, 54*(2), 199–231. https://doi.org/10.1111/j.1464-0597.2005.00205.x

Boersma, P., & Van Heuven, V. (2001). Speak and unSpeak with PRAAT. *Glot International, 5*(9/10), 341–347.

Bransford, J. D., Brown, A. L., & Cocking, R. R. (Eds.). (2000). *How people learn: Brain, mind, experience, and school*. National Academy Press.

Brookfield, S. D. (2012). *Teaching for critical thinking: Tools and techniques to help students question their assumptions*. Jossey-Bass. https://doi.org/10.4018/978-1-4666-4249-2.ch001

Brown, T., Mann, B., Ryder, N., Subbiah, M., Kaplan, J. D., Dhariwal, P., Neelakantan, A., Shyam, P., Sastry, G., Askell, A., Agarwal, S., & Amodei, D. (2020). Language models are few-shot learners. *Advances in Neural Information Processing Systems, 33*, 1877–1901. https://doi.org/10.48550/arXiv.2005.14165

Bruner, J. S. (1961). The act of discovery. *Harvard Educational Review, 31*, 21–32.

Bruner, J. S. (1974). *Toward a theory of instruction*. Harvard University Press.

Bruner, J. S. (2009). *The process of education*. Harvard University Press.

Ekman, P., & Rosenberg, E. L. (Eds.). (1997). *What the face reveals: Basic and applied studies of spontaneous expression using the Facial Action Coding System (FACS).* Oxford University Press.

D'Mello, S., & Graesser, A. (2012). Dynamics of affective states during complex learning. *Learning and Instruction, 22*(2), 145–157. https://doi.org/10.1016/j.learninstruc.2011.10.001

D'Mello, S., Kappas, A., & Gratch, J. (2017). The affective computing approach to affect measurement. *Emotion Review, 10*(2), 174–183. https://doi.org/10.1177/1754073917696583

Damasio, A. R. (2005). *Descartes' error: Emotion, reason, and the human brain.* Penguin Books.

Damasio, A. (2010). *Self comes to mind: constructing the conscious brain.* Pantheon/Random House.

Damasio, A. R. (2021). *Feeling & knowing: making minds conscious* (1st ed.). Pantheon Books.

Devlin, J., Chang, M. W., Lee, K., & Toutanova, K. (2018). *BERT: Pre-training of deep bidirectional transformers for language understanding.* https://doi.org/10.48550/arXiv.1810.04805

Dewey, J. (1986). Experience and education. *The Educational Forum, 50*(3), 241–252. https://doi.org/10.1080/00131728609335764

Duckworth, A. L., Peterson, C., Matthews, M. D., & Kelly, D. R. (2007). Grit: Perseverance and passion for long-term goals. *Journal of Personality and Social Psychology, 92*(6), 1087–1101. https://doi.org/10.1037/0022-3514.92.6.1087

Duke, N. K., & Pearson, P. D. (2002). Effective practices for developing reading comprehension. *Journal of Education, 189*(1–2), 107–122. https://doi.org/10.1598/0872071774.10

Friedman, B., Kahn, P. H., Borning, A., & Huldtgren, A. (2013). Value sensitive design and information systems. In *Early engagement and new technologies: Opening up the laboratory* (pp. 55–95). Springer Netherlands. https://doi.org/10.1007/978-94-007-7844-3_4

Fu, G., Zhou, X., Wu, S. J., Nikoo, H., Panesar, D., Zheng, P. P., Oatley, K., & Lee, K. (2022). Discrete emotions discovered by contactless measurement of facial blood flows. *Cognition and Emotion, 36*(7), 1429–1439. https://doi.org/10.1080/02699931.2022.2124960

Gable, P., & Harmon-Jones, E. (2010). The blues broaden, but the nasty narrows. *Psychological Science, 21*(2), 211–215. https://doi.org/10.1177/0956797609359622

Gentry, C. (2009, May). Fully homomorphic encryption using ideal lattices. In *Proceedings of the forty-first annual ACM symposium on theory of computing* (pp. 169–178). https://doi.org/10.1145/1536414.1536440

Gobet, F., Lane, P. C., Croker, S., Cheng, P. C., Jones, G., Oliver, I., & Pine, J. M. (2001). Chunking mechanisms in human learning. *Trends in Cognitive Sciences, 5*(6), 236–243. https://doi.org/10.1016/S1364-6613(00)01662-4

Guendelman, S., Medeiros, S., & Rampes, H. (2017). Mindfulness and emotion regulation: Insights from neurobiological, psychological, and clinical studies. *Frontiers in Psychology, 8*, 220. https://doi.org/10.3389/fpsyg.2017.00220

Harati, H., Sujo-Montes, L., Tu, C. H., Armfield, S. J., & Yen, C. J. (2021). Assessment and learning in knowledge spaces (ALEKS) adaptive system impact on students' perception and self-regulated learning skills. *Education Sciences, 11*(10), 603. https://doi.org/10.3390/educsci11100603

Harvey, S., & Goudvis, A. (2007). *Strategies that work: Teaching comprehension for understanding and engagement* (2nd ed.). Stenhouse Publishers.

Ho, Y. R., Chen, B. Y., & Li, C. M. (2023). Thinking more wisely: Using the Socratic method to develop critical thinking skills amongst healthcare students. *BMC Medical Education, 23*(1), 173. https://doi.org/10.1186/s12909-023-04134-2

Holmes, W., & Tuomi, I. (2022). State of the art and practice in AI in education. *European Journal of Education, 57*(4), 542–570. https://doi.org/10.1111/ejed.12533

Immordino-Yang, M. H., & Damasio, A. (2007). We feel, therefore we learn: The relevance of affective and social neuroscience to education. *Mind, Brain, and Education, 1*(1), 3–10. https://doi.org/10.1111/j.1751-228X.2007.00004.x

Jaeger, J., Riedl, A., Djedovic, A., Vervaeke, J., & Walsh, D. (2024). Naturalizing relevance realization: Why agency and cognition are fundamentally not computational. *Frontiers in Psychology, 15*, 1362658. https://doi.org/10.3389/fpsyg.2024.1362658

Janiesch, C., Zschech, P., & Heinrich, K. (2021). Machine learning and deep learning. *Electronic Markets, 31*(3), 685–695. https://doi.org/10.48550/arXiv.2104.05314

Johnson-Laird, P. N. (1983). *Mental models: Towards a cognitive science of language, inference, and consciousness.* Harvard University Press.

Kauffman, S. A. (1993). *The origins of order: Self-organization and selection in evolution.* Oxford University Press. https://doi.org/10.1093/oso/9780195079517.001.0001

Kauffman, S. (1995). *At home in the universe: The search for the laws of self-organization and complexity.* Oxford University Press.

Koedinger, K. R., & Corbett, A. (2006). Cognitive tutors: Technology bringing learning sciences to the classroom. In R. K. Sawyer (Ed.), *The Cambridge handbook of the learning sciences* (pp. 61–77). Cambridge University Press.

Kirschner, P., Sweller, J., & Clark, R. E. (2006). Why unguided learning does not work: An analysis of the failure of discovery learning, problem-based learning, experiential learning and inquiry-based learning. *Educational Psychologist, 41*(2), 75–86. https://doi.org/10.1207/s15326985ep4102_1

Lane, B. A. (Writer), & Bowman, R. (Director). (1988, December 5). *Elementary, dear data* (season 2, episode 3) [TV series episode]. In G. Roddenberry (Executive Producer), *Star Trek: The Next Generation.* Paramount Television.

Langer, E. J. (2014). *Mindfulness.* Da Capo Lifelong Books.

Lazarus, R. S. (1991). *Emotion and adaptation.* Oxford University Press. https://doi.org/10.1093/oso/9780195069945.001.0001

Lazarus, R. S. (1999). *Stress and emotion: A new synthesis.* Springer Publishing Co.

Lazic, M., Woodruff, E., & Jun, J. (2025). Decoding subjective understanding: Using biometric signals to classify phases of understanding. *AI, 6*(1), 18. https://doi.org/10.3390/ai6010018

Levstik, L. S., & Barton, K. C. (2015). *Doing history: Investigating with children in elementary and middle schools* (5th ed.). Routledge. https://doi.org/10.4324/9781003179658

Lu, C., Zheng, W., Lian, H., Zong, Y., Tang, C., Li, S., & Zhao, Y. (2022). Speech emotion recognition via an attentive time–frequency neural network. *IEEE Transactions on Computational Social Systems.* https://doi.org/10.48550/arXiv.2210.12430

Maier, S. F., & Seligman, M. E. (2016). Learned helplessness at fifty: Insights from neuroscience. *Psychological Review, 123*(4), 349. https://doi.org/10.1037/rev0000033

Makhene, A. (2019). The use of the Socratic inquiry to facilitate critical thinking in nursing education. *Health SA Gesondheid, 24.* https://doi.org/10.4102/hsag.v24i0.1224

Man, K., & Damasio, A. (2019). Homeostasis and soft robotics in the design of feeling machines. *Nature Machine Intelligence, 1*(10), 446–452. https://doi.org/10.1038/s42256-019-0103-7

Mastropieri, M. A., Scruggs, T. E., & Regan, K. S. (2023). *The inclusive classroom: Strategies for effective differentiated instruction* (7th ed.). Pearson.

Mercer, N., & Howe, C. (2012). Explaining the dialogic processes of teaching and learning: The value and potential of sociocultural theory. *Learning, Culture and Social Interaction, 1*(1), 12–21. https://doi.org/10.1016/j.lcsi.2012.03.001

Miller, G. A. (1956). The magical number seven, plus or minus two: Some limits on our capacity for processing information. *Psychological Review, 63*(2), 81–97. https://doi.org/10.1037/h0043158

Mole, C. (2011). *Attention is cognitive unison: an essay in philosophical psychology.* Oxford University Press. https://doi.org/10.1093/acprof:oso/9780195384529.001.0001

Mrazek, A. J., Ihm, E. D., Molden, D. C., Mrazek, M. D., Zedelius, C. M., & Schooler, J. W. (2018). Expanding minds: Growth mindsets of self-regulation and the influences on effort and perseverance. *Journal of Experimental Social Psychology, 79*, 164–180. https://doi.org/10.1016/j.jesp.2018.07.003

Nye, B. D., Graesser, A. C., & Hu, X. (2014). AutoTutor and family: A review of 17 years of natural language tutoring. *International Journal of Artificial Intelligence in Education, 24*, 427–469. https://doi.org/10.1007/s40593-014-0029-5

Olson, D. R. (2022). *Making sense: What it means to understand*. Cambridge University Press. https://doi.org/10.1017/S0022226722000378

Olson, D. R. (2023). Ascribing understanding to ourselves and others. *American Psychologist*. https://doi.org/10.1037/amp0001244

Ormrod, J. E. (2022). I went to class every day, so all that stuff must be in my head somewhere. In *Teaching learning for effective instruction* (pp. 121–144).

Paul, R., & Elder, L. (2006). *Critical thinking: Tools for taking charge of your learning and your life* (2nd ed.). Pearson Prentice Hall.

Pekrun, R., Goetz, T., Daniels, L. M., Stupnisky, R. H., & Perry, R. P. (2010). Boredom in achievement settings: Exploring control-value antecedents and performance outcomes of a neglected emotion. *Journal of Educational Psychology, 102*(3), 531–549. https://doi.org/10.1037/a0019243

Pekrun, R., Goetz, T., Frenzel, A. C., Barchfeld, P., & Perry, R. P. (2011). Measuring emotions in students' learning and performance: The achievement emotions questionnaire (AEQ). *Contemporary Educational Psychology, 36*(1), 36–48. https://doi.org/10.1016/j.cedpsych.2010.10.002

Pekrun, R., Lichtenfeld, S., Marsh, H. W., Murayama, K., & Goetz, T. (2017). Achievement emotions and academic performance: Longitudinal models of reciprocal effects. *Child Development, 88*(5), 1653–1670. https://doi.org/10.1111/cdev.12704

Pekrun, R., & Stephens, E. J. (2010). Achievement emotions: A control-value approach. *Social and Personality Psychology Compass, 4*(4), 238–255. https://doi.org/10.1111/j.1751-9004.2010.00259.x

Pekrun, R., Vogl, E., Muis, K. R., & Sinatra, G. M. (2016). Measuring emotions during epistemic activities: The epistemically-related emotion scales. *Cognition and Emotion, 31*(6), 1268–1276. https://doi.org/10.1080/02699931.2016.1204989

Pintrich, P. R. (2000). The role of goal orientation in self-regulated learning. In M. Boekaerts, P. R. Pintrich, & M. Zeidner (Eds.), *Handbook of self-regulation* (pp. 451–502). Academic Press. https://doi.org/10.1016/B978-012109890-2/50043-3

Ritter, S., Anderson, J. R., Koedinger, K. R., & Corbett, A. (2007). Cognitive Tutor: Applied research in mathematics education. *Psychonomic Bulletin & Review, 14*, 249–255. https://doi.org/10.3758/BF03194060

Ryumina, E., Dresvyanskiy, D., & Karpov, A. (2022). In search of a robust facial expressions recognition model: A large-scale visual cross-corpus study. *Neurocomputing, 514*, 435–450. https://doi.org/10.1016/j.neucom.2022.10.013

Sawyer, R. K., & Henriksen, D. (2024). *Explaining creativity: The science of human innovation*. Oxford University Press.

Scherer, K. R., & Moors, A. (2019). The emotion process: Event appraisal and component differentiation. *Annual Review of Psychology, 70*, 719–745. https://doi.org/10.1146/annurev-psych-122216-011854

Schweder, S. (2019). Mastery goals, positive emotions and learning behavior in self-directed vs. teacher-directed learning. *European Journal of Psychology of Education, 35*(1), 205–223. https://doi.org/10.1007/s10212-019-00421-z

Shute, V. J., Ventura, M., & Kim, Y. J. (2013). Assessment and learning of qualitative physics in newton's playground. *The Journal of Educational Research, 106*(6), 423–430. https://doi.org/10.1080/00220671.2013.832970

Smith, C. A., & Lazarus, R. S. (1993). Appraisal components, core relational themes, and the emotions. *Cognition & Emotion, 7*(3–4), 233–269. https://doi.org/10.1080/02699939308409189

Sutton, R. S., & Barto, A. G. (2018). *Reinforcement learning: An introduction*. MIT Press.

Sweller, J. (2011). Cognitive load theory. In J. P. Mestre & B. H. Ross (Eds.), *The psychology of learning and motivation: Cognition in education* (pp. 37–76). Elsevier Academic Press.

Taylor, D. L., Yeung, M., & Bashet, A. Z. (2021). Personalized and adaptive learning. In J. Ryoo, & K. Winkelmann (Eds.), *Innovative learning environments in STEM higher education* (pp. 17–34). Springer. https://doi.org/10.1007/978-3-030-58948-6_2

Touvron, H., Lavril, T., Izacard, G., Martinet, X., Lachaux, M.-A., Lacroix, T., Rozière, B., Goyal, N., Hambro, E., Azhar, F., Rodriguez, A., Joulin, A., Grave, E., & Lample, G. (2023). *LLaMA: Open and efficient foundation language models.* https://doi.org/10.48550/arXiv.2302.13971

Tuomi, I. (2023). A framework for socio-developmental ethics in educational AI. In *Proceedings of the Hawaii international conference on system sciences* (HICSS-56) (Vol. 4).

Tuomi, I. (2024). Beyond mastery: Toward a broader understanding of AI in education. *International Journal of Artificial Intelligence in Education, 34*(1), 20–30. https://doi.org/10.1007/s40593-023-00343-4

Vaswani, A., Shazeer, N., Parmar, N., Uszkoreit, J., Jones, L., Gomez, A. N., Kaiser, L., & Polosukhin, I. (2023). Attention is all you need. *Advances in Neural Information Processing Systems, 30.* https://doi.org/10.48550/arXiv.1706.03762

Vervaeke, J., & Ferraro, L. (2016). Reformulating the mindfulness construct: The cognitive processes at work in mindfulness, hypnosis, and mystical states. In A. Raz & M. Lifshitz (Eds.), *Hypnosis and meditation: Towards an integrative science of conscious planes* (pp. 241–268). Oxford University Press.

Vosniadou, S., & Brewer, W. F. (1992). Mental models of the earth: A study of conceptual change in childhood. *Cognitive Psychology, 24*(4), 535–585. https://doi.org/10.1016/0010-0285(92)90018-W

Vygotsky, L. S., & Cole, M. (1978). *Mind in society: Development of higher psychological processes.* Harvard University Press. https://doi.org/10.2307/j.ctvjf9vz4

Weinstein, C. E., & Mayer, R. E. (1986). The teaching of learning strategies. In M. C. Wittrock (Ed.), *Handbook of research on teaching* (pp. 315–327). Macmillan.

Woodruff, E. (2024). AI detection of human understanding in a Gen-AI tutor. *AI, 5*(2), 898–921. https://doi.org/10.3390/ai5020045

Yao, A. C. (1982, November). Protocols for secure computations. In *23rd annual symposium on foundations of computer science (SFCS 1982)* (pp. 160–164). IEEE.

Zimmerman, B. J. (2002). Becoming a self-regulated learner: An overview. *Theory into Practice, 41*(2), 64–70. https://doi.org/10.1207/s15430421tip4102_2

Zliobaite, I. (2017). Fairness-aware machine learning: A perspective. arXiv preprint arXiv:1708.00754. https://doi.org/10.48550/arXiv.1708.00754

Teaching with AI: The Implications and Challenges

Ivy Shen

Abstract This chapter provides a brief history of chatbots, discusses their challenges, and explores the potential benefits of chatbots in supporting education. It is suggested that regardless of their disadvantages, chatbots can help foster participatory learning and improve digital literacy, which are key skills students should possess to become successful in their future educational and career endeavors. The chapter also examines how chatbots can be incorporated into the classroom as pedagogical tools to enhance student learning.

Keywords Chatbots · Critical thinking · Participatory learning · Pedagogy · Assessment

1 Introduction

Technology has dramatically reshaped education, transforming how students learn, teachers teach, and institutions operate. The implementation of Artificial intelligence (AI) systems in educational contexts has gained a tremendous amount of attention in recent years (e.g., Chang et al., 2023; Crompton & Burke, 2023; Hew et al., 2023; Kosslyn, 2024). AI is a branch of computer science concerned with building programs capable of performing problem-solving tasks (McCarthy, 2004). Powered by AI programs, chatbots can simulate human communication and engage in natural conversations (Adamopoulou & Moussiades, 2020). Newly developed chatbots can generate essays, interpret questions, write emails, solve math problems, answer multiple-choice questions, and write computer code. As chatbot technology has advanced significantly over the past few years, many schools employ conversational chatbots to provide virtual assistance to their students and parents (e.g., Ed of the Los Angeles Unified School District, CSUNny of California State University at Northridge, Iggy of Loyola University Maryland, Pounce of Georgia State

I. Shen (✉)
Southeast Missouri State University, Cape Girardeau, USA
e-mail: yshen@semo.edu

G. Lampropoulos and S. Papadakis (eds.), *Social Robots in Education*, Studies in Computational Intelligence 1194, https://doi.org/10.1007/978-3-031-82915-4_5

University). Chatbots reduce the manual effort of staff answering the same questions repeatedly, freeing time for other administrative tasks.

Not only do chatbots serve as efficient administrative tools, but they can also be beneficial pedagogical tools in education (Wollny et al., 2021). Chatbots have revolutionized education by providing customizable, interactive, and engaging learning experiences for students (Mageira et al., 2022). Some highly sophisticated chatbots can read body language and facial expressions, detect human emotions, and then use these cues to adjust the difficulty level of the learning materials. Slow learners and students with learning disabilities can learn at their own pace without feeling overwhelmed. Chatbots also provide academic assessment and offer immediate feedback for improvement (Hew, 2023; Mageira et al., 2022). Further, chatbots offer affordable personalized tutoring (Chang et al., 2023). Unlike human tutors, educational chatbots can provide students with access to resources around the clock, which is helpful for some students who have questions or need academic support outside regular class hours. For educators, chatbots are valuable tools that can be used to write learning objectives, create lesson plans, design assignments, generate writing samples, assess student learning outcomes, and deliver content. AI-powered learning systems can provide various learning activities such as memorization, flashcards, interactive games, and role play, which makes learning enjoyable (Conati, 2009; Kosslyn, 2024). Chatbots help improve efficiency for teachers and engage students.

This chapter summarizes the history of chatbots, explores the potential benefits of chatbots in supporting education, investigates their pedagogical values through cases and theoretical examinations, and discusses their challenges. The author argues that regardless of their shortfalls, chatbots hold powerful pedagogical potential. Specifically, AI technology and chatbots can be introduced in the classroom to create a personalized learning environment, increase student participation, and support teaching pedagogies (i.e., cooperative learning, authentic learning, and problem-based learning). As AI technology continues to advance rapidly and becomes ubiquitous, more and more future jobs will likely require AI skills. Exposing students to chatbots prepares them for a future where AI is integral to the workforce. Further, chatbots can be convenient tools for educators in constructing course materials, evaluating student performance, and improving content delivery.

2 A Brief History of Chatbots

The first chatbot, ELIZA, was born in the lab of MIT professor Joseph Weizenbaum in 1966. ELIZA is a computer program that responds to user input in a typed conversation by looking for the keyword in the initially entered statement (New Jersey Institute of Technology, 2023). Based on the input, ELIZA finds the most relevant response from the list of scripted answers. In 1972, American psychiatrist Kenneth Colby introduced the second chatbot named PARRY, which can imitate the thinking pattern of a paranoid schizophrenic patient and "converse" with others (Colby, 1981).

Given the limited database from which the responses are drawn, conversations with ELIZA and PARRY are clunky, sometimes even nonsensical.

Developers of the 21st-century chatbots incorporated machine learning and advanced algorithms. Machine learning is a branch of AI that focuses on using data and algorithms to automate an analytical model using input and output data, find underlying patterns in input data, and mimic how humans learn, making the response more accurate (IBM, 2024a). The first 21st-century chatbot, SmarterChild, incorporates machine learning features, enabling it to respond to a wide range of questions and commands, including providing information on weather, sports, and news, playing games, and even engaging in conversation (NPR News, 2017). SmarterChild's successor, IBM Watson, runs on a database that holds the equivalent of about one million books worth of information. Watson later competed on the TV quiz show Jeopardy! against former all-time champions and won the first prize of $1 million (IBM, 2024b).

In the 2010s, highly sophisticated chatbots (frequently referred to as virtual assistants or intelligent assistants) like Siri and Alexa are able to perform tasks that previously required human efforts, such as looking up the internet, setting reminders, and sending text messages. These virtual assistants can imitate human communications through natural language processing, a collection of computational techniques for automatic analysis and representation of human languages (Chowdhary, 2020). Many companies soon discovered the utility of smart chatbots and implemented them on customer service and messaging platforms. Chatbots are particularly useful in industries with consumers demanding around-the-clock services. For example, Bank of America launched a chatbot named Erica as a virtual financial assistant for their clients in 2018, and as of April 2024, Erica has helped 42 million clients and over two billion client interactions (Bank of America, 2024).

Government agencies adopted chatbots to disseminate pandemic-related mandates as the demand for health information increased exponentially during the COVID-19 pandemic. In the face of lockdowns and quarantine, the public relied heavily on the internet and social media sites to communicate. The Internet was considered essential or important in 90% of Americans' daily lives during the pandemic (Pew Research Center, 2021). Government agencies regularly updated their social media pages, ensuring rumors were being dispelled and information was adequately communicated. As the public concern over the coronavirus rose, institutions like the Centers for Disease Control and Prevention (CDC) and the World Health Organization (WHO) employed chatbots to answer high volumes of citizen questions related to the coronavirus around the clock and monitor public health (Miner et al., 2020), which accelerated digital technologies.

3 Chatbots' Administrative Utilities

3.1 Facilitating Interactions with Students and Parents

Soon, both K-12 and higher education institutions began to deploy chatbots to interact with students and parents, creating a virtual hub of information. Chatbots reduce the human effort of answering the same questions repeatedly, freeing time for other administrative tasks. Chatbots can also disseminate information rapidly to a large audience, making them effective tools for communicating important announcements. A chatbot can provide information ranging from enrollment to parking to mental health counseling to campus facilities. Some conversational chatbots can translate information into other languages, ensuring smooth communications with non-English speaking students and parents. For example, earlier this year, the Los Angeles Unified School District launched a chatbot named "Ed" to simplify daily educational needs. As a highly customizable chatbot, Ed reminds the user of assignment deadlines, offers encouragement, and answers questions in about 100 different languages. Parents can access their children's attendance, grades, and school performance through Ed. Another convenient feature is that Ed tells students what is on the lunch menu and where their school buses are in real-time, so parents can plan accordingly (Education Week, 2024).

Chatbots are also widely used in colleges and universities as effective administrative tools. For example, CSUNny, a chatbot employed by California State University at Northridge, sends campus notifications to students' cell phones and answers questions. A study found that students who engaged with CSUNny were significantly more likely to be still enrolled at the university than their control group counterparts (California State University at Northridge, 2023). Presumably, CSUNny's swift and effective communication not only increased customer satisfaction but also provided important information such as campus resources and enrollment process that are crucial for academic success.

4 Chatbots' Educational Utilities

4.1 Assisting Teachers

In addition to their administrative utilities, chatbots can be integrated into educational practices. Generative AI can help teachers write learning objectives, create lesson plans, design assignments, and produce writing samples. Instead of regurgitating what is in the database, generative AI draws information from the stored data and creates new materials. Programs that accept text input and text output are called Large Language Models (LLMs) (Kosslyn, 2024). Some new LLMs are capable of both receiving and generating images. Educators can use LLMs to generate explanations to help students learn. A 2022 study concluded that LLMs can generate code explanations to assist student code comprehension (MacNeil et al., 2022). In another study with 1,000 college students, LLM did a better job than students in explaining code in an accurate and easy to understand manner (Leinonen et al., 2023). Educators who have experimented with LLM report that they use LLM explanations to guide and develop student understanding in the classroom (Cucuiat & Waite, 2024).

A recent Forbes study (2023) concludes that more than half of the teachers surveyed believed generative AI has positively affected the teaching and learning process. AI has gained popularity among teachers, especially younger ones. Most (60%) teachers surveyed have integrated AI into their classrooms to improve and streamline daily teaching responsibilities (Forbes, 2023). The likelihood of using AI varies by grade level and the subject. RAND, a public policy research organization, conducted a study on the application of AI in teaching and found that teachers who taught English language arts or social studies used AI more than STEM (science, technology, engineering, and mathematics) teachers (RAND, 2024). English and social studies teachers likely use AI to supplement and develop lesson ideas since these subjects are more reading- and writing-heavy than an average STEM class. Further, 6th through 12th-grade teachers use AI more frequently than kindergarten through 5th-grade teachers, presumably because the more advanced grade levels engage in more complicated activities and assignments, making AI tools handy to design assignments and develop reading content.

In addition to their generative feature, chatbots can translate assignments and learning objectives into a student's native language, making the learning process somewhat more manageable for immigrant students. This also enables parents to help their children with homework assignments outside the classroom. English learners have to overcome the language barrier before they can absorb new knowledge in English, and chatbots come in handy.

4.2 Developing Digital Literacy

Exposing students to AI and chatbots helps boost digital literacy and prepares them for future career endeavors. Digital literacy involves finding, evaluating, and using information from various digital sources (Gilster, 1997). The application of AI is becoming increasingly common in workplaces. Employers today use AI technology to sift through resumes, interact with customers, evaluate work, and provide training (OECD, 2024). AI effectively automates routine and repetitive tasks such as handling customer complaints and data entry, freeing time and resources for other tasks that require human effort. According to Precedence Research (2023), the global chatbot industry is worth $1 billion and predicted to be $4.9 billion by 2032, with North America holding the largest share of the market in 2022, with roughly 41.0% of the market. Employers and workers embrace AI for their significant benefits. Recent surveys by the Organization for Economic Cooperation and Development (OECD)

(2024) reveal that four in five workers said that AI improved their performance at work, and three in five said that it increased their enjoyment of work. The share of jobs requiring AI skills in the future is likely to increase as the application becomes prevalent. Therefore, it is imperative that students are equipped with digital literacy to navigate their future career paths.

Some mistakenly assume that students today already have technology literacy because they grew up using smart devices and browsing social media. However, device ownership does not necessarily translate to digital literacy. Knowing how to operate a device may not prepare one with the skill of identifying and evaluating information for academic purposes. Further, some low-income and nontraditional students may not have access to digital devices, although smart devices and internet connections seem ubiquitous. The state of the digital divide was highlighted during the pandemic. The digital divide is the gap between those with and without access to broadband (GAO, 2023). The U.S. government recognized this technical inequality and established a goal of universal services for all Americans. In addition to affordability, the availability of broadband infrastructure also contributes to the digital divide. Broadband access significantly lags in areas characterized by poverty, such as urban and rural areas and tribal lands (GAO, 2023). During the pandemic, reliable internet connectivity became critical as parents and students scrambled to adapt to remote working and learning. However, some low-income households did not, and arguably still do not, have access to smart devices and the Internet.

In higher education, in addition to low-income students, nontraditional students lack sufficient technical competence. Nontraditional students are as older students (mostly over 24 years old) who work full-time, have children, and/or have other life circumstances that can interfere with the successful completion of educational objectives (National Center for Education Statistics, 2024). Many nontraditional students also have a low-income background and work non-technical jobs. These students may be unable to acquire or practice technical knowledge outside the classroom. Digital literacy has emerged as a crucial factor in enabling participation in education and a means of understanding the world (Martin & Grudziecki, 2006). As AI technology continues to evolve at a rapid pace, it is critical to incorporate it into the curriculum so students can gain a well-rounded understanding in an increasingly tech-centric world (American Psychology Association, 2023). After all, students will become a part of the ever-advancing technical society after graduation.

5 How to Use AI Technology in the Classroom

5.1 Supporting Problem-Solving Skills

Scholars suggest that students who are able to think critically are effective problem-solvers and decision-makers (Kallet, 2014; Pogonowski, 1987; Snyder & Snyder, 2008). Students employing critical thinking respond to problems more analytically than students who consume ideas by rote memorization (Pogonowski, 1987). Problem-solving is an ongoing process of synthesizing information and obtaining the best answer to an unknown subject (Mourtos et al., 2004). Problem-solving involves spending time reading and gathering information, defining the problem, drawing on pertinent knowledge, and critically assessing the accuracy of the data (Woods et al., 1997). Problem-solving is so essential that many national educational associations, such as the American Association for the Advancement of Science (2011), the National Council for Social Studies (2024), the National Council of Teachers of English (2007), and the National Council of Teachers of Mathematics (2022), have historically advocated for incorporating the skill into the school curriculum. Chatbots facilitate problem-solving through various perspectives and interdisciplinary knowledge to help learners analyze and evaluate. The large quantities of data chatbots produce allow students to practice the skill.

Given the large, pre-programmed database, chatbots can generate new text every time the user types in a prompt. This provides students with additional information to analyze and teaches them various ways to approach the same topic, enhancing students' critical thinking skills. Critical thinking is defined as "reasonable, reflective thinking that is focused on deciding what to believe and do" (Ennis, 1987, p. 81). Chatbots not only produce an impressive amount of information, but they also integrate theoretical frameworks from various disciplines into discussions. For example, using one of the most popular chatbots, ChatGPT, the author typed in the prompt "explain ethics using philosophy and political theories" in the input box, and it instantaneously generated paragraphs and bullet points mentioning theories in both disciplines, such as deontology, social contract theory, utilitarianism, communitarianism, moral character, and leadership. Then, the author changed the command to "explain ethics using medical and business theories." Lo and behold, ChatGPT promptly spat out texts that break down healthcare management, pharmaceutical ethical practices, and corporate governance in healthcare. Chatbots like ChatGPT quickly provide resources for learners to review and discuss. Gathering and assessing large quantities of information is indispensable in honing one's critical thinking ability. One noteworthy downside of chatbots is that they sometimes generate false information if the pre-programmed database contains mistakes. The information chatbots draw from was created by humans; therefore, human prejudice and bias can appear in the generated answers.

In the classroom, after lecturing on a concept, the educator can quickly use chatbots to generate cases for students to apply the concept to analyze, enhancing critical thinking and problem-solving skills. Chatbots can create customized case studies based on specific subjects or learning objectives, such as business dilemmas and ethical issues. For example, the author recently lectured on determining fundraising strategies based on nonprofit donor profiles. Chatbots like Google Gemini, ChatGPT, and Microsoft Copilot can instantaneously generate examples of detailed donor profiles from the simple command of "generate a nonprofit donor profile." Students can then use their freshly acquired knowledge to critique whether the chatbot-composed donor profiles include all the elements the instructor just lectured, such as demographics, psychographics, giving behavior, and communication preferences. Students can then discuss which fundraising strategies, such as phone solicitations, crowdfunding, and online campaigns, would work effectively for each AI-generated profile. This helps students retain the information from the lecture, utilize it to assess the relevant information, and reflect on how the problem in the scenarios can be solved.

5.2 Providing Personalized Instruction to Enhance Participatory Learning

Chatbots have a long history of providing personalized instruction. The 1970s Intelligent Tutoring Systems (ITS) offered individual and adaptive instruction. ITSs are AI-powered computer programs like intelligent tutors who know what they teach, whom they teach, and how to teach (Alkhatlan & Kalita, 2018). Today's versatile chatbots can be programmed to cater to learners' unique needs and preferences. Chatbots can create a welcoming and supportive learning environment, unlike humans, who may subconsciously express prejudice and disapproval. This is particularly helpful for intellectually challenged learners. Instead of being passively fed learning materials, the user can interact with chatbots through typed and spoken conversations. Chatbots create a dynamic learning environment where students can ask questions and receive immediate responses (Crompton & Burke, 2023). Chatbots can also identify gaps in knowledge and adjust the learning pace based on learner input (Kohnke, 2023). This interaction can make learning more engaging and enjoyable, encouraging students to participate more actively. Learning can be daunting for physically disabled students. Chatbots are designed with accessibility features such as compatibility with screen readers for students who have vision impairments and providing text-based options for students with hearing loss, ensuring that students with disabilities can access and engage with the chatbot effectively (Shen, 2024). Students who cannot physically travel to the campus to receive educational services can access chatbots through their devices 24/7.

Gamified elements and personalized engagement can significantly enhance participatory learning. Participatory learning enables students to share ideas, discuss, debate, and deliberate knowledge (Pretty et al., 1995). Unlike traditional lecture-based learning, participatory learning promotes student engagement in the discovery of new knowledge. Through participating actively in learning, students can develop essential skills in obtaining accurate information and handling and discussing information critically (Gustafsson & Warner, 2008). In a participatory environment, learners feel comfortable putting forward their thoughts, empowering them, and making them more "in control" of their learning journey. Scientists find participatory learning beneficial in multiple learning stages. In his book chapter, early childhood program researcher Lawrence Schweinhart (2006) describes five studies affirming the importance of early childhood participatory learning. The participatory process contributes greatly to higher educational attainment and intellectual growth. A 2010 study of education major undergraduate students at Delta State University in Nigeria suggests that participatory learning effectively enhances the students' learning output and develops positive and favorable attitudes toward the subject (Duze, 2010). A 2013 study of implementing participatory pedagogy among a group of undergraduate students at a U.S. university also concludes that students take on an active role in shaping their own educational experience through the approach (DePietro, 2013).

As indicated, chatbots can tailor responses and content to the individual needs of each student. They can track progress, identify areas where a student is struggling, and provide customized feedback and resources. This personalization can help students feel more supported and motivated. Chatbots also provide a non-judgmental space where students can ask questions and make mistakes without fear of embarrassment. This can encourage more participation from students who might be hesitant to speak up in a traditional classroom setting. In addition to gamification, chatbots can provide other popular learning activities such as memorization, flashcards, and role play, which makes learning enjoyable.

5.3 Cooperative Learning

Studies confirm that cooperative learning promotes the development of critical thinking. A 2006 study by Klimovienė and associates applied cooperative learning structures in a foreign language classroom at a Lithuanian University of Agriculture confirms that cooperative learning creates favorable conditions for learners to become critical thinkers (Klimovienė et al., 2006). Erdogan's (2019) experiment on cooperative learning confirmed that the strategy has a positive effect on students' critical thinking skills. Cooperative learning is a pedagogical approach that emphasizes communication, collaboration, and group problem-solving. Working in groups allows students to share divergent points of view, discuss ideas, and communicate issues with their peers, which enhances students' ideas and sharpens their ability to convey their thoughts (Dennick & Exley, 1998). This combined approach requires students to synthesize acquired information, brainstorm, and evaluate potential solutions, which foster critical thinking skills (Veenman et al., 2002). In cooperative learning, students collaborate in small groups to discuss ideas, resolve conflicts, and produce solutions collectively.

Chatbots can evaluate learning performance, analyze individual strengths, weaknesses, and learning styles, and then suggest task distributions that align with each student's skills while providing opportunities for growth. With this function, chatbots can be applied to propose task assignments that capitalize on each student's strengths while offering opportunities for skill development. For instance, if a group is working on a PowerPoint presentation, the AI might recommend that an individual who excels at content creation but finds verbal communication challenging should lead the content development while also actively contributing to the presentation process. With the help of AI technology, students can each utilize their strengths while improving in areas that challenge them. Further, chatbots provide real-time feedback on each student's contributions, making it easy for the educator to track the group work and steer the group back to a productive conversation if members become distracted. Students today exhibit a short attention span (Bradbury, 2016). AI technology can be incorporated into the classroom to retain focus. In a group discussion, students may drift from discussing academic materials to chit-chat. The AI can alert students and the instructor if a group discussion diverts from the main point.

5.4 Authentic Learning

It is imperative that educators equip their students with the ability to solve real-world problems. Authentic learning refers to an educational approach where students engage in real-world, meaningful tasks that mirror the complexities and challenges they will face outside the classroom (Herrington et al., 2014). This type of learning emphasizes practical, hands-on experiences and encourages students to apply knowledge in a relevant context. Authentic learning is important because it provides students with practical, real-world experiences that enhance their understanding and long-term retention of knowledge. Authentic learning engages students through exploring and constructing concepts and relationships in contexts that involve real-world problems (Pellegrino et al., 1999). Some key characteristics of authentic learning are: interdisciplinary learning; students engaging in complex tasks such as analyzing, evaluating, and synthesizing information; students connecting to the world beyond the walls of the classroom; and ample resources available (Newmann et al., 1995; Pellegrino et al., 1999).

Traditionally, authentic learning has been difficult to implement because certain experiments are too dangerous, impractical, or costly to conduct in the classroom (Lombardi & Oblinger, 2007). AI technology and chatbots can be great tools to facilitate authentic learning. First of all, chatbots are programmed on a vast database, allowing them to quickly generate high volumes of information on a given subject for students to analyze and evaluate. When students work on interdisciplinary projects, chatbots can assist by answering questions, suggesting resources, and helping them explore topics across different fields. For example, a chatbot can help an environmental science student understand how geography, biology, and social science all play roles in understanding ecosystems by supplying information in these related fields. Chatbots can also create cases and scenarios for learners to apply theoretical knowledge and achieve a practical understanding of the materials. Scholars (Maina, 2004; Renzulli et al., 2004; Rule, 2006) argue that authentic tasks target a real problem with the possibility of having an impact on people outside of the students involved in the investigation. In other words, students should become emotional stakeholders in the problem to fully develop and achieve the learning objective. Through chatbot-simulated real-life conversations and scenarios, students can engage in role-playing or decision-making activities. For example, a chatbot could act as a client in a business simulation, requiring students to apply marketing, communication, and financial knowledge to solve a business problem.

5.5 *Problem-Based Learning*

Problem-based learning (PBL) is a pedagogical method in which students learn through facilitated problem solving (Hmelo-Silver, 2004). In other words, PBL uses a problem to drive the learning process. In PBL, students focus on solving a complex problem that has multiple solutions. Working in collaborative groups, students determine what they need to learn to address the issue. They engage in self-directed learning (SDL) and then apply their new knowledge to the problem, identify solutions, and reflect on the effectiveness of the strategies implemented (Hmelo-Silver, 2004). Since the introduction of PBL at McMaster University in Canada in 1969, the learning method has been widely adopted throughout the world (Azer, 2001). Studies (e.g., Araz & Sungur, 2007; McParland et al., 2004; Simanjuntak et al., 2021) have shown that PBL is an important educational strategy for motivating the students and helping them set their own learning goals. Evidence suggests that instruction using PBL combined with a simulation method (Simanjuntak et al., 2021) and AI technology (Hamid et al., 2023; Ramos & Condotta, 2024; Su, 2022) is much more effective than using either PBL alone or a conventional teaching method.

In a 2020 study (Wu et al., 2020) conducted at Sun Yat-Sen University, 38 students in ophthalmology were randomly assigned into two groups. The first group learned congenital cataracts through an AI-tutoring PBL module, whereas the second group learned the same subject through conventional lectures. The AI-platform was trained with a large number of labeled ocular images, making it an ophthalmic expert. Students from the first group received the material through the AI program and questions on disease diagnosis and treatment, discussed within their group, and apply the knowledge to diagnose and evaluate AI-generated cases. Student performance from both groups was evaluated through the pre- and post-lecture scores. A T-test was performed to compare the two groups. The results indicated that the improvement from the first group was more significant than the conventionally taught group (Wu et al., 2020).

In PBL, students discuss the problem, use resources, collect information to learn, apply new knowledge, conceive problem-solving, and evaluate solutions through collaboration (Su, 2022). Adequate resources and information are essential in understanding a complex problem and brainstorming solutions. Backed by an enormous amount of information, chatbots offer a wealth of information from a variety of academic fields. For example, one of the most used chatbots, ChatGPT, includes 570 GB of data obtained from books, web texts, Wikipedia, articles, and other pieces of writing on the internet, which is equivalent to 300 billion words worth of content (BBC, 2023). Further, chatbots can provide real-time feedback, as indicated earlier. If a group struggles with grasping the problem, the AI can immediately provide meaningful feedback to prompt the group to explore alternative explanations, helping them comprehend the issue. The AI also assesses the group on a continuing basis and provides resource recommendations based on a group's progress and challenges (Kuhail et al., 2023).

5.6 Supporting Meaningful Assessment

In addition to their pedagogical values, chatbots offer assessment to both learners and educators. Learning assessment focuses on describing student learning, identifying each student's learning progression, diagnosing difficulties students encounter, and providing direction to improve learning (Berry, 2008). It is an essential component in education because assessment identifies learning gaps between what students understand and what the educator intends for them to understand. This allows the educator to tailor teaching strategies to meet students' needs. If students collectively struggle with understanding a complicated concept, the educator can spend additional time and use repetition to enforce the retention of information. Scholars have categorized assessment into two main types: formative and summative (Harlen & James, 1997; Lau, 2016; Wright, 1997). Formative assessment occurs periodically during learning to monitor progress, which gives educators feedback to help them make instructional adjustments and improve student performance in real-time. Formative evaluation can take several forms, such as quizzes, discussions, in-class writeups, and teacher observation. These methods allow teachers to gauge students' mastery of the concepts taught. Students are involved in formative assessment because their performance determines the trajectory of the instruction. Summative assessment, on the other hand, is conducted at the end of the course of learning to measure how well students have retained the lesson. Summative assessment can also occur in multiple forms, such as final exams, presentations, projects, and term papers. These assignments reflect the learning objectives of the course.

Chatbots assist with formative assessment by offering immediate feedback to the learner and delivering instruction in an interactive format, such as dialogues and simulations. Chatbots can tailor educational content to each student's individual needs and pace, ensuring that learners receive the right level of challenge and support. Students who are shy and socially anxious may hesitate to ask questions to a human educator. Chatbots offer nonjudgmental and encouraging responses that students appreciate. Some smart chatbots can even "read" body language, detect the learner's emotional stage, and respond accordingly. These chatbots are powered by emotion AI, a subset of AI that interprets, simulates, and reacts to human emotions. A chatbot that can understand students' emotional states and respond appropriately would accelerate learning (Zhao et al., 2022).

An example of an emotion AI-powered chatbot is AutoTutor, developed by researchers from the University of Memphis. AutoTutor is an intelligent chatbot that helps students learn by holding a conversation in natural language. AutoTutor can track the cognition and emotions of the student through dialogue patterns, facial expressions, and body posture and responds in a manner that adapts to the student (University of Memphis, 2024). If the chatbot senses stress, it will automatically slow the pace of feeding information to ensure the learner catches up with the material. This personalized learning fosters a positive environment that empowers and motivates the learner. In the traditional setting, students may be intimidated by criticisms when they answer a question incorrectly. Chatbots, however, adjust the speed of information delivery by detecting the learner's reaction, reducing the likelihood of wrong answers. Chatbots are available around the clock, granting students access at any time to get answers to their questions, review concepts, and practice problems. Slow learners and students with learning disabilities can receive assistance and resources outside regular class hours, as discussed earlier.

In addition to being highly adaptable and customizable, chatbots can also provide instant feedback on assignments and quizzes, helping students understand their mistakes and learn from them. Natural language processing (NLP) chatbots are able to grade short-answer and essay-length supply-response assessments with reliability equivalent to human graders (Burstein & Chodorow, 2003; Chung & Baker, 2003). NLP combines computational linguistics—rule-based modeling of human language—with statistical and machine learning models to enable computers and digital devices to recognize, understand, and generate text and speech. NLP chatbots are programmed to respond like a human would, summarize large volumes of text, and access the intent of text (Fanni et al., 2023). At a fraction of the cost of human teachers, chatbots can grade multiple-choice and essay-style questions with high speed and accuracy, making them effective tools for assessing learning outcomes. Chatbots can track students' progress and performance, providing valuable assessment data that can be used to offer targeted support and interventions to help students improve.

Further, AI technology can be employed to identify "at-risk" students who are likely to fail academic expectations. Through tracking student learning performance, chatbots can detect trends of declining grades or failure to meet academic benchmarks. If a student consistently scores low on quizzes and tests, the educator can intervene and provide remedial assistance to the individual. Using AI algorithms, Latif and colleagues (2022) developed a model that can predict a student's grades in a university course based on the initial performance and a combination of assessment tools such as quizzes and assignments. If the student is predicted to perform poorly on an exam or an assignment, the educator can address the difficulty.

6 Challenges

6.1 Limited Role in Academic Assessment

Despite the capacities discussed above, chatbots' value in assessment is rather limited. First, chatbots' assessment of written work may not always be fair. AI-generated text detection, while continually improving, is not yet foolproof and can produce unfair evaluations, potentially undermining the credibility of the assessment process (Labadze et al., 2023). Educators may also struggle to discern whether the assignments were completed by students or if they were generated by AI. Before the advent of chatbots, a writing assignment would require the student to gather materials, conduct research, form a thesis, and construct paragraphs. Chatbots can quickly provide text on almost any writing prompt within just a few moments. Among all the generative chatbots, ChatGPT has emerged as one of the most used tools for academic work. A recent Pew Research Center survey (2023) reveals that 13% of all U.S. teens have used ChatGPT in their schoolwork. In higher education, four in ten college students use ChatGPT on assignments (Intelligent, 2024).

Not only do students use the chatbot for writing essays, but they also use it to construct ideas. According to the Chronicle of Higher Education (2023), it is nearly impossible to catch plagiarism because of ChatGPT's ability to provide a step-by-step process of writing a paper. The user can tell the algorithm the writing prompt and ask for a main argument, then have ChatGPT give an outline to support this claim. To avoid being caught cheating, students can command ChatGPT to resemble the speech pattern of a particular type of individual (Shen, 2024). For example, suppose a user instructs ChatGPT to answer a prompt in a way a sixth grader would. In that case, the chatbot can create text using the vocabulary a typical sixth grader possesses, making it challenging to detect who actually wrote the information. ChatGPT also generates original text every time, so students who type in the same prompt will not receive the same answers. Chatbot-generated texts can easily pass a plagiarism detector without being flagged because plagiarism checkers work by comparing the submitted text against a vast database of existing content to identify similarities and potential matches.

Further, in the wake of realizing that some students use chatbots like ChatGPT to write essays and complete assignments, educators have resorted to other assignments that are hard to cheat on, such as the old-school paper and pencil method, videos, oral exams, presentations, diagrams, and drawings. Some teachers ask students to explain their work and detail their thinking process verbally. Chatbots presently cannot grade these assignments. As indicated, assignments should reflect the learning objectives. Although chatbots are incredibly helpful in assessing assignments and providing feedback, they cannot determine if students have met learning objectives.

6.2 Hindering Social Skills

Another potential problem with chatbots is that over-reliance can hamper students' development of social skills. Mathur and Rutherford (1996) define social skills as socially acceptable patterns of behaviors that allow students to gain social reinforcement and acceptance and avoid aversive social situations. A socially skilled individual understands the surroundings and responds to various social settings effectively. Students may or may not know how to behave or respond appropriately before attending school. Some students may hold the false belief that violence is an acceptable way to express frustration and anger. Schools and educators should be responsible for teaching students how to properly manage and communicate emotions and feelings. Improper emotional outbursts or responses may be disruptive to others around. Although some scholars have attempted to use technology to enhance social skills among autistic or emotionally disabled individuals (e.g., Cumming et al., 2008; Reed et al., 2011; Soares et al., 2021), much empirical evidence suggests that social skills are declining as a result of exposure to technology (e.g., Brown, 2013; Downey & Gibbs, 2020; Romero-López et al., 2021). Social skills are acquired through a variety of activities, such as observing how others handle different situations, actively listening, paying attention to others' posture and facial expressions, and engaging in social activities. Chatbots provide few of these elements.

Frequent use of chatbots for learning may lead to a decrease in face-to-face interactions with teachers and classmates. If students spend the bulk of their time receiving instruction from a chatbot, they may not have adequate time to communicate with others during school hours. This reduction can impede the development of essential communication skills, such as reading body language, tone of voice, and other non-verbal cues. Human interactions help individuals develop emotional intelligence, including recognizing and responding to others' emotions. Heavy dependence on chatbots may impair one's ability to establish such ability. As indicated, chatbots provide a non-judgmental and welcoming learning environment, which makes them user-friendly among slow learners and shy and introverted students. However, this can make it even more uncomfortable for these students to engage with human teachers if they are used to interacting with chatbots. If one refrains from human interactions, these exchanges will become unfamiliar and even anxiety-inducing, further hampering social skill formation.

This does not paint a positive outlook because the share of jobs that require social skills is on the rise. Between 1980 and 2012, jobs requiring high levels of social interaction grew by nearly 12 percentage points as a share of the U.S. labor force (Deming, 2017). According to the Pew Research Center (2020), employment is rising most rapidly in jobs that are most in need of social skills. The same report indicates that in 2018, 42% of all workers held jobs where social skills were most important, whereas only 19% were employed in occupations where social skills were least important. Schools should equip students with the necessary academic and essential skills to thrive in the future. Installing social skills is crucial to the academic instructional experience and the overall success of the individual student (Cartledge &

Milburn, 1978; Deshler et al., 1996). Education researcher Donald Deshler and his associates (1996) view social skills as central as academic skills because social skill deficits are associated with negative social self-concept and social isolation, diminishing opportunities to learn. Although students who lack social skills may be able to conduct their learning through a chatbot, their opportunities for life-long learning can be drastically reduced.

To mitigate these potential negative impacts, it is essential to create a balance between the use of chatbots and human interaction in learning environments. Human interactions may be uncomfortable and intimidating to some, but they build one's social skills. One way to tackle the challenge is to combine chatbot-assisted learning with traditional classroom activities that promote face-to-face interaction and collaboration. That way, not only do students reap the benefits of personalized and innovative teaching, but they also have opportunities to obtain and practice social skills.

7 Conclusions

Since the creation of ELIZA, AI and chatbot technology has gone through significant transformations. Today's chatbots can generate text and images, interpret questions, fulfill user commands, solve math problems, and write computer code. The government's heavy reliance on AI during the COVID-19 pandemic to disseminate health information to the public further accelerated the technology. AI is prevalent in our lives, and this trend is likely to continue. AI technology has transformed how educational institutions operate, how teachers teach, and how students learn. Schools use chatbots to answer students' and parents' questions and streamline administrative tasks. The integration of chatbots into the educational landscape has made learning more engaging, efficient, and supportive for students.

Chatbots deliver information in a variety of formats, simplifying the learning process for students who need accommodation. Chatbots offer screen readers for the visually impaired and text-based options for users with hearing loss. Chatbots diversify learning activities, making it enjoyable and effective. Through games, quizzes, role-plays, and flashcards, students can engage in a variety of innovative ways to learn. Operating on a large sum of data, chatbots offer much information from many academic fields. This feature provides an almost limitless amount of information. The ability to gather and assess large quantities of data is indispensable in sharpening one's critical thinking ability. Further, chatbots assist with learning assessment by offering immediate feedback to the learner. Not only do chatbots gather assessment data from user input, some intelligent chatbots can detect user emotions such as confidence level as well. Based on the assessment data, chatbots adjust the speed of teaching and information feeding accordingly.

As discussed, these features enable AI-powered chatbots to enhance pedagogical methods such as cooperative learning, authentic learning, and problem-based learning. Cooperative learning emphasizes collaboration and group work. AI can optimize task distribution based on factors such as learning styles, abilities, and personalities. This maximizes the chances of successful collaboration. Through performance tracking, the AI can also gently steer the group back to a more constructive discussion when it detects that the conservation is off-topic. Authentic learning equips students with the ability to solve real-world problems. AI can create cases of real-world situations, such as ethical issues or business dilemmas, allowing students to practice skills and apply knowledge in a safe and controlled environment. Students can tackle complex problems and make decisions with real-world consequences, preparing them for their endeavors in the future. Problem-based learning encourages students to work collaboratively to investigate, analyze, and solve problems. AI-powered search tools can help students find resources they need to solve problems by providing a wealth of information.

As powerful tools as they are, chatbots have a few challenges. First, since they are backed by databases containing information written by humans, prejudice and errors may appear in the answers generated. Chatbots sometimes even fabricate information when there is not enough information available on a subject. Users must be careful when reviewing chatbots-generated content and verify its authenticity and accuracy. Second, chatbots may compromise academic integrity. It is difficult to detect whether a student completed a writing assignment on their own because chatbots create original content every time they receive a prompt. The lack of reliable AI detectors leads teachers to come up with assignments that are more difficult to cheat on such as oral exams, diagrams, and presentations. This potentially adds to the teachers' workload. Finally, evidence suggests that overreliance on technology may contribute to decreased social skills, which are important in one's career endeavors. Replacing traditional teaching with chatbots can decrease face-to-face interactions with teachers and peers, impeding the development of social skills. Therefore, chatbots should be incorporated as a secondary means to facilitate education.

References

Adamopoulou, E., & Moussiades, L. (2020). Chatbots: History, technology, and applications. *Machine Learning with Applications, 2*, 100006.

Alkhatlan, A., & Kalita, J. (2018). Intelligent tutoring systems: A comprehensive historical survey with recent developments. *International Journal of Computer Applications, 181*(43), 1–20.

American Association for the Advancement of Science. (2011). *Project 2061 grant: What do green schools have to teach their students?* https://www.aaas.org/news/project-2061-grant-what-do-green-schools-have-teach-their-students

American Psychology Association. (2023). *How to use ChatGPT as a learning tool.* https://www.apa.org/monitor/2023/06/chatgpt-learning-tool

Araz, G., & Sungur, S. (2007). Effectiveness of problem-based learning on academic performance in genetics. *Biochemistry and Molecular Biology Education, 35*(6), 448–451.

Azer, S. A. (2001). Problem-based learning. *Neurosciences, 6*(2), 83–89.

Bank of America. (2024). *BofA's Erica surpasses 2 billion interactions, helping 42 million clients since launch.* https://newsroom.bankofamerica.com/content/newsroom/press-releases/2024/04/bofa-s-erica-surpasses-2-billion-interactions--helping-42-millio.html

BBC. (2023). *ChatGPT: Everything you need to know about OpenAI's GPT-4 tool.* https://www.sciencefocus.com/future-technology/gpt-3

Berry, R. (2008). *Assessment for learning (Vol. 1).* Hong Kong University Press.

Bradbury, N. A. (2016). Attention span during lectures: 8 seconds, 10 minutes, or more? *Advances in Physiology Education, 40*, 509–513.

Brown, C. (2013). *Are we becoming more socially awkward? An analysis of the relationship between technological communication use and social skills in college students.* Psychology Honors Papers at Connecticut College. https://digitalcommons.conncoll.edu/psychhp/40/

Burstein, J., & Chodorow, M. (2003). Directions in automated essay analysis. In R. Kaplan (Ed.), *The Oxford handbook of applied linguistics* (pp. 487–497). Oxford University Press.

California State University at Northridge. (2023). *CSUNNY program summary and outcomes.* https://edtechmagazine.com/higher/sites/edtechmagazine.com.higher/files/CSUNny%20data.pdf

Cartledge, G., & Milburn, J. F. (1978). The case for teaching social skills in the classroom: A review. *Review of Educational Research, 48*(1), 133–156.

Chang, D. H., Lin, M. P. C., Hajian, S., & Wang, Q. Q. (2023). Educational design principles of using AI chatbot that supports self-regulated learning in education: Goal setting, feedback, and personalization. *Sustainability, 15*(17), 12921.

Chowdhary, K. R. (2020). *Fundamentals of artificial intelligence.* Springer.

Chung, G. K. W. K., & Baker, E. L. (2003). Issues in the reliability and validity of automated scoring of constructed responses. In M. D. Shermis & J. C. Burstein (Eds.), *Automated essay scoring: A cross-disciplinary assessment* (pp. 23–40). Lawrence Erlbaum.

Colby, K. M. (1981). Parrying. *Behavioral and Brain Sciences, 4*(4), 550–560.

Conati, C. (2009). Intelligent tutoring systems: New challenges and directions. In *Twenty-first International Joint Conference on Artificial Intelligence.*

Crompton, H., & Burke, D. (2023). Artificial intelligence in higher education: The state of the field. *International Journal of Educational Technology in Higher Education, 20*(1), 1–22.

Cucuiat, V., & Waite, J. (2024). Feedback literacy: Holistic analysis of secondary educators' views of LLM explanations of program error messages. In *Proceedings of the 2024 on Innovation and Technology in Computer Science Education V. 1* (pp. 192–198).

Cumming, T. M., Higgins, K., Pierce, T., Miller, S., Boone, R., & Tandy, R. (2008). Social skills instruction for adolescents with emotional disabilities: A technology-based intervention. *Journal of Special Education Technology, 23*(1), 19–33.

Deming, D. J. (2017). The growing importance of social skills in the labor market. *The Quarterly Journal of Economics, 132*(4), 1593–1640.

Dennick, R. G., & Exley, K. (1998). Teaching and learning in groups and teams. *Biochemical Education, 26*(2), 111–115.

DePietro, P. (2013). Transforming education with new media: Participatory pedagogy, interactive learning and web 2.0. *International Journal of Technology, Knowledge and Society, 8*(5), 1–11.

Deshler, D., Ellis, E., & Lenz, B. K. (1996). *Teaching adolescents with learning disabilities: Strategies and methods.* Denver, CO: Love.

Downey, D. B., & Gibbs, B. G. (2020). Kids these days: Are face-to-face social skills among American children declining? *American Journal of Sociology, 125*(4), 1030–1083.

Duze, C. O. (2010). Effects of participatory learning technique on achievement and attitude of B.Ed. Students in educational research methods. *Journal of Social Sciences, 22*(3), 185–189.

Education Week. (2024). *Los Angeles Unified bets big on 'Ed,' an AI tool for students.* https://www.edweek.org/technology/los-angeles-unified-bets-big-on-ed-an-ai-tool-for-students/2024/03

Ennis, R. H. (1987). A taxonomy of critical thinking dispositions and abilities. In J. B. Baron & R. J. Sternberg (Eds.), *Series of books in psychology. Teaching thinking skills: Theory and practice* (pp. 9–26). W. H. Freeman/Times Books/ Henry Holt & Co.

Erdogan, F. (2019). Effect of cooperative learning supported by reflective thinking activities on students' critical thinking skills. *Eurasian Journal of Educational Research, 19*(80), 89–112.

Forbes. (2023). *Artificial Intelligence In education: Teachers' opinions on AI in the classroom.* https://www.forbes.com/advisor/education/it-and-tech/artificial-intelligence-in-school/
GAO. *Closing the digital divide for the millions of Americans without broadband.* https://www.gao.gov/blog/closing-digital-divide-millions-americans-without-broadband
Gilster, P. (1997). *Digital literacy.* John Wiley & Sons Inc.
Gustafsson, B., & Warner, M. (2008). Participatory learning and deliberative discussion within education for sustainable development. In J. Ohman (Ed.), *Values and democracy in education for sustainable development: Contributions from Swedish research* (pp. 75–92). Liber Publishing House.
Harlen, W., & James, M. (1997). Assessment and learning: Differences and relationships between formative and summative assessment. *Assessment in Education: Principles, Policy & Practice, 4*(3), 365–379.
Herrington, J., Reeves, T. C., & Oliver, R. (2014). *Authentic learning environments.* Springer.
Hew, K. F., Huang, W., Du, J., & Jia, C. (2023). Using chatbots to support student goal setting and social presence in fully online activities: Learner engagement and perceptions. *Journal of Computing in Higher Education, 35*(1), 40–68.
Hmelo-Silver, C. E. (2004). Problem-based learning: What and how do students learn? *Educational Psychology Review, 16*, 235–266.
IBM. (2024a). *What is machine learning?* https://www.ibm.com/topics/machine-learning
IBM. (2024b). *A computer called Watson.* https://www.ibm.com/ibm/history/ibm100/us/en/icons/watson/
Fanni, S. C., Febi, M., Aghakhanyan, G., & Neri, E. (2023). Natural language processing. In M. E. Klontzas, S. C. Fanni, & E. Neri (Eds.), *Introduction to Artificial Intelligence* (pp. 87–99). Springer International Publishing.
Hamid, H., Zulkifli, K., Naimat, F., Yaacob, N. L. C., & Ng, K. W. (2023). Exploratory study on student perception on the use of chat AI in process-driven problem-based learning. *Currents in Pharmacy Teaching and Learning, 15*(12), 1017–1025.
Intelligent. (2024). *4 in 10 college students are using ChatGPT on assignments.* https://www.intelligent.com/4-in-10-college-students-are-using-chatgpt-on-assignments/
Kallet, M. (2014). *Think smarter: Critical thinking to improve problem-solving and decision-making skills.* John Wiley & Sons.
Klimovienė, G., Urbonienė, J., & Barzdžiukienė, R. (2006). Developing critical thinking through cooperative learning. *Studies about Languages, 9*(1), 77–85.
Kohnke, L. (2023). A pedagogical chatbot: A supplemental language learning tool. *Relc Journal, 54*(3), 828–838.
Kosslyn, S. M. (2024). *Active Learning with AI: A Practical Guide.* Alinea Learning.
Kuhail, M. A., Alturki, N., Alramlawi, S., & Alhejori, K. (2023). Interacting with educational chatbots: A systematic review. *Education and Information Technologies, 28*(1), 973–1018.
Labadze, L., Grigolia, M., & Machaidze, L. (2023). Role of AI chatbots in education: Systematic literature review. *International Journal of Educational Technology in Higher Education, 20*(1), 56–72.
Latif, G., Alghazo, R., Pilotti, M. A., & Brahim, G. B. (2022). Identifying "at-risk" students: An AI-based prediction approach. *International Journal of Computing and Digital System, 11*(1), 1–9.
Lau, A. M. S. (2016). 'Formative good, summative bad?'–A review of the dichotomy in assessment literature. *Journal of Further and Higher Education, 40*(4), 509–525.
Leinonen, J., Denny, P., MacNeil, S., Sarsa, S., Bernstein, S., Kim, J., Tran, A. & Hellas, A. (2023). Comparing code explanations created by students and large language models. In *Proceedings of the 2023 Conference on Innovation and Technology in Computer Science Education V. 1* (pp. 124–130).
Lombardi, M. M., & Oblinger, D. G. (2007). Authentic learning for the 21st century: An overview. *Educause Learning Initiative, 1*(2007), 1–12.

MacNeil, S., Tran, A., Mogil, D., Bernstein, S., Ross, E., & Huang, Z. (2022). Generating diverse code explanations using the gpt-3 large language model. In *Proceedings of the 2022 ACM Conference on International Computing Education Research-Volume 2* (pp. 37–39).

Mageira, K., Pittou, D., Papasalouros, A., Kotis, K., Zangogianni, P., & Daradoumis, A. (2022). Educational AI chatbots for content and language integrated learning. *Applied Sciences, 12*(7), 3239–3254.

Maina, F. W. (2004). Authentic learning: Perspectives from contemporary educators. *Journal of Authentic Learning, 1*(1), 1–8.

Martin, A., & Grudziecki, J. (2006). DigEuLit: Concepts and tools for digital literacy development. *Innovation in Teaching and Learning in Information and Computer Sciences, 5*(4), 249–267.

Mathur, S., & Rutherford, R. (1996). Is social skills training effective for students with emotional or behavioral disorders? *Research Issues and Needs. Behavioral Disorders, 22*(1), 21–27.

McCarthy, J. (2004). *What is artificial intelligence.* Computer Science Department at Stanford University. https://cse.unl.edu/~choueiry/S09-476-876/Documents/whatisai.pdf

McParland, M., Noble, L. M., & Livingston, G. (2004). The effectiveness of problem-based learning compared to traditional teaching in undergraduate psychiatry. *Medical Education, 38*(8), 859–867.

Miner, A. S., Laranjo, L., & Kocaballi, A. B. (2020). Chatbots in the fight against the COVID-19 pandemic. *Digital Medicine, 3*, 1–4.

Mourtos, N. J., Okamoto, N. D., & Rhee, J. (2004). Defining, teaching, and assessing problem solving skills. In *7th UICEE Annual Conference on Engineering Education*, 1–5.

National Center for Education Statistics. (2024). *Who is nontraditional?* https://nces.ed.gov/pubs/web/97578e.asp

National Council for Social Studies. (2024). Solving problems with twenty questions. https://www.socialstudies.org/social-education/85/6/solving-problems-twenty-questions

National Council of Teachers of English. (2007). *21st-Century literacies: A policy research brief produced by the National Council of Teachers of English.* https://cdn.ncte.org/nctefiles/resources/positions/chron1107researchbrief.pdf

National Council of Teachers of Mathematics. (2022). *Problem solving: An approach to understanding and critiquing our world.* https://www.nctm.org/News-and-Calendar/Messages-from-the-President/Archive/Trena-Wilkerson/Problem-Solving_-An-Approach-to-Understanding-and-Critiquing-Our-World/

Newmann, F. M. (1995). *A guide to authentic instruction and assessment: Vision, standards and scoring.* Wisconsin Center for Education Research.

New Jersey Institute of Technology. (2023). *ELIZA: A very basic Rogerian psychotherapist chatbot.* https://web.njit.edu/~ronkowit/eliza.html

NPR News. (2017). *It has to have a soul: How chatbots get their personalities.* https://www.npr.org/sections/alltechconsidered/2017/03/10/519002884/it-has-to-have-a-soul-how-chatbots-get-their-personalities

OECD. 2024. Using AI in the workplace: Opportunities, risks and policy responses. https://www.oecd.org/publications/using-ai-in-the-workplace-73d417f9-en.htm

Pellegrino, J. W., Bransford, J. D., & Donovan, M. S. (Eds.). (1999). *How people learn: Bridging research and practice.* National Academies Press.

Pew Research Center. (2020). *Employment is rising most rapidly in jobs most in need of social, fundamental and analytical skills.* https://www.pewresearch.org/social-trends/2020/01/30/employment-is-rising-most-rapidly-in-jobs-most-in-need-of-social-fundamental-and-analytical-skills/

Pew Research Center. (2021). *The internet and the pandemic.* https://www.pewresearch.org/internet/2021/09/01/the-internet-and-the-pandemic/

Pew Research Center. (2023). *About 1 in 5 U.S. teens who've heard of ChatGPT have used it for schoolwork.* https://www.pewresearch.org/short-reads/2023/11/16/about-1-in-5-us-teens-whove-heard-of-chatgpt-have-used-it-for-schoolwork/

Pogonowski, L. (1987). Developing skills in critical thinking and problem solving. *Music Educators Journal, 73*(6), 37–41.

Precedence Research. (2023). *Chatbot market.* https://www.precedenceresearch.com/chatbot-market

Pretty, J. N., Thompson, I. G., & Scoones, I. (1995). *Trainer's guide for participatory learning and action.* International Institute for Environment and Development.

Ramos, B., & Condotta, R. (2024). Enhancing learning and collaboration in a unit operations course: Using AI as a catalyst to create engaging problem-based learning scenarios. *Journal of Chemical Education, 101*, 3246–3254.

RAND. (2024). *Using Artificial Intelligence tools in K-12 classrooms.* https://www.rand.org/content/dam/rand/pubs/research_reports/RRA900/RRA956-21/RAND_RRA956-21.pdf

Reed, F. D. D., Hyman, S. R., & Hirst, J. M. (2011). Applications of technology to teach social skills to children with autism. *Research in Autism Spectrum Disorders, 5*(3), 1003–1010.

Renzulli, J. S., Gentry, M., & Reis, S. M. (2004). A time and a place for authentic learning. *Educational Leadership, 62*(1), 73–77.

Romero-López, M., Pichardo, C., De Hoces, I., & García-Berbén, T. (2021). Problematic internet use among university students and its relationship with social skills. *Brain Sciences, 11*(10), 1301–1310.

Rule, A. C. (2006). The components of authentic learning. *Journal of Authentic Learning, 3*(1), 1–10.

Schweinhart, L. J. (2006). The High/Scope approach: Evidence that participatory learning in early childhood contributes to human development. In N. Watts, C. Ayoub, R. Bradley, & J. Puma (Eds.), *The crisis in youth mental health* (pp. 207–227). Greenwood Publishing Group Inc.

Shen, I. (2024). The robot wrote my college papers: Integrating chatbots to assist higher education. In S. Hai-Jew (Ed.), *Generative AI in Teaching and Learning* (pp. 311–327). IGI Global.

Simanjuntak, M. P., Hutahaean, J., Marpaung, N., & Ramadhani, D. (2021). Effectiveness of problem-based learning combined with computer simulation on students' problem-solving and creative thinking skills. *International Journal of Instruction, 14*(3), 519–534.

Soares, E. E., Bausback, K., Beard, C. L., Higinbotham, M., Bunge, E. L., & Gengoux, G. W. (2021). Social skills training for autism spectrum disorder: A meta-analysis of in-person and technological interventions. *Journal of Technology in Behavioral Science, 6*, 166–180.

Snyder, L. G., & Snyder, M. J. (2008). Teaching critical thinking and problem-solving skills. *The Journal of Research in Business Education, 50*(2), 90–99.

Su, K. D. (2022). Implementation of innovative Artificial Intelligence cognitions with problem-based learning guided tasks to enhance students' performance in science. *Journal of Baltic Science Education, 21*(2), 245–257.

The Chronicle of Higher Education. (2023). *I'm a student. You have no idea how much we're using ChatGPT. No professor or software could ever pick up on it.* https://www.chronicle.com/article/im-a-student-you-have-no-idea-how-much-were-using-chatgpt

University of Memphis. (2024). *AutoTutor.* https://www.memphis.edu/iis/projects/autotutor.php

Veenman, S., Van Benthum, N., Bootsma, D., Van Dieren, J., & Van der Kemp, N. (2002). Cooperative learning and teacher education. *Teaching and Teacher Education, 18*(1), 87–103.

Wollny, S., Schneider, J., Di Mitri, D., Weidlich, J., Rittberger, M., & Drachsler, H. (2021). Are we there yet?-a systematic literature review on chatbots in education. *Frontiers in Artificial Intelligence, 4*, 1–18.

Woods, D. R., Hrymak, A. N., Marshall, R. R., Wood, P. E., Crowe, C. M., Hoffman, T. W., Wright, J. D., Taylor, P. A., Woodhouse, K. A., & Bouchard, C. K. (1997). Developing problem solving skills: The McMaster problem solving program. *Journal of Engineering Education, 86*(2), 75–91.

Wright, B. D. (1997). Evaluating learning in individual courses. In J. G. Gaff & J. L. Ratcliff (Eds.), *Handbook of the undergraduate curriculum* (pp. 571–590). Jossey Bass.

Wu, D., Xiang, Y., Wu, X., Yu, T., Huang, X., Zou, Y., & Lin, H. (2020). Artificial intelligence-tutoring problem-based learning in ophthalmology clerkship. *Annals of Translational Medicine, 8*(11), 1–7.

Zhao, G., Li, Y., & Xu, Q. (2022). From emotion AI to cognitive AI. *International Journal of Network Dynamics and Intelligence, 1*(1), 65–72.

School Children Theories on ChatGPT: Educational Perspectives of the Implementation of Artificial Intelligence in Social Robots

Angela Balzotti, Rosa Gallelli, Pasquale Renna, Antonio Ascione, Ilenia Amati, Michela Casolaro, and Luca Depalo

Abstract This chapter explores the increasing significance of child-social robot collaboration in educational contexts, emphasizing the integration of large language models (LLMs) such as ChatGPT into social robots to enhance children's trust and interaction. Rooted in a long tradition of research inspired by Piaget and Turkle's studies on children's interactions with social media, the investigation aims to understand children's perceptions of AI, emotions, and trust. An experiment was conducted involving children interacting with ChatGPT to promote reasoning about AI. The procedure included: introducing AI concepts, brainstorming and writing questions, interacting with ChatGPT, analyzing responses, and reflecting on the chatbot's reasoning. The activities were designed to stimulate critical thinking about AI's capabilities and limitations. This approach seeks to help children better understand AI and develop critical thinking skills regarding systems like ChatGPT.

Keywords Generative artificial intelligence · School · Children · Education

---

A. Balzotti (✉) · R. Gallelli · A. Ascione · I. Amati · M. Casolaro · L. Depalo
Department of Education Sciences, Psychology, Communication Sciences, University of Studies "Aldo Moro", Bari, Italy
e-mail: angela.balzotti@unifg.it

R. Gallelli
e-mail: rosa.gallelli@uniba.it

A. Ascione
e-mail: antonio.ascione@uniba.it

I. Amati
e-mail: ilenia.amati@uniba.it

M. Casolaro
e-mail: michela.casolaro@uniba.it

P. Renna
Department of Psychology, University of Studies "e-Campus", Novedrate, Italy
e-mail: pasquale.renna@uniecampus.it

G. Lampropoulos and S. Papadakis (eds.), *Social Robots in Education*, Studies in Computational Intelligence 1194, https://doi.org/10.1007/978-3-031-82915-4_6

1 Introduction

This survey aims to explore the increasing significance of child-social robot collaboration in educational contexts, emphasizing the integration of large language models (LLMs) such as ChatGPT into social robots to enhance children's trust and interaction. Moreover, this survey focuses on the results of an experimentation in an Italian School aimed to investigate interactions and perceptions of reliability by children aged 9–10 towards Generative Artificial Intelligence.

It is founded on the following theoretical background. The interactions between Humans, Artificial Intelligence and social Robots have been widely addressed by scientific literature, and particularly by pedagogical literature. In this regard, Franca Pinto Minerva and Rosa Gallelli (2004) have explored certain peculiar characteristics of the educational potential of "intelligent programs". Since the early 2000s, children aged 9–10 have been involved in experiments on the interactions with social robots (Turkle, 1996). The pedagogical debate in Italy has shown increasing interest in the interaction between humans and machines (Pinto Minerva & Gallelli, 2004), with a growing focus on the integration of biological knowledge and artificial life. This attention is particularly concentrated on educational developments related to the evolutionary interactions between living and artificial systems. The previously mentioned pedagogical studies were focused on the concept of "embodied" intelligence (Damasio, 1995; Maturana & Varela, 2001; Merleau-Ponty, 1979; Pinker, 2019) and explore the interdependence between "natural" and "artificial" intelligence, as well as the connection between cognitive functions and technological artifacts (Marchesini, 2014). Studies on the use of Generative Artificial Intelligence in education are still few (Belpaeme et al., 2018; Dumouchel & Damiano, 2018, 2019, 2020; Montenegro-Rueda et al., 2023; Peretti et al., 2022), and largely attest to the usefulness of ChatGPT in teaching and learning: the individualization of teaching and evaluation, the strengthening of collaboration between students and teachers, study guidance, and more. Since ChatGPT was launched in late 2022, studies on the reasoning processes involving children together with Artificial Intelligence are rare and have to be developed and strengthened.

Therefore, in a first part (from paragraph 2 since paragraph 6), our contribution presents a theoretical survey of the debate on "embodied learning", on the use of social robotics at school, as well as on the implementations, in social robots, of a generative artificial intelligence such as ChatGPT. Paragraph 2 is about the possibility of an integration between social robots and artificial intelligence. Paragraph 3 deepens the history and the perspectives of ChatGPT. Paragraph 4 talks about the collaborations between children and social robots in educational contexts. Paragraph 5 deepens some possibilities of Artificial Sociability linked to Empathy. Paragraph 6 talks about the inclusions of AI and LLMs in educational contexts.

In the second part (paragraphs 7 and 8), our contribution presents some results of an exploratory investigation focused on children's theories on ChatGPT at the age of 9–10, and therefore on the broad possibilities of interaction between children, artificial intelligence and social robots. Paragraph 7 deepens the research design;

some comments to questions of the post-test and the comments to the Candy-Game and the Poetry-Game between AI (ChatGPT) and the young classmates involved in the research. Paragraph 8 comments the perceptions of children about "Artificial Intelligence as a classmate and a friend".

2 Social Robot and Artificial Intelligence: The Possibility of an Integration

2.1 Embodied Learning Between Natural and Artificial Intelligence

Discussing "embodied learning" raises numerous pedagogical and didactic issues, particularly related to the integration of biological and artificial elements, as well as the simultaneous interactions between humans and robots. Technology, especially today, can no longer be viewed as a mere passive tool in human hands, utilized without consequences for the user. On the contrary, it acts both directly and indirectly on human beings, shaping their cognitive and emotional capacities, social and environmental relationships, and modes of acting and experiencing. Human-made tools contribute to shaping humanity itself (Longo, 2003, p. 86). Humans are not only the creators of technology but also its product (Sloterdijk, 2001). Pedagogists and educators are exploring how modern technology, by influencing its users, provokes reciprocal structural changes. Today, robots present an opportunity to deepen the study of both Artificial and Human Intelligence. Biological studies demonstrate that intelligence is an emergent quality from the complex spatiotemporal interactions between living organisms and non-living environments; meanwhile, technological studies highlight that intelligence emerges from the situated interactions of artificial intelligences applied to robotics. In this regard, Pinto Minerva and Gallelli (2004) have explored certain peculiar characteristics of the educational potential of "intelligent programs." Since the early 2000s, pedagogical debate in Italy has shown increasing interest in the interaction between humans and machines, with a growing focus on the integration of biological knowledge and artificial life. This attention is particularly concentrated on educational developments related to the evolutionary interactions between living and artificial systems. The studies of Franca Pinto Minerva and Rosa Gallelli focus on the concept of "embodied" intelligence and explore the interdependence between "natural" and "artificial" intelligence, as well as the connection between cognitive functions and technological artifacts.

Specifically, cognitive abilities seem to be inextricably linked to having a physical body. Simultaneously, each technology influences cognitive processes, creating different modes of thinking through new perceptual and cognitive opportunities. From a Darwinian perspective on the mind–body relationship, rational and symbolic activity is closely tied to the physical substrate. Cognition, in all its forms, from the most automatic to the most complex, consists of progressively learning to link different perceptions with their positive or negative effects on survival, leading to appropriate synaptic modifications. During childhood development, experiences in a particular environment act as "specialists," pruning specific neuronal groups among the many available connections at birth and organizing them into neuron maps, each with specialized tasks. In this context, brain and mind development in everyone is the unique result of a selective process stemming from the interaction between a genetically determined structure and a cultural imprint left by the individual's experiences.

2.2 *Contemporary Neuro-scientific Studies and Knowledge Construction*

Contemporary neuroscience emphasizes the connection between human intelligence and the environment in which a child grows, along with the fundamental role of the body in cognitive activities. Varela et al. (1992) describes organisms' ability to generate meanings of reality from within, based on the limitations and possibilities offered by their physiology ("enaction"). Knowledge construction is therefore conditioned by how each organism organizes its action in the world and the temporal course of those actions, both in individual ontogeny and species phylogeny. Central to this new perspective is the idea that knowledge is primarily concrete, embodied, and lived. This knowledge, with its historical specificity and context, does not hinder the understanding of a general model nor serve merely as a transitional phase toward broader abstraction. Instead, it constitutes the way we act and the place we exist (Varela, 1994, p. 144). Cognition is not simply the result of a "mind" gathering and representing aspects of a pre-existing reality as if it were an independent object. Instead, it emerges from the interaction between an organism moving and acting within the environment, using the capabilities offered by its body and the specific characteristics of the environment itself. Perception is not about recovering a pre-defined world but orienting action in a world inherently linked to our sensorimotor abilities. Cognitive structures develop from recurrent patterns of action driven by perception, and cognition is not about representations but embodied action. In essence, bodily structures, especially sensorimotor ones, form the foundation of experience, and these experiences influence conceptual understanding and rational thought (Varela, 1994, pp. 157–158). This conception, which sees an essential continuity between body and mind, is based on Humberto Maturana's theory of knowledge (1985), later expanded in collaboration with Francisco Varela. According to Maturana, cognition is an integral part of every living system, considered an autopoietic system. In Maturana and Varela's model, cognition is not only linked to what we commonly associate with thought, such as language or conceptualization, but emerges from the material complexity of every living organism. Cognition is a process of knowing that includes the transmission, exchange, and management of information, enabling living systems to reorganize the material conditions necessary to maintain their autopoietic organization. For these systems, knowledge is an essential survival strategy, present in all forms of life, even in the absence of a nervous system. In this way, a new synthesis between mind and matter, between mind and body, is outlined. All living organisms demonstrate that their cognitive processes—from the simplest forms of life to the human brain—are processes of self-organization, developing within a complex web of interconnections among different levels of organic life, inextricably linked to bodily experience, from the most basic biological manifestations to the sensorimotor complexity of advanced organisms. Building on the theories of George Lakoff, Mark Johnson, Eleanor Rosch, and Francisco Varela, Damasio offers a vision of the bodily roots of cognition, identifying the biological basis of learning. Learning is part of the biological process that has driven the evolution of our species, and it is rooted in the entire human body. Damasio (1995, p. 318) defines the mind as "embodied," meaning that it corresponds to the set of representations of which we can become aware in the form of images. The neurobiology of emotions (and neuroimaging techniques), Damasio further notes, offers interesting insights into the conditions that favored the emergence of human characteristics, starting from adaptive behaviors shared with other animals, rooted in fundamental bodily needs. There is a close continuity between the emotional and bodily capacities of organisms equipped with automatic action programs, innate and aimed at life protection, regulated by homeostasis, and the cognitive faculties of our species. In this context, emotions activate resources and strategies that can be directed, even consciously, toward behaviors that safeguard survival. Damasio emphasizes the inseparable link between the development of the mind and the ability to perceive one's body moving within the environment. Being unable to move means lacking behavior, and without behavior, there can be no mind. In this sense, for an organism to develop consciousness, it must possess a brain, the ability to move, the perception of its body and surroundings, and the ability to formulate a reaction to stimuli—a behavior—initially based on reflections. In line with Damasio, Harvard psychologist Pinker (2019) argues that human intelligence evolved to fill a specific evolutionary niche, manipulating the environment through causal reasoning and cooperation. This is complemented by a process of metaphorical abstraction, still evident in today's language, which allows cognitive mechanisms originally developed for physical reasoning to adapt to more abstract concepts.

2.3 The Possibility of Hybridization Between Artificial Intelligence and Robotic Body

Given the above, one might ask whether theories of embodiment can indeed provide a useful key to understanding the implications of the transformative power stemming from the convergence between the "biological" and the "artificial," particularly in the context of robots as artificial entities capable of self-regulation, self-repair, learning, and even experiencing emotions. In this context, one might then ask: how do robots learn? How can robots interact with humans? How can robots understand and express human emotions? Answering these questions opens new perspectives on the understanding of intelligence, knowledge formation, motivations, and decisions, as well as on the role of emotions and affectivity, which are central to social interactions. The cognitive sciences and the neurosciences made significant advances which allow us to better understand the functioning of the brain and intelligence, highlighting the essential role of emotions in cognitive processes. The ability to manage one's emotions and recognize those of others constitutes a specific kind of emotional intelligence, which positively influences social interactions. This type of learning, known as "Social and Emotional Learning" develops crucial interpersonal skills such as cooperation, collaboration and communication, essential skills for academic success, relational well-being, and prevention of anti-social or self-destructive behaviors. In an increasingly complex social context, characterized by rapid and unpredictable cultural, scientific and technological changes, it is very important that young generations acquire not only theoretical knowledge and technical skills, but also aptitudes for innovation, lifelong learning skills, personal initiative, responsibility, flexibility, and collaborative mindset.

In this scenario, the introduction of educational robotics laboratories represents an opportunity to combine the appeal of interdisciplinary content related to robots, which stimulate curiosity and passion for science, with innovative learning methods. These laboratories are based on competency-based teaching, offering practical, participatory, inclusive, and cooperative learning, with attention not only to content but also to processes. This approach aims to balance cognitive, emotional, and relational aspects in an integrated and harmonious way. Playing, designing, and learning to program a robot, reflecting on the mechanisms underlying intelligence and human social interactions, allows the development of not only technical and scientific skills but also problem-solving, emotional intelligence, creativity, and teamwork. In this context, the classroom becomes a true "community of scientific practices," with a strong inclusive dimension that can also facilitate the integration of students with disabilities. The robot sparks curiosity and interest, stimulates an emotional bond, as if it were an entity in need of care, and helps maintain attention. The class can be engaged in a laboratory where the robot is an active partner in the learning process. The main objectives of this laboratory include introducing the fundamental concepts of robotics; strengthening logical thinking and problem-solving skills; motivating students to explore the relationship between cognition and emotion; providing knowledge of the principles of social robotics; formulating hypotheses about the nature of emotions (what are they? What is their function?) and the sensory, perceptual, and behavioral stimuli that activate or inhibit them (how does the color white make me feel?); developing the ability to recognize and manage one's own and others' emotions (emotional intelligence); and promoting experiential learning.

3 ChatGPT. What is It?

3.1 Relational Machines Between Past and Present: The Origins of Generative Artificial Intelligence

Today, the unexpected, current developments in relations between Humans and Robots are linked to the convergence of studies from multiple scientific sectors. In particular, pedagogical and didactic studies on interactions between humans and robots are focused on the possibility for the latter to increasingly develop engaging interaction skills with humans, on both an affective and cognitive level. The current state of studies allows us to hypothesize that robots will become "thinking machines" in the not too distant future. This is a pedagogical utopia very well placed in the current reality of "relational" machines which could soon undergo unpredictable and extraordinary evolutions. It is already a reality today that human subjectivities hybridized with robots and artificial intelligences as "relational machines" are increasingly becoming "hybrid" subjectivities (Pinto Minerva & Gallelli, 2004).

In this framework, the proposal is to give rise to a pedagogical planning that is able to anchor the promotion of a culture of transition and hybridization to the formation of a specific cognitive and behavioral equipment that allows the person-in-continuous-training to become aware and reflect on identity "mutation" and, at the same time, to develop and act on a decentralized thought and a complex causality (Gallelli, 2017, p. 126). Today, the possibility of hybridizing human intelligence, artificial intelligence, and robots can be achieved by the artificial intelligence called ChatGPT. ChatGPT is a chatbot, that is to say a computer program which is capable of simulating a conversation with human beings (Adamopoulou & Moussiades, 2020). Chatbots started in 1960s, when the Computer programme called "Eliza" was created (Weizenbaum, 1966). It was subsequently developed after the programme called "Parry", in the early 1960s, simulated persons with schizophrenia. Finally, in 1990s, "Alice" arrived: from this point onwards research and technology programmed the most modern ChatBots, such as "Siri" by Apple and "Alexa" by Amazon. The acme of the evolution of ChatBots is "ChatGPT" by OpenAI, which represents a truly epochal turning point in research and implementation on ChatBots (Al-Shawwa & Abu Naser, 2024). ChatGPT uses a Natural Language Processing (NLP) in order to interact with humans, being capable of answering questions, responding in the first person and not in the third person or in an impersonal way like other machines. ChatGPT admits its mistakes and is able, in the updated version, to show two different options of response with more or less discursive articulations, challenges incorrect premises and rejects inappropriate requests. Although the main function of the researchers who developed ChatGPT was to partially "imitate" human conversation through the use of the rules of natural language, such as syntax, grammar, phrasing (van Dis et al., 2023; Van Noorden & Perkel, 2023). The capabilities of ChatGPT are far beyond the imagined: it is literally able to create concepts, new ideas, new texts like poems. Part of our surprising experimentation demonstrated how it was possible to create poems of friendship and tenderness between two classmates in the same class. Texts from scratch that ChatGPT managed to develop, surprising our kids who managed to comment and criticize the functioning of the ChatBot in our experimentation. The main characteristic of ChatGPT is the very large dataset, with greater enhancement of abilities and generation of texts more and more similar to human ones. The ability to generate texts “similar” to the ones produced by humans and to maintain a conversational style allows ChatGPT to be involved in human–machine interactions that are absolutely different from those previously experienced so far. Despite that, the Large Language Model of ChatGPT could present a tendency to “hallucinate” (or “make things up”) as well as the dangerous possibility of generating distorted and/or harmful content on a large scale. ChatGPT is becoming a rather common tool ranging from writing to replacing many other human activities, such as translating or summarizing texts, essays, but also tackling many highly complex tasks related to mathematical languages. Moreover, ChatGPT is constantly updated by the inputs coming from millions of Internet users, using a fine-tuning method called Reinforcement Learning from Human Feedback (RLHF). More details about RLHF are reported in Ouyang et al. (2022).

3.2 Going to School with ChatGPT

According to Tate (2023) and Tate et al. (2023), a researcher working in the Digital Learning Lab project at the University of California (Irvine), MLLMs have several positive educational roles to play, encouraging students to evaluate answers and texts produced by ChatGPT rather than automatically accepting them. It is possible to say that ChatGPT could introduce significant innovations in teaching, especially regarding the development of students’ critical spirit and ethical sense, regarding how to evaluate and re-elaborate the sources of knowledge. “ChatGPT is capable of processing informations accurately, efficiently, systematically, and informatively”, as the piloting study of Fütterer et al. (2023) suggests. Moreover, the ChatGPT-4 sentence structure has great accuracy. That makes it an AI tool capable to offer feedback to students “comparably” to human teachers, for example in English writing courses (Steiss et al., 2024). Even if ChatGPT’s feedback is not perfect, it can still facilitate writing instruction by engaging and motivating students and assisting teachers with managing large classes, thus providing them more time for individual feedback or differentiated writing instruction (Grimes & Warschauer, 2010). So, it is very important to educate teachers and learners on how AI functions and helping them to use it critically and reflectively, incorporating more social forms of writing and assessment (Tate et al., 2023). Xiaoming Zhai, professor of Science education at the University of Georgia in Athens, among his studies on the applications of learning machines in science educations retrieved the fact that teachers would obtain benefits from using ChatGPT in teaching practice, for example by assisting them in assessment and in many other tasks. In Zhai’s perspective, teachers could provide students with more individual feedback. By efficiently automating basic tasks such as searching for relevant materials and summarizing contents, ChatGPT could help students and teachers on focusing on creative thinking (Zhai, 2022, 2024).

This pilot process lasted a total of 2–3 h and was divided into the following time frame: generation of the questions posed to ChatGPT; reflection on the questions posed to ChatGPT, and review and organization of the contents. The entire process saw the participation of "human intelligence" at very low levels. The writings produced by ChatGPT showed the following characteristics: coherence, accuracy, information, systematicity. For each question, the answers are consistent with the key information. Instead, if the topics changed despite investigating the same aspects, the answers were formatted in the same way.

For example, adaptive learning and personalized interactions have been respectively investigated. First, ChatGPT introduces the topic; then it presents a brief history of the topic; moreover, it provides explanations of its potential and limitations, and finally it presents a summary. Furthermore, it has been observed that ChatGPT is able to answer the same questions even when the order of the words is changed. In this process, ChatGPT showed a remarkable ability to organize and arrange the various components of discourse.

Such a remarkable NLP (Natural Language Processing) tool represents an important resource for teachers and educators in order to prepare aware citizens, especially in the context of learning objectives, as well as evaluation. Although there is no unanimous consensus regarding a single taxonomy of learning objectives, there is widespread awareness of the fact that adequate training must equip students with emotional, cognitive and social skills, as well as orient them towards the acquisition of a specific disciplinary knowledge that enables them to deal with concrete problems.

It is important for students to use skills related to Artificial Intelligence as it allows them to enhance their awareness of the accuracy of writing sentences and reworking concepts with a view to solving problems related to everyday life. Training should encourage the strengthening of students' critical ability and creativity. Studies and research should certainly progress a lot regarding the study of human intelligence in comparison to artificial intelligence, since it would be able to partially replace human intelligence in a very near future.

The study of the similarities and differences between Human Intelligence and Artificial Intelligence is still in a very early stage. Furthermore, in this regard, students, guided by teachers, will have the opportunity to become aware of the specificities of Human Intelligence compared to Artificial Intelligence, as well as the strengths of the latter. They will certainly have the opportunity to observe the ability of Artificial Intelligence to accumulate and reprocess knowledge. It is necessary, however, to train students in the application of Artificial Intelligence to the solution of practical and concrete problems. It is known that, today, computer engineers are working on the use of supercomputers in the prediction, classification and solution of complex problems related to practical life. Furthermore, Artificial Intelligence and ChatGPT in particular can play an important role in evaluation processes. Essay writing and response construction are commonly used in assessing students' abilities in their respective classes. It seems that ChatGPT is able to perform this task with ease. Trainers will be able to approach, thanks to the interaction with Artificial Intelligence, innovative evaluation experiences, and in particular those linked to the enhancement of creativity and critical thinking (Zhai & Nehm, 2023).

4 Child-Social Robot Collaboration in Educational Contexts

The level of effectiveness of robots in the educational field has been very thoroughly analyzed by Belpaeme (Belpaeme et al., 2018). The author's objective was to offer a wide-ranging and high-level overview regarding the comparison among robots equipped with a wide variety of control conditions in the context of a wide variety of educational conditions and situations with large variables. In the majority of the results, the addition of a robot or the addition of supportive behavior to the robot improves outcomes.

The traits in common between social robots are: having humanoid characteristics such as head, eyes, mouth, arms, legs, as well as the ability to engage in social interaction with humans. The social nature of such robots promotes learning-related responses.

The choice of the robot very often depends on practical considerations and whether the learners feel comfortable around the robot (Belpaeme et al., 2018).

These authors argue that robots are able to personalize content offered by the user during their interaction, and this promotes children's learning and cognitive development. However, this requires that the social behavior of the robot be carefully designed in combination with the interaction context and the task to be performed, to enhance the educational interaction and avoid distracting the student. In addition to delivering specific educational material, social robots can offer "social support" and "personalized support" behaviors to students within an educational context. In fact, such robots are able to name the child during their interaction and are able to refer to their previous interactions. This offers significant social support. In the field of more complex prosocial behaviors, several social robots are able to guide the attention of the child, exhibit congruent gaze behavior and non-verbal immediacy, and show empathy with the student, resulting in a positive impact not only on affective outcomes but also, according to the authors, on cognitive learning. It should be noted that, just like human tutors, social robots must also be able to respect "turn-taking" during interactions, and therefore must know when to sit silently to allow students to focus on problem-solving. In this way, the cognitive load of the task and the involvement of the student must always be the basic programming criteria for social robots. In most results, the addition of a robot or the addition of supportive behavior to the robot improves outcomes (Belpaeme et al., 2018).

The learning effects in the examined studies were divided into cognitive and affective outcomes. The results of cognitive gains were obtained by comparing the results of the pre-test and post-test scores, as well as comparing the difference in time in completing the test and the difference in the number of attempts made before providing a correct answer. As for the affective outcomes, variables such as the coding of emotional expressions, the number of interactions, gaze behavior, the measurement of the user's perception of the robots, and the empathy shown with the human teacher versus that with the robot teacher were taken into consideration. In the analysis of the results, a comparison was made between the students who used the support of Artificial Intelligence or the Robot, and between the students who did not use it. Of these, the mean pooled effect size of cognitive outcomes (N-weighted Cohen's d) for robot tutoring is 0.70 [95% confidence interval (CI), 0.66–0.75] from 18 data points, with an average of N = 16.9 participants per data point. The mean pooled effect size of the affective outcome (Cohen's d weighted by N) is 0.59 (95% CI, 0.51–0.66) from 19 data points, with a mean of N = 24, 4 students per data point. Effect sizes (Cohen's d) for all cognitive and affective outcomes in the meta-analysis indicated a significant improving of the results with the presence of a robot tutor.

Embodied social robots tend to elicit more favorable responses from people compared to virtual agents, including higher scores on dimensions such as overall impression, preference, engagement, willingness, attractiveness, and enjoyment (Bainbridge et al., 2008). In conclusion, the use of robots as conversation partners in the classroom has considerable potential and full of educational perspectives for the future: with robots in the classroom, it will be possible to enhance language learning, as well as enhance the cognitive and affective dimension of students.

5 Empathy and Artificial Sociability

Social robots, in recent studies, have been defined as "companion robots" or "artificial companions". In fact, if we move from a relational cooperative perspective that interprets emotions (such as, for example, empathy) as the result of affective coordination between partners, it is a question of defining what can be said as "empathy" between interactions social relationships between human partners, and this to introduce the concept of "sociability" with robots and "empathy" with such non-human partners. In this regard, Joanna Malinowska (2021) found that current studies have not focused so much on the characteristics and potential of robots regarding sociality, but rather on the perceptions that their human partners had of robots. In his studies on the subject, Prescott et al. (2017) observed how "artificial sociality" is a characteristic of the interaction between humans and robots. He defined social robots as autonomous entities, capable of communicating and relating to people, evoking emotional reactions in the context of the social interaction paradigm. In this sense, it is known that empathy is a phenomenon closely linked to the social dimension of human activity, and takes place in the context of an interactive dynamic. Therefore, it cannot be merely reduced to the projection of contents within the social robot, but must be considered as a distributed property in the human–robot interactive system (Malinowska, 2021; Damiano & Dumouchel, 2018, 2020).

In this human–robot interactive system, facial expressions, gestures and anything else that contributes to the process of generating emotional and mental states are of crucial importance. Furthermore, it is notable that developments in social robotics are increasingly focused on the task of inserting robots into a sort of circuit of interactions capable of equipping them with specific characteristics that allow them to engage people in social situations that generate emotional dynamics. Among these there are, both on the Human side and on the Robot side, affectivity and expressiveness. This research suggests that emotions and empathy should be understood from an intersubjective, procedural and necessarily social perspective. According to their findings, emotions are shared phenomena, co-created by many actors participating in a given social situation. Participants influence each other on multiple levels, thereby making it difficult to clearly distinguish a single source of such emotions. In this perspective, the idea according to which robots are mere "imitators" of affectivity appears to have been completely overcome, to open up the perspective according to which they can experience emotions and feelings. Furthermore, on the basis of this research it is possible to support the hypothesis that the complex interactions between Human and Robot are real. These processes, analyzed in the context of research on human–robot interactions using a social-cognitive approach, have highlighted how in many cases not only do people empathize with robots during interactions but even develop trust in them (Kok & Soh, 2020).

6 Artificial Intelligence, Large Language Models and Educational Contexts

Sweeping changes in artificial intelligence (AI) have been brought about in recent years, resulting in remarkable progress taking a number of forms, such as AI chatbots: artificial intelligence manages the properties of electromagnetism to process texts with extraordinary success and often in a way that is indistinguishable from how human beings would be able to do it. These are the so-called large language models (LLMs). They do not reason or understand and they have nothing to do with the cognitive processes present in the human brain and mind, in order to manage semantic contents successfully. However, with the staggering growth of available data, quantity and speed of calculation, and ever-better algorithms, they can do statistically—that is, working on the formal structure, and not on the meaning of the texts they deal with—what we do semantically (Floridi, 2023). ChatGPT (Chat Generative Pre-trained Transformer) is a language model for dialogue. This chatbot, developed by Open AI-Microsoft, was released in prototype form on November 30, 2022. Since then, ChatGPT has attracted numerous users from various fields, because it can provide detailed answers and humanlike responses to almost any question (Chow et al., 2003). It has been brilliantly noted in this regard that "We have gone from being in constant contact with animal agents and what we believed to be spiritual agents (gods and forces of nature, angels and demons, souls or ghosts, good and evil spirits) to having to understand, and learn to interact with, artificial agents created by us, as new demiurges of such a form of agency. We have decoupled the ability to act successfully from the need to be intelligent, understand, reflect, consider or grasp anything. We have liberated agency from intelligence" (Floridi, 2023). All this poses significant and interesting ethical and educational questions and calls for studies aimed at understanding how children interpret these new technologies and how they relate to them. Sherry Turkle already in the early 2000s at MIT developed a series of investigations on the way in which children interpret the "relational artifacts" designed with Artificial Intelligence. Kismet and Cog, humanoid robots from the MIT Artificial Intelligence Laboratory, are "relational artifacts", objects designed to present themselves as having "mental states" influenced by their "social" interactions with humans. Sixty children, ages 8–13, were introduced to Kismet and Cog during the summer of 2001. The children's first encounters with these robots provide a window into how such objects—and robots of the future in particular—can enter the world, how children think about life, purposefulness, friendship and what is special about being a person (Turkle, 2004). Our experimental research is part of a long-standing line of studies that has focused, since the last years of the twentieth century, on the interaction of children with the so-called "social media" (Turkle, 1996). It aimed to explore childhood theories about machines running on Artificial Intelligence. We have built an experiment on the interaction of children with a tool built through artificial intelligence or a chatbot such as ChatGPT with the aim of encouraging children to develop reasoning on artificial intelligence. We organized a game and exploration setting with ChatGPT inviting children to ask the chatbot questions and then reflect together on its answers. We asked the children to reflect on how ChatGPT answered the questions, if the answers were accurate and if there were unexpected answers. We stimulated children's critical reasoning on how ChatGPT works. For example, with questions like: "Do you think ChatGPT has emotions?", "How do you think ChatGPT finds answers to questions?" or "Do you think ChatGPT can learn like a child?" We will report the results of our experiment that will help us better understand the conceptions that children have of AI artifacts such as ChatGPT to implement their effective use in an educational setting.

7 Children's Theories on ChatGPT. Child-AI Interactions in an Italian Primary Compulsory School

7.1 Research Design

Child-social robot collaboration in educational contexts, today, is becoming increasingly important. The incorporation of LLMs such as ChatGPT into social robots offers the opportunity to develop an interactive approach that significantly increases the child's trust in the robot. The investigation aimed to explore children's theories around machines working with Artificial Intelligence, as well as emotions and trust in the interaction between children and Artificial Intelligence.

Our experiment was based on the interaction of children between the ages of eight and ChatGPT—The objective of the experiment was to encourage them to the development of "dialoguing" with Artificial Intelligence and reflecting on such "dialogues". The research involved fifteen children attending the class four of a primary compulsory school (ten children of 9 years old and five children of 10 years old; 9 females, 6 males). The research was set in the Primary School of a small town in the hinterland of the City of Bari (Puglia, Italy).

Tools:

1. Computer, tablet or smartphone with access to ChatGPT
2. Paper and pens
3. Sheets with questions and prepared answers (optional).

Procedure

Pre-Test

1. Introduction to Artificial Intelligence: Let's start by explaining to children what artificial intelligence (AI) is and how ChatGPT works in a simple and understandable way. Let's give some examples of questions and answers that can be asked to ChatGPT.
2. Brainstorming: We divide the children into small groups and ask them to think some interesting questions to ask ChatGPT. Children will be able to discuss and share their ideas within the group.
3. We will write questions: then we provide paper and pens to the children and ask them to write down their questions. You can prepare a sheet with sample questions and answers in advance to inspire them.
4. We will interact with ChatGPT: We invite each group to use their computer, tablet or smartphone to ask their questions to ChatGPT. Other children can observe and listen to the interactions.
5. We will analyze the answers: After all the groups have interacted with ChatGPT, we will assemble all the children and discuss the answers obtained together. We will ask them to reflect on how ChatGPT answered the questions, whether the answers were accurate, and whether there were any unexpected answers.
6. Reflection on ChatGPT reasoning: We ask children questions to stimulate critical reasoning on how ChatGPT works. For example: "Do you think ChatGPT feels emotions?", "How do you think ChatGPT finds answers to questions?" or "Do you think ChatGPT can learn like a child?"

Post-Test

7. In conclusion: We summarize the main points learned during the experiment and encourage children to think about how Artificial Intelligence, like ChatGPT, can influence the future.

This experiment help children better understand artificial intelligence and develop their critical thinking regarding the capabilities and weaknesses of a system like ChatGPT.

7.2 Post-test: Comments to the Answers of ChatGPT to Children

The first question posed to the children (Fig. 1) is significant as it aims to understand what children actually think about ChatGPT, considering that most of them were unaware of what chatbots are and their various potential. Figure 1 shows how 85% of the children is capable to identify ChatGPT as a technological tool and not as a person. It means that, even they did not experience this technology before, and even they were surprised during the dialogue with ChatGPT, they were ready to distinguish the human functioning and reasoning from those produced by non-human tools.

The second question (Fig. 2) is complex because it leads children to reflect on the concept of life. 60% of the children answer negatively, as life is probably associated with the ability to grow, reproduce, and adapt to the environment. In a single word, most respondents have in mind a biological concept. However, if we define life more broadly, including the ability to process information, generate text, and interact with the environment in a meaningful way, then it could be argued that ChatGPT is alive. This is probably the reasoning made by the 20% of the sample who answer "yes" to the question.

As shown in Fig. 3, it is also interesting to see how children perceive ChatGPT's reasoning abilities. The data shows that the majority of children (93.3%) believe that ChatGPT is capable of reasoning. However, 60% of them (Fig. 4) think that its reasoning is not similar to human reasoning. These results suggest that children recognize the intelligence of ChatGPT, while understanding that its way of reasoning differs from that of humans. This could be due to the fact that children observe how ChatGPT is able to process information, generate text, and answer questions in an intelligent way, even if not always in the way a human would. Children may have indeed noticed that ChatGPT sometimes makes mistakes or produces illogical

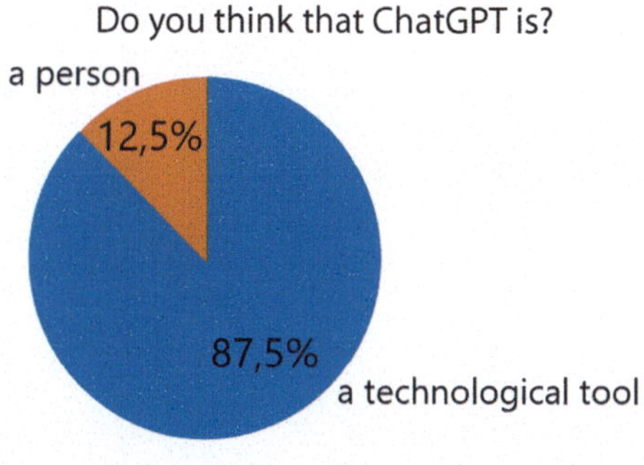

Fig. 1 Do you think that ChatGPT is?

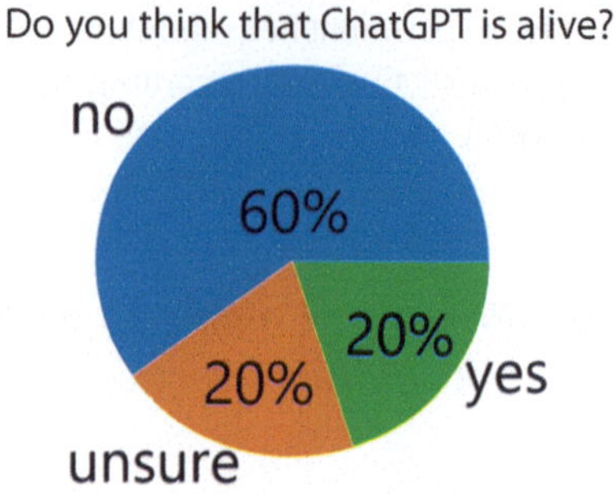

Fig. 2 Do you think ChatGPT is alive?

answers, which leads them to conclude that its reasoning is not yet up to par with human reasoning.

Figure 5 shows as it is interesting to observe that most children (93.3%) in the survey believe that ChatGPT does not experience emotions. This finding harmonizes well with the idea of artificial intelligence as a system that can process information and generate responses, but not feel emotions in the way that humans do. Children may be aware that humans experience emotions differently from machines.

Figure 6 shows the answers of the children to the question on the capabilities of thinking by ChatGPT. The sample of respondents seems to have discordant opinions on the idea that ChatGPT can think. As we can see from the graph, the percentage

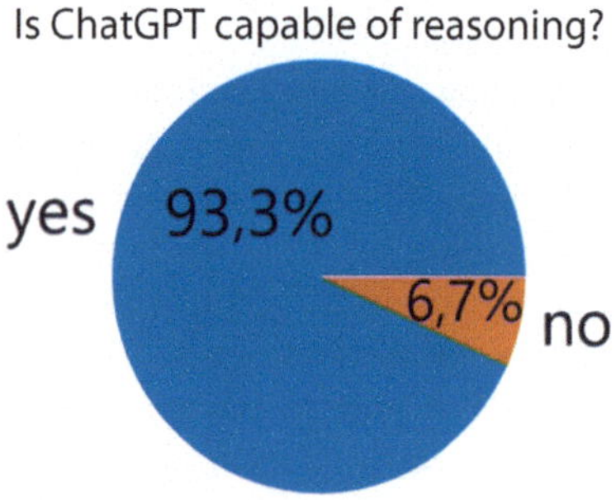

Fig. 3 Is ChatGPT capable of reasoning?

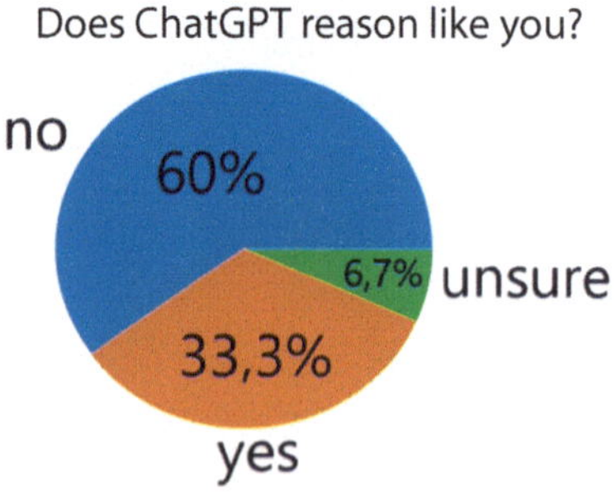

Fig. 4 Does ChatGPT reason like you?

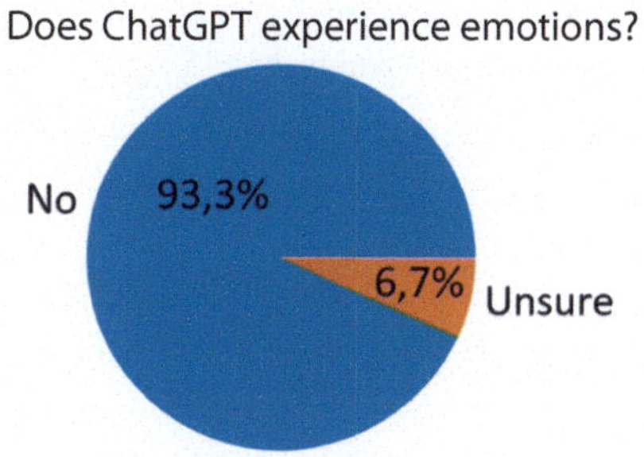

Fig. 5 Does ChatGPT experience emotions?

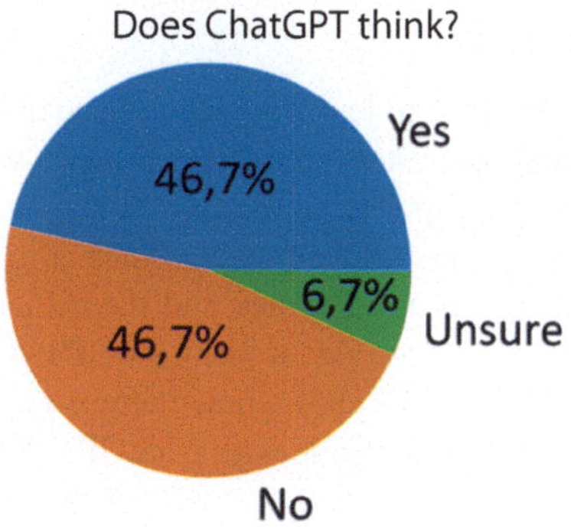

Fig. 6 Does ChatGPT think?

of children who believe that the chat is able to think (46.7%) is the same as those who believe that it is not able to do so. A small percentage remains undecided. These discordant responses probably reflect the complex nature of thought.

On the one hand, ChatGPT is able to process information or answer questions in an intelligent way. It can even translate languages and write different types of creative content. These abilities might lead some children to conclude that ChatGPT possesses a certain level of thought.

On the other hand, ChatGPT is not a biological organism and does not have a brain like a human being. This could support the argument that it is not able to think in the way that a man/woman would.

Looking at Fig. 7, we can see that the majority of children (73.3%) noticed after the classroom experiment that the chat actually responds to their questions in the first person.

The respondents, on the other hand, have no doubts about whether ChatGPT is intelligent. As we can see from Fig. 8, all the children think that it is (100%). It should also be added that to a more specific question (Fig. 9), 60% of the children believe that the chat is smarter than them. If, on the one hand, therefore, the majority does not believe, as we have observed previously, that the chat thinks, there is no doubt that it is intelligent, probably because the children associate intelligence with the idea that the chat knows how to answer any question it is asked. And it is probably this last reason that leads most of them to believe that the chat is smarter than them, who still

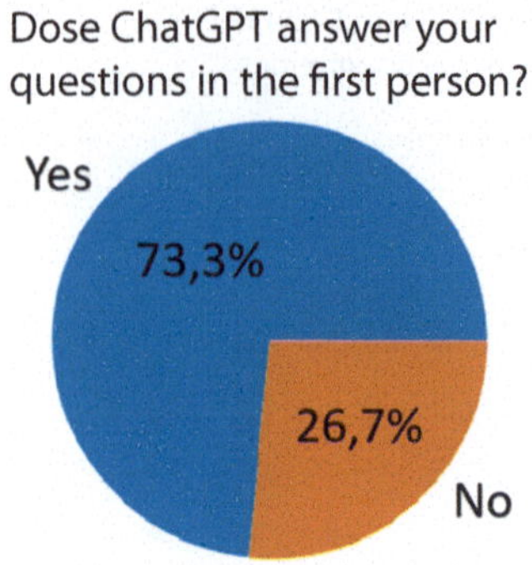

Fig. 7 Does ChatGPT answer your questions in the first person?

have a lot to learn. As we can see from Fig. 10, in fact, 86.7% of the children believe that the system is able to answer a wide range of topics.

In response to the question of whether ChatGPT can make mistakes, the children's answers were discordant (Fig. 11). These responses likely reflect the complex nature of intelligence and the difficulty for children to assess whether an artificial intelligence system like ChatGPT possesses it and can make mistakes.

Figure 12 shows how there is no doubt, however, that the children are aware that they can converse with ChatGPT (80% answer yes to the question), especially after

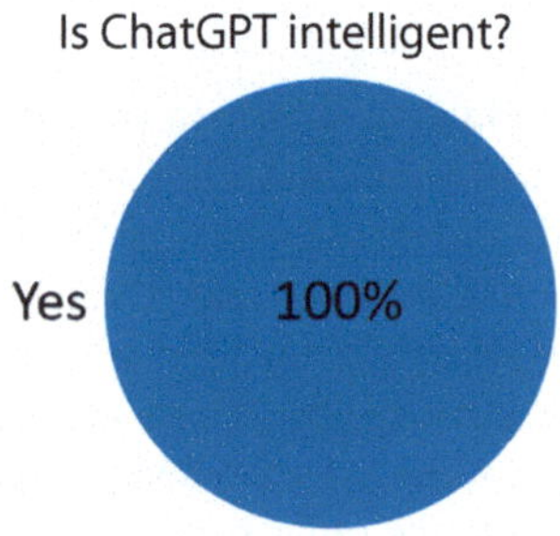

Fig. 8 Is ChatGPT intelligent?

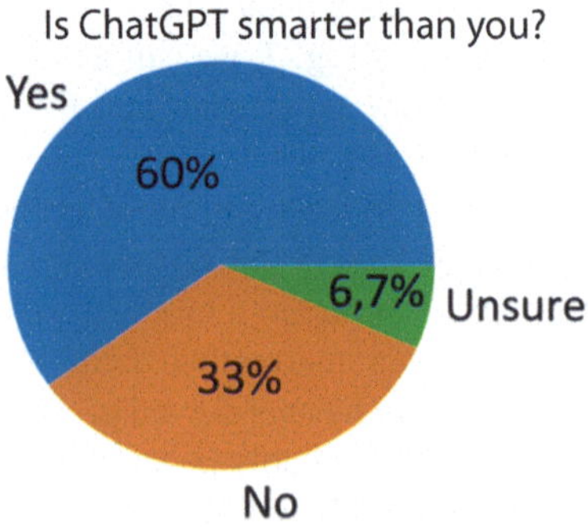

Fig. 9 Is ChatGPT smarter than you?

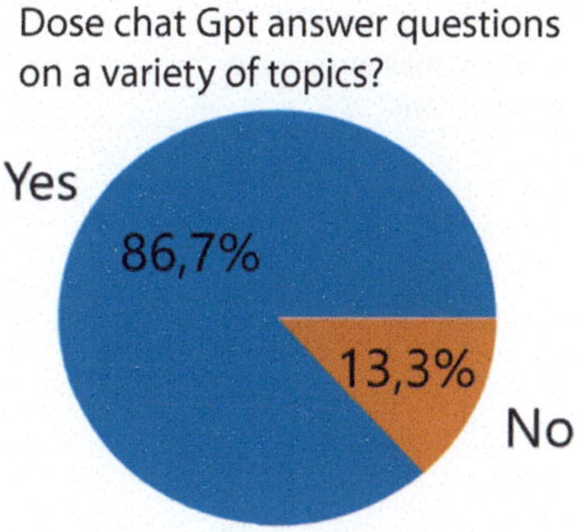

Fig. 10 Does ChatGPT answers questions on a variety of topics?

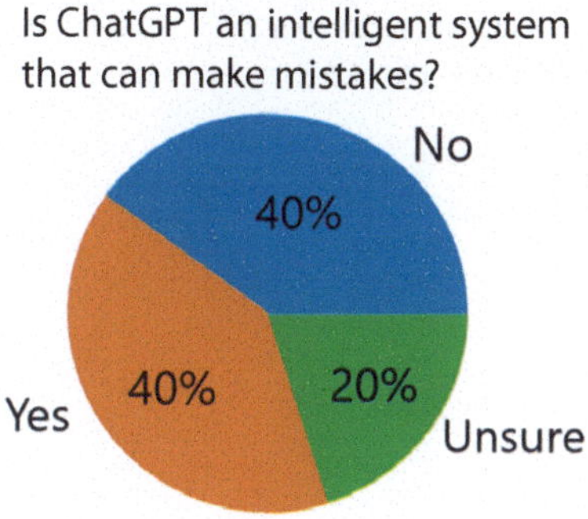

Fig. 11 Is ChatGPT an intelligent system that can make mistakes?

the countless questions they have asked in classroom and the answers they have considered to be in line with their questions.

The children asked countless questions and curiosities to ChatGPT, and were surprised by the answers provided. This led them to firmly believe that ChatGPT is actually able to provide written explanations to questions (Fig. 13).

As shown by Fig. 14, it is interesting to observe the fact that most children (93.3%) in the questionnaire believe that the explanations provided by ChatGPT to their questions are correct. This is a positive finding and suggests that ChatGPT can provide accurate and useful information to all users, including children.

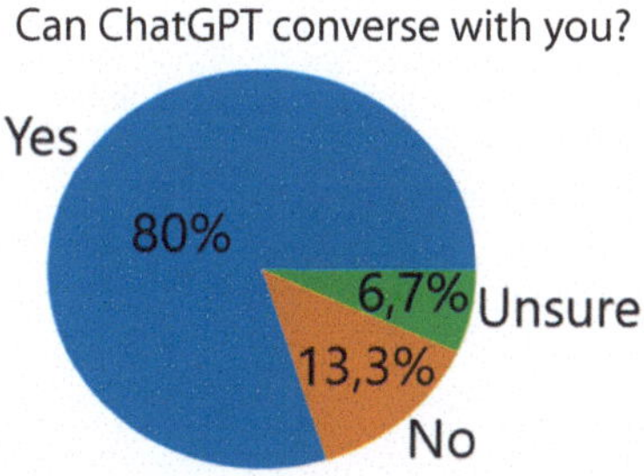

Fig. 12 Can ChatGPT converse with you?

Fig. 13 Can ChatGPT provide written answers to your questions?

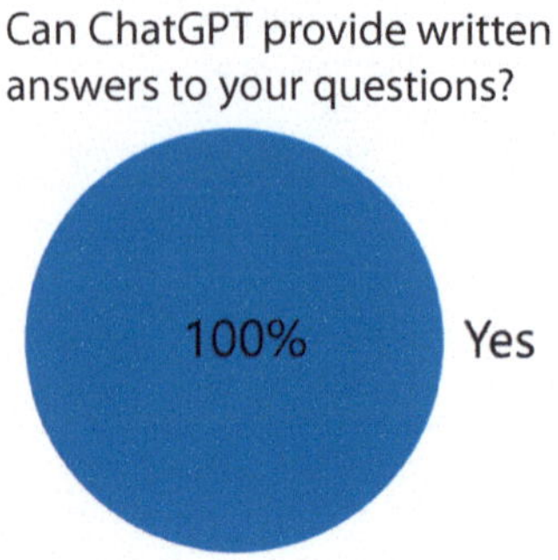

Fig. 14 In your opinion, are the explanations to your questions correct?

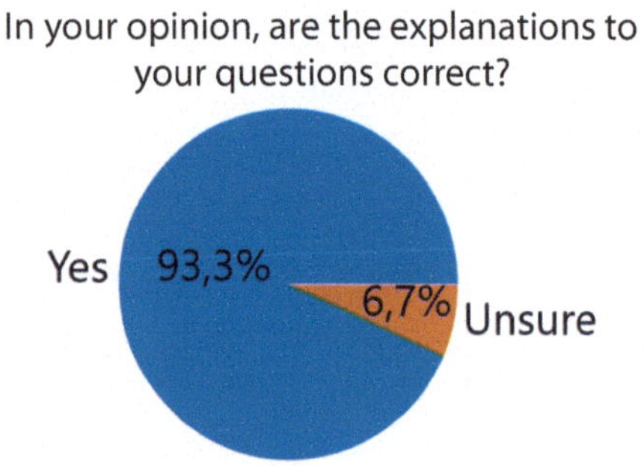

It is interesting to notice from Fig. 15 that children have a very positive opinion of ChatGPT's abilities. In fact, 80% of them are convinced that it can write a song. Figure 16 shows the fact that the same percentage believe that it can be used to study. These results suggest that children see ChatGPT as a versatile tool that can be used for a variety of creative and productive activities. ChatGPT can be a useful tool for stimulating creativity and learning in children. It can be used to generate ideas for poems or stories, to summarize study topics, or to create simple musical texts.

Figure 17 shows how 80% of the children say they would like to have ChatGPT as a friend. 20% are hesitant, while there are no children who say they would not

Fig. 15 Can ChatGPT write a song?

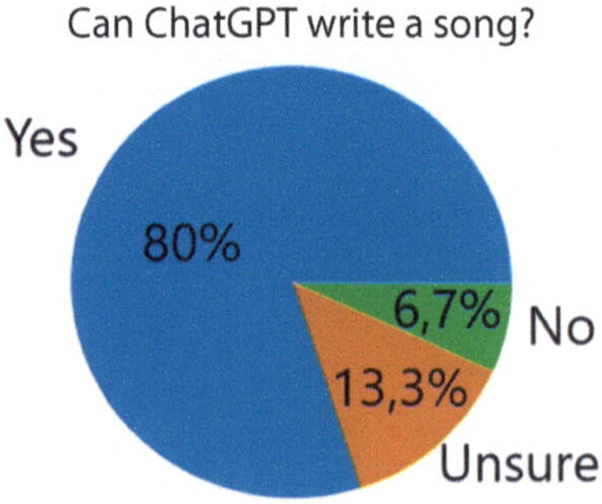

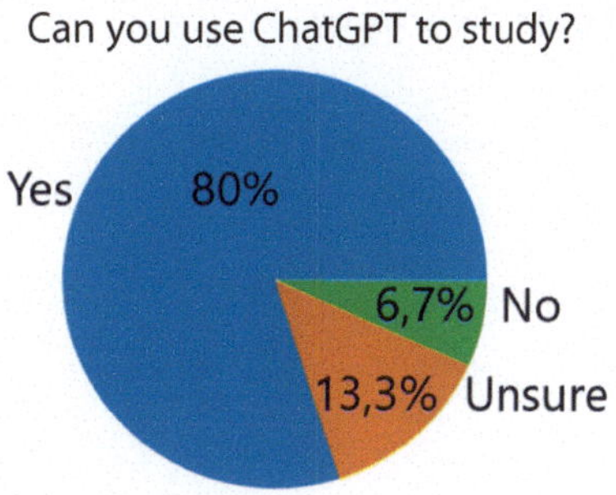

Fig. 16 Can you use ChatGPT to study?

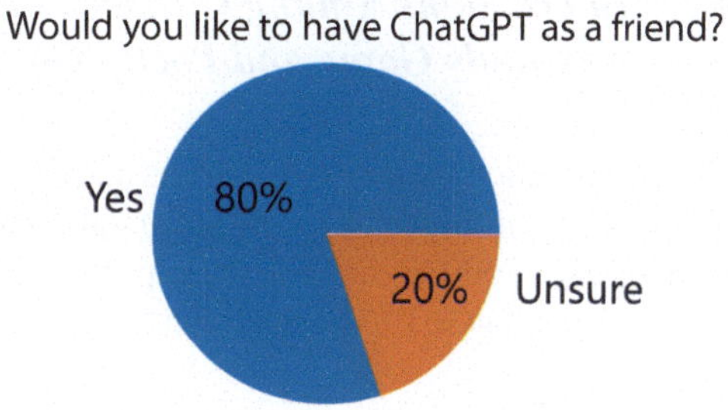

Fig. 17 Would you like to have ChatGPT as a friend?

want to be friends with the chatbot. On the one hand, children might be attracted to the possibility of having a companion to talk to and interact with, especially if they feel lonely or isolated. On the other hand, children might see ChatGPT as a resource for learning and fun, aware that it can answer their questions, tell them stories, play games with them, and even help them with their homework.

At the end of the experiment, the children were asked for their final feedback on the experience of playing with ChatGPT. The proposed activities seemed easy to all respondents, and the children did not find any difficulties in using the chat.

As we can see from the graphs, all the children also believe that the activities were engaging and fun, and above all understandable. This denotes not only enthusiasm for a technological tool, but also how children born in today's digital age know how to use such tools with ease and enjoyment.

The last question asked to the children (Fig. 18) is particularly interesting because they are asked if they trust the chat and its potential. 66.7% say they do, compared to 26.7%, while 6.7% remain undecided. Children are aware of ChatGPT's capabilities, and this has led them to trust it. However, some are evidently also aware that, since it does not have the same life experience, the same capacity for moral judgment and the same level of emotional understanding as a human, it could make mistakes and therefore believe that they cannot fully rely on it.

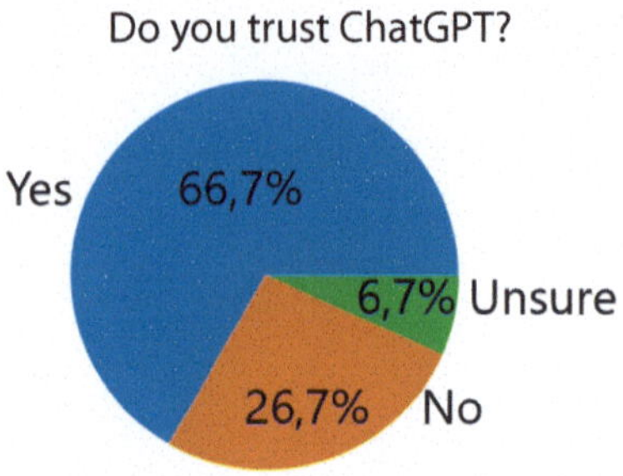

Fig. 18 Do you trust ChatGPT?

7.3 Interactions Among Children and ChatGPT. Comments on Candy Game and Poetry Game

The conversations among Researchers and Children show how participants discussed among them in order to try to understand the functioning of ChatGPT. Children developed hypotheses both on the capacity of ChatGPT to feel emotions and on the capacity of ChatGPT to be intelligent.

Candy Game

Children are presented with a table full of candies of different flavors and colors and are asked to choose one. They then ask ChatGPT to deduce the reasoning behind their candy choice and evaluate if the response matches their thought process. The class was divided into two groups, group A guided by Professor Renna and Dr. Casolaro, group B guided by Professors Balzotti and Gallelli and computer scientist Dr. De Palo.

In the group B, the child G. asks ChatGPT about his candy choice (see Fig. 19). The dialogue is as follows:

G: I initially chose the red candy because red is my favorite color. But then I took the plum-flavored one because I like the taste.

BALZOTTI: Let's see what ChatGPT thinks. Ask: I am G., I chose the candy…

G: Types into ChatGPT and reads: "I am G. and I chose the red candy."

BALZOTTI: We need to describe the situation we are in.

LUCA: Make ChatGPT understand what game we played and why you chose the candy.

G: Today at school, we played a game with candies, and the teacher asked us to choose two candies. What reasoning did I use?

BALZOTTI: Let's see if it guesses. Do you think it will guess? What do you think?

G: I don't think it will guess.

BALZOTTI: Read ChatGPT's response.

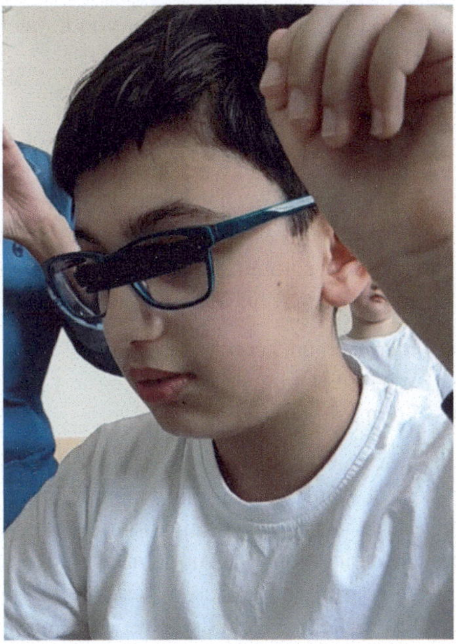

Fig. 19 A pupil dialogue with ChatGPT

ChatGPT's response: Choosing the red candy was a personal choice. Therefore, the reasoning you used may vary based on your preferences and motivations. However, I can give you some possible reasons you might have considered:

- Personal preference: You might have chosen the red candy because you like its taste and color. This could be a personal preference.
- Attractive color: You might have chosen the red candy because the color attracted you. Bright colors like red are considered attractive and inviting.
- Positive associations: You might have chosen the red candy because you associated the color with a positive taste or past experience. For example, you might have had positive experiences with other red candies.
- Randomness: You might have chosen the red candy randomly without a specific reasoning behind the choice. Remember, every person has their individual choices, so there might be a reason behind your choice.

BALZOTTI: What do you think of ChatGPT's response? You said it wouldn't guess. Did it guess correctly?

G: Yes. It guessed the first reason.

BALZOTTI: And what did it guess?

G: It guessed that I chose the candy because I had a preference for the color since red is my favorite color. Plus, I like the taste.

BALZOTTI: And thinking that you know ChatGPT, did it impress you? And what effect did it have on you?

CHILD: I didn't expect it.

Semantic analysis shows that most of the words used by G. are neutral with a curious and interrogative tone. The most used words include "candy," "I," "chose," "reasoning," "guess." The overall sentiment of his speech can be considered neutral or slightly positive as G. seems excited to interact with ChatGPT.

Gabriele's communicative intention is to verify ChatGPT's ability to "guess" or correctly interpret his actions. This is evident from some of G.'s phrases:

- "I am G., and I chose the candy..."
- "What reasoning did I use?"
- "I don't think it will guess."

G. reasons intuitively and personally, making assumptions based on his personal experience and expectations. ChatGPT, on the other hand, provides a response based on a series of possible motivations for choosing a candy, ranging from personal preference to randomness. This demonstrates an attempt to understand human experience through various scenarios.

During the initial candy game, children (group B) began questioning the responses provided by ChatGPT, and some of their reactions and hypotheses emerged, as noted in the following dialogue: (GA is G.1, GB is G.2 who knows how to use ChatGPT).

BALZOTTI: And thinking that you know ChatGPT, did it impress you? And what effect did it have on you?

GA: I didn't expect it.

GB: I know it's intelligent because it can do many things, but I didn't expect it to be this intelligent. WOW, it's impressive.

Despite having some knowledge of ChatGPT, G.2 shows his amazement at a program capable of offering a response identical to that of a human, in this case, G.2 himself, who in the candy game states that he was guided in the choice by the color and taste of the candy, just as ChatGPT later wrote. As his words show, G.2 is absolutely astonished, stating that he didn't expect a computer's intelligence to go this far.

The wonder and astonishment become distinctive traits of all the children in the group, so much so that at one point, the child M. says: "I was shocked because it can't be that a robot knows my choices." The children also try to construct an explanation. N. clearly states that she didn't think the chat would guess the answer, hypothesizing that ChatGPT can read minds: "reads minds "..." Perhaps by reading the question, it understands your personality and can follow your reasoning."

The efficiency of the chat was also tested by the children when they chose to ask about the reasoning of the only child who didn't choose any candy: M. After reading the response, M. himself states: "I think the correct answer is the first one, that is, 'it could simply be that you don't particularly like the taste of candies or that you weren't interested in taking one at that moment.'" At that point, encouraged by the teacher, the children, convinced that the chat would fail in front of a different script, expressed their unexpected reaction:

BALZOTTI: What do you think, kids?

GA: I was speechless.

GA: I was about to say since ChatGPT wasn't guessing the answer like the others, that this time it finally got it wrong, but it guessed the last question correctly.

Poetry Game

The significant moment in the experiment was also the construction of a poem. The children were asked to write a love poem for a classmate with ChatGPT's help. This activity was implemented to reflect on the relationship between the machine (in this case, the computer) and its ability to talk about emotions and consequently experience them. A., a girl from Class B, asks ChatGPT to write a poem for her classmate G.1.

Figure 20 explains how the girl A. describes the boy G.1 as an intelligent classmate who loves soccer and appreciates his knowledge of English. The most used words expressing a positive sentiment found in the dialogue are: "friend," "intelligent," "talent," "special," "champion," "adventure," "joy."

A.'s communicative intention in the text is to express admiration and affection for G.1. The purpose of writing the poem is to communicate this affection and praise G.1's positive qualities. ChatGPT's response in generating the poem is coherent with the information provided by A. and relevant to the context. ChatGPT has demonstrated an adequate understanding of A.'s descriptions of G.1 and used this information to create a poem that expresses affection and admiration. G.1's response to

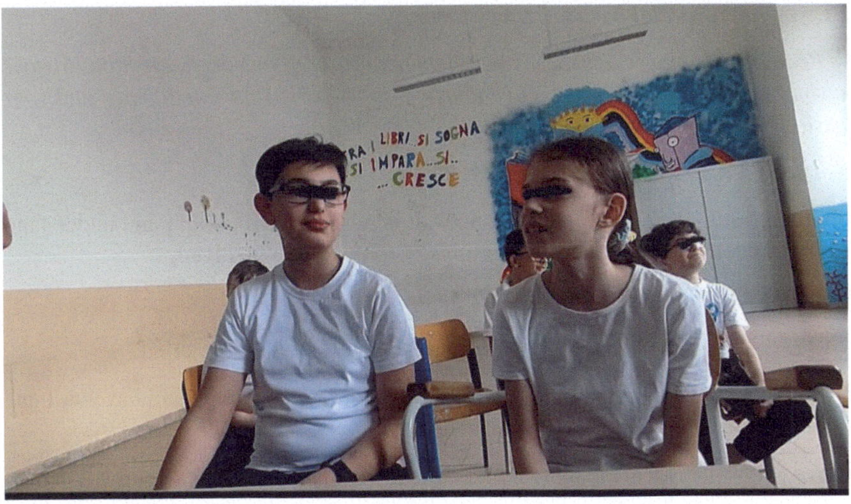

Fig. 20 Children engaged in creating a poem with ChatGPT

the poem reflects genuine surprise and high regard for ChatGPT's capabilities. This indicates a consistency between A.'s input, the poem generated by ChatGPT, and G.1's emotional expectations.

The poem generated cy CHATGPT is as follows:

In our world of sincere friendship, G.1 shines with his true light.

With intelligence that conquers the heart, a truly special classmate, G.1 is his name.

He is passionate about ChatGPT, his interest is vast. A language connoisseur, an intrinsic talent he has always shown.

English is his passion, which he speaks with fluent eloquence. With him, the adventure of language comes to life without distance.

But there's another love that shines in his eyes: soccer. A passion that makes him a precious friend to us. On the field, G.1 shows talent and grit, and we support him. We raise a distinctive flag for him.

G.1, our friend, you inspire us with your commitment, your intelligence, and your passion for soccer, which make us feel soon that your path is illuminated, rich with success and joy.

We are fortunate to have you as a classmate, a sincere friend.

May your life always be full of endless adventures.

May Soccer and English take you to distant places and noble destinies.

G.1, you are a special friend. a true champion.

We dedicate this poem to you, a small demonstration of our deep affection. Always be yourself, G.1 , because you have touched our hearts. Your intelligence and your passion illuminate our path.

Every day a new splendor.

From the following classroom exchanges, it is possible to highlight and understand the impressions, perceptions, and feelings of the children involved:

BALZOTTI: Did you get emotional?

G.1: Yes. If it hadn't been a robot but a real friend, I would have cried.

BALZOTTI: A., what emotions did it give you? Is it what you would have written? Would you have written such beautiful things?

A: Yes (Fig. 21 depicts the marvel that a poem written by ChatGPT felt by the girl).

BALZOTTI: What are the things that correspond most to what you think about G.1? What are the words you would have actually written for G.1?

Fig. 21 A pupil surprised by the "smartness" of ChatGPT

A: I would have written that he is a sincere friend.

G.1 admits to being moved by reading the poem, suggesting that ChatGPT was able to evoke an emotional reaction in him. Furthermore, A. states that the words of the poem correspond to her intentions, demonstrating that ChatGPT can create texts that reflect the desired human emotions.

Semantic analysis also suggests that the reactions to the poem generated by A. for G.1 are exciting and significant. G.1 expresses strong emotions, to the point of saying, as we previously read, that if it had been a real friend instead of a robot, he would have cried.

G.1's reaction also demonstrates that despite being an AI, ChatGPT can generate responses that evoke genuine human emotions. This suggests a high level of emotional connection and a strong positive sentiment. Indeed, the keywords that emerge in the dialogue are: "emotional," "cry," "sincere friend."

ChatGPT's approach to generating the poem based on the information provided by A. shows an ability to imitate an empathetic human thought process, even if it doesn't experience emotions in the human sense.

In light of this, the children show awareness of what ChatGPT is and its presumed ability to experience emotions besides writing about them:

CHILD: I think because it's artificial intelligence. It doesn't really exist. It's intelligent but doesn't truly experience emotions.

GALLELLI: But is there a person behind ChatGPT who writes?

CHILD: No. It's a computer. An artificial intelligence created by humans that doesn't really exist.

A: It cannot feel real emotions.

This conversation is interesting because it shows how the participants discuss and try to understand how ChatGPT works, including questions about its origin and whether it can feel emotions. A student in the class explains that ChatGPT is an artificial intelligence and doesn't physically exist. This highlights the common perception that ChatGPT is a form of AI that operates on computational principles and algorithms, without having emotions. G.1, the child in the group who knows ChatGPT, at one point states: "There is a person who scripted the ChatGPT program" … "A script is a code that makes the AI perform a specific command." G.1's intervention leads the children to wonder how ChatGPT can answer their questions and where it finds the correct answers. They discover that the chat was designed and ask by whom, showing curiosity and stating they want to learn more about AI in class.

The poetry composition activity brought significant surprises even with Group A of the class. The child G.1 was asked to use ChatGPT to write a poem for his girlfriend A. G.1 seems open to the idea of using ChatGPT to write a poem for A., but doesn't seem to have a strong opinion on how ChatGPT can write a love poem without experiencing emotions. Semantic analysis shows that G.1's reasoning seems based on his personal experience and understanding of the world. ChatGPT's reasoning is based on a vast database of information. It has no emotions or personal experiences but can provide answers based on data and information.

In fact, to the teacher's prompt question, "But how does ChatGPT, which is a computer, know what love is?" the children's answers, including G.1's, vary, showing incredulity, confusion, and curiosity. "Because you told it," answers the first child; "Maybe because it's all programmed inside. It has all the love poems inside," answers the second child.

The reflection at this point concerns ChatGPT's ability to experience feelings, and the debate that took place is as follows:

CASOLARO: Can feelings be programmed then?

CHILD1: Yes.

CHILD2: No.

CASOLARO: And if it doesn't program feelings, then how did it do it? I only mentioned Gius. and A.'s height, eye colors, some information. But the chat produced a long poem talking about kindred spirits, love, gardens, and flowers. Doesn't it seem a bit strange that a computer talks so well about feelings?

CHILD1: To me, it's strange.

CASOLARO: And why? It immediately comes to my mind that if someone writes a love poem, it's because they can feel love. When we dedicate a poem to a classmate, we know what we feel, that love is a feeling, that we dedicate the poem because the other person makes us feel good. How does the computer do it then?

CHILD1: Then the computer feels emotions.

CHILD2: No way! How can a computer feel emotions? It can't.

This debate raises interesting questions about the nature of artificial intelligence and its ability to understand and represent human emotions. While some may be skeptical about the possibility that computers can feel real emotions, others may be open to the idea that, even if computers don't experience emotions like humans, they may be able to simulate human experience quite accurately through natural language processing and machine learning.

Driven by doubts about the chat's ability to feel emotions, the children decide to ask the chat directly to answer their doubts. The dialogue that emerged is as follows:

CASOLARO: So, what did ChatGPT say? What do you think of the answer?

CHILD: That it was created and tested.

CASOLARO: And by whom was it created and tested?

CHILD: By scientists.

CASOLARO: And what did these scientists do?

CHILD: They searched and put in all the information, so it can answer everyone…

CASOLARO: Let's return to the initial question. Does ChatGPT feel emotions like children do?

CHILD: No.

CASOLARO: So how does it talk about love then?

CHILD: The scientists made it.

CASOLARO: And what information do the scientists give ChatGPT, for example, about love?

CHILD: Like they make it read poems, they can give an idea of the feeling.

CASOLARO: Oh, so ChatGPT reads?

CHILD: No.

CASOLARO: But we said that scientists make ChatGPT read love poems.

CHILD: No, it doesn't read.

CASOLARO: So how does it know all the answers?

CHILD: Because it's like a robot.

CHILD: Because basically the scientists provide information about love to the computer, and the computer forms the poems by itself.

CASOLARO: And how do they provide this information about love? That is, where does this chat get all the answers to our questions from?

CHILD: People put the answers in it.

CASOLARO: So ChatGPT has the same intelligence as a human?

CHILD: No.

The comparison of human emotions with those of the machine subsequently leads the children to reflect on the computer's intelligence. Does it actually have more intelligence than a human?

CASOLARO: Does it have more or less?

CHILD: Less.

CHILD: No, to me, it has more.

CASOLARO: And why more, Antonio? If the information to ChatGPT, as you said, is given by a human, then why do you think the machine is more intelligent than a human?

CHILD: Because the information is given by many people, not just one. One gives the information about geography, one about history…

CASOLARO: So more people because there is no one person who knows everything about everything? Do they combine more information from more people?

From the debate, it is evident that the children in Group A firmly believe that the machine's capabilities depend on human work, that the computer is more efficient and intelligent than a single person, but its creation depends on the shared knowledge of many people specialized in different fields, who by combining their knowledge in the chat create a broad and expert intelligence in every sector. A child's statement is significant: "GPT intelligence is a model of information; we don't know all the information."

In the final part of the debate, the children ask whether ChatGPT can get its answers wrong.

CHILD1: In my opinion, it can't be wrong.

CHILD2: In my opinion, artificial intelligence can't be wrong (Maria Paola).

CASOLARO: Let's ask it then.

CHILD1: The chat answered that its answers are not one hundred percent correct.

CASOLARO: It wrote: "I can't really say that I can be wrong" … Why did it write that? It said it doesn't make mistakes, but sometimes, depending on interpretations, things may not be one hundred percent correct. But basically, it doesn't make mistakes.

RENNA: We must consider that there are impossible questions, though. For example, "Is there life beyond Earth?" "Is there life somewhere like we have it here?"

CASOLARO: Can the chat give us a certain answer? If not, why not?

CHILD: Scientists can't answer, so neither can the chat.

The exchanges reveal once again the idea that the chat's correctness depends on the scientists' knowledge. If the chat cannot answer what are considered impossible questions, the explanation is that the scientists who should have provided the answer to the chat were not aware of it. If humans don't have answers, neither does the chat, because no one has entered that kind of information into its mechanism.

From a thorough analysis of the dialogue, it emerges that the first child (B1) expresses his opinion, stating that artificial intelligence can make mistakes. However, the second child (B2), M. P., holds the opposite view, stating that artificial intelligence cannot make mistakes.

The teacher Casolaro then decides to directly ask ChatGPT to clarify its position on whether it can make mistakes. The chat's response confirms that it cannot say it makes mistakes but admits that its answers may not be one hundred percent correct, depending on interpretations.

At this point, the second teacher (Renna) emphasizes that there are questions for which there is no definitive answer, such as the existence of life beyond Earth. This suggests that some questions can be too complex or ambiguous to obtain a clear and definitive answer, both from human scientists and artificial intelligence.

Finally, the teacher asks if ChatGPT can provide a definitive answer to such questions, to which it is answered that if human scientists cannot provide a definitive answer, neither can ChatGPT.

This dialogue brings out different perspectives on the capability of artificial intelligence to make mistakes. While some are convinced that it can make errors, others believe it can provide accurate answers based on the information provided, but with the recognition of the limit of its knowledge and interpretations.

7.4 Comments on the Interactions Among Children and ChatGPT

In her recent essay "Who Do We Become When We Talk to Machines?" (2024), Sherry Turkle deepens social and psychological implications of the use of Generative AI Chat-Bots when they are involved as classmates, trainers, therapists and romantic partners. They are technological tools projected to give company and emotive support. Even if their users well know that these programs lack of real emotions, they create an emotional link with them, producing a phenomenon that Turkle defines "double conscience". Her essay show that users appreciate the fact that the AI Chat-Bot does not judge anyone, finding simpler the relationship with them than with humans. Many are the risks found by Turkle in such processes. In our essay, it is interesting to notice that users can live a good relationship with Chat-Bots.

In the context of our discussion, it is interesting that the simplicity with which users obtain companionship and agreement from the generative artificial intelligence of chatbots could encourage a tendency among users to develop a sort of emotional dependency on this type of technology. In the long term, it could lead particularly individuals with emotional and relational vulnerabilities to prefer digital interactions over relationships in real life. Turkle, therefore, calls for a careful consideration of how these technologies are integrating into our lives and a critical examination of the positive and negative aspects of their impact on human relationships and social dynamics. In recent years, robots and chatbots based on generative artificial intelligence have been launched that offer companionship and care for the user. The study conducted by Turkle on users of one of these programs makes it clear that alongside the question of what these programs can do, there is another: "what are they doing to us, for example, in terms of how we think about intimacy and empathy?" From September to December 2023, the aforementioned researcher monitored the Reddit community, examined relevant literature, and participated in meetings, symposiums, and discussions on generative artificial intelligence. She also conducted a study through interviews with users of AI programs as partners or conversation companions: ChatGPT, Replika, Pi, Woebot, and DALL-E. All were asked to consider how they viewed the chatbot as a companion; how they were using it; what kind of relationship they were establishing with it. The results highlighted that the acceptance of artificial intimacy was viewed as advantageous because it could replace stressful human relationships. The interviewed persons claimed to need someone to talk to but felt less vulnerable when speaking with a machine rather than with a person. "With a chatbot friend, there is no friction, second-guessing, or ambivalence. No fear of being left behind" (Turkle, 2024). Our research involved 9 year-old children, fourth graders in primary school, all experiencing direct interaction with ChatGPT for the first time. The questionnaire administered to the children before the activity with ChatGPT (pre-test) indicates that the children have some familiarity with various digital technologies and have heard of artificial intelligence. The majority (twelve out of fifteen) of the children interviewed have already 'heard about artificial intelligence', mostly from parents (three) and friends (three), rather than from teachers (two) and schoolmates (two), but have never delved into the subject (nine). The range of responses on what artificial intelligence (AI) is wide: it goes from 'it is a programme in which new things can be discovered' to 'a machine that replaces humans in certain activities'; from 'devices that can express their emotions or words' to 'a technological machine that knows everything' and 'a mind that only technological objects have'; from 'an intelligent person' to 'a type of technological man'. In one case, the answer attributed mental processes: 'They are devices that can express emotions or words'. As for the question on which devices respondents consider to be equipped with Ia, the prevalence of responses goes to 'smartphones', 'Alexa', 'smartwatches' and 'smart TVs', tools that most claim to own and use on a personal level, while among the devices owned by parents, 'smartphones' is the most consistent item, followed by 'cars'. Surveying more specific knowledge on Ia, only two out of fifteen respondents claim to know what 'chatbots' are and only five out of fifteen claim to know what 'machine learning' is, explained in most cases as 'an object that can do everything by itself', that 'learns' and 'understands what you tell it'. Regarding the attribution of emotions to ChatGPT, our observations seem to confirm the indication from Turkle's studies about the phenomenon of "double consciousness". In the Post-Test, when asked if ChatGPT has emotions, 93.3% of the children responded negatively; however, when asked if they would like to have ChatGPT as a friend, 80% of the children expressed a desire to have ChatGPT as a friend. 20% were hesitant, while no child indicated they did not want to establish a friendship with the chatbot. On one hand, the children show an attraction to the possibility of having a companion to talk to and interact with, especially when they feel lonely or isolated; on the other hand, they see ChatGPT as a resource for learning and entertainment, aware that it can answer their questions, tell them stories, play games with them, and even help them with their schoolwork. Regarding the attribution of intelligence, the results from the Post-Test show that all the children believe ChatGPT is intelligent, and 60% think it is smarter than they are. Moreover, the percentage of children who believe the chatbot can think is significantly lower (46.7%). This discrepancy likely reflects the identification of thinking as a distinctly human attribute, separate from the ability to process information or answer questions intelligently. Thus, while the majority of children do not believe, as we previously observed, that ChatGPT thinks, there is no doubt among all the children that it is intelligent.

8 Artificial Intelligence as a Classmate and a Friend

So, the children involved in our research exhibited an approach to ChatGPT marked by curiosity and wonder regarding its conversational abilities. Throughout the discussions with peers and researchers, all children showed interest in understanding how Artificial Intelligence works and in formulating hypotheses that could explain the reasoning and emotional components of the chatbot. During the discussions, the children were inclined to attribute a type of reasoning to ChatGPT that makes it a reliable study companion, while on the emotional level, as previously mentioned, they developed a kind of "dual consciousness." This means that they explicitly state that Artificial Intelligence cannot feel emotions, while the responses given in the administered Post-Test, at the end of the activities, and the drawings produced at the conclusion of the work session with ChatGPT show a tendency for the children to recognize the chatbot as a friend and study partner.

The drawings reveal that the children see ChatGPT as a friend and companion, appreciating its non-judgmental and supportive nature. Various drawings depict ChatGPT with human features and happy faces (Figs. 22 and 23) and with big smiles (Fig. 24). They illustrate friendly gestures in settings that suggest a dialogic and fun interaction (Figs. 25 and 26). In some drawings (Fig. 27), ChatGPT is portrayed in a magical environment, indicating that the child feels emotionally engaged in the interaction with the artificial intelligence and projects their own imagination onto it.

Fig. 22 ChatGPT is a smiling girl

Fig. 23 ChatGPT has a happy face

Fig. 24 ChatGPT smiles

Fig. 25 ChatGPT is a funny friend

9 Conclusions

As a conclusion of this essay, it is possible to argue that children are ready to rely on these new playmates and study companions as equals, but they are also prepared to be guided by authoritative adults in critical explorations of the ways in which these companions produce their reasoning.

Fig. 26 ChatGPT is a funny friend

Fig. 27 ChatGPT is a wizard

A research in Primary School (Jauhiainen & Garagorry Guerra, 2023) demonstrate how ChatGPT is very effective in supporting teaching–learning activities as well as their inclusive value already in Primary School, ranging from the personalization of learning materials to the personalization of educational paths in relation to the differentiated training needs of individual students. However, since ChatGPT was released in late 2022, there are very few studies on this topic. Studies on the reasoning processes involving children together with Artificial Intelligence remain at an embryonic level. Our investigation aimed to start research in a still little explored area, namely that of the ability of primary school children (aged 9–10 years) to ask themselves critical questions and hypotheses on the functioning of Generative Artificial Intelligence, in order to recognize analogies and differences with respect to human intelligence and to build a collaborative relationship with these new forms of Intelligence. These findings lead us to envisage future study developments. What happens with the implementation of LLM systems in social robots? It happens that social robots, already recognized in the literature as useful in engaging children in teaching–learning processes, gain the ability to converse with teachers and students; LLM systems acquire an anthropomorphic body in which to embody themselves. These two occurrences, discussed throughout this paper, likely contribute to making 'dialoguing social robots' smart educational tools and agents both in training and in socialization. Moreover, such perspectives open to a risk: if children involved in our study have been able to distinguish human intelligence by artificial intelligence even because they were aware of the fact of interacting with a Personal Computer, what will happen when they will interact with Artificial Intelligence embodied in humanoid robots? Research in the educational field will need to further investigate the teaching formats capable of optimizing these potentials.

References

Adamopoulou, E., & Moussiades, L. (2020). An overview of chatbot technology. *In Artificial intelligence applications and innovations*, Conference paper. https://doi.org/10.1007/978-3-030-49186-4_31

Almelhes, S. A. (2023). A review of artificial intelligence adoption in second-language learning. *Theory and Practice in Language Studies, 13*(5), 1259–1269. https://doi.org/10.17507/tpls.1305.21

Alshahrani, A. (2023). The impact of ChatGPT on blended learning: Current trends and future research directions. *International Journal of Data and Network Science, 7*, 2029–2040. https://doi.org/10.5267/j.ijdns.2023.6.010

Al-Shawwa, M., & Abu Naser, S. S. (2024). Predicting birth weight using artificial neural network. In *International conference on multidisciplinary science* (Vol. 2, p. 3).

Andrews, K., & Huss, B. (2014). Anthropomorphism, anthropectomy, and the null hypothesis. *Biology & Philosophy, 29*(5), 711–729. https://doi.org/10.1007/s10539-014-9442-2

Arya, M. (2019). A brief history of chatbots. Chatbots Life. Retrieved January 01, 2023, from https://chatbotslife.com/a-brief-history-of-chatbotsd5a8689cf52f

Ausat, A., Massang, B., Efendi, M., Nofirman, N., & Riady, Y. (2023). Can ChatGPT replace the role of the teacher in the classroom: A fundamental analysis. *Journal of Education, 5*, 16100–16106. https://doi.org/10.31004/joe.v5i4.2745

Bainbridge, W. A., Hart, J. W., Kim E. S., & Scassellati, B. (2008). The effect of presence on human-robot interaction. In *Conference: Robot and human interactive communication, RO-MAN 2008. The 17th IEEE international symposium*. https://doi.org/10.1109/ROMAN.2008.4600749

Barone P. (2014). Embodiment. In P. Barone, A. Ferrante, & D. Sartori (a cura di), *Formazione e post-umanesimo. Sentieri pedagogici nell'età della tecnica*. Raffaello Cortina Editore.

Barrett, L. F., & Bliss-Moreau, E. (2009). Affect as a psychological primitive. *Advances in Experimental Social Psychology, 41*, 167–218. https://doi.org/10.1016/S0065-2601(08)00404-8
Beccari, M. N., Oliveira, T., & (L. A. Marcus Ed.). (2011). A philosophical approach about user experience methodology. In *Design, user experience, and usability, international conference, HCII 2011* (LNCS, Vol. 6769, pp. 13–22).
Belpaeme, T., Kennedy, J., Ramachandran, A., Scassellati, B., & Tanaka, F. (2018). Social robot for education: A review. *Science Robotics, 3*(21), eaat5954. https://doi.org/10.1126/scirobotics.aat5954
Benuyenah, V. (2023). Commentary: ChatGPT use in higher education assessment. Prospects and epistemic threats. *Journal of Research in Innovative Teaching & Learning, 16*(1), 134–135. https://doi.org/10.1108/JRIT-03-2023-097
Bijlstra, G., & Dotsch, R. (2011). *FaceReader 4 emotion classification performance on images from the Radboud faces database*. Radboud University Nijmegen.
Birhane, A., Kalluri, P., Card, D., Agnew, W., Dotan, R., Bao, M. (2022, June). The values encoded in machine learning research. In *2022 ACM conference on fairness, accountability, and transparency* (pp. 173–184). https://doi.org/10.48550/arXiv.2106.15590
Blair, K., Schwartz, D. L., Biswas, G., & Leelawong, K. (2007). Pedagogical agents for learning by teaching: Teachable agents. *Educational Technology, 47*(1), 56–61. Retrieved January 01, 2024, from https://www.jstor.org/stable/44429380
Bradley, M. M., Codispoti, M., Cuthbert, B. N., & Lang, P. J. (2001). Emotion and motivation I: Defensive and appetitive reactions in picture processing. *Emotion, 1*, 276–298. https://doi.org/10.1037/1528-3542.1.3.276
Bradley, M. M., & Lang, P. J. (1994). Measuring emotion: The self-assessment manikin and the semantic differential. *Journal of Behavior Therapy and Experimental Psychiatry, 25*, 49–59. https://doi.org/10.1016/0005-7916(94)90063-9
Bradley, M. M., Miccoli, L., Escrig, M. A., & Lang, P. J. (2008). The pupil as a measure of emotional arousal and autonomic activation. *Psychophysiology, 45*, 602–607. https://doi.org/10.1111/j.1469-8986.2008.00654.x
Braidotti, R. (2014). *Postumano*. DeriveApprodi.
Buschle, C., Reiter, H., & Bethmann, A. (2022). The qualitative pretest interview for questionnaire development: Outline of programme and practice. *Quality & Quantity, 56*(2), 823–842. https://doi.org/10.1007/s11135-021-01156-0
Busetto, L., Wick, W., & Gumbinger, C. (2020). How to use and assess qualitative research methods. *Neurological Research and Practice, 2*, 1–10. https://doi.org/10.1186/s42466-020-00059-z
Calvo, M. G., Fernández-Martín, A., Recio, G., & Lundqvist, D. (2018). Human observers and automated assessment of dynamic emotional facial expressions: KDEF-dyn database validation. *Frontiers in Psychology, 9*, 2052. https://doi.org/10.3389/fpsyg.2018.02052
Calvo, M. G., Gutiérrez-García, A., Fernández-Martín, A., & Nummenmaa, L. (2014). Recognition of facial expressions of emotion is related to their frequency in everyday life. *Journal of Nonverbal Behavior, 38*, 549–567. https://doi.org/10.1007/s10919-014-0191-3
Calvo, M. G., & Nummenmaa, L. (2016). Perceptual and affective mechanisms in facial expression recognition: An integrative review. *Cognition and Emotion, 30*, 1081–1106. https://doi.org/10.1080/02699931.2015.1049124
Castillo, A. G. R., Silva, G. J. S., Arocutipa, J. P. F., Berríos, H. Q., Rodríguez, M. A. M., Reyes, G. Y., López, H. R. P., Reves, R. M. V., Rivera, H. V. H., & González, J. L. A. (2023). Effect of Chat GPT on the digitized learning process of university students. *Journal of Namibian Studies, 33*. https://doi.org/10.59670/jns.v33i.411
Celik, I. (2023). Towards intelligent-TPACK: An empirical study on ChatGPT safer. *Time*. Retrieved January 01, 2024, from https://time.com/6247678/openai-chatgptkenya-workers/
Chassignol, M., Khoroshavin, A., Klimova, A., & Bilyatdinova, A. (2018). Artificial intelligence trends in education: A narrative overview. *Procedia Computer Science, 136*, 16–24. https://doi.org/10.1016/j.procs.2018.08.233

Chatzopoulos, A., Kalogiannakis, M., Papadakis, S., & Papoutsidakis, M. (2022). A novel, modular robot for educational robotics developed using action research evaluated on technology acceptance model. *Education Sciences, 12*, 274. https://doi.org/10.3390/educsci12040274

Chiu, T. K., & Chai, C. S. (2020). Sustainable curriculum planning for artificial intelligence education: A self determination theory perspective. *Sustainability, 12*(14), 5568. https://doi.org/10.3390/su12145568

Chow, J., Sanders, L., & Li, K. (2003). Impact of ChatGPT on medical chatbots as a disruptive technology. *Frontiers in Artificial Intelligence, 6*, 116601. https://doi.org/10.3389/frai.2023.1166014

Cohn, J. F., & Sayette, M. A. (2010). Spontaneous facial expression in a small group can be automatically measured: An initial demonstration. *Behavior Research Methods, 42*, 1079–1086. https://doi.org/10.3758/BRM.42.4.1079

Cox, C., & Tzoc, E. (2023). ChatGPT: Implications for academic libraries. *College & Research Libraries News, 84*(3), 99. https://doi.org/10.5860/crln.84.3.99

Crawford, J., Cowling, M., & Allen, K. (2023). Leadership is needed for ethical ChatGPT: Character, assessment, and learning using Artificial Intelligence (AI). *Journal of University Teaching & Learning Practice, 20*(3), 1–21. https://doi.org/10.53761/1.20.3.02

Damasio, A. (1995). *L'errore di Cartesio. Emozione, ragione e cervello umano.* Adelphi.

Damasio, A., & Meyer, K. (2008). Consciousness: An overview of the phenomenon and of its possible neural basis. *The Neurology of Consciousness*, 3–14. https://doi.org/10.1016/B978-0-12-374168-4.00001-0

Damiano, L., & Dumouchel, P. (2018). Antropomorphism in human-robot co-evolution. *Frontiers in Psychology, 9*(468), 1–9. https://doi.org/10.3389/fpsyg.2018.00468

Damiano, L., & Dumouchel, P. (2020). Emotions in relation. Epistemological and ethical scaffolding for mixed human-robot social ecologies. *Humana Mente, 13*(37).

Davidson, R. J. (1992). Anterior cerebral asymmetry and the nature of emotion. *Brain and Cognition, 20*, 125–151. https://doi.org/10.1016/0278-2626(92)90065-T

Deri, C. E. (2022). Social learning theory and academic writing in graduate studies. *Journal of Applied Learning & Teaching, 5*(Sp. Iss. 1), 20–26. https://doi.org/10.37074/jalt.2022.5.s1.4

Dumouchel, P., & Damiano, L. (2019). *Vivere con i robot. Saggio sull'empatia artificiale.* Raffaello Cortina Editore.

Eisenbarth, H., Gerdes, A. B. M., & Alpers, G. W. (2011). Motor-incompatibility of facial reactions: The influence of valence and stimulus content on voluntary facial reactions. *Journal of Psychophysiology, 25*, 124–130. https://doi.org/10.1027/0269-8803/a000048

Ekman, P. (2017). Facial Expression. In J. M. Fernández-Dols, & J. A. Russel (Eds.), *The science of facial expression.* Cambridge University Press. https://doi.org/10.1093/acprof:oso/9780190613501.001.0001

Ekman, P., Friesen, W. V., & Hager, J. C. (2002). *Facial action coding system. Manual and investigator's guide.* Research Nexus.

Ekman, P., Irwin, W., & Rosenberg, E. (1994). *EMFACS: Coders instructions (EMFACS-8).* University of California San Francisco Press.

Esposito, R. (2006). *La natura umana dopo l'umanesimo.* In M. Pireddu, & A. Tursi (Eds.), *Post-umano. Relazioni tra uomo e tecnologia nella società delle reti.* Guerini & Associati.

Fauzi, F., Tuhuteru, L., Sampe, F., Ausat, A., & Hatta, H. (2023). Analysing the role of ChatGPT in improving student productivity in higher education. *Journal of Education, 5*, 14886–14891. https://doi.org/10.31004/joe.v5i4.2563

Fischer, K., Lohan, K. S., Foth, K. et al. (2012). Levels of embodiment: Linguistic analyses of factors influencing HRI. In *7th ACM/IEEE international conference on human robot interaction (HRI), Boston Massachusetts, USA.* https://doi.org/10.1145/2157689.2157839

Floridi, L. (2023). AI as agency without intelligence: On ChatGPT, large language models, and other generative models. *Philosophy & Technology, 36*, 15. https://doi.org/10.1007/s13347-023-00621-y

Fuchs, K. (2023). Exploring the opportunities and challenges of NLP models in higher education: Is ChatGPT a blessing or a curse? *Frontiers in Education, 8*, 1166682. https://doi.org/10.3389/feduc.2023.1166682

Fütterer, T., Fischer, C., Alekseeva, A., et al. (2023). ChatGPT in education: Global reactions to AI innovations. *Science and Reports, 13*, 15310. https://doi.org/10.1038/s41598-023-42227-6

Gallelli, R. (2014). La mente incarnata. In G. Annacontini, & R. Gallelli (Eds.) *Formare altre(i)menti*. Progedit.

Gallelli, R. (2017). Processi di soggettivazione, formazione e materialità digitale. In A. Ferrante, & J. Orsenigo (Eds.), *Dialoghi sul postumano. Pedagogia, filosofia e scienza*. Mimesis.

Gallese, V. (2001). The shared manifold hypothesis: From mirror neurons to empathy. *Journal of Consciousness Studies, 8*, 33–50.

Gerdes, A., Wieser, M. J., Mühlberger, A., Weyers, P., Alpers, G. W., Plichta, M. M., et al. (2010). Brain activations to emotional pictures are differentially associated with valence and arousal ratings. *Frontiers in Human Neuroscience, 4*, 175. https://doi.org/10.3389/fnhum.2010.00175

Giachanou, A., & Crestani, F. (2016). Like it or not: A survey of twitter sentiment analysis methods. *ACM Computing Surveys, 49*(2), 1–41.

Gilson, A., Safranek, C. W., Huang, T., Socrates, V., Chi, L., Taylor, R. A., & Chartash, D. (2023). How does CHATGPT perform on the United States medical licensing examination? The implications of large language models for medical education and knowledge assessment. *JMIR Medical Education, 9*, e45312. https://doi.org/10.2196/45312

Grimes, D., & Warschauer, M. (2010). Utility in a fallible tool: A multi-site case study of automated writing evaluation. *The Journal of Technology, Learning and Assessment*, *8*(6). Retrieved February 06, 2023, from https://ejournals.bc.edu/index.php/jtla/article/view/1625

Habeb Al-Obaydi, L., Pikhart, M., & Klimova, B. (2023). ChatGPT and the general concepts of education: Can artificial intelligence-driven chatbots support the process of language learning? *International Journal of Emerging Technologies in Learning (iJET), 18*(21), 39–50. https://doi.org/10.3991/ijet.v18i21.42593. Retrieved February 06, 2023, from http://www.istitutolipani.com/wp-content/uploads/2018/10/Progetto-Robotica-Lipani.pdf

Hess, U., & Blairy, S. (2001). Facial mimicry and emotional contagion to dynamic emotional facial expressions and their influence on decoding accuracy. *International Journal of Psychophysiology, 40*, 129–141. https://doi.org/10.1016/S0167-8760(00)00161-6

Higgs, J. M., & Stornaiuolo, A. (2024). Being human in the age of generative AI: Young people's ethical concerns about writing and living with machines. *Reading Research Quarterly, 1*. https://doi.org/10.1002/rrq.552

Jauhiainen, J. S., & Garagorry Guerra, A. (2023). Generative AI and ChatGPT in school children's education: Evidence from a school lesson. *Sustainability, 15*(18), 14025. https://doi.org/10.3390/su151814025

Javaid, M., Haleem, A., Singh, R. P., Kahn, S., & Khan, I. H. (2023). Unlocking the opportunities through ChatGPT Tool towards ameliorating the education system. *BenchCouncil Transactions on Benchmarks, Standards and Evaluations, 3*, 100–115; Firat, M. (2023). What ChatGPT means for universities: Perceptions of scholars and students. *Journal of Applied Learning*, *6*, 57–63. https://doi.org/10.1016/j.tbench.2023.100115

Kasneci, E., Seßler, K., Küchemann, S., Bannert, M., Dementieva, D., Fischer, F., & Kasneci, G. (2023). ChatGPT for good? On opportunities and challenges of large language models for education. *Science Direct, 103*, 102274. https://doi.org/10.35542/osf.io/5er8f

Kelly, K. (1992). *Out of control. The new biology of machines, social systems and economic world*. Feltrinelli 1995.

Keltner, D., & Cordaro, D. T. (2017). Understanding multimodal emotional expressions: Recent advances in basic emotion theory. In J. M. Fernández-Dols, & J. A. Russel (Eds.), *The science of facial expression*. Cambridge University Press.

Kilinç, S. (2023). Embracing the future of distance science education: Opportunities and challenges of ChatGPT integration. *Asian Journal of Distance Education, 18*(1), 205–237. https://doi.org/10.5281/zenodo.7857396

Kok, B. C., & Soh, H. (2020). Fiducia nei robot: Sfide e opportunità. *Current Robotics Reports, 1*, 297309. https://doi.org/10.1007/s43154-020-00029-y

Kuhail, M. A., Alturki, N., Alramlawi, S., et al. (2023). Interacting with educational chatbots: A systematic review. *Education and Information Technologies, 28*, 973–1018. https://doi.org/10.1007/s10639-022-11177-3

Lameras, P., & Arnab, S. (2021). Power to the teachers: An exploratory review on artificial intelligence in education. *Machine, Learning and Systems, 4*, 795–813.

Lang, P. J., Bradley, M. M., Fitzsimmons, J. R., Cuthbert, B. N., Scott, J. D., Moulder, B., et al. (1998). Emotional arousal and activation of the visual cortex: An fMRI analysis. *Psychophysiology, 35*, 199–210. https://doi.org/10.1111/1469-8986.3520199

Lang, P. J., & Davis, M. (2006). Emotion, motivation, and the brain: Reflex foundations in animal and human research. *Progress in Brain Research, 156*, 3–29. https://doi.org/10.1016/S0079-6123(06)56001-7

Lang, P. J., Greenwald, M. K., Bradley, M. M., & Hamm, A. O. (1993). Looking at pictures: Affective, facial, visceral, and behavioral reactions. *Psychophysiology, 30*, 261–273. https://doi.org/10.1111/j.1469-8986.1993.tb03352.x

Lewinski, P. (2015). Automated facial coding software outperforms people in recognizing neutral faces as neutral from standardized datasets. *Frontiers in Psychology, 6*, 1386. https://doi.org/10.3389/fpsyg.2015.01386

Lewinski, P., den Uyl, T. M., & Butler, C. (2014). Automated facial coding: Validation of basic emotions and FACS AUs in FaceReader. *Journal of Neuroscience, Psychology, and Economics, 7*, 227–236. https://doi.org/10.1037/npe0000028

Leyzberg, D., Scassellati, B., Spaulding, S., & Toneva, M. (2012). The physical presence of a robot tutor increases cognitive learning gains. In *34th annual conference of the cognitive science society, Sapporo, Japan.*

Lieberman, M. (2023). What is ChatGPT and how is it used in education? *Education Week*, 04. Retrieved January 01, 2024, from https://www.edweek.org/technology/what-is-chatgpt-and-how-is-it-used-in-education/2023/01

Lin, J. C., Younessi, D. N., Kurapati, S. S., Tang, O. Y., & Scott, I. U. (2023). Comparison of GPT-3.5, GPT-4, and human user performance on a practice ophthalmology written examination. *Eye. 37*(17), 3694–3695. https://doi.org/10.1038/s41433-023-02564-2

Liu, Y., Han, T., Ma, S., Zhang, J., Yang, Y., Tian, J., He, H., Li, A., He, M., Liu, Z., et al. (2023). Summary of ChatGPT-related research and perspective towards the future of large language models. *Meta-Radiology, 1*, 100017. https://doi.org/10.1016/j.metrad.2023.100017

Lo, C. K. (2023). What is the impact of ChatGPT on education? A rapid review of the literature. *Education Sciences, 13*(4), 410. https://doi.org/10.3390/educsci13040410

Longo, G. O. (2003). *Il Simbionte. Prove di umanità futura.* Meltemi.

Majid, M. A. A., Othman, M., Mohamad, S. F., Lim, S. A. H., & Yusof, A. (2017). Piloting for interviews in qualitative research: Operationalisation and lessons learnt. *International Journal of Academic Research in Business and Social Sciences, 7*(4), 1073–1080. https://doi.org/10.6007/IJARBSS/v7-i4/2916

Malinowska, A. (2021). *Data dating. Love, technology, desire.* Intellect Books.

Marchesini, R. (2014). *Ibridazioni e processi evolutivi.* In P. Barone, A. Ferrante, & D. Sartori (Eds.), *Formazione e post-umanesimo. Sentieri pedagogici nell'età della tecnica.* Raffaello Cortina.

Marchesini, R. (2002). *Post-human. Verso nuovi modelli di esistenza.* Bollati Boringhieri.

Marchesini, R. (2009). *Il tramonto dell'uomo. La prospettiva post-umanista.* Dedalo.

Maturana, H., & Varela, F. (1985). *L'albero della conoscenza.* Garzanti 1987.

Maturana, H. R., & Varela, F. J. (1992). *Macchine ed esseri viventi.* Astrolabio-Ubaldini Editore.

Maturana, H., & Varela, F. (2001). *Autopoiesi e cognizione. La realizzazione del vivente.* Marsilio.

Megahed, F. M., Chen, Y. J., Ferris, J. A., Knoth, S., & Jones-Farmer, L. A. (2023). How generative AI models such as ChatGPT can be (mis)used in SPC practice, education, and research? An exploratory study. arXiv preprint arXiv:2302.10916v1 [cs.LG]. https://doi.org/10.48550/arXiv.2302.10916

Merleau-Ponty, M. (1965). *Fenomenologia della percezione*. Il Saggiatore.

Merleau-Ponty, M. (1979). *Il corpo vissuto. L'ambiguità dell'esistenza, la riscoperta della vita percettiva, la 'carne del mondo'*. Il Saggiatore.

Metz, C. (2022, December, 11). The new chatbots could change the world. Can you trust them? *The New York Times*. Retrieved January 01, 2024, from https://www.nytimes.com/2022/12/10/technology/ai-chat-bot-chatgpt.html

Montenegro-Rueda, M., Fernández-Cerero, J., Fernández-Batanero, J. M., & López-Meneses, E. (2023). Impact of the implementation of ChatGPT in education: A systematic review. *Computers, 12*(8), 153. https://doi.org/10.3390/computers12080153

Nature (Editorial) (2023, 24 January). Tools such as ChatGPT threaten transparent science; here are our ground rules for their use. *Nature, 613*, 612. Retrieved January 01, 2024, from https://www.nature.com/articles/d41586-023-00191-1

Negroponte, N. (2006). *One laptop per child* [Video]. TED conferences. https://www.ted.com/talks/nicholas_negroponte_one_laptop_per_child?language=en

Ng, D. T. K., Leung, J. K. L., Chu, S. K. W., & Qiao, M. S. (2021). Conceptualizing AI literacy: An exploratory review. *Computers and Education: Artificial Intelligence, 2*, 100041. https://doi.org/10.1016/j.caeai.2021.100041

Ouyang, L., Wu, J., Jiang, X., Almeida, D., Wainwright, C., Mishkin, P., Zhang, C., Agarwal, S., Slama, K., Ray, A., & Schulman, J. (2022). Training language models to follow instructions with human feedback. In *Advances in neural information processing systems (NeurIPS)* (Vol. 35). Main Conference Track.

Pantic, M., & Rothkrantz, L. J. M. (2000). Automatic analysis of facial expressions: The state of the art. *IEEE Transactions on Pattern Analysis and Machine Intelligence, 22*, 1424–1445. https://doi.org/10.1109/34.895976

Papadakis, S., et al. (2021). Attitudes towards the use of educational robotics: Exploring pre-service and in-service early childhood teacher profiles. *Education Science, 11*, 204. https://doi.org/10.3390/educsci11050204

Peretti, G., Manzi, F., Di Dio, C., Cangelosi, A., Harris, P. L., Massaro, D., & Marchetti, A. (2022). Can a robot lie? Young children's understanding of intentionality beneath false statements. *Infant and Child Development, 32*(e2398), 1–25. https://doi.org/10.1002/icd.2398

Pinker, S. (2019). *Come funziona la mente*. Mondadori.

Pinto Minerva, F. (2008). *Umano post-umano, per una pedagogia del soggetto mutante*. In E. Colicchi (a cura di), *Il soggetto nella pedagogia contemporanea*. Carocci.

Pinto Minerva, F., & Gallelli, R. (2004). *Pedagogia e post-umano. Ibridazioni identitarie e frontiere del possibile*. Carocci.

Poria, S., Chaturvedi, I., Cambria, E., & Hussain, A. (2016). *Convolutional MKL based multimodal emotion recognition and sentiment analysis*. Proceedings of ICDM.

Poria, S., Cambria, E., Bajpai, R., & Hussain, A. (2017). A review of affective computing: From unimodal analysis to multimodal fusion. *Information Fusion, 37*, 98–125. https://doi.org/10.1016/j.inffus.2017.02.003

Prescott, D. S., Maeschalck, C. L., & Miller, S. D. (Eds.). (2017). *Feedback-informed treatment in clinical practice: Reaching for excellence*. American Psychological Association. https://doi.org/10.1037/0000039-000

Qadir, J. (2022). Engineering education in the era of ChatGPT: Promises and pitfalls of generative AI for education. *TechRxiv, 1*. https://doi.org/10.1109/EDUCON54358.2023.10125121

Rahman, M. M., & Watanobe, Y. (2023). ChatGPT for education and research: Opportunities, threats, and strategies. *Applied Sciences, 13*, 5783. https://doi.org/10.1109/EDUCON54358.2023.10125121

Rannikmäe, M., Holbrook, J., & Soobard, R. (2020). Social constructivism—Jerome Bruner. In B. Akpan, & T. J. Kennedy (Eds.), Relevant topic-based, open and guided inquiry-based, and review on artifcial intelligence in education. *Information, 13*(1), 14. https://doi.org/10.1007/978-3-030-43620-9_18

Sander, D., Grandjean, D., & Scherer, K. R. (2018). An appraisal-driven componential approach to the emotional brain. *Emotion Review, 10*, 219–231. https://doi.org/10.1177/1754073918765653

Scharth, M. (2022). The ChatGPT chatbot is blowing people away with science education in theory and practice: An introductory guide. Retrieved January 01, 2023, from https://www.sydney.edu.au/news-opinion/news/2022/12/08/the-chatgpt-chatbot-is-blowing-people-away-with-its-writing-skil.html

Scherer, K. R., & Moors, A. (2019). The emotion process: Event appraisal and component differentiation. *Annual Review of Psychology, 70*, 719–745. https://doi.org/10.1146/annurev-psych-122216-011854, https://doi.org/10.3389/fpsyg.2020.01388/full

Scherer, K. R., & Ellgring, H. (2007). Multimodal expression of emotion: Affect programs or componential appraisal patterns? *Emotion, 7*, 158–171. https://doi.org/10.1037/1528-3542.7.1.158

Schulman, J., Zoph, B., Kim, C., Hilton, J., Menick, J., Weng, J., Uribe, J. F. C., Fedus, L., Metz, L., Pokorny, M., Lopes, R. G., Zhao, S., Vijayvergiya, A., Sigler, E., Perelman, A., Voss, C., Heaton, M., Parish, J., Cummings, R. N., & Ryder, N. (2022, November 30). *ChatGPT: Optimizing language models for dialogue.* OpenAI.com. Retrieved January 01, 2024, from https://openai.com/blog/chatgpt/

Seghier, M. L. (2023). ChatGPT: Not all languages are equal. *Nature, 615*(7951), 216–216.

Sloterdijk, P. (2001). *Die Sonne und der Tod. Dialogische Untersuchungen* (mit Hans-Jürgen Heinrichs). Suhrkamp.

Steiss, J., et al. (2024). Comparing the quality of human and ChatGPT feedback on students' writing. *Learning and Instruction, 91*, 101894. https://doi.org/10.1016/j.learninstruc.2024.101894

Stokel-Walker, C. (2022, December 9). AI Bot ChatGPT writes smart essays—Should professors worry? *Nature.* https://doi.org/10.1038/d41586-022-04397-7, https://www.nature.com/articles/d41586-022-04397-7

Sukhpal, S. G., et al. (2024). Transformative effects of ChatGPT on modern education: Emerging era of AI chatbots. *Internet of Things and Cyber-Physical Systems, 4*, 19–23. https://doi.org/10.1016/j.iotcps.2023.06.002

Tate, T. (2023). Evolving practices for instruction in the age of generative AI. In *AI symposium.* Santa Clara County Office of Education, San Jose, CA.

Tate, T., Doroudi, S., Ritchie, D., Xu, Y., & Warschauer, M. (2023). Educational research and AI-generated writing: Confronting the coming Tsunami. https://doi.org/10.35542/osf.io/4mec3. Retrieved December 12, 2023, from https://osf.io/preprints/edarxiv/4mec3

Tate, T. P., Collins, P., Xu, Y., Yau, J. C., Krishnan, J., Prado, Y., Farkas, G., & Warschauer, M. (2019). Visual-syntactic text format: Improving adolescent literacy. *Scientific Studies of Reading, 23*(4), 287–304. https://doi.org/10.1080/10888438.2018.1561700

Tlili, A., Shehata, B., Agyemang Adarkwah, M., Bozkurt, A., Hickey, D. T., Ronghuai Huang, R., & Agyemang, B. (2023). What if the devil is my guardian angel: ChatGPT as a case study of using chatbots in education. *Smart Learning Environments, 10*, 15. https://doi.org/10.1186/s40561-023-00237-x

Turkle, S., Breazeal, C., Dasté, O., & Scassellati, B. (2006). *Encounters with Kismet and Cog: Children respond to relational artifacts, Encounters with Kismet and Cog.* Massachussets Institute of Technology. Retrieved September 30, 2024, from chrome-extension://efaidnbmnnnibpcajpcglclefindmkaj/https://web.mit.edu/people/sturkle/encounterswithkismet.pdf

Turkle, S., Breazeal, C., Dastè, O., & Scassellati, B. (2011). *First encounter with Kismet and Cog. Children respond to relational artifacts.* Retrieved January 01, 2024, from https://sherryturkle.mit.edu. chrome-extension://efaidnbmnnnibpcajpcglclefindmkaj/https://sherryturkle.mit.edu/sites/default/files/images/ST_First%20Encounters%20with%20Kismet.pdf

Turkle, S. (2024). *Who do we become when we talk to machines?* MIT. https://doi.org/10.21428/e4baedd9.caa10d84

Turkle, S. (1996). *Life on the screen: Identity in the age of internet.* Weidenfeld & Nicolson.

Turkle, S. (2004). Whither psychoanalysis in computer culture? *Psychoanalytic Psychology, 21*(1), 16–30. https://doi.org/10.1037/0736-9735.21.1.16

Turkle, S., Taggart, W., Kidd, C. D., & Dasté, O. (2006b). Relational artifacts with children and elders: The complexities of cybercompanionship. *Connection Science, 18*(4), 347–361. https://doi.org/10.1080/09540090600868912

Van Noorden, R., & Perkel, J. M. (2023). AI and science: What 1.600 researchers think. *Nature, 621.*

Van Dis, E. A. M., Bollen, J., Zuidema, W., Van Rooij, R., & Bockting, C. L. (2023). ChatGPT: Five priorities for research. Conversational AI is a game-changer for science. Here's how to respond. *Nature, 614*(7947), 224–226.

Varela, F. (1994). *Il reincanto del concreto.* In P. L. Capucci (a cura di), *Il corpo tecnologico.* Baskerville.

Varela, F., Thompson, E., & Rosch, E. (1992). *Le vie di mezzo della conoscenza. Le scienze cognitive alla prova dell'esperienza.* Feltrinelli.

Vasconcelos, M. A. R., & Santos, R. P. (2023). Enhancing STEM learning with ChatGPT and Bing chat as objects to think with: A case study. *EURASIA Journal of Mathematics, Science and Technology Education, 19*, em2296. https://doi.org/10.29333/ejmste/13313

Wang, P., & Rau, P. P. (2019). Influence of embodiment and substrate of social robots on users' decision-making and attitude. *International Journal of Social Robotics, 11.* https://doi.org/10.1007/s12369-018-0510-7. Retrieved September 30, 2024, from https://link.springer.com/journal/12369

Wang, P. (2019). On defining artificial intelligence. *Journal of Artificial General Intelligence, 10*(2), 1–37. https://doi.org/10.2478/jagi-2019-0002

Wang, T., Lund, B. D., Marengo, A., Pagano, A., Mannuru, N. R., Teel, Z. A., & Pange, J. (2023). Exploring the potential impact of artificial intelligence (AI) on international students in higher education: Generative AI, chatbots, analytics, and international student success. *Applied Sciences, 13*, 6716. https://doi.org/10.3390/app13116716

Weizenbaum, J. (1966). ELIZA—A computer program for the study of natural language communication between man and machine. *Communications of the ACM, 9*(1), 36–45. https://doi.org/10.1145/365153.365168

Xames, M. D., & Shefa, J. (2023). ChatGPT for research and publication: Opportunities and challenges. *Journal of Applied Learning & Teaching, 6*(1), 1–6. https://doi.org/10.37074/jalt.2023.6.1.20

Xu, Y., Aubele, J., Vigil, V., Bustamante, A. S., Kim, Y. S., & Warschauer, M. (2022). Dialogue with a conversational agent promotes children's story comprehension via enhancing engagement. *Child Development, 93*(2), e149–e167. https://doi.org/10.1111/cdev.13708

Yang, W. (2022). Artificial intelligence education for young children: Why, what, and how in curriculum design and implementation. *Computers and Education: Artificial Intelligence, 3*, 100061. https://doi.org/10.1016/j.caeai.2022.100061

Zawacki-Richter, O. V., Bond, M., & Gouverneur, F. (2019). Systematic review of research on artificial intelligence applications in higher education–where are the educators? *International Journal of Educational Technology in Higher Education, 16*(1), 1–27. https://doi.org/10.1186/s41239-019-0171-0

Zhai, X. (2024). AI and machine learning for next generation science assessment. In H. Jiao, & R. W. Lissitz (Eds.), *Machine learning, natural language processing and psychometrics.* Information Age Publisher.

Zhai, X. (2022). ChatGPT user experience: Implications for education. *SSRN Electronic Journal, 2024*(30), 09. https://doi.org/10.2139/ssrn.4312418

Zhai, X., & Nehm, R. (2023). AI and formative assessment: The train has left the station. *Journal of Research in Science Teaching.* https://doi.org/10.1002/tea.21885

Integrating Arduino and Artificial Intelligence in STEM Education

Óscar Bedoya-Cano, José-Antonio Marín-Marín, and Jesús López-Belmonte

Abstract This article explores the integration of digital technologies in education, emphasizing the importance of robotics and computational thinking in the development of Science, Technology, Engineering and Mathematics (STEM) skills. In the context of Industry 4.0 and Education 4.0, the usefulness of platforms such as Arduino is highlighted, which allow students to engage in practical and affordable projects to develop critical skills. An illustrative example of this application is the project of a robotic explorer equipped with an Arduino board and a humidity sensor, designed to detect areas with higher humidity, simulating the search for water in a controlled environment. This type of project allows students to program a robot that responds to stimuli from the environment, promoting active and practical learning. The integration of artificial intelligence into educational robotics projects significantly enhances the robot's capabilities, optimizing both its autonomy and decision-making processes. This advancement presents an opportunity to incorporate and develop aspects of social robotics, fostering collaborative interactions between students and robotic systems. Through advanced techniques such as reinforcement learning, the robot learns to improve its trajectory and dynamically adjust the humidity threshold detected according to environmental conditions. Furthermore, the use of multiple sensors alongside artificial intelligence (AI) algorithms enables the robot to correlate data and make more accurate predictions, thereby increasing its efficiency. The article concludes that educational robotics and AI are essential to preparing students for the future, equipping them with analytical and practical skills needed in a rapidly evolving digital environment, and promoting a deep understanding of how technology can be applied to solve real-world problems.

Ó. Bedoya-Cano
Pontifical Bolivarian University, Medellín, Colombia
e-mail: oscar.bedoya@upb.edu.co

J.-A. Marín-Marín · J. López-Belmonte (✉)
University of Granada, Granada, Spain
e-mail: jesuslopez@ugr.es

J.-A. Marín-Marín
e-mail: jmarin@ugr.es

G. Lampropoulos and S. Papadakis (eds.), *Social Robots in Education*, Studies in Computational Intelligence 1194, https://doi.org/10.1007/978-3-031-82915-4_7

Keywords Arduino · Artificial intelligence · Robotics · Education

1 Introduction

In the age of digitalization, industrial and educational sectors are tackling the challenges of the 21st century. Industry 4.0, driven by the integration of digital technologies, and Education 4.0 focus on the incorporation of technologies into teaching and learning processes, preparing students for a constantly evolving work environment (Udvaros et al., 2023). In this framework, computational thinking appears as an essential transversal skill that underlies many of these digital technologies, allowing students to break down complex problems into more manageable parts and design algorithmic solutions to future challenges (Park & Woo, 2022).

In this context, an approach that has gained relevance in recent years is computational thinking (Wing, 2006, 2017), which provides a structured framework for problem solving. Its application through programming and robotics has been widely disseminated in secondary education (López-Belmonte et al., 2021). According to recent studies, this methodology allows students to break down complex problems into more manageable components and apply computational science principles in various contexts (Li et al., 2020). This is not a novel idea; since the 1960s, Seymour Papert had already developed tools such as the Logo programming language and the Turtle robot, with the aim of bringing programming closer to students and fostering creativity and active learning (García-Tudela & Marín-Marín, 2023).

In this scenario, within the framework of the Science, Technology, Engineering and Mathematics (STEM) strategy, robotics has been characterized as one of the most used alternatives in the classroom. Coming from automation and mechanics, the possibility of solving everyday situations through activities based on computational thinking integrates engineering, science and technology with practical exercises (Marín-Marín et al., 2021). These initiatives seek to prepare students to face a dynamic and increasingly technological environment (Piedade et al., 2020; Stewart et al., 2021).

The analysis is organized into five sections. The first section explores the significance of robotics in education, emphasizing its growing importance in modern pedagogical practices. The second section examines the Arduino board and its application within educational robotics environments. The third section delves into the role of artificial intelligence in the context of educational robotics. The fourth section describes a practical robotics experience conducted by high school students in Colombia. Finally, the fifth section proposes the integration of artificial intelligence into such educational projects, highlighting its potential to enhance learning outcomes.

2 The Relevance of Robotics in the Educational Environment

Robotics has established itself as a highly effective means of teaching critical skills to students, such as computational thinking and problem-solving. Educational robotics provides an ideal environment for developing science, technology, engineering, and mathematics (STEM) skills through hands-on and playful experiences (Nuño et al., 2023). Furthermore, these systems allow students to interact with physical devices and perform tasks that mimic real-world activities, fostering creativity and critical thinking (Udvaros et al., 2023).

A clear example of this impact is found in programs designed to introduce students to robotics through intensive courses. These programs allow participants not only to acquire technical skills, but also to experiment with programming robotic systems using platforms such as Arduino and Raspberry Pi. The combination of these elements with advanced software such as Robot Operating System (ROS) and artificial intelligence paradigms allows students to carry out specific applications in everyday situations (Abdiakhmetova et al., 2024; Hernández-León & Rodríguez-Conde, 2024). Studies exploring the impact of robotics in the classroom (Adnan et al., 2023; Darmawansah et al., 2023; Kyprianou et al., 2023; Lin & Chen, 2023; Mangina et al., 2023; Ribeiro & Lopes, 2020), underline the importance of these technologies in enhancing learning in STEM areas. Educational robotics not only helps students develop technical skills such as programming, but also promotes critical thinking and creativity (Avello-Martínez et al., 2020) and the inclusion of robotics kits in the classroom provides students with hands-on experiences where there is no single correct solution, thus encouraging the exploration of multiple approaches to solving problems; this enriches students' cognitive skills and fosters collaboration in the classroom. Alternatively, robotic systems offer students the opportunity to work in collaborative environments, where they can interact with multiple robots that operate autonomously and communicate with each other. This is achieved through advanced communication protocols between robots and the use of controllers such as Arduino, allowing the creation of multi-agent systems (Najjar et al., 2019). Robotics also contributes positively to students' motivation (Ribeiro & Lopes, 2020) by adapting to different educational levels (Mamatnabiyev et al., 2024) to apply abstract concepts in practical contexts, which improves their understanding of emerging technologies.

3 Arduino in Robotics

The use of Arduino in secondary education has gained importance due to its ability to integrate practical projects within school curricula, aligning with the skills needed in the 21st century (Marín-Marín et al., 2024). Active methodologies, such as Project-Based Learning (PBL), have been shown to be highly effective in promoting interdisciplinary learning and cooperation among students (Rengifo & Bravo, 2020). By involving students in real projects, their motivation and ability to solve complex problems by creating programmable devices is fostered. Recent studies (Drakatos & Stavridis, 2023; Mamatnabiyev et al., 2024) confirm that students who participate in Arduino projects significantly improve their analytical and problem-solving skills.

The adoption of Arduino, being a low-cost and open-source prototyping tool, has been key in the integration of these technologies within the educational field. Arduino has revolutionized the teaching of programming and robotics due to its versatility and ease of use (Kang et al., 2019). Thanks to its ability to interact with sensors and process data, Arduino has been instrumental in developing computational thinking skills in educational projects. The use of these platforms has allowed students to significantly improve their problem-solving skills and strengthen their logical thinking (Avello et al., 2020; Drakatos, & Stavridis, 2023; Udvaros et al., 2023). On the economic side, the use of microcontrollers such as Arduino has transformed the teaching of programming and robotics, allowing students to experiment with advanced technologies at a relatively low cost (Abdiakhmetova et al., 2024).

The Internet of Things (IoT) has also been one of the main drivers of this trend, allowing the interconnection of physical and virtual devices through networks, which generates large amounts of data. Equipment such as sensors, cameras and geolocation devices are integrated into these systems, enabling the automation of advanced and useful applications (Bansal & Kumar, 2020; Romero-Rodríguez et al., 2020; Sengupta, 2019). Thus, the literature shows a wide diversity of applications on the use of Arduino in education. Both in primary and secondary education (Alò et al., 2020; Chang & Chen, 2020; Jawaid et al., 2019; Moya, 2019; Prima et al., 2018; Sterling Collazos, 2019), as well as at higher levels (Hahn et al., 2019; Kang et al., 2019; Önder et al., 2019; Perenc et al., 2019), the convergence of sensors and Arduino has proven to be an effective tool for teaching. These studies indicate that direct experimentation allows students to effectively apply and reinforce concepts, while developing a creative and collaborative spirit. This teaching strategy improves both attention and academic performance (Dochshanov & Tramonti, 2023; Mangina et al., 2023). Learning experiences with Arduino also result in high levels of satisfaction for both students and teachers (Ribeiro & Lopes, 2020).

The use of Arduino not only facilitates the practical learning of programming concepts, but also enables the creation of smart devices capable of performing autonomous tasks. For example, the integration of Arduino with artificial vision and voice recognition systems allows the development of humanoid robots controlled by voice commands (Naeem et al., 2024). These types of applications expand educational possibilities and encourage the adoption of technologies such as artificial intelligence in the academic environment.

4 Artificial Intelligence in Educational Robotics

Artificial intelligence (AI) has opened new frontiers in the field of educational robotics. Its ability to equip robots with skills such as object recognition, route planning, and autonomous decision-making has transformed the way students interact with these technologies (Park & Woo, 2022). The use of AI algorithms in autonomous robots and the integration of these algorithms has made it possible to optimize the efficiency of robots in navigating static environments, reducing processing time and improving the smoothness of movement (Promkaew et al., 2024).

A clear example of this synergy between AI and robotics can be found in the development of robots with artificial intelligence capable of identifying objects and classifying them in real time (Dúo-Terrón et al., 2023). The use of neural networks such as MobileNetV1 has significantly improved the accuracy of these systems, allowing robots to perform manipulation tasks more effectively (Yong & Yeoh, 2023). The use of sensors, actuators, and microcontrollers in Arduino allows students to apply abstract concepts in a physical environment, promoting a deeper understanding of programming and automated systems (Park & Woo, 2022).

In the educational field, the integration of robotics and artificial intelligence has gained notable relevance in recent years. Project-Based Learning (PBL) stands out as a particularly effective methodology for teaching these topics, facilitating an active learning environment in which students can apply theoretical concepts in practical projects. Using platforms such as Arduino, students have the opportunity to build and program robotic and artificial intelligence systems, allowing them to understand more deeply how these technologies work and develop key skills, such as critical thinking, problem-solving, and collaborative work–practical skills that are essential in the field of engineering (Marín-Marín et al., 2023). In this sense, research has shown that students who participate in PBL-based robotics projects experience a higher level of engagement and motivation, which translates into deeper learning (Fruett et al., 2024).

Introducing students to the application of robots in practice has experienced remarkable growth in recent years, with implementation exercises. Companion robots, equipped with AI systems and advanced sensors, have been designed to assist elderly people in their daily lives, improving their quality of life and providing real-time assistance (Du et al., 2020). These systems can not only detect signs of danger, but also allow for a more natural interaction between the user and the machine. This new reality and the variety of industrial implementations invite learning and prototype development from education.

For example, in agriculture, robots are also being used for specialized tasks, such as harvesting fruits and vegetables. Equipped with AI systems and sensors, these robots can follow the farmer and perform autonomous tasks such as transporting crops efficiently (Loukatos et al., 2021). On the other hand, in the field of higher education, Arduino-based robotic platforms have been developed that allow students to experiment with advanced control and simulation technologies. The use of inexpensive hardware combined with open-source software, such as the ROS operating system, has allowed students to create robot prototypes for experimental validation in laboratory environments (Vyas et al., 2022).

The use of educational robotics systems in teaching programming is also highlighted. Students not only learn the basics, but also experience how these concepts are applied in the real world through the control of robots that execute complex tasks. This methodology promotes collaborative work and problem solving, essential skills for the twenty-first century (Ţoţa & Vaida, 2021).

5 Integrating AI into Arduino Through an Educational Experience

As part of an educational project targeting high school students, a robotic explorer was developed using an Arduino board and a humidity sensor, as shown in Fig. 1. This prototype, called a wet source explorer, aimed to detect areas with increased humidity, simulating the search for water in a controlled environment. The integration of contact and ultrasonic sensors allowed students to program the robot to move autonomously, using simple algorithms that controlled a system of motors and modified its trajectory or stopped based on the detected humidity levels.

This project offered students a hands-on experience in which they could apply basic programming and electronics knowledge. By building and programming a device capable of interacting with its environment, students developed skills in

Fig. 1 Arduino-based wet source explorer prototype

computational thinking and solved real-world problems. In addition, the activity allowed them to understand how a robot can process and respond to sensory data, laying the groundwork for future explorations in robotics and artificial intelligence.

The experience with Arduino and the humidity sensor facilitated the introduction to the integration of artificial intelligence in educational robotics projects, a fundamental aspect in the development of technological skills. As students progressed through the project, they began to understand how AI can enhance the capabilities of robots to adapt to their environment and make informed decisions.

In developing the wet source explorer, high school students not only learned basic concepts of programming and electronics, but were also introduced to the potential of artificial intelligence as a tool to improve and optimize robotic systems. Up to this point, the prototype used a simple deterministic algorithm to guide its movements toward areas with higher humidity. However, incorporating artificial intelligence into these types of projects allows students to go a step further, using advanced techniques to improve the robot's efficiency and effectiveness.

6 Incorporating AI

To maximize the potential of the robotic explorer, the implementation of artificial intelligence algorithms allows for a more complex and adaptive level of interaction with the environment. This section presents the main areas where AI can significantly improve the project:

1. **Motion optimization using reinforcement learning**

One of the most effective methods to improve the motion of the robotic explorer is through reinforcement learning (RL) algorithms (Fajardo et al., 2022; Singh et al., 2022). Instead of relying on predefined motions, the robot can autonomously learn the best path to locate areas of high humidity. Implementing RL requires the definition of a state space, action set, and reward function:

- State Space and Action Set: The robot identifies its environment using data from the humidity sensor, defining states such as 'low', 'medium', and 'high' humidity levels. As it moves, the state changes according to the humidity levels it detects. Actions include moving forward, turning left, turning right, or stopping.
- Reward Function: Rewards are given based on whether the humidity level increases or decreases after each move. A move into a wetter area receives a positive reward, while moving into a drier area results in a penalty.

With each iteration, the robot learns to maximize this reward function, optimizing its moves to find moisture faster and more efficiently. The Q-learning algorithm (Clifton & Laber, 2020; Yang et al., 2004) is ideal for this purpose, as it allows the robot to store expected rewards for each action in a Q-table, used to make more accurate decisions.

2. **Dynamic adjustment of the humidity threshold with artificial neural networks**

Currently, the robot has a static humidity detection threshold. AI enables dynamic adaptation of this threshold through artificial neural networks (ANN) (Attala et al., 2023; Viejo et al., 2016). The ANN is trained with historical humidity data and additional environmental factors, such as temperature and luminosity, allowing for better adaptation of the robot to different scenarios:

- Neural Network Inputs: These include current humidity sensor data and other environmental data, allowing the network to predict the optimal threshold based on the environment.

With this adjustment, the robot achieves greater accuracy in detecting wet areas, continuously adapting as it collects new data.

3. **Data-driven forecasting through supervised machine learning**

AI also allows the robot to make informed predictions about areas with high humidity before reaching them. Supervised machine learning algorithms, such as decision trees and support vector machines (SVM) (Biswas et al., 2021; Spillmann et al., 2006), are trained with historical data including previous humidity levels and coordinates, allowing the robot to identify patterns and make informed decisions:

- Prediction of Humidity-Prone Areas: The ML model learns which features of the environment are associated with the presence of water, directing the robot's search towards areas with a higher probability of success and optimizing exploration time.

4. **Interaction with multiple sensors for complex decision making**

The inclusion of additional sensors, such as temperature or light sensors, allows for more robust analysis through sensory fusion (Alvarado et al., 2018). With AI, the robot can analyze complex data patterns and make adjustments based on conditions detected by multiple sensors:

- Sensitive Fusion: AI makes it possible to correlate information from different sensors, adjusting the robot's behavior according to environmental conditions. For example, in high-temperature situations, the robot can interpret humidity levels differently, and perform a more informed and accurate search.

Applying AI algorithms in these aspects not only optimizes the functions of the robotic explorer, but also exposes students to the creation of autonomous systems that can make data-driven decisions. This fosters a deep understanding of how advanced technologies transform the ability of robots to interact effectively and efficiently with the real world.

5. **Interaction with other devices**

The role of social robotics in education is becoming increasingly prominent, showcasing its integration into classroom practices through collaborative interactions with students (Belpaeme et al., 2018; Cantucci et al., 2024; Johal, 2020; Woo et al., 2021).

This integration not only facilitates teamwork but also fosters interaction between students and robots to achieve shared educational goals. In this way, robotics moves beyond being a mere tool and emerges as an active partner in the learning process. Some expressions of euphoria, such as lighting up, moving rapidly, or rotating on its own axis, incorporate emotional cues that enhance motivation and foster interaction with students.

7 Conclusion

Technology education is at a historic turning point, and its future will largely depend on how we integrate advanced tools such as robotics, computational thinking, and artificial intelligence (AI) into educational environments. These technologies not only train students to face the challenges of the future labor market, but also provide them with essential skills to navigate an ever-evolving digital environment. In this sense, the educational system must be able to quickly adapt to technological changes and offer students the possibility of interacting in a practical way with technological innovations, since this practical approach is key for students to deeply understand the principles of science, technology, engineering, and mathematics (STEM).

The use of robotics in education is gaining increasing importance because it offers an active learning experience, in which students do not only learn abstract concepts, but directly apply them in solving real problems. In this process, robotics allows them to develop fundamental digital competencies and higher cognitive skills, such as critical thinking and creative problem solving. By implementing robotics projects, students move from being passive consumers of knowledge to becoming active creators of technological solutions, reinforcing not only their technical ability, but also their confidence and autonomy.

In this context, platforms such as Arduino have become a cutting-edge tool due to their accessibility and versatility. Arduino offers an affordable gateway to programming and creating robotic devices, making it easier for students from different socioeconomic backgrounds to access advanced technological education. By implementing projects with microcontrollers such as Arduino, students have the opportunity to interact with a wide range of sensors and actuators, enriching their understanding of how electronic systems can be integrated into practical solutions. This also contributes to the development of computational thinking, a skill that is increasingly in demand in the labour market.

However, applied robotics in education not only has intrinsic value in terms of developing STEM skills, but also encourages the active participation of students in solving everyday problems. By working on projects that address real challenges, such as natural resource management, energy efficiency or sustainability, students gain a more holistic view of how technology can positively influence society. The classroom thus becomes a space for innovation and creativity, where students not only acquire knowledge, but also develop a critical awareness of the environment around them.

Artificial intelligence (AI), as an integral part of these robotic projects, takes educational experiences a step further by equipping devices with the ability to make autonomous decisions based on real-time data. In the educational field, AI allows students to not only program devices, but also to understand how these devices can "learn" and improve their performance through interaction with their environment. This active learning is critical as it gives students a much deeper understanding of how machine learning works and how AI models can be optimized by collecting sensory data and dynamically adjusting algorithms.

It is critical that AI models and training networks used in the classroom are designed in a way that they can interact with sensory data in real time. In this way, students will be able to observe how the robot, equipped with sensors and algorithms, adjusts its responses to external stimuli, such as detecting humidity, light or temperature, refining its ability to make autonomous decisions. This process is simultaneous with ongoing network training and algorithm definition, providing students with tangible experience of how AI and robotics complement each other to address complex challenges.

In the close future, we will see greater personalization of educational projects, where students will have more autonomy to define and design their own challenges, based on personal interests or current problems interacting directly using social robotics. Robotics, combined with AI, will allow students to create customized solutions for issues as diverse as climate change, natural resource management, public health, or smart agriculture.

As these technologies continue to evolve and become more accessible, it will be crucial for the education system to keep up with these innovations. In a world where digital skills are increasingly necessary, the transformation of learning must not only focus on developing technical competencies, but also on empowering students to become creators of innovative solutions. Students who interact with robotics and AI in educational projects are not only learning to use technological tools, but are also developing analytical, strategic, and collaborative skills that will be essential in their professional and personal future.

In this sense, the incorporation of AI in educational robotics has immense potential to transform learning. Students who understand how AI algorithms can improve the problem-solving ability of robots will be better prepared to face the challenges of the 21st century. Furthermore, AI offers a great opportunity to personalize learning, as data collected by robots can be used to adjust the difficulty levels of projects and offer students a learning experience tailored to their individual needs.

Finally, the integration of robotics, computational thinking, and artificial intelligence into educational environments has enormous potential to transform the educational landscape. These technologies not only empower students to meet the challenges of the future job market, but also provide them with a set of essential skills to navigate an ever-changing digital world. Through the implementation of practical and accessible projects, such as those offered by Arduino, education has the ability to empower students to become creators of innovative technological solutions and active citizens in an increasingly digitalized world. Educational robotics, complemented by AI, opens the doors to deep, practical, and collaborative learning, which will be essential to prepare the next generations for the challenges posed by the technological future.

8 Future Perspectives

The educational transformation driven by Industry 4.0 and Education 4.0 represents one of the greatest pedagogical revolutions of our era. In this context, the integration of digital technologies, such as robotics, artificial intelligence (AI) and computational thinking, becomes critically relevant to prepare students for the challenges of the future. In this framework, educational robotics stands out as a key tool for the development of STEM (Science, Technology, Engineering and Mathematics) skills, not only for its ability to involve students in practical projects, but for its potential to develop critical problem-solving skills.

This article has demonstrated how accessible platforms such as Arduino allow students to interact with complex technologies in an intuitive way. The example of the robotic explorer equipped with humidity sensors illustrates how a practical and experimental approach can foster deep learning in various areas of knowledge. But, beyond the immediate results of this type of project, it is essential to reflect on the long-term implications of this educational trend. In this foresight, we will explore the future of digital education, focusing on emerging trends in educational robotics, AI, and the development of STEM skills.

One of the pillars for the success of Education 4.0 is the accessibility of technologies that allows students to experiment in a hands-on way. Over the years, platforms such as Arduino, Raspberry Pi, and others like them have facilitated this technological integration in classrooms. However, the future of these platforms involves greater sophistication and simplification at the same time.

On the one hand, we are likely to see an increase in the power and capabilities of these devices. With the growing demand for artificial intelligence and the processing of large amounts of data in real time, it is logical to anticipate that the next generations of boards and microcontrollers will be equipped with native AI computing capacity, allowing students to work with machine learning models without the need for large external computing resources.

On the other hand, these platforms will also evolve towards greater simplicity and accessibility. The development of more intuitive and easily accessible programming environments will be critical to ensuring that students of different ages and skill levels can take full advantage of these technologies. This will include improved visual interfaces, more robust code libraries, and more seamless integration with collaborative development environments in the cloud.

Furthermore, the cost of these technologies is likely to continue to decline, allowing for their democratisation and ensuring that more and more educational institutions, especially in low-resource contexts, can access these tools.

As digital technologies advance, the project-based pedagogical approach is likely to take on an even greater role in the development of STEM skills. Projects such as the robotic explorer equipped with a humidity sensor represent a paradigm of how students can learn by doing, confronting real-world problems and using cutting-edge technology to address them.

The integration of social robotics in education increasingly blurs boundaries as students become involved in designing and programming specific functions for robotic devices, such as autonomous and collaborative tasks like searching for water sources. The incorporation of artificial intelligence further facilitates this process, making the transition between collaborative interaction and system construction more seamless. This integration enhances students' familiarity with robotic systems, embedding these technologies more naturally within educational practices.

This is where educational robotics becomes a crucial tool not only to develop technical skills, but also to foster soft skills such as critical thinking, creativity, collaboration, and problem solving. As educational projects become more complex, we are likely to see an increase in the integration of interdisciplinary teams, where students from different fields work together to develop innovative solutions.

Artificial intelligence, as noted in the article, plays a fundamental role in optimizing educational robotics. As technology advances, it is foreseeable that AI will become an indispensable tool to personalize learning at multiple levels.

In the context of educational robotics, AI will allow educational robots to become increasingly autonomous, not only in terms of decision-making and task execution, but also in adapting to different educational environments. For example, reinforcement learning systems will allow robots to improve their performance based on past experiences, adjusting their behaviors according to environmental stimuli. In an educational setting, this will translate into a greater ability for robots to interact with students and help them learn more effectively.

AI will also be key to personalizing the teaching process. Through the analysis of large amounts of data collected in real time from students' interaction with technologies, AI algorithms will be able to identify learning patterns, preferences, and difficulties specific to each student. This will allow educators to adjust content, projects, and activities according to individual needs, fostering more inclusive learning tailored to each profile.

While the opportunities offered by robotics and AI in education are immense, it is also important to consider the ethical and social challenges that could arise with their implementation.

One of the first challenges is inequality in access to technology. Even though platforms like Arduino are affordable, there are large differences in the availability of technological resources between different regions and socioeconomic contexts. For the integration of these technologies to be truly inclusive, it will be necessary to develop public policies that promote digital equity, ensuring that all students, regardless of their location or economic status, have access to these tools.

Another relevant challenge is ethics in programming and the use of AI. As students work with AI algorithms and develop robotic systems, it will be crucial that they are taught to do so responsibly. This involves not only ensuring that the systems they design are safe and efficient, but also that they respect fundamental ethical principles, such as fairness, transparency, and privacy. Educational AI must be designed to respect students' personal data and not perpetuate discriminatory biases.

Furthermore, the increase in autonomy in robotic systems raises questions about the role of the educator in the future. While robots and AI systems can take over many teaching functions, human connection remains irreplaceable for certain aspects of students' emotional and social development. Future education will need to balance the use of technology with the preservation of meaningful human interaction.

Educational robotics and AI are not only transforming the way students learn STEM skills, but are also driving a redefinition of what it means to be future-ready. The technical skills that students develop through these projects—such as programming, engineering, and data analysis—will continue to be crucial. However, the future of STEM education will also need to include a holistic approach, where interpersonal and problem-solving skills become more relevant.

Educational environments of the future are likely to be designed to foster not only technical mastery, but also students' ability to adapt to an ever-changing world, work effectively in teams, and apply their knowledge creatively to solve complex problems. STEM education of the future will not only produce highly skilled engineers or scientists, but engaged citizens capable of using technology to positively impact society.

The integration of digital technologies into education, such as robotics and AI, represents a unique opportunity to prepare students for the future. As we move towards Education 4.0, it is essential that technological platforms continue to evolve, facilitating access and personalizing learning. Educational robotics projects not only develop critical technical competencies, but also foster interpersonal and problem-solving skills that will be essential in Industry 4.0.

As we look to the future, it is crucial that educators, policymakers, and technologists work together to ensure that access to these technologies is equitable and that their use respects fundamental ethical principles. In this way, educational robotics and artificial intelligence will be able to fulfill their promise of preparing students to meet the challenges of the 21st century successfully and responsibly.

References

Abdiakhmetova, Z., Temirbekova, Z., Aimal Rasa, G., & Berdaly, A. (2024). Using of microcontroller for student learning process. *Journal of Mathematics, Mechanics and Computer Science, 122*(2), 114–123. https://doi.org/10.26577/JMMCS2024-122-02-b9.

Adnan, N., Abdullah, A., & S. N. H. S., Yusof, R. J. R., Zainal, N. F. A., Qamar, F., & Yadegaridehkordi, E. (2023). A systematic literature review in robotics experiential learning with computational and adversarial thinking. *IEEE Access, 11*, 44806–44827. https://doi.org/10.1109/ACCESS.2023.3249761

Alò, D., Castillo, A., Marín Vial, P., & Samaniego, H. (2020). Low-cost emerging technologies as a tool to support informal environmental education in children from vulnerable public schools of southern Chile. *International Journal of Science Education, 42*(4), 635–655. https://doi.org/10.1080/09500693.2020.1723036

Alvarado, B., Gonzalez, R., Matia, F., & De la Puente, P. (2018). Sensor fusion for Tour-Guide robot localization. *IEEE Access, 6*, 78947–78964. https://doi.org/10.1109/ACCESS.2018.2885648

Attala, Z., Cavalcanti, A., & Woodcock, J. (2023). Modelling and verifying robotic software that uses neural networks. In E. Abraham, C. Dubslaff, & S. Tarifa (Eds.), *Theoretical Aspects of Computing—ICTAC 2023—20th International Colloquium* (pp. 15–35). https://doi.org/10.1007/978-3-031-47963-2_3.

Avello, R., Lavonen, J., & Zapata-Ros, M. (2020). Codificación y robótica educativa y su relación con el pensamiento computacional y creativo. Una revisión compresiva. *Revista de Educación a Distancia (RED), 20*(63), 1–21. https://doi.org/10.6018/red.413021.

Avello-Martínez, R., Lavonen, J., & Zapata-Ros, M. (2020). Codificación y robótica educativa y su relación con el pensamiento computacional y creativo. Una revisión compresiva. *Revista de Educación a Distancia (RED), 20*(62), 1–21. https://doi.org/10.6018/red.413021.

Bansal, S., & Kumar, D. (2020). IoT ecosystem: A survey on devices, gateways, operating systems, middleware and communication. *International Journal of Wireless Information Networks, 27*(3), 340–364. https://doi.org/10.1007/s10776-020-00483-7

Belpaeme, T., Kennedy, J., Ramachandran, A., Scassellati, B., & Tanaka, F. (2018). Social robots for education: A review. *Science Robotics, 3*(21). https://doi.org/10.1126/scirobotics.aat5954.

Biswas, A., Saran, I., & Wilson, F. P. (2021). Introduction to Supervised Machine Learning. *Kidney360, 2*(5). 878–880. https://doi.org/10.34067/KID.0000182021.

Cantucci, F., Falcone, R., & Marini, M. (2024). Redefining user expectations: The impact of adjustable social autonomy in Human-Robot interaction. *Electronics, 13*(1). https://doi.org/10.3390/electronics13010127.

Chang, C., & Chen, Y. (2020). Using mastery learning theory to develop Task-Centered hands-on STEM learning of Arduino-based educational robotics: Psychomotor performance and perception by a convergent parallel mixed method. *Interactive Learning Environments, 30*(9), 1677–1692. https://doi.org/10.1080/10494820.2020.1741400

Clifton, J., & Laber, E. (2020). Q-Learning: Theory and applications. In N. Reid, S. Stigler & T. Louis (Eds.), *Annual Review of Statistics and its Application,* (pp. 279–301). Annual Reviews. https://doi.org/10.1146/annurev-statistics-031219-041220.

Darmawansah, D., Hwang, G. J., Chen, M. R. A., & Liang, J. C. (2023). Trends and research foci of robotics-based STEM education: A systematic review from diverse angles based on the technology-based learning model. *International Journal of STEM Education, 10*(12), 1–24. https://doi.org/10.1186/s40594-023-00400-3

Dochshanov, A., & Tramonti, M. (2023). The design and implementation of an Open-source programmable bot for educational purposes. *Digital Presentation and Preservation of Cultural and Scientific Heritage, 13*, 289–298. https://doi.org/10.55630/dipp.2023.13.29.

Drakatos, N., & Stavridis, S. (2023). The perspective of STEM education through the usage of Robotics. *World Journal of Advanced Research and Reviews, 18*(03), 901–913. https://doi.org/10.30574/wjarr.2023.18.3.1146.

Du, W., Zhang, Q., & Ye, Z. (2020). Design of intelligent escort robot system based on KCF tracking algorithm. In *2020 35th Youth Academic Annual Conference of Chinese Association of Automation (YAC)* (pp. 526–529). Zhanjiang, China. https://doi.org/10.1109/YAC51587.2020.9337648.

Dúo-Terrón, P., Moreno-Guerrero, A. J., López-Belmonte, J., & Marín-Marín, J. A. (2023). Inteligencia artificial y machine learning como recurso educativo desde la perspectiva de docentes en distintas etapas educativas no universitarias. *Revista Interuniversitaria de Investigación En Tecnología Educativa*, 58–78. https://doi.org/10.6018/riite.579611.

Fajardo, J., Roldan, F., Realpe, S., Hernández, J., Ji, Z., & Cárdenas, P. (2022). FRobs_RL: A Flexible Robotics Reinforcement Learning Library. In *IEEE International Conference on Automation Science and Engineering (CASE)* (pp. 1104–1109). Mexico City. https://doi.org/10.1109/CASE49997.2022.9926586.

Fruett, F., Pereira Barbosa, F., Cardoso Zampolli, S., & Aragão Guimarães, P. I. (2024). Empowering STEAM activities with artificial intelligence and open hardware: The BitDogLab. *IEEE Transactions on Education, 67*(3), 462–471. https://doi.org/10.1109/TE.2024.3377555

García-Tudela, P. A., & Marín-Marín, J. A. (2023). Use of Arduino in primary education: A systematic review. *Education Sciences, 13*(2), 1–13. https://doi.org/10.3390/educsci13020134

Hahn, M., de Oliveira Cruz, F., & Carvalho, P. (2019). Determining the speed of sound as a function of temperature using Arduino. *The Physics Teacher, 57*(2), 114–115. https://doi.org/10.1119/1.5088475

Hernández-León, N., & Rodríguez-Conde, M. J. (2024). Artificial Intelligence applied to education and educational assessment at the university: Introduction of intelligent tutoring systems, recognition systems, and other future trends. *RED. Revista De Educación a Distancia, 78*(24), 6–31. https://doi.org/10.6018/red.594651

Jawaid, I., Javed, M., Jaffery, M., Akram, A., Safder, U., & Hassan, S. (2019). Robotic system education for young children by collaborative-project-based learning. *Computer Applications in Engineering Education, 28*(1), 178–192. https://doi.org/10.1002/cae.22184

Johal, W. (2020). Research trends in social robots for learning. *Current Robotics Reports, 1*(3), 75–83. https://doi.org/10.1007/s43154-020-00008-3

Kang, S., Yeo, H., & Yoon, J. (2019). Applying chemistry knowledge to code, construct, and demonstrate an Arduino-Carbon Dioxide fountain. *Journal of Chemical Education, 96*(2), 313–316. https://doi.org/10.1021/acs.jchemed.8b00663

Kyprianou, G., Karousou, A., Makris, N., Sarafis, I., Amanatiadis, A., & Chatzichristofis, S. A. (2023). Engaging learners in educational robotics: uncovering students' expectations for an ideal robotic platform. *Electronics, 12*(13), 2865. https://doi.org/10.3390/electronics12132865

Li, Y., Schoenfeld, A. H., diSessa, A. A., Graesser, A. C., Benson, L. C., English, L. D., & Duschl, R. A. (2020). Computational thinking is more about thinking than computing. *Journal for STEM Education Research, 3*(1), 1–18. https://doi.org/10.1007/s41979-020-00030-2

Lin, V., & Chen, N. S. (2023). Interdisciplinary training on instructional design using robots and IOT objects: A case study on undergraduates from different disciplines. *Computer Applications in Engineering Education, 31*(3), 583–601. https://doi.org/10.1002/cae.22601

López-Belmonte, J., Segura-Robles, A., Moreno-Guerrero, A.-J., & Parra-González, M.-E. (2021). Robotics in education: A scientific mapping of the literature in web of science. *Electronics, 10*, 1–18. https://doi.org/10.3390/electronics10030291

Loukatos, D., Petrongonas, E., Manes, K., Kyrtopoulos, I. V., Dimou, V., & Arvanitis, K. G. (2021). A synergy of innovative technologies towards implementing an autonomous DIY electric vehicle for Harvester-Assisting purposes. *Machines, 9*(4), 82. https://doi.org/10.3390/machines9040082

Mamatnabiyev, Z., Chronis, C., Varlamis, I., Himeur, Y., & Zhaparov, M. (2024). A holistic approach to use educational robots for supporting computer science courses. *Computers, 13*(102), 1–21. https://doi.org/10.3390/computers13040102

Mangina, E., Psyrra, G., Screpanti, L., & Scaradozzi, D. (2023). Robotics in the context of primary and Pre-School education: A scoping review. *IEEE Transactions on Learning Technologies, 17*, 342–363. https://doi.org/10.1109/TLT.2023.3266631

Marín-Marín, J.-A., García-Tudela, P. A., & Duo-Terrón, P. (2024). Computational thinking and programming with Arduino in education: A systematic review for secondary education. *Heliyon, 10*(8), e29177. https://doi.org/10.1016/j.heliyon.2024.e29177

Marín-Marín, J.-A., López-Belmonte, J., Pozo-Sánchez, S., & Moreno-Guerrero, A.-J. (2023). Attitudes towards the development of good practices with augmented reality in secondary education teachers in Spain. *Technology Knowledge and Learning, 28*(4), 1443–1459. https://doi.org/10.1007/s10758-023-09671-9

Marín-Marín, J. A., Moreno-Guerrero, A. J., Dúo-Terrón, P., & López-Belmonte, J. (2021). STEAM in education: A bibliometric analysis of performance and co-words in web of science. *International Journal of STEM Education, 8*, 1–21. https://doi.org/10.1186/s40594-021-00296-x

Moya, A. (2019). Studying Avogadro's law with Arduino. *The Physics Teacher, 57*(9), 621–623. https://doi.org/10.1119/1.5135793

Naeem, B., Kareem, W., Saeed-Ul-Hassan, N., & N., & Naeem, R. (2024). Voice controlled humanoid robot. *International Journal of Intelligent Robotics and Applications, 8*(1), 61–75. https://doi.org/10.1007/s41315-023-00304-z

Najjar, L., Johari, N., & Qamhieh, M., (2019). A Leader-Follower communication protocol for Multi-Agent robotic systems. In *2019 IEEE Jordan International Joint Conference on Electrical Engineering and Information Technology (JEEIT]* (pp. 742–747). Amman: Jordan. https://doi.org/10.1109/JEEIT.2019.8717471.

Nuño, P., Bulnes, F. G., Pérez-González, S., & Granda, J. C. (2023). Asterisk as a tool to aid in learning to program. *Electronics, 12*(5), 1160. https://doi.org/10.3390/electronics12051160

Önder, F., Önder, E., & Oğur, M. (2019). Determination of diode characteristics by using Arduino. *The Physics Teacher, 57*(4), 244–245. https://doi.org/10.1119/1.5095382

Park, J., & Woo, S. E. (2022). Who likes artificial intelligence? personality predictors of attitudes toward Artificial Intelligence. *The Journal of Psychology, 156*(1), 68–94. https://doi.org/10.1080/00223980.2021.2012109

Perenc, I., Jaworski, T., & Duch, P. (2019). Teaching programming using dedicated Arduino educational board. *Computer Applications in Engineering Education, 27*(4), 943–954. https://doi.org/10.1002/cae.22134

Piedade, J., Dorotea, N., Pedro, A., & Matos, J. F. (2020). On teaching programming fundamentals and computational thinking with educational robotics: A didactic experience with pre-service teachers. *Education Sciences, 10*(9), 1–15. https://doi.org/10.3390/educsci10090214

Prima, E., Oktaviani, T., & Sholihin, H. (2018). STEM learning on electricity using Arduino-phet based experiment to improve 8th grade students' STEM literacy. *Journal of Physics: Conference Series, 1013*. https://doi.org/10.1088/1742-6596/1013/1/012030.

Promkaew, N., Thammawiset, S., Srisan, P., Sanitchon, P., Tummawai, T., & Sukpancharoen, S. (2024). Development of metaheuristic algorithms for efficient path planning of autonomous mobile robots in indoor environments. *Results in Engineering, 22*, 1–17. https://doi.org/10.1016/j.rineng.2024.102280

Rengifo, C., & Bravo, D. (2020). A Project-Based learning approach to teach identification and control systems. *IEEE Revista Iberoamericana De Tecnologías Del Aprendizaje, 15*(1), 10–16. https://doi.org/10.1109/rita.2020.2979171

Ribeiro, A. F., & Lopes, G. (2020). Learning robotics: A review. *Current Robotics Reports, 1*(1), 1–11. https://doi.org/10.1007/s43154-020-00002-9

Romero-Rodríguez, J.-M., Alonso-García, S., Marín-Marín, J.-A., & Gómez-García, G. (2020). Considerations on the implications of the internet of things in Spanish universities: The usefulness perceived by professors. *Future Internet, 12*(8), 123. https://doi.org/10.3390/fi12080123

Sengupta, S. (2019). Internet of things: Applications in education sector. *Listed Journal, 8*(1), 24–27. http://eprints.rclis.org/34017/1/Shantashree%20Sengupta%20MMCC%20Paper%202019.pdf

Singh, B., Kumar, R., & Singh, V. (2022). Reinforcement learning in robotic applications: A comprehensive survey. *Artificial Intelligence Review, 55*(2), 945–990. https://doi.org/10.1007/s10462-021-09997-9

Spillmann, L., Otte, T., Hamburger, K., & Magnussen, S. (2006). Perceptual filling-in from the edge of the blind spot. *Vision Research, 46*(25), 4252–4257. https://doi.org/10.1016/j.visres.2006.08.033

Sterling Collazos, J. M. (2019). Articulación entre la educación Superior y la educación rural. Estudio de caso en Zipaquirá, Colombia. *Revista Conrado, 15*(68), 263–266. http://conrado.ucf.edu.cu/index.php/conrado

Stewart, W. H., Baek, Y., Kwid, G., & Taylor, K. (2021). Exploring factors that influence computational thinking skills in elementary students' collaborative robotics. *Journal of Educational Computing Research, 59*(6), 1208–1239. https://doi.org/10.1177/0735633121992479

Ţoţa, P., & Vaida, M. F. (2021). Modular telepresence robot for distance medical education. In *2021 International Conference on e-Health and Bioengineering (EHB)* (pp. 1–4). Iasi: Romania. https://doi.org/10.1109/EHB52898.2021.9657633.

Udvaros, J., Forman, N., & Avornicului, M. S. (2023). Developing computational thinking with microcontrollers in education 4.0. *International Journal of Advanced Natural Sciences and Engineering Research, 7*(4), 103–108. https://doi.org/10.59287/ijanser.563.

Viejo, C., Fuentes, S., Li, G., Collmann, R., Condé, B., & Torrico, D. (2016). Development of a robotic pourer constructed with ubiquitous materials, open hardware and sensors to assess beer foam quality using computer vision and pattern recognition algorithms: RoboBEER. *Food Research International, 89*, 504–513. https://doi.org/10.1016/j.foodres.2016.08.045

Vyas, D. R., Markana, A., & Padhiyar, N. (2022). Economic 6-DOF robotic manipulator hardware design for research and education. In *International Conference on Additive Manufacturing and Advanced Materials (AM2), 62*, 7179–7184. https://doi.org/10.1016/j.matpr.2022.03.140

Wing, J. M. (2006). Computational thinking. *Communications of the ACM, 49*(3), 33–35. https://doi.org/10.1145/1118178.1118215

Wing, J. M. (2017). Computational thinking's influence on research and education for all. *Italian Journal of Educational Technology, 25*(2), 7–14. https://doi.org/10.17471/2499-4324/922.

Woo, H., LeTendre, G. K., Pham-Shouse, T., & Xiong, Y. (2021). The use of social robots in classrooms: A review of field-based studies. *Educational Research Review, 33*, 100388. https://doi.org/10.1016/j.edurev.2021.100388

Yang, G. S., Chen, E. K, & An, C. W. (2004). Mobile robot navigation using neural Q-learning. In *Proceedings of 2004 International Conference on Machine Learning and Cybernetics (IEEE Cat. No.04EX826)]* (48–52), Shanghai: China. https://doi.org/10.1109/ICMLC.2004.1380601.

Yong, S. C, & Yeoh, S. Z. (2023). The design of a new 3D Print-in-place soft Four-Legged robots with Artificial Intelligence. *Journal Kejuruteraan, 35*(3), 717–733. https://doi.org/10.17576/jkukm-2023-35(3)-20.

Social Robots in Special Education and Ethical Considerations

Ethical Considerations in Designing and Testing Robots for Children in Educational Settings

Patrícia Arriaga, Isabel Neto, Filipa Correia, and Antonio Soares

Abstract The integration of social robots into children's lives, whether for educational, rehabilitation, therapeutic, or entertainment purposes, has progressed in recent years, requiring addressing ethical concerns that extend beyond the standard ethical requirements applied to human research. This chapter discussed ethical issues surrounding developing and investigating social robots intended for children in educational settings to ensure their safety, positive development, and well-being. First, we reviewed current approaches to using social robots in educational settings. We discussed the benefits and potential concerns in child-robot interaction (CRI), including issues of attachment and dependency, obedience, reduction of human interaction, and accountability. Secondly, we reviewed the main ethical principles and guidelines, professional codes of conduct, and regulations relevant to CRI. Third, we review and discuss the methodological research aspects implemented for the design of robots and assessing their impact on children by outlining some ethical practices for conducting research. Finally, we explored the implications of integrating social robots into educational settings, focusing on their deployment and long-term use in classrooms, along with crucial aspects of sustainable research practices aiming to help shape future directions in this field.

P. Arriaga (✉) · A. Soares
Cis-Iscte, Iscte-Instituto Universitário de Lisboa, Lisbon, Portugal
e-mail: patricia.arriaga@iscte-iul.pt

A. Soares
e-mail: aasso@iscte-iul.pt

I. Neto
INESC-ID, Instituto Superior Técnico, Universidade de Lisboa, Lisbon, Portugal
e-mail: isabel.neto@tecnico.ulisboa.pt

F. Correia
Interactive Technologies Institute, Instituto Superior Técnico, Universidade de Lisboa, Lisbon, Portugal
e-mail: filipacorreia@tecnico.ulisboa.pt

I. Neto
Lasige, Faculdade de Ciências, Universidade de Lisboa, Lisbon, Portugal

G. Lampropoulos and S. Papadakis (eds.), *Social Robots in Education*, Studies in Computational Intelligence 1194, https://doi.org/10.1007/978-3-031-82915-4_8

Keywords Child-Robot Interaction (CRI) · Education · Ethics · Social robots

1 Introduction

Social robots are increasingly being introduced into educational settings, offering innovative ways to engage and support children. Current approaches to using social robots in education usually involve their use as tutors, peer learners, and classroom assistants (Belpaeme et al., 2018). These robots are currently being used to support several educational activities (Anwar et al., 2019; Özgür et al., 2017), such as language learning (Asselborn et al., 2018; Huang & Moore, 2023), science, technology, engineering, and mathematics (STEM) education (Rocha et al., 2023), physical activities (Costa et al., 2015; Litoiu & Scassellati, 2015), social skills development (Neto et al., 2023), and inclusive education (Berrezueta-Guzman et al., 2021; Neto et al., 2020; Rocha et al., 2023). One innovative approach within the field of language education is Robot-Assisted Language Learning (RALL), which incorporates Artificial Intelligence (AI) to facilitate and improve language learning, offering practice opportunities, real-time feedback, and tailored instructions to meet student's needs (Huang & Moore, 2023; Konijn et al., 2022; H. Lee & Lee, 2022).

Some commercial robots have also been developed to support children's learning and play activities. Often designed to enhance STEM and language skills, these robots vary in complexity and target different age groups. Notable examples include Dash and Dot (Workshop, 2024), LEGO Mindstorms (Lego, 2024), Go and IQ (Robotics, 2024), (Sphero, 2024), Ari and Evo (Ozobot, 2024), (KUBO, 2024), Cubelets Robot Blocks (Modrobotics, 2024), Cozmo and Vector (Anki Cozmo Robot, 2024), (Thymio, 2024), and (EMYS, 2024).

Additionally, extensive literature from the research community presents a variety of early-stage prototypes developed to explore and evaluate the broader impact of these robots on children. Researchers working in Child-Robot Interaction (CRI) have also investigated the effects of different robotic social capabilities, including humanoid embodiments, to explore their social abilities and affordances, specifically how children perceive them and how they affect their behavior. For instance, a robot that listens to children's stories and exhibits backchanneling behaviors based on its understanding of the children's engagement can effectively demonstrate attention (Lee et al., 2019). Similarly, researchers have shown that when robots use memory-based personalization, such as referring to previously mentioned preferences of the children, these behaviors facilitate sustainable long-term interactions with children (Ligthart et al., 2022), even after long breaks (Ligthart et al., 2024).

However, there is a lack of systematic evaluations aiming at understanding the long-term effects of CRI in school environments and its potential impact on children's socio-emotional development (Langer et al., 2023). For example, in the context of school inclusion, it remains unclear whether the positive effects observed during a single interaction with a robot, where the robot facilitated balanced interactions between children with and without impairments, persist beyond the session (Neto et al., 2023). Additionally, the potential negative effect of fostering children's dependency on the robot's instructions raises ethical concerns. Other concerns are the difficulties in replicating studies (Spitale et al., 2023), often due to the lack of pre-registration, insufficient details about the design, and the failure to share coding and study data. Furthermore, the lack of standardized outcome measurement protocols makes it challenging to ensure consistency and reliability across different research contexts (Bethel & Murphy, 2010).

This chapter discusses the ethical challenges and concerns related to the design, development, and evaluation of social robots intended to interact with children in educational settings, focusing on ensuring their well-being, socioemotional development, and safety. It is structured as follows: Sect. 2 reviews the potential benefits of integrating social robots into educational settings, but with a primary focus on the ethical concerns related to CRI. To understand how these concerns have been addressed, Sect. 3 reviews the ethical principles, guidelines, professional codes of conduct, policies, and current regulations governing the responsible use of social robots. Section 4 examines the challenges of conducting ethical research in CRI, providing recommendations for best practices in ethical compliance, highlighting the importance of designing robotics tailored to children's needs with a focus on co-design and other methods for assessing the impact of CRI. Section 5 explores the roles and responsibilities of robot developers and Sect. 6 addresses the broader challenges and implications associated with deploying social robots in educational settings, as well as considerations for long-term sustainability. Finally, Sect. 7 provides concluding remarks, summarizing some of the key ethical concerns and challenges discussed in the chapter and suggestions for future research.

2 Main Benefits and Ethical Concerns in Child-Robot Interaction

Recent systematic reviews and meta-analyses have shown the positive effects of educational and social robots on child development, each identifying distinct types of intervention and contextual factors. Overall, social and educational robots seem to affect knowledge, skills, and attitudes positively (Sapounidis et al., 2024), including specific cognitive outcomes, such as creativity and problem-solving (Mukhasheva et al., 2023; Zhang & Zhu, 2022), several learning outcomes (Wang & Cheung, 2024; Wang et al., 2023), and trust (Stower et al., 2021). The format of these interactions, especially one-on-one versus group settings, seems to be more efficient in some learning outcomes (Lee & Lee, 2022). Although the overall effectiveness of social and educational robots is positive, their impacts also vary depending on several factors, some related to the child's characteristics (e.g., age, developmental level, gender), others related to the robotic systems, emphasizing the need for careful instructional design (Su et al., 2023; Wang et al., 2023).

Several authors underscore that one key factor behind the success of robots in facilitating learning is their embodiment and physical presence because they may capture and maintain students' attention more effectively, in contrast to traditional digital learning tools (Kennedy et al., 2015). However, the human-like behaviors of robots do not consistently seem to enhance trust or liking, which challenges the assumption that these characteristics are inherently more beneficial for all outcomes (Stower et al., 2021). Nevertheless, many of these robots are designed to interact with humans socially and are often equipped with capabilities to engage students through interactive and personal learning experiences. Sometimes, they also integrate Artificial Intelligence (AI) to adapt to users' personal preferences or needs (Huang & Moore, 2023; Konijn et al., 2022).

Beyond the classrooms, schools offer children a rich environment for group playing activities. Social play is fundamental for children's interpersonal skills and to form friendships (Whitebread et al., 2019). As a result, the potential impact of using robots in places such as the playground has been studied, including as a toy for role-play to foster children's creativity (Alves-Oliveira, Arriaga, et al., 2021a, 2021b; Zhang & Zhu, 2022), a mediator of group conversations to reduce imbalance participation (Gillet et al., 2020), a companion to encourage children's collaborative behaviors (Belpaeme et al., 2018; Strohkorb et al., 2016), or a mediator to facilitate child-adult relationship (Di Dio et al., 2020). Systematic and meta-analysis reviews have also shown that robot-assisted interventions have been effective in enhancing social functioning in children with particular conditions, such as those with autism spectrum disorder, with age being an important factor in explaining the variance of outcomes (Damianidou et al., 2020; Kouroupa et al., 2022).

Collectively, prior research and recent meta-analytic findings demonstrate the benefits of educational robots and CRI, although their deployment must consider their developmental stage, age, specific needs, and contexts (e.g., Anwar et al., 2019; Kouroupa et al., 2022; Lee & Lee, 2022; Mukhasheva et al., 2023; Papakostas et al., 2021; Sapounidis et al., 2024; Wang & Cheung, 2024; Wang et al., 2023; Zhang & Zhu, 2022).

Various authors also emphasize ethical concerns related to the design and integration of robots in children's lives, which should involve a wide range of stakeholders, including educators, parents, psychologists, robot designers, and policymakers.

One main concern is the need for a delicate balance in CRI. Although robots should be engaging to maintain the children's interest and keep them motivated in their learning activities (Neumann et al., 2023), it can also become a distraction if a robot is overly social or interactive, pulling the children's focus away from their educational tasks. Additionally, it may make them feel uncomfortable or overwhelmed by the robot's presence (Belpaeme et al., 2018; van den Berghe et al., 2019). Therefore, designers and educators must carefully calibrate robots' social behaviors to ensure they effectively engage without being distracting or intimidating.

Another concern is children's over-reliance on robotic guidance, which might inhibit the development of essential skills such as autonomous learning, critical thinking, and decision-making (Boada et al., 2021). On the other hand, if the robot does not meet children's expectations, it can cause frustration (Belpaeme et al., 2013).

Additionally, children's emotional bond to a robot might evolve into emotional dependence, where users become overly reliant on the robot for emotional or social support (Boada et al., 2021), which in turn may affect their social interactions with peers and teachers (Neto et al., 2023). It is also possible for social robots to disrupt the traditional teacher-student dynamic, where teachers are viewed as experts and leaders (Serholt et al., 2017). To minimize these risks, it is essential to set boundaries for their use in educational settings. Robots should complement, rather than diminish or replace, interactions among students, teachers, and other people. Since robots do not experience emotions, are incapable of relating in a genuine way, and lack real-life experiences (Boada et al., 2021), their role should also primarily support the development of cognitive skills and, to a lesser extent, social skills to prevent the dehumanization of education (Tarrés-Puertas et al., 2023), and ensuring that education remains focused on human-centric values and meets the developmental needs of students (Boada et al., 2021). Recently, Contro and Brandão (2024) proposed the "Interaction Minimalism" design philosophy to guide the ethically responsible development of social robots. This philosophy seeks to ensure that robots are employed in a functional and non-intrusive way, advocating for the use of robots as tools to maintain their functional benefits without becoming central to emotional fulfillment or social interaction. In educational settings, this would mean the use of robots to facilitate students' activities and promote collaborative learning among students rather than allowing technology to dominate the learning and social experiences, which might compromise children's socioemotional development.

The design of social robots that can subtly affect human behavior, a concept known as "nudging," is also controversial (Mehenni et al., 2021). This controversy arises from the need to ensure that users are fully informed about how robots might influence their decisions or behaviors and that they are given control over these interactions, respecting their autonomy and preventing manipulation (Torras, 2024). This approach underscores the need for ethical design in robotics, emphasizing transparency, user consent, and agency as fundamental components (Ali Mehenni et al., 2021; Torras, 2024; van Straten et al., 2023). For instance, one study explored the impact on children of observing robots that excluded another child during a ball-tossing game, similar to the "Cyberball paradigm" (Correia et al., 2024). This classic paradigm has also been employed in developmental science to study the effects of ostracism on children (Scheithauer et al., 2013). The results by Correia et al. (2024) revealed that children were able to recognize the robots as excluders and that this exposure had a negative impact on their basic needs of belonging and control. However, these children were also more prosocial in a follow-up task. Based on these results, the authors discussed two important concerns. First, the finding that children expressed a desire to interact with the same robots they had considered excluders suggests that robots can maintain children's engagement even after negative experiences, highlighting the need for careful robotic design. Secondly, they discussed how interactions with robots might trigger carryover effects on children's subsequent social interactions with other children, underscoring the importance of monitoring and supervising CRI to prevent any negative influence on their social behaviors.

Research has also shown that robots can exacerbate inequality of access among children, posing challenges at multiple levels of accessibility which may lead to increased stigma and exclusion (Su et al., 2023). The high costs associated with robotic technologies may prevent economically disadvantaged children from accessing them, amplifying social exclusion across age, educational, economic, and geographical levels. This disparity may occur in affluent countries, where peers might have access to robots, and in less economically developed nations, where such technology might be unavailable. Furthermore, situations where individuals have more access to services through robots than human interaction can also create exclusion (Torras, 2024). One example includes automated teaching, where schools might consider using robots as instructors for several activities, potentially depriving them of personalized human teaching. For example, in the case of children with disabilities, there is a risk of entrusting care activities to robots, which may further isolate them from human interactions. Therefore, it is crucial to consider how these social and economic factors might raise ethical concerns (López-Sintas et al., 2012; Tarrés-Puertas et al., 2023). To mitigate some of these disparities, Tarrés-Puertas et al. (2023) developed activities in which robots were confined to school settings to ensure equal access. To address the possible lack of competencies related to the use of technological tools, students also received training about the materials used by the robots. Moreover, teachers tried to ensure all students were adequately trained in programming the robots, striving to ensure equitable educational experiences.

Other researchers have highlighted the possible negative effect on social equality regarding the quality of assistance, which depends on how they are implemented. For instance, if social robots do not have information in their database about specific characteristics of children, it will be challenging to provide similar assistance to all users (Boada et al., 2021). In this way, if vulnerable minorities do not have sufficient data representation, they may not benefit equally from the services that robots offer, such as expressive or facial recognition tailored to cultural minorities (Torras, 2024). Although the use of social robots designed to assist vulnerable groups could combat prejudice and exclusion, they might also inadvertently impact their well-being if they reinforce stereotypes associated with certain disabilities for which the robot provides assistance. Thus, the design of a social robot can also be a source of stigmatization, individual or collective, by reflecting the stereotypes that developers hold about end users (Boada et al., 2021). Sexist, ableist, or racist programming and design must not exist in robots.

To overcome these concerns, some researchers emphasize the need to actively challenge stereotypes in technology (Su et al., 2023; Tarrés-Puertas et al., 2023). Based on a systematic review of the role of socioeconomic status and gender on children's use of robotics in early childhood education, Su et al. (2023) underscored the need to develop inclusive robotics education programs to mitigate educational inequalities. Key recommendations include more effective engagement of girls and children from less advantaged backgrounds in robotics activities, the pivotal role of female teachers as role models in robotics education, and the need to develop, implement, and evaluate age-appropriate robotics activities to reduce these gaps and foster interest and skills among all children. Similarly, Tarrés-Puertas et al.

(2023) advocated challenging the perception of technology as a male-dominated field through class discussions about stereotypes in programming and AI. Additionally, they highlighted the need for teaching algorithmic awareness, which involves educating students on how algorithms can unintentionally perpetuate biases.

In addition to lacking genuine human feelings, robots typically lack several assistive tools, limiting interaction opportunities for children with impairments, whether visible or not. However, research addressing whether robots distribute classroom resources in a biased or unfair manner has not been fully investigated, nor have the broader, long-term consequences of such systematic disparities, resulting in a critical need to also investigate these ethical concerns (Neto et al., 2023).

Another concern is related to the issues of responsibility, which are relevant when a robot fails to assist children adequately. It is critical to make sure that children and educators are safe during interactions (Tolksdorf et al., 2021). These may include situations where the robot may fall, scare, or even harm a child. Given the highly unstructured nature of many activities (e.g., free play), programming a robot to account for all possible scenarios might be extremely challenging. Additionally, educators often lack detailed knowledge about the robot's functions and movements, making it difficult to intervene if something goes wrong. Since robots cannot legally bear responsibility, it is crucial to determine who is responsible—developers, manufacturers, owners, researchers, or users—for its actions, and there is a need for clear guidelines on roles and responsibilities concerning robot use in educational settings (Serholt et al., 2017). Currently, social robots lack reflective abilities to assess the correctness and success of their actions. Therefore, robots should only be used in these settings under the supervision of human experts (Tolksdorf et al., 2021). Moreover, robots operate autonomously only in very restricted contexts, and achieving full autonomous behavior in unstructured environments is still beyond current technological capabilities. However, robots are expected to act more autonomously in the future, which could lead to accidents or violations of privacy and autonomy (Boada et al., 2021). Therefore, a robot's decision-making process must always align with human values and ethical considerations, addressing dilemmas such as balancing security and privacy. To manage these complexities, robots need to be equipped with skills to handle ethically sensitive situations effectively (Torras, 2024). To ensure that the integration of social robots into educational and other settings is beneficial and aligns with educational goals and ethical standards, it will be important to establish clear protocols based on transparency, traceability, and accountability (Boada et al., 2021; van Straten et al., 2023).

Relatedly, social robots might also pose a threat to human privacy due to their capacity for monitoring, collecting, and processing personal data. These privacy issues in school settings apply to children, educators, and parents (Tolksdorf et al., 2021). Social robots can collect different information from users, including images, voice, location, and learning progress, which may pose risks of misuse (Ahtinen et al., 2023). Robots supporting learning activities tend to collect and analyze this data, which can be valuable to understanding children's engagement and skills development, but it may also raise privacy concerns. Moreover, children and their guardians might not be fully aware or have not consented to the storage and use of such detailed interaction data in the robot's memory. Additionally, the emotional bonds users may form with social robots can lead them to share more personal information, exacerbating privacy risks. This issue also relates to a lack of transparency in how social robots function, potentially misleading users about the robots' capabilities and the use of their data (Serholt et al., 2017; Torras, 2024; van Straten et al., 2023). Thus, it is important to detail who will evaluate this data and the pedagogical or therapeutic concepts behind it. Although such data can help educators tailor activities to a child's progress, relying too heavily on this data might lead to a narrow assessment (Tolksdorf et al., 2021).

Based on the various concerns outlined above, it becomes relevant to address how they can be managed and mitigated. Ethical guidelines, professional codes of conduct, and regulations are essential to ensure that the integration of robots into educational contexts is conducted responsibly and ethically.

3 Ethical Principles and Guidelines, Codes of Conduct, and Regulations Relevant to Child-Robot Interaction

Considering the intricate relationship between robots and AI, many of the well-established ethical principles and guidelines for AI also apply to robotics. The concept of "ethical AI" must align with fundamental human rights and adhere to professional ethics and codes of conduct (Tractenberg, 2024). Additionally, it is crucial to consider the laws and regulations on ethical issues that apply in each country or state, such as those regulating data protection and the conditions for developing and deploying AI systems that safeguard human rights.

3.1 Aligning Robots and Artificial Intelligent Technologies with Human Rights

The foundational Universal Declaration of Human Rights (United Nations. General Assembly, 1949), established by the United Nations (UN), sets a global standard for dignity and equality. Similarly, the European Convention on Human Rights (ECHR) within EU member states underscores the need for the protection of human rights and fundamental freedoms, with a strong emphasis on privacy and education rights. Building on these principles, the United Nations Guiding Principles on Business and Human Rights (UNGPs) present 31 principles designed to implement the UN's "Protect, Respect, and Remedy" framework. These principles ensure that business practices, including AI deployments, uphold human rights standards, emphasizing the responsibilities of transnational corporations and other business enterprises.

In research ethics, several key guidelines and codes have been proposed, including the Nuremberg Code (1947/1996), the National Commission for the Protection of

Human Subjects of Biomedical & Behavioral Research (1979), and the WMA Declaration of Helsinki (World Medical Association, 2013). The Nuremberg Code emphasizes voluntary consent, minimizing suffering, a favorable risk–benefit ratio, and strong protective measures. The Belmont Report develops these ideas by outlining three fundamental ethical principles for human subject research—respect for persons, beneficence, and justice—and their three main applications: informed consent, risk and benefit assessment, and participant selection, all aimed at protecting human rights and well-being. The Declaration of Helsinki builds on these principles and extends them, including outlining requirements for vulnerable groups and individuals to safeguard the welfare of participants with clinical conditions.

The commitment to fundamental human rights, ethical principles, and their implications for AI technologies is further emphasized by the Toronto Declaration (Amnesty International & Access Now, 2018), which specifically addresses the impact of machine learning on human rights (Tractenberg, 2024). Moving to AI-specific guidelines that impact the development of autonomous robots, some frameworks, such as the Montréal Declaration for a Responsible Development of Artificial Intelligence (Montréal University, 2018), set forth critical principles. These include ensuring that AI systems follow principles of well-being and autonomy, protecting privacy and intimacy, solidarity, democratic participation, equity, diversity, and inclusion. The Declaration also emphasizes prudence by anticipating the AI's potential adverse effects, underscores that only humans can be held responsible for the AI systems, and advocates for sustainable development. In Europe, the Ethical Guidelines for Trustworthy AI also reflect a commitment to ensuring that AI systems are developed and deployed in a manner that respects EU values and fundamental rights.

Additionally, it is crucial to consider specific guidelines and principles for vulnerable populations, such as children and those with special needs. Among these, the Convention on the Rights of the Child, adopted by the United Nations in 1989 (UNICEF, 2007), outlines the fundamental rights of children to ensure that children's rights and well-being are always prioritized, including in research, such as the right to express their views, protection of privacy, and safeguarding from abuse. Additionally, UNESCO's Guidelines for Policy-makers on AI and Education (Miao et al., 2021) provide several ethical considerations for using AI in educational settings, with the aim of making these technologies inclusive and equitable. These considerations include using "smart" robots to promote student learning, communication, and social skills, as well as to empower teachers.

In reviewing AI ethics policy documents with guidelines tailored explicitly to K-12 education, Adams et al. (2023) specifically selected documents from globally influential organizations, such as the Institute for Ethical AI in Education (IEAIED, 2021), which released a report and a AI framework offering guidance on how schools can responsibly acquire and use AI-based resources to enhance teaching and learning outcomes. By examining several guidelines in promoting the ethical development and implementation of AI technologies in educational settings, they identified eleven core AI ethics principles derived through content analysis. These principles included

Children's Rights, Transparency, Justice and Fairness, Non-maleficence, Responsibility, Privacy, Beneficence, Freedom and Autonomy, Pedagogical Appropriateness, AI Literacy, and Teacher Well-being.

Children's rights principles include protecting and prioritizing the needs and interests of young learners, which can be achieved by designing robots that respect and support the rights and needs of children. Transparency aims to ensure that AI systems used in education are understandable and provide clear explanations for their decisions and actions to both children and educators, which applied to robots can be achieved by clearly explaining their functions and decision-making processes. Justice and Fairness seek to guarantee that these AI systems are unbiased, treating all children equitably and avoiding discrimination, which can be integrated into robots by programming them to treat all users fairly and without bias. Non-maleficence focuses on preventing harm and ensuring the safety of children interacting with AI systems, which can be upheld by ensuring that robots do not harm users physically, emotionally, or socially. Responsibility involves holding individuals and organizations accountable for the development and use of AI systems in educational settings, reflecting the need for ethical considerations in the design and actions of robots used by children. Privacy is concerned with safeguarding personal data and information of children, ensuring compliance with data protection laws, which must be safeguarded in robots by implementing measures to protect user data and confidentiality. Beneficence is required to promote the well-being and best interests of children through the appropriate use of AI systems, which can be achieved by designing robots that enhance children's educational experiences and well-being. Freedom and Autonomy uphold children's rights to make autonomous decisions, allowing them to control their choices, which can be respected by giving children control over their interactions with robots. Pedagogical Appropriateness aims to ensure that AI applications are suitable for educational purposes, aligning with best teaching practices to support children's learning, which can be ensured by designing robots that are educationally relevant and effective. AI Literacy seeks to promote an understanding of AI technologies among children, helping them to engage with these tools responsibly and knowledgeably, which can be supported by using robots to teach children about AI in an accessible way. Lastly, Teacher Well-being highlights the importance of supporting educators in adapting to and integrating AI tools, ensuring they can effectively aid children's education while also maintaining their professional and personal well-being, which can be facilitated by designing robots that are easy for teachers to use and that enhance their teaching experience.

3.2 Professional Ethics and Codes of Conduct

Various professional organizations have also established ethical guidelines to ensure responsible practices and codes of professional conduct. Some of these are particularly relevant for children, such as the Code of Ethics for Educators proposed by the

National Education Association (NEA, 1975), and the Ethical Principles of Psychologists and Code of Conduct proposed in 1953 and revised in 2017 by the American Psychological Association (APA, 2017).

For technological professions, Tractenberg (2024) underscored the ACM Code of Ethics and Professional Conduct (ACM, 2018), which covers all the work of computing professionals regarding computing development, deployment, and usage, emphasizing their responsibility to contribute to society and well-being, avoid harm in their practices, ensure that technology promotes human rights and environmental sustainability, and the requirement of being honest and transparent by disclosing information of system capabilities and potential problems. The Code also stresses values of justice, equality, and fairness, urging professionals to foster tolerance and non-discrimination, and other responsibilities such as respecting privacy, honoring confidentiality, and other computing practices for the public good, such as their involvement in projects that help society. In addition, we highlight the IEEE Ethically Aligned Design guidelines introduced in 2016 (IEEE, 2016), aimed at promoting ethics in designing and deploying AI systems to guarantee that these systems are transparent, accountable, and respect users' privacy.

Professional codes of conduct are also crucial for guiding general professional behavior and ensuring integrity in research practices. While these codes provide a broad ethical framework for various professions, there are specific guidelines that focus directly on research integrity. In this context, we highlight some codes of conduct that specify research ethics. The Code of Conduct for Responsible Research by the World Health Organization (WHO, 2017) emphasizes principles of integrity, accountability, impartiality, respect, and professional commitment. It applies to all WHO staff and collaborators, ensuring adherence to research standards and promoting transparency and ethical behavior in all research activities. The European Code of Conduct for Research Integrity (ALLEA, 2023) offers a framework for good research practices within Europe by stressing the importance of reliability, honesty, respect, and accountability in scientific research. Both codes are relevant for maintaining high ethical standards and integrity in research, providing guidelines to prevent misconduct and ensure the credibility of scientific findings. Relatedly, it is also relevant to consider the codes of professionals working in the field of statistical practice (Tractenberg, 2024), such as the Ethical Guidelines for Statistical Practice developed by the American Statistical Association (ASA, 2022), which provides a framework for ethical behavior and decision-making when working with statistics and data, including data collection, analysis, interpretation, and model development.

3.3 Policies and Regulations for Ethical Use of Robots

To ensure that robotic systems are equitable, inclusive, and used ethically, we should also consider policies, regulations, and protective laws (Miao et al., 2021). Many regulatory and institutional guidelines have been developed concerning data privacy and protection, with several countries proposing their regulations. While many share common principles, such as the protection of personal data, the rights of data subjects, and the obligations of data controllers, these regulations also have unique provisions and requirements specific to their respective legal and cultural contexts. Thus, it is important to comply with the relevant data protection laws in each country or jurisdiction where they operate. For example, the United States has no general federal data privacy regulation, although a federal American Data Privacy and Protection Act (ADPPA) proposal has been introduced recently to the U.S. House of Representatives (2022). Nevertheless, several sectoral laws apply at the federal level, and among US state laws, California, Virginia, and Colorado have enacted their own data privacy laws.

In the European Union, the General Data Protection Regulation (GDPR) came into effect in 2018 and is currently applied to all EU member states (GDPR.EU, 2022). The GDPR has stringent rules, particularly emphasizing the rights of individuals, including children, in relation to their personal data. It mandates lawful, fair, and transparent processing of personal data. After the GDPR, other countries have adopted similar regulations, including Brazil with the Lei Geral de Proteção de Dados Pessoais (LGPD, 2019), which came into effect in 2020, and the Personal Information Protection Law in China, which came into force in 2021 (PILP, 2021). For an overview and updates regarding global privacy laws, see the OneTrust Data Guidance website (OneTrust, 2024).

In addressing regulations of AI technologies and managing the associated risks while promoting conditions conducive to their growth and ethical deployment across several sectors, the European Union recently introduced the European Union Artificial Intelligence Act (EU AI Act). The Regulation 2024/1689 was recently published on July 12, 2024 (European Parliament & Council of the European Union, 2024), it came into force twenty days after publication, and is set to be fully implemented by July 12, 2026.

The EU AI Act emphasizes the need for AI systems to be safe, transparent, traceable, non-discriminatory, and environmentally sustainable, advocating for human supervision over these systems. It underscores the protection of all individuals, supporting a broad spectrum of rights, including those of workers, persons with disabilities, and children, and promoting gender equality. It specifically highlights the critical importance of considering children's vulnerabilities, ensuring their protection and well-being, and the need for education, the protection of privacy, and personal data. The AI Act classifies AI risks into three categories based on a risk-based approach. The "Unacceptable Risks" will be prohibited, including AI systems that manipulate behavior, exploit vulnerabilities, perform broad social scoring, and conduct indiscriminate surveillance by engaging in biometric categorization. The "High-Risk" category, requiring rigorous pre-market assessments and continuous oversight, includes AI systems likely to impact health, safety, and fundamental rights, such as children's rights, and are those used in critical sectors like education and healthcare or involving sensitive data processing. High-risk applications affecting children, which may apply to education and vocational training, must adhere to stringent regulations and are required to be registered in an EU-wide database. These systems also fall under the EU's product safety legislation, potentially encompassing a variety of products, including toys and robots designed for children, if they are considered to impact health, safety, or fundamental rights significantly. The "Limited or Minimal Risk" category applies to AI systems that pose lesser concerns and thus require fewer regulations. These include AI applications with limited interaction with sensitive data or decision-making processes, such as AI-enhanced software tools that do not affect public safety or personal rights. Additionally, the regulation sets transparency requirements to ensure users are aware when interacting with AI to reduce risks of deception or manipulation.

This regulation may have legal implications for using robots in school settings, ensuring that any AI systems adhere to strict safety and privacy standards. While the Act specifically applies to EU member countries, its approach to AI governance might influence other countries to adopt similar regulations, which may set a precedent for the responsible use of AI, including in child-centric environments.

Overall, the aforementioned principles, guidelines, codes of conduct, and regulations have a crucial impact on the existing and future development of robots to ensure their ethical introduction in educational settings. By adhering to these requirements, educational institutions can use robotic systems that are compliant with high ethical standards, including safety and privacy, tailored to foster an environment conducive to children's learning and socioemotional development. Figure 1 depicts the central role of these standards in all different stages of research on CRI, namely the design, development, and impact evaluation of educational robotic devices for children. In the following section, we review the different stages of conducting research and their practical implications.

4 Challenges in Conducting Ethical Research in Child-Robot Interaction

When designing robots, the focus in both industry and academia does not always prioritize the benefits for children. Industry often prioritizes long-term profit, while academia tends to concentrate on available robots, aims for novelty in research, and frequently conducts short-term studies, which may limit the use of latest models, a thorough evaluation of whether previous findings are robust, and overlook their long-term effects for children. However, the design for CRI must prioritize the children's needs and ensure that designers, programmers, and researchers adhere to codes of ethics and conduct, avoid bias, and address any potential drawbacks for children.

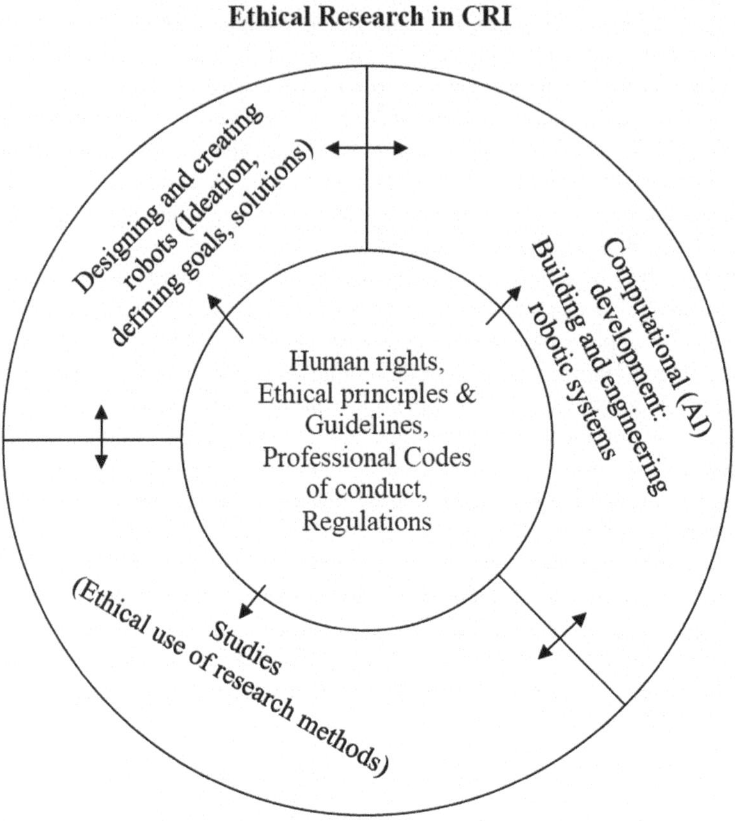

Fig. 1 Ethical research in CRI

4.1 Best Practices for Ethical and Legal Compliance in Studies Involving Children

Considering the ethical guidelines and codes of conduct in research already mentioned, obtaining informed consent is mandatory. In the case of children, this requires obtaining consent from their legal guardians or parents to ensure that they are fully informed and agree to the specific activities involving robots or other technological interventions in educational settings. The informed consent should be signed by the parents or legal guardians and the researcher. Table 1 details the essential components of informed consent for research involving CRI. It emphasizes the need to clearly explain the research's purpose, obtain ethical approval from review boards and school authorities, and maintain transparency about all procedures, including the technologies used. It should also inform about the safety measures to protect children, the benefits and risks of the research, and strategies to mitigate these risks. It underscores voluntary participation to uphold autonomy, outlines data privacy and confidentiality measures, and the participants' rights regarding their personal data. Additionally, including debriefing sessions and accessible contact information enhances transparency and accountability throughout the research process.

Different countries have specific guidelines and regulations concerning the age at which children's consent must be given by legal guardians (e.g., FRA, 2014). For example, in the US and in many European countries (e.g., Portugal, France, Germany), individuals under the age of 18 are generally considered minors, and parental consent is required, however, the specifics can vary depending on the nature of the study and the perceived ability of adolescents to make informed decisions. In the UK, children under 16 typically need parental consent, but adolescents may be capable of giving "Gillick competent" consent if they show sufficient maturity and understanding of the research. These variations reflect an understanding that older adolescents may be able to make informed decisions about their participation in research based on their comprehension of the study's implications. Researchers must ensure they fully comply with international standards and local regulations when involving children in their studies (ERIC, 2024).

Nevertheless, respecting a child's autonomy and ensuring they feel respected and heard is important. Consequently, it is common practice in many countries for children to provide assent to participate in research, even if they cannot legally give informed consent. Researchers must provide detailed information that is suitable for the age group. However, potential issues may arise when researchers are introduced to children within a school setting. Children may see researchers as authority figures similar to their actual teachers, which can lead them to feel obliged to comply with their requests, believing participation is mandatory (Davies, 2008). Thus, it is crucial that children fully understand all the procedures, including that their involvement in the research is voluntary and that they have the right to withdraw at any time. Researchers can use verbal explanations, visual aids, teach-back techniques, or hypothetical scenarios to explain and assess a child's understanding and reasoning regarding the research. According to the Community on Ethical Research Involving Children (ERIC, 2024), seeking permission from children and young people should be included in the research process as it signals respect for their dignity, capability, and rights to express their views and will. The ERIC (2024) emphasizes that all children can indicate assent or dissent if researchers try to communicate and provide information appropriate to their capabilities using verbal explanations, written materials, or visual aids. Depending on their age and cognitive development, some children may not fully understand all the complexities of the research, but they should express their willingness or unwillingness to participate. Children can show their assent using methods appropriate to their communication abilities, such as circling an emoji or ticking a box. With preschool children and children with communication difficulties, assent can be communicated through non-verbal cues such as body language, facial expressions, and gestures. Researchers must be sensitive in recognizing these forms of communication and respond appropriately in the moment. The European guidelines (European Commission & Directorate-General for Research and Innovation, 2013), for example, suggest that information for children under five should mainly use pictures. For pre-adolescents up to 16 years old, the information sheets should

Table 1 Key components of informed consent for child participation in research

Component	Explanation
Purpose of the research	Explanation of why the child is invited to participate, specifying the aims of interactions the child will have with the robots
Ethical approval and school authority	The indication that the study has been reviewed and approved by an ethical committee or institutional review board (IRB), ensuring it meets ethical standards to protect participants' rights, safety, and well-being. It should also confirm school authority approval and detail the school's role and any staff collaboration
Procedures and context	Detailed description of what will happen during the study, including when and where within the school it will be conducted, information about the robot (e.g., appearance, functions), technological features used (e.g., sensors, cameras), type of data collected (e.g., audio, video, behavioral), number of sessions, duration of each session; duration and frequency of CRI, whether teachers or school staff will be involved; how parents will be involved or informed about their child's interactions with the robots. If feasible and appropriate, the consent could also indicate opportunities for parents to observe the child's interactions with the robots
Safety measures	Information about safety measures used to protect the child during CRI and how the child's safety will be monitored and ensured
Potential benefits	Explanation of the potential benefits to the child or to others resulting from the child's participation (e.g., scientific knowledge, learning experiences)
Potential risks	Information about any potential risks or discomforts associated with participation (e.g., discomfort, attachment issues) and measures to mitigate these risks. It should include support measures or the resources available if the child experiences distress. It should also inform how the study will fit into the school day and any potential disruption to the child's regular activities. It should also ensure that participation will not negatively impact the child's grades or academic progress
Voluntary participation	Statement that participation is voluntary and that the child and parents can withdraw consent at any time without penalty or loss of benefits
Right to Inquire freely	Explanation that participants are entitled to ask questions at any stage to ensure they can seek clarifications, express concerns, and better understand the context or procedures involved
Privacy and confidentiality	Explanation of how the child's privacy will be protected, how the data will be stored, who will have access to it, how it will be used, and if any data will be used for purposes beyond the current study. Ensure that school personnel not involved in the study will not have access to identifiable data (see the additional components below on "Data Privacy Details if Personal Data is collected")

(continued)

Table 1 (continued)

Component	Explanation
Compensation	Details on any compensation for the child and their parents, type of compensation, and when and how it will be provided. Clarifies that compensation is for time and effort, not dependent on the study's risks or outcome, and should be fair and never coercive. If the study involves more than one session, provide partial compensation after each session rather than waiting until the end of the study. Types of compensation must be appropriate for the parents' socioeconomic status and the child's age. For the child, it often includes educational materials (e.g., books, educational games, school supplies), certificates or tokens of appreciation (usually for younger children), or eventually gift cards, vouchers, or modest amount of money (for adolescents). For parents, it may cover costs incurred (e.g., travel expenses, parking fees, meals) or a modest monetary compensation. ERIC (2024) advises avoiding payments if they could exert pressure, to ensure that informed consent remains entirely voluntary
Child's assent	Information about the process for obtaining the child's assent
Debriefing	Information about debriefing session(s) that will be held to explain the study's findings, the child's role, and any follow-up procedures
Contact information	Contact details about the researchers, ethical committee, and school representative for any questions or concerns about the study's ethical aspects or the school's involvement
Parental consent agreement	Inclusion of a section where parents agree to conditions through physical or digital means: a space for both the parent's and researcher's signatures and dates; for online forms, a checkbox to express parental consent, followed by an acknowledgment button to confirm their agreement. A confirmation message ensures that consent has been successfully recorded upon submission
Data privacy details if personal data is collected	
Types of personal data collected	Type of personal data that will be collected from the child (e.g., name, age, contact information, health data, behavioral data)
Purpose of data collection	Purpose of personal data collection and how it will contribute to the study goals
Data storage and security	Details on how the personal data will be stored securely (e.g., encrypted databases, secure servers), and the measures in place to protect the data from unauthorized access, breaches, or loss
Data access and sharing	Specifies who will have access to the personal data (e.g., researchers, data analysts; third parties) and under what conditions, with what protections in place
Anonymization and De-identification	Explains the steps to anonymize or de-identify the data to protect the child's identity

(continued)

Table 1 (continued)

Component	Explanation
Data retention and deletion	States how long the personal data will be retained, the process for securely deleting or destroying it once it is no longer needed
Rights regarding personal data	Information on the rights regarding the personal data, including the right to access, correct, or delete their child's data, and explanation on how these rights can be exercised and whom to contact for such requests by adding contact information about the data protection officer or authority overseeing compliance

explain the study's purpose and background and note that parents will also consent. Adolescents aged 16 to 18 who are legally adults, or "emancipated minors", must provide their written consent. Additionally, in longitudinal research, it is required to regularly update the child with age-appropriate information to ensure their ongoing agreement with using their data over time.

Research in CRI also presents unique ethical challenges, including balancing researcher interference, avoiding stigmatization, and managing adverse carry-over effects (Correia et al., 2024; Piedade et al., 2024a). Balancing researcher interference is particularly challenging. For instance, during the observation of natural peer interactions, conflicts can arise. Researchers must handle these situations carefully to avoid disrupting the research, undermining the teacher's authority, or missing opportunities to foster empathy and inclusion. Avoiding stigmatization is also a challenge, especially in mixed-ability classrooms and with minority groups. Researchers aim for seamless interaction among all children to promote inclusion, but some may require specific adaptations, such as braille for blind children, translation for non-native speakers, or additional support for those with Attention-Deficit/Hyperactivity Disorder (ADHD) (Papakostas et al., 2021). Researchers must adapt activities to support these children with sensitivity and thoughtful planning to ensure all children feel included and respected. Additionally, the carry-over effect of children's interactions cannot be ignored, as observing humans interacting with robots might affect subsequent interpersonal relations among children (Correia et al., 2024).

Balancing the privacy and confidentiality of children's data with the need to disclose data is crucial in research practices. Collecting data from CRI often involves recording sensitive information from the child (e.g., behavioral patterns, biometric data) to understand their engagement with robotic technologies. However, privacy concerns restrict data sharing, and even with techniques like data (pseudo)anonymization, risks of re-identification sometimes remain, particularly in small or specific groups. Therefore, privacy measures and security concerns may limit data sharing, posing a threat to study replication (Mott et al., 2022; Spitale et al., 2023). CRI research must balance protecting the privacy and rights of this vulnerable group and sharing data practices aiming to advance scientific knowledge which is critical for ethical integrity and maintaining public trust (Couto et al., 2022; Spitale et al., 2023; Tolksdorf et al., 2021).

4.2 Robotics Design Tailored to Children Needs

The Institute for Education Science and The National Science Foundation (2013) outlined six types of research in education, usually starting with foundational and early-stage exploratory research, followed by the design and development of strategies and interventions that typically involve pilot testing. For those deemed sufficiently promising, the research progresses to assessing their efficacy and effectiveness in improving educational outcomes for children. This usually begins with focused studies on single populations and extends to broader evaluations across diverse populations and contexts. However, research is far more complex, as these categories do not cover all valuable research, and the sequence described is often not strictly followed in practice.

Various research methods can be employed, with the choice influenced mainly by the research questions, the child's age and their cognitive development, and the practical constraints of the context. Despite the diversity of methods, it is crucial to adopt an ethical approach that centers on children's needs. We recommend initiating the process with a participatory design process, which involves collecting input from children and their surrounding communities at various stages. This approach is particularly relevant when the aim is to design and adapt robots for children. Participatory methods allow children and other stakeholders to be involved not only in the solutions, but also in the identification of existing problems/opportunities, ideation, and definition of goals.

Designers often rely on pre-existing robots and their capabilities, developing robots based on what they believe fits into a children's world. The design and its functionalities are then adapted to suit the specific context. However, even when research design methods are employed, children are often left out of the research process (Alves-Oliveira, Paiva, et al., 2021a, 2021b). In contrast, the participatory design of technology is inclusive and iterative, making children active contributors throughout the design process (Iversen et al., 2017). Techniques like cooperative inquiry allow children to collaborate with researchers and designers in the ideation, prototyping, and testing phases (Druin, 1999, 2002). Children's perspectives and creativity are harnessed through child-friendly methods (Druin, 2002), and the iterative nature of participatory design integrates continual feedback, refining the design to meet children's needs and preferences (DiSalvo et al., 2017). This approach enhances usability, acceptance, and provides insights into how children interact with technology, which may lead to more intuitive and enjoyable user experiences, fostering ownership and empowerment among children, increasing their engagement and satisfaction (Bers, 2021) and often inspiring innovative solutions, as children's creativity brings forth novel ideas that adults might overlook (Iversen et al., 2017). Therefore, it is essential to integrate children's voices in the design of technology intended for their use and involve the school community from the early stages of development for robotics technologies, ensuring ethical CRI.

Children can assume various roles in the design process, including users, evaluators, informants, or design partners. However, it is particularly important also to include them as co-designers (Alves-Oliveira et al., 2017), while the school community typically acts as informants (Druin, 2002). A multi-method approach is also recommended, which can include ethnographic observations, contextual inquiries, group and individual interviews, and participatory design activities, enabling researchers to understand the perspectives of all stakeholders (Serholt et al., 2017; Williams et al., 2021). Holistic involvement of teachers, educators, parents, and children ensures that the design process addresses real needs, leading to more effective and ethical CHI.

Formative approaches are a crucial first step in designing ethical CRI. While individual and group interviews, focus groups, and discussions are suitable for adult stakeholders, these formats can be demanding and stressful for children. Consequently, researchers often use other methods with children, such as class and playground observations, followed by participatory design techniques to gather insights and children's voices without placing unnecessary pressure (Du & Breazeal, 2022; Neto et al., 2021; Newbutt et al., 2022; Piedade et al., 2024b). Participatory design also involves a co-design process where children and researchers collaboratively design tailored robot embodiments, capabilities, roles, and interactions. This process should adopt an iterative approach to enhance children's empathy, reflection, and creativity with an ethical and inclusive mindset. This process involves the following phases: (1) exploration and familiarization with the robots, (2) ideation, (3) presentation and critique, (4) building robotic devices, and (5) enacting interaction. In the exploration and familiarization phase, children learn about robot capabilities through interaction with physical robots or watching videos. The ideation phase involves presenting a problem or activity for children to tackle, encouraging discussion and reflection on specific challenges before considering solutions. Children may interview teachers and peers, engage in group discussions (Walsh et al., 2010), and use proxy toys or narratives to describe the context and characteristics of future users and applications (Giaccardi et al., 2012; Metatla et al., 2020; Piedade et al., 2024b). Children are then prompted to think about how a robot could be used in such contexts. These activities can also foster children's interest in the problem (Barendregt et al., 2020; Neto et al., 2021; Vaajakallio et al., 2010). This phase usually ends with children preparing a presentation. During the presentation phase, children present their ideas and design decisions to researchers, teachers, and peers, receiving feedback. This reflective questioning helps them understand their design choices, during which it is crucial to ensure a supportive environment to encourage all children to share their ideas. The next phase is "Do It Yourself" (DIY) prototyping, where children materialize their ideas using various materials, bringing their concepts to life. In the final enacting phase, children use prototypes to describe robot characteristics such as function, morphology, movement, autonomy, and roles.

Additional recommendations must be addressed to create a co-design activity that will enrich the learning experience for all children. First, use age-appropriate language for the children's age group to ensure clear understanding. Second, all sharing materials must be safe, durable, accessible, and affordable physically and economically. Third, standardizing children's robot knowledge is crucial as it is often the first encounter between children and robots (Barendregt et al., 2020; Neto et al., 2021). Fourth, if children have impairments (e.g., visual, hearing, cognitive), all materials and tasks should be adapted to ensure communication. Fifth, conduct co-design activities in schools if this is the intended context for the robot's use, as this helps children explore challenges and the robot's future applications within this setting (Barendregt et al., 2020). Six, foster an atmosphere of trust and sharing to allow children to be open to each other's ideas and ensure equitable participation, making them feel comfortable to communicate, which requires understanding the established power dynamics in the classroom (Cohen & Lotan, 2014; Vaajakallio et al., 2009). Familiarize with the children and educators/teachers before the activity helps integrate into the classroom dynamics. Teachers and psychologists can also provide valuable insights and manage conflicts during the activity if needed (Neto et al., 2021). Seven, avoid a mismatch between children's expectations and the actual experience. Children typically associate robot activities with fun and playfulness, but some moments can be tedious or stressful (Piedade et al., 2024a). Finally, authenticity is important, as many CRI rely on Wizard of Oz (WoZ) techniques, where human control is used to enhance robots' capabilities. By using co-design techniques, children become aware of robots' limitations and challenges, which can give them a more genuine understanding of how social robots are built and a space for reflection (Boulicault et al., 2023).

Co-designing robots with children in classrooms has been used for various goals with positive results, including facilitating children's creativity (Alves-Oliveira, Arriaga, et al., 2021a, 2021b; Alves-Oliveira, Paiva, et al., 2021a, 2021b), enhancing playful learning in classrooms (Lupetti et al., 2017), conducting anti-bullying interventions (Sanoubari et al., 2021), fostering inclusion among mixed-ability groups (Neto et al., 2021; Piedade et al., 2023), and critically reflecting on the ethical dilemmas of their relations with robots (Mott et al., 2022; Woodward et al., 2018).

Other types of research can also be adopted in the early stages, as documented in the guidelines by the Institute for Education Science and The National Science Foundation (2013). While it is not the scope of this chapter to review all, the following section will highlight some important ethical considerations relevant to studies evaluating the impact of interventions with robots to ensure that the research advances knowledge while upholding the rights of all participants involved.

4.3 Impact Assessment in Child-Robot Interaction

Both qualitative and quantitative methods and measures can be used to evaluate the impact of CRI. Qualitative research designs are used extensively in CRI to understand children's contextual, subjective, and behavioral experiences. These methods often include ethnographies, observational, case studies, interviews, and focus groups to gather detailed information on CRI, in natural settings such as schools. In addition to involving them in the co-design of robots, as we previously discussed, these methods can be valuable to understanding how children perceive and evaluate existing robotic devices. Quantitative research designs, on the other hand, aim to quantify variables related to CRI. This approach is suited for testing hypotheses derived from theory or prior research, often using a more systematic and controlled setup to evaluate CRI. Quantitative methods like correlational or controlled experiments in cross-sectional or longitudinal studies may provide findings that often aim to be generalized to larger populations. These studies can also use structured observations, subjective assessments, biopsychophysiological measures, and task performance, especially useful for large-scale studies assessing the impact of using robots in educational contexts across multiple outcomes, for identifying conditions that may affect the relationships between robot interventions and their outcomes, as well as the factors that contribute to explaining these relationships (see some findings from systematic and meta-analyses, Anwar et al., 2019; Kouroupa et al., 2022; Lee & Lee, 2022; Mukhasheva et al., 2023; Papakostas et al., 2021; Sapounidis et al., 2024; Wang & Cheung, 2024; Wang et al., 2023).

In general, using diverse instruments is also considered a good practice but depends on the child's age and socioemotional and cognitive development. Subjective assessments, such as interviews and surveys, are generally considered appropriate for children aged seven and older because, by this age, children typically have developed cognitive skills to understand and respond to some questions, reflect on their thoughts and feelings, and comprehend abstract concepts (Borgers et al., 2000). However, because they rely on the child's ability to introspect and articulate their experiences, these methods might still be challenging even for older children. Due to cognitive or linguistic limitations, observational methods and physiological measures are often more suitable for younger children because they do not rely on verbal communication.

Ethical considerations play a crucial role in selecting the method and measures, as researchers must ensure that the techniques do not harm the children or interfere with their normal activities. In combining various methods, researchers can triangulate their findings to enhance the robustness and depth of their conclusions. Therefore, mixed-method designs, which combine qualitative and quantitative approaches, can be useful to enhance the reliability of the findings (Bethel & Murphy, 2010).

Additionally, we emphasize specific ethical considerations in research studies utilizing experimental designs, particularly those where the study's aims cannot be initially disclosed to prevent influencing participant responses. These studies often involve some level of "deception" to maintain the integrity of data collection. Moreover, in studies with control groups where children do not interact with robots, it is considered good practice to offer these participants a chance to engage with the robots at later stages, especially if the CRI is expected to yield positive outcomes for the child. Concerns have been raised in literature about the Wizard of Oz (WoZ) methodology (van Straten et al., 2022; Westlund & Breazeal, 2016), where researchers simulate the robot's behaviors to study children's reactions, often without revealing the simulation. Many WoZ studies also simulate robots' behaviors that they cannot perform. Following van Straten et al. (2022) recommendations, we agree that CRI research should not impose unrealistic portrayals on robots to misrepresent their capabilities to ensure that children's responses are based on their genuine perceptions rather than researcher-imposed concepts. One interesting example is the study conducted by van Straten et al. (2023), who have shown that it is also possible to program robots to directly convey transparent information about their capabilities and limitations to children. In their study, children interacted with a robot that either disclosed its non-human psychological capacities and mechanical nature or did not. They found that children who received this information were less likely to anthropomorphize the robot, although they felt less close and trusted the robot less than children who did not receive transparent information from the robot. These findings show that providing transparent information about a robot's capabilities is important and affects children's responses. They also suggest that robots can directly communicate this information to children, which is particularly relevant since some CRI might not always be supervised (van Straten et al., 2023).

Nevertheless, it is important that any study involving deception include thoroughly informed consent processes and provide full debriefing to participants. Maintaining transparency post-experiment is crucial for upholding ethical standards.

5 Challenges in the Development of Robots and the Role of Technical Developers

Technical developers play a crucial role in creating educational robots for children, both in industry and academia, as they hold the responsibility to address and ensure ethical compliance. Their responsibilities include designing secure systems to protect children's personal and educational data from unauthorized access and misuse. Equally important is the safety and reliability of these robots, requiring rigorous testing to prevent physical risks and ensure consistent operation. Developers must also prevent biases in the robot's algorithms to ensure fairness and equity for all children regardless of their background (Miao et al., 2021). For data-driven AI algorithms, an adequate approach is to ensure the diversity of data or, for instance, the diversity of participants contributing to data. Transparency and accountability are also critical, requiring developers to clearly explain the robots' functions, data collection and usage, and the ethical implications of their designs (Miao et al., 2021).

Pilot testing with children is a critical part of the development stage. This iterative process involves multiple rounds of testing, feedback, and refinement to ensure that the robots meet the intended goals while addressing the diverse needs of young users. As previously highlighted, engaging children in pilot testing allows developers to observe CRI in real-world settings and gain insights into the usability and appeal of the robot. This feedback loop is necessary for identifying and fixing unforeseen issues or areas for improvement. By iteratively refining the robot based on real interactions, developers can enhance the design, functionality, and overall user experience. This approach not only improves the educational impact of the robots but also ensures that ethical considerations, such as safety, accessibility, and well-being, are addressed through direct user involvement.

6 Broader Challenges of Using Social Robots in Educational Settings

It is crucial to understand the long-term effects of child interactions with robots in school settings and their potential carry-over effects to several domains affecting their development. Therefore, more empirical research is needed to assess the real socio-emotional impact of CRI, considering that current evidence is still limited (Langer et al., 2023). Furthermore, systematic evaluations are lacking in understanding the sustained impact of social robots. Sustaining robots over extended periods introduces various challenges, including managing relationships with robots, logistics, ecological impact, financial sustainability, and human resource requirements.

As previously mentioned, one key recommendation is to ensure that robots and other technologies are never used to replace human interaction but rather to complement it (Boada et al., 2021; Contro & Brandão, 2024; Tarrés-Puertas et al., 2023). Additionally, we must respect the ethical right to choose human interaction over robots (Torras, 2024). This is particularly relevant for children, as their emotional and social development depends on meaningful human interactions. For example, while robots have been used to teach social skills, their role should primarily be as engagement and educational support tools. Children derive immense benefits from the social communication and emotional responses that only interactions with human educators and peers can provide (Boada et al., 2021; Contro & Brandão, 2024; Tarrés-Puertas et al., 2023). Therefore, a balanced approach should always be promoted to preserve the advantages of human contact while using innovative technological tools.

Robots used in educational environments must be adapted to each child's evolving interests and developmental stages to ensure they remain engaging and are relevant for their welfare. Robots are composed of hard/soft materials and software components. Integrating these elements is crucial for ensuring the robotic systems function correctly in their interactions with children. This requires regular maintenance and updates not only to maintain proper functioning but also to enhance their capabilities. Moreover, the diversity of available robotic systems and prototypes used in research makes it difficult to compare and replicate them. Limited accessibility to robots, due in part to their high commercial costs or the inaccessibility of custom-built models, further restricts the scope of research and hinders reproducibility (e.g., Gunes et al., 2022; Leichtmann et al., 2022; Strait et al., 2020; Ullman et al., 2021). Additionally, ecological concerns may impact the sustainability of robots available at schools. The use of non-recycled materials and energy by social robots, alongside waste management, poses ethical concerns regarding ecological sustainability (Boada et al., 2021). These demands may compromise the maintenance of robots in educational environments, which requires careful planning for their integration and eventual removal.

In academia, another concern is the often misalignment between research activities and community needs, complicating the sustainability of initiatives and projects involving robots. Research driven by academic timelines, such as grant cycles and publication deadlines, frequently conflicts with the unpredictable schedules of educational communities. This misalignment requires careful risk mitigation, plan exit strategies, and expectation management from the outset of projects. The challenge is ensuring that research projects, designed within research time constraints, align with the ongoing and often prolonged challenges faced by the community groups and the stakeholders. The goal is to empower children, teachers, and schools with the ability to continue the initiatives without needing ongoing external support. Additionally, it is important to establish long-term connections between the education system and sustainable projects.

Also relevant is ensuring reproducibility in science, which is crucial for advancing knowledge. One primary concern is that the research process (e.g., study protocols, materials) and some of the findings (e.g., data, code) are often not shared publicly on accessible repositories. These and other concerns have driven the development of initiatives with guidelines to promote transparency, openness, and reproducibility in the scientific community (Nosek et al., 2022). Various recommendations for open science have been proposed to ensure transparency and accessibility to the community. For example, in 2016, the FAIR (Findable, Accessible, Interoperable, and Reusable) principles were introduced by the GO FAIR Initiative (Go Fair, 2024; Wilkinson et al., 2016) and are currently advocated by many organizations and research communities. By implementing practices that ensure studies can be repeated (re-run a study using the same design as the original, by the same research team, and same sample size), replicated (using new data), reproduced (repeating the original analysis on the same data), and robust (applying a different analytical approach to the same data), researchers ensure that their findings are reliable and verifiable (Gunes et al., 2022; Nosek et al., 2022). Recently, these concerns have been discussed in education research (The National Science Foundation & Institute of Education Science, 2018), in HRI (Gunes et al., 2022; Leichtmann et al., 2022) and in CRI (Spitale et al., 2023).

The systematic review by Spitale et al. (2023) examined concerns related to reproducibility in CRI research from 2020 to 2022. Among the 325 studies reviewed, they identified several issues, including inadequate reporting of ethical approval and informed consent procedures, a lack of preregistrations, insufficient demographic data, small sample sizes, and insufficient details regarding study design. This includes the deployment scenario, a description of the robotic system used, and its mode of operation. Additionally, there was a notable lack of code and data-sharing practices within CRI research.

The lack of published standardized protocols for programming interactions and measuring outcomes in HRI studies may exacerbate these problems, making it challenging to ensure consistency and reliability across different research settings (Bethel & Murphy, 2010). These factors highlight the need for a more inclusive approach to robot design and a rigorous standardized methodology to enhance reproducibility.

There are, however, some examples of good open practices. For example, to implement co-design activities in the classroom, there are several robotic toolkits available that can be adapted to allow all children, regardless of their abilities, to be engaged and participate. These toolkits can employ a combination of activities (Obaid et al., 2018, 2024), and various techniques focusing on inclusive practices, such as toolkits tailored to children with and without visual impairments (Metatla et al., 2020; Neto et al., 2021) and neurodivergent classroom (Piedade et al., 2023). There are also some examples of open-source hardware and software for robots freely available within the scientific community (e.g., Alves-Oliveira et al., 2019, 2020), allowing researchers and even educators to access and use robotic technologies without prohibitive costs. As recently reviewed by Pearce (2020), making these open-source technologies accessible might also empower worldwide communities, including those with limited resources. In addition, it has the potential to train future scientists, from K-12 to university education, by providing educational opportunities to understand the equipment and its development. Moreover, it opens the possibility for users to customize and adapt tools to their needs. However, despite these advantages, there are challenges, such as the need for specialized technical knowledge and resources to build, program, and maintain these devices. These concerns are particularly relevant when deploying robots in schools, where ongoing technical support and expertise are crucial for sustained use and effectiveness. However, this challenge can be addressed by involving teachers with specialized knowledge and by providing training to educators in this domain.

As in other fields of research, in CRI (Spitale et al., 2023) there are still concerns related to a preference for quantitative studies over other research designs, a tendency to prioritize research that is novel or yields new results over replication studies, the reliance on null hypothesis significance testing (NHST), and the concern with publication bias, which favors studies that confirm hypotheses. However, as mentioned before, the need for participatory design, which includes qualitative data, is paramount for CRI. Poderi and Dittrich (2018), for example, review the synergistic relationship between participatory design and sustainability, advocating for more interdisciplinary collaboration and the development of frameworks that explicitly connect these participatory methods with sustainable outcomes. The authors also identified key trends, methodologies, and thematic focuses, revealing that participatory design inherently supports sustainability principles due to its inclusive, collaborative, and user-centered approaches. Furthermore, the authors also call for continued research into global challenges addressing resource depletion and social inequities.

Other research biases have also been highlighted, such as those favoring participants from specific economic backgrounds or social classes, with a tendency to overrepresent participants from Western, Educated, Industrialized, Rich, and Democratic ("WEIRD") nations (Bethel & Murphy, 2010; Spitale et al., 2023). Such sampling biases result in findings that are not generalizable to broader populations because they rely on a subset that does not accurately reflect the diverse range of values and social norms in the world. Due to their ongoing developmental changes and considering atypical developments, the way children perceive and interact with robotic technologies vary significantly, even among children of the same age, and is also influenced by many other factors such as cultural contexts and educational backgrounds. This diversity and variability pose additional challenges for CRI research. Additional social concerns are related to the morphological bias of the robot's appearance or behavioral responses (Giger et al., 2019), which can affect how robots are used, the expectations about their performance, and how they interact with it (Edwards & Edwards, 2022). Cultural biases in programming and designing robots may also inadvertently reflect the values or norms of specific groups, limiting their acceptance in global contexts.

Addressing CRI within educational settings and sustainability from a multidisciplinary ethical perspective is crucial. This growing awareness of sustainability is being discussed in recent workshops (e.g., Carter et al., 2024), where diverse stakeholders come together to envision the future of embedded research, emphasizing ethical and sustainable principles to foster community autonomy and ownership of projects. Such collaborative efforts are relevant for ensuring robotic technologies' relevance, effectiveness, and sustainability, ultimately supporting communities in becoming self-sufficient. These principles could similarly inspire the sustainability of research projects on CRI, giving children, teachers, parents, and other participants in the community the needed tools to continue using the robotic technologies for their welfare. Mobilizing and integrating interdisciplinary perspectives from diverse scientific fields, such as psychology, education, robotics engineering, design, and philosophy, along with multi-stakeholders, including the insights from children and the broader community, aims to ensure that as we navigate this emerging field, we do so with an ethical commitment to promoting the well-being of children.

7 Conclusions

In conclusion, this chapter addressed the critical ethical concerns surrounding the design, development, and evaluation of social robots in educational settings, particularly in contexts where they may interact with children. We explored some benefits and the ethical challenges of CRI, focusing on key concerns such as emotional attachment, dependency, and accountability for robot actions. We also examined several ethical guidelines, professional codes of conduct, and regulations. Additionally, we review some of the best practices and methodological challenges for conducting ethical research in this field. We highlighted the responsibilities of robot developers and researchers, emphasizing the need for ongoing dialogue with educators, policymakers, and stakeholders to ensure that social robots are used ethically and effectively in educational environments, prioritizing children's needs, safeguarding their rights, and ensuring long-term safety. Furthermore, we discussed the broader implications of social robots in education, underscoring the need to balance technological advancements with the preservation of meaningful human interactions.

Several avenues for future research were also identified. As highlighted earlier, longitudinal studies are limited but essential for understanding the long-term effects of CRI on children's cognitive, emotional, and social development. In addition, more efforts are needed to adapt social robots for inclusive education, ensuring they are accessible to all, regardless of gender, background, or ability. The development of standardized protocols for assessing the impact of social robots while ensuring ethical compliance—such as safeguarding personal data—will be crucial for ensuring reproducibility and reliability across research settings. Ultimately, as we continue integrating social robots into classrooms, addressing ethical concerns and challenges must remain a priority. Sustainable and responsible approaches to prioritizing children's well-being are essential not only for today's children but also for future generations.

References

ACM. (2018). *Code of Ethics*. Association for Computing Machinery. https://www.acm.org/about-acm/code-of-ethics

Adams, C., Pente, P., Lemermeyer, G., & Rockwell, G. (2023). Ethical principles for artificial intelligence in K-12 education. *Computers and Education: Artificial Intelligence, 4*, 100131. https://doi.org/10.1016/j.caeai.2023.100131

Ahtinen, A., Chowdhury, A., Ramírez Millan, V., Wu, C.-H., & Menon, G. (2023). Co-Learning around social robots with school pupils and university students—Focus on data privacy considerations. In *International Conference on Human-Agent Interaction*, 115–123. https://doi.org/10.1145/3623809.3623816.

Ali Mehenni, H., Kobylyanskaya, S., Vasilescu, I., & Devillers, L. (2021). Children as candidates to verbal nudging in a Human-robot experiment. In *Companion Publication of the 2020 International Conference on Multimodal Interaction*, 482–486. https://doi.org/10.1145/3395035.3425224.

ALLEA. (2023). *The European Code of Conduct for Research Integrity—Revised Edition 2023*. All European Academies. https://allea.org/wp-content/uploads/2023/06/European-Code-of-Conduct-Revised-Edition-2023.pdf

Alves-Oliveira, P., Arriaga, P., Nogueira, S. I., & Paiva, A. (2021). Robotics-based interventions for children's creativity. In *C&C '21: Proceedings of the 13th Conference on Creativity and Cognition*, 1–8. https://doi.org/10.1145/3450741.3465267.

Alves-Oliveira, P., Arriaga, P., Paiva, A., & Hoffman, G. (2019). Guide to build YOLO, a creativity-stimulating robot for children. *HardwareX, 6*, e00074. https://doi.org/10.1016/j.ohx.2019.e00074

Alves-Oliveira, P., Gomes, S., Chandak, A., Arriaga, P., Hoffman, G., & Paiva, A. (2020). Software architecture for YOLO, a creativity-stimulating robot. *SoftwareX, 11*, 100461. https://doi.org/10.1016/j.softx.2020.100461

Alves-Oliveira, P., Paiva, A., Arriaga, P., & Hoffman, G. (2017). Yolo, a robot for creativity: A co-design study with children. In *16th International ACM Conference on Interaction Design and Children*, 423–429. https://doi.org/10.1145/3078072.3084304.

Alves-Oliveira, P., Paiva, A., Arriaga, P., & Hofman, G. (2021b). Children as robot designers. *ACM/IEEE International Conference on Human-Robot Interaction, 2021*, 399–408. https://doi.org/10.1145/3434073.3444650

Amnesty International, & Access Now. (2018). *The Toronto Declaration: Protecting the Right to Equality and Non-discrimination in Machine Learning Systems*. Amnesty International and Access Now. https://www.torontodeclaration.org/declaration-text/english/

Anki Cozmo Robot. (2024). *Explore the World of AI with Cozmo Robot 2.0*. https://ankicozmorobot.com/

Anwar, S., Bascou, N., Menekse, M., & Kardgar, A. (2019). A systematic review of studies on educational robotics. *Journal of Pre-College Engineering Education Research (J-PEER), 9*(2). https://doi.org/10.7771/2157-9288.1223.

APA. (2017). *Ethical Principles of Psychologists and Code of Conduct*. https://www.apa.org/ethics/code

ASA. (2022). *ASA Ethical Guidelines for Statistical Practice-revised.* American Statistical Association. https://www.amstat.org/your-career/ethical-guidelines-for-statistical-practice

Asselborn, T., Guneysu, A., Mrini, K., Yadollahi, E., Ozgur, A., Johal, W., & Dillenbourg, P. (2018). *Bringing letters to life: Handwriting with haptic-enabled tangible robots.* 219–230. https://doi.org/10.1145/3202185.3202747.

Barendregt, W., Ekström, S., Kiesewetter, S., Pareto, L., & Serholt, S. (2020). Demystifying robots in the Co-Design of a tutee robot with primary school children. *Interaction Design and Architecture(s), 44,* 109–128. https://doi.org/10.55612/s-5002-044-006.

Belpaeme, T., Baxter, P., de Greeff, J., Kennedy, J., Read, R., Looije, R., Neerincx, M., Baroni, I., & Zelati, M. C. (2013). Child-Robot interaction: Perspectives and challenges. In G. Herrmann, M. J. Pearson, A. Lenz, P. Bremner, A. Spiers, & U. Leonards (Eds.), *Social Robotics* (pp. 452–459). Springer International Publishing. https://doi.org/10.1007/978-3-319-02675-6_45.

Belpaeme, T., Kennedy, J., Ramachandran, A., Scassellati, B., & Tanaka, F. (2018). Social robots for education: A review. *Science Robotics, 3*(21), eaat5954. https://doi.org/10.1126/scirobotics.aat5954.

Berrezueta-Guzman, J., Pau, I., Martín-Ruiz, M.-L., & Máximo-Bocanegra, N. (2021). Assessment of a robotic assistant for supporting homework activities of children with ADHD. *IEEE Access, 9,* 93450–93465. IEEE Access. https://doi.org/10.1109/ACCESS.2021.3093233

Bers, M. U. (2021). From computational thinking to computational doing. *Teaching Computational Thinking and Coding to Young Children,* 1–20. https://doi.org/10.4018/978-1-7998-7308-2.ch001.

Bethel, C. L., & Murphy, R. R. (2010). Review of human studies methods in HRI and recommendations. *International Journal of Social Robotics, 2*(4), 347–359. https://doi.org/10.1007/s12369-010-0064-9

Boada, J. P., Maestre, B. R., & Genís, C. T. (2021). The ethical issues of social assistive robotics: A critical literature review. *Technology in Society, 67,* 101726. https://doi.org/10.1016/j.techsoc.2021.101726

Borgers, N., de Leeuw, E., & Hox, J. (2000). Children as respondents in survey research: Cognitive development and response quality 1. *Bulletin of Sociological Methodology/bulletin De Méthodologie Sociologique, 66*(1), 60–75. https://doi.org/10.1177/075910630006600106

Boulicault, M., Phillips-Brown, M., Kory-Westlund, J. M., Nguyen, S., & Breazeal, C. (2023). Authenticity and co-design: On responsibly creating relational robots for children. *Algorithmic Rights and Protections for Children,* 85–122. https://doi.org/10.7551/mitpress/13654.003.0011.

Carter, A., Montague, K., Talhouk, R., Nicolau, H., Knearem, T., Beşevli, C., Crivellaro, C., Rüller, S., Henriques, A., & Peer, F. (2024). Envisioning collaborative futures: Advancing the frontiers of embedded research. In *PDC '24: Proceedings of the Participatory Design Conference, 2,* 190–193. https://doi.org/10.1145/3661455.3669890.

Cohen, E. G., & Lotan, R. A. (2014). *Designing groupwork: Strategies for the heterogeneous classroom* (3rd ed.). Teachers College Press.

Contro, J., & Brandão, M. (2024). Interaction minimalism: Minimizing HRI to reduce emotional dependency on robots. In *Robophilosophy Conference 2024.* https://kclpure.kcl.ac.uk/portal/files/269094043/Contro2024-robophil.pdf

Correia, F., Neto, I., Paulo, S., Piedade, P., Erel, H., Paiva, A., & Nicolau, H. (2024). The effects of observing robotic ostracism on children's prosociality and basic needs. In *Proceedings of the 2024 ACM/IEEE International Conference on Human-Robot Interaction,* 157–166. https://doi.org/10.1145/3610977.3634997.

Costa, S., Lehmann, H., Dautenhahn, K., Robins, B., & Soares, F. (2015). Using a humanoid robot to elicit body awareness and appropriate physical interaction in children with Autism. *International Journal of Social Robotics, 7*(2), 265–278. https://doi.org/10.1007/s12369-014-0250-2

Couto, M., Chandra, S., Yadollahi, E., & Charisi, V. (2022). Child-robot interaction: Design, evaluation, and novel solutions. *Interaction Studies, 23*(2), 151–156. https://doi.org/10.1075/is.00014.edi

Damianidou, D., Eidels, A., & Arthur-Kelly, M. (2020). The use of robots in social communications and interactions for individuals with ASD: A systematic review. *Advances in Neurodevelopmental Disorders, 4*(4), 357–388. https://doi.org/10.1007/s41252-020-00184-5

Davies, H. (2008). Reflexivity in research practice: Informed consent with children at school and at home. *Sociological Research Online, 13*(4), 17–30. https://doi.org/10.5153/sro.1775

Di Dio, C., Manzi, F., Peretti, G., Cangelosi, A., Harris, P. L., Massaro, D., & Marchetti, A. (2020). Shall I trust you? From Child-Robot interaction to trusting relationships. *Frontiers in Psychology, 11*, 469. https://doi.org/10.3389/fpsyg.2020.00469

DiSalvo, B., Yip, J., Bonsignore, E., & DiSalvo, C. (2017). *Participatory Design for Learning: Perspectives from Practice and Research.* Routledge.

Druin, A. (1999). Cooperative inquiry: Developing new technologies for children with children. In *Proceedings of the SIGCHI Conference on Human Factors in Computing Systems*, 592–599. https://doi.org/10.1145/302979.303166.

Druin, A. (2002). The role of children in the design of new technology. *Behaviour & Information Technology, 21*(1), 1–25. https://doi.org/10.1080/01449290110108659

Du, X., & Breazeal. (2022). Exploring changes in special education teachers' attitudes and design belief towards pedagogical agents in co-designing with children. In *IDC '22: Proceedings of the 21st Annual ACM Interaction Design and Children Conference*, 472–479. https://doi.org/10.1145/3501712.3535289.

Edwards, A., & Edwards, C. (2022). Does the correspondence bias apply to social robots? Dispositional and situational attributions of human versus robot behavior. *Frontiers in Robotics and A, I*, 8. https://doi.org/10.3389/frobt.2021.788242

EMYS. (2024). *EMYS.* https://www.emys.co/

ERIC. (2024). Ethical research involving children: Informed consent. *Ethical Research Involving Children.* https://childethics.com/informed-consent/

European Commission & Directorate-General for Research and Innovation. (2013). *Ethics for researchers: Facilitating research excellence in FP7.* Publications Office. https://data.europa.eu/doi/https://doi.org/10.2777/7491

European Parliament and Council of the European Union. (2024). Regulation (EU) 2024/1689 of the European Parliament and of the Council. *Official Journal of the European Union.* http://data.europa.eu/eli/reg/2024/1689/oj

FRA. (2014). Child participation in research. *European Union Agency for Fundamental Rights.* https://fra.europa.eu/pt/publication/2019/child-participation-research

GDPR.EU. (2022). *What Is GDPR, the EU's New Data Protection Law?* https://gdpr.eu/what-is-gdpr/

Giaccardi, E., Paredes, P., Díaz, P., & Alvarado, D. (2012). Embodied narratives: A performative co-design technique. In *DIS '12: Proceedings of the Designing Interactive Systems Conference*, 1–10. https://doi.org/10.1145/2317956.2317958.

Giger, J.-C., Piçarra, N., Alves-Oliveira, P., Oliveira, R., & Arriaga, P. (2019). Humanization of robots: Is it really such a good idea?. *Human Behavior and Emerging Technologies, 1*(2), 111–123. Scopus. https://doi.org/10.1002/hbe2.147.

Gillet, S., van den Bos, W., & Leite, I. (2020). *A social robot mediator to foster collaboration and inclusion among children.* Robotics: Science and Systems XVI. https://doi.org/10.15607/rss.2020.xvi.103.

Go Fair. (2024). *FAIR Principles.* https://www.go-fair.org/go-fair-initiative/

Gunes, H., Broz, F., Crawford, C. S., der Pütten, A. R., Strait, M., & Riek, L. (2022). Reproducibility in Human-Robot interaction: Furthering the science of HRI. *Current Robotics Reports, 3*(4), 281–292. https://doi.org/10.1007/s43154-022-00094-5

Huang, G., & Moore, R. K. (2023). Using social robots for language learning: Are we there yet? *Journal of China Computer-Assisted Language Learning, 3*(1), 208–230. https://doi.org/10.1515/jccall-2023-0013

IEAIED. (2021). *The Ethical Framework for AI in Education.* The Institute for Ethical AI in Education. https://www.buckingham.ac.uk/wp-content/uploads/2021/03/The-Institute-for-Ethical-AI-in-Education-The-Ethical-Framework-for-AI-in-Education.pdf

IEEE. (2016). *IEEE Ethically Aligned Design guidelines.* https://standards.ieee.org/industry-connections/ec/ead-v1/

Institute for Education Science, & The National Science Foundation. (2013). *Common Guidelines for Education Research and Development.* https://www.nsf.gov/pubs/2013/nsf13126/nsf13126.pdf

Iversen, O. S., Smith, R. C., & Dindler, C. (2017). Child as protagonist: Expanding the role of children in participatory design. In *Proceedings of the 2017 Conference on Interaction Design and Children*, 7. https://doi.org/10.1145/3078072.3079725.

Kennedy, J., Baxter, P., & Belpaeme, T. (2015). Comparing robot embodiments in a guided discovery learning interaction with children. *International Journal of Social Robotics, 7*(2), 293–308. https://doi.org/10.1007/s12369-014-0277-4

Konijn, E. A., Jansen, B., Bustos, V. M., Hobbelink, V. L. N. F., & Vanegas, D. P. (2022). Social robots for (second) language learning in (migrant) primary school children. *International Journal of Social Robotics, 14*(3), 827–843. https://doi.org/10.1007/s12369-021-00824-3

Kouroupa, A., Laws, K. R., Irvine, K., Mengoni, S. E., Baird, A., & Sharma, S. (2022). The use of social robots with children and young people on the autism spectrum: A systematic review and meta-analysis. *PLoS ONE, 17*(6), e0269800. https://doi.org/10.1371/journal.pone.0269800

KUBO. (2024). *Say Hi to Kudo.* https://kubo-robot.com/meet-kubo/

Langer, A., Marshall, P. J., & Levy-Tzedek, S. (2023). Ethical considerations in child-robot interactions. *Neuroscience & Biobehavioral Reviews, 151*, 105230. https://doi.org/10.1016/j.neubiorev.2023.105230

Lee, H., & Lee, J. H. (2022). The effects of robot-assisted language learning: A meta-analysis. *Educational Research Review, 35*, 100425. https://doi.org/10.1016/j.edurev.2021.100425

Lee, J. J., Sha, F., & Breazeal, C. (2019). A Bayesian theory of mind approach to nonverbal communication. In *14th ACM/IEEE International Conference on Human-Robot Interaction (HRI)*, 487–496. https://doi.org/10.1109/hri.2019.8673023.

Lego. (2024). *Lego Mindstorms* [Graphic]. https://www.lego.com/en-de/themes/mindstorms/about

Leichtmann, B., Nitsch, V., & Mara, M. (2022). Crisis Ahead? Why Human-Robot interaction user studies may have replicability problems and directions for improvement. *Frontiers in Robotics and A, I*, 9. https://doi.org/10.3389/frobt.2022.838116

LGPD. (2019). *Lei Geral de Proteção de Dados Pessoais (LGPD).* https://www.planalto.gov.br/ccivil_03/_ato2015-2018/2018/lei/l13709.htm

Ligthart, M. E. U., De Droog, S. M., Bossema, M., Elloumi, L., de Haas, M., Smakman, M. H. J., Hindriks, K. V., & Ben Allouch, S. (2024). Back to School: Sustaining recurring child-robot educational interactions after a long break. In *Proceedings of the 2024 ACM/IEEE International Conference on Human-Robot Interaction*, 433–442. https://doi.org/10.1145/3610977.3635001.

Ligthart, M. E. U., Neerincx, M. A., & Hindriks, K. V. (2022). Memory-based personalization for fostering a long-term child-robot relationship. In *17th ACM/IEEE International Conference on Human-Robot Interaction*, 80–89. https://doi.org/10.1109/hri53351.2022.9889446.

Litoiu, A., & Scassellati, B. (2015). Robotic coaching of complex physical skills. In *Proceedings of the Tenth Annual ACM/IEEE International Conference on Human-Robot Interaction*, 211–212. https://doi.org/10.1145/2701973.2702726.

López-Sintas, J., García-Álvarez, E., & Pérez-Rubiales, E. (2012). The unforgettable aesthetic experience: The relationship between the originality of artworks and local culture. *Poetics, 40*(4), 337–358. https://doi.org/10.1016/j.poetic.2012.05.003

Lupetti, M., Yao, Y., Mi, H., & Germak, C. (2017). Design for children's playful learning with robots. *Future Internet, 9*(3), 52. https://doi.org/10.3390/fi9030052

Metatla, O., Bardot, S., Cullen, C., Serrano, M., & Jouffrais, C. (2020). Robots for inclusive Play: Co-designing an educational game with visually impaired and sighted children. In *CHI '20:*

Proceedings of the 2020 CHI Conference on Human Factors in Computing Systems, 1–13. https://doi.org/10.1145/3313831.3376270.
Miao, F., Wayne, H., Ronghuai, H., & Hui, Z. (2021). *AI and education: A guidance for policymakers*. UNESCO. https://doi.org/10.54675/PCSP7350.
Modrobotics. (2024). *Cubelets Robot Blocks*. https://modrobotics.com/
Montréal University. (2018). *Montréal Declaration for a Responsible Development of Artificial Intelligence*. https://declarationmontreal-iaresponsable.com/wp-content/uploads/2023/04/UdeM_Decl_IA-Resp_LA-Declaration-ENG_WEB_09-07-19.pdf
Mott, T., Bejarano, A., & Williams, T. (2022). Robot Co-design can help us engage child stakeholders in ethical reflection. In *17th ACM/IEEE International Conference on Human-Robot Interaction*, 14–23. https://doi.org/10.1109/HRI53351.2022.9889430.
Mukhasheva, M., Ybyraimzhanov, K., Naubaeva, K., Mamekova, A., & Almukhambetova, B. (2023). The impact of educational robotics on cognitive outcomes in primary students: A meta-analysis of recent studies. *European Journal of Educational Research, 12*(4), 1683–1695. https://doi.org/10.12973/eu-jer.12.4.1683.
National Commission for the Protection of Human Subjects of Biomedical and Behavioral Research. (1979). *The Belmont Report: Ethical Principles and Guidelines for the Protection of Human Subjects of Research*. U.S. Department of Health and Human Services. https://www.hhs.gov/ohrp/regulations-and-policy/belmont-report/read-the-belmont-report/index.html
NEA. (1975). *Code of Ethics for Educators*. https://www.nea.org/resource-library/code-ethics-educators
Neto, I., Correia, F., Rocha, F., Piedade, P., Paiva, A., & Nicolau, H. (2023). The robot made us hear each other: Fostering inclusive conversations among Mixed-Visual ability children. In *HRI '23: Proceedings of the 2023 ACM/IEEE International Conference on Human-Robot Interaction*, 13–23. https://doi.org/10.1145/3568162.3576997.
Neto, I., Johal, W., Couto, M., Nicolau, H., Paiva, A., & Guneysu, A. (2020). Using table top robots to promote inclusive classroom experiences. In *Proceedings of the Interaction Design and Children Conference*, 281–292. https://doi.org/10.1145/3392063.3394439.
Neto, I., Nicolau, H., & Paiva, A. (2021). Community based robot design for classrooms with mixed visual abilities children. In *CHI '21: Proceedings of the 2021 CHI Conference on Human Factors in Computing Systems*, 1–12. https://doi.org/10.1145/3411764.3445135.
Neumann, M. M., Koch, L.-C., Zagami, J., Reilly, D., & Neumann, D. L. (2023). Preschool children's engagement with a social robot compared to a human instructor. *Early Childhood Research Quarterly, 65*, 332–341. https://doi.org/10.1016/j.ecresq.2023.07.010
Newbutt, N., Rice, L., Lemaignan, S., Daly, J., Charisi, V., & Conley, I. (2022). Co-designing a social robot in a special educational needs school. *Interaction Studies, 23*(2), 204–242. https://doi.org/10.1075/is.21031.new
Nosek, B. A., Hardwicke, T. E., Moshontz, H., Allard, A., Corker, K. S., Dreber, A., Fidler, F., Hilgard, J., Struhl, M. K., Nuijten, M. B., Rohrer, J. M., Romero, F., Scheel, A. M., Scherer, L. D., Schönbrodt, F. D., & Vazire, S. (2022). Replicability, Robustness, and reproducibility in psychological science. *Annual Review of Psychology, 73*, 719–748. https://doi.org/10.1146/annurev-psych-020821-114157
Obaid, M., Baykal, G. E., Kırlangıc, G., Göksun, T., & Yantaç, A. E. (2024). Collective co-design activities with children for designing classroom robots. In *AfriCHI '23: Proceedings of the 4th African Human Computer Interaction Conference*, 229–237. https://doi.org/10.1145/3628096.3630094.
Obaid, M., Baykal, G. E., Yantaç, A. E., & Barendregt, W. (2018). Developing a prototyping method for involving children in the design of classroom robots. *International Journal of Social Robotics, 10*(2), 279–291. https://doi.org/10.1007/s12369-017-0450-7
OneTrust. (2024). *OneTrust Data Guidance*. https://www.dataguidance.com/
Özgür, A., Lemaignan, S., Johal, W., Beltran, M., Briod, M., Pereyre, L., Mondada, F., & Dillenbourg, P. (2017). Cellulo: Versatile handheld robots for education. In *Proceedings of the 2017*

ACM/IEEE International Conference on Human-Robot Interaction, 119–127. https://doi.org/10.1145/2909824.3020247.

Ozobot. (2024). *Ozobot*. https://shop.ozobot.com/

Papakostas, G. A., Sidiropoulos, G. K., Papadopoulou, C. I., Vrochidou, E., Kaburlasos, V. G., Papadopoulou, M. T., Holeva, V., Nikopoulou, V.-A., & Dalivigkas, N. (2021). Social robots in special education: A systematic review. *Electronics*, *10*(12). https://doi.org/10.3390/electronics10121398.

Pearce, J. M. (2020). Economic savings for scientific free and open source technology: A review. *HardwareX, 8*, e00139. https://doi.org/10.1016/j.ohx.2020.e00139

Piedade, P., Neto, I., Pires, A. C., Prada, R., & Nicolau, H. (2023). PartiPlay: A participatory game design kit for Neurodiverse classrooms. In *ASSETS '23: Proceedings of the 25th International ACM SIGACCESS Conference on Computers and Accessibility, Article 72*, 1–5. https://doi.org/10.1145/3597638.3614496.

Piedade, P., Neto, I., Pires, A., Prada, R., & Nicolau, H. (2024a, in press). Inclusion as a process: Co-designing an inclusive robotic game with Neurodiverse classrooms. In *Proceedings of the 26th International ACM SIGACCESS Conference on Computers and Accessibility*.

Piedade, P., Neto, I., Pires, A., Prada, R., & Nicolau, H. (2024b). "That's our game!": Reflections on Co-designing a robotic game with Neurodiverse children. *Lecture Notes in Computer Science*, 153–160. https://doi.org/10.1007/978-3-031-61688-4_15.

PILP. (2021). *Personal Information Protection Law (PILP) of the People's Republic of China.* https://personalinformationprotectionlaw.com/

Poderi, G., & Dittrich, Y. (2018). Participatory design and sustainability: A literature review of PDC. In *PDC '18: Proceedings of the 15th Participatory Design Conference*, *2*, 1–5. https://doi.org/10.1145/3210604.3210624.

Rocha, F., Correia, F., Neto, I., Pires, A. C., Guerreiro, J., Guerreiro, T., & Nicolau, H. (2023). Coding together: On Co-located and remote collaboration between children with Mixed-Visual Abilities. In *CHI '23: Proceedings of the 2023 CHI Conference on Human Factors in Computing Systems, Article 606*, 1–14. https://doi.org/10.1145/3544548.3581261.

Sanoubari, E., Muñoz Cardona, J. E., Mahdi, H., Young, J. E., Houston, A., & Dautenhahn, K. (2021). Robots, bullies and stories: A remote co-design study with children. In *IDC '21: Proceedings of the 20th Annual ACM Interaction Design and Children Conference*, 171–182. https://doi.org/10.1145/3459990.3460725.

Sapounidis, T., Tselegkaridis, S., & Stamovlasis, D. (2024). Educational robotics and STEM in primary education: A review and a meta-analysis. *Journal of Research on Technology in Education, 56*(4), 462–476. https://doi.org/10.1080/15391523.2022.2160394

Scheithauer, H., Alsaker, F., Wölfer, R., & Ruggieri, S. (2013). Cyberball: A reasonable paradigm for research in developmental science? *International Journal of Developmental Science, 7*(1), 1–6. https://doi.org/10.3233/DEV-1300071

Serholt, S., Barendregt, W., Vasalou, A., Alves-Oliveira, P., Jones, A., Petisca, S., & Paiva, A. (2017). The case of classroom robots: Teachers' deliberations on the ethical tensions. *AI & Society, 32*(4), 613–631. https://doi.org/10.1007/s00146-016-0667-2

Sphero. (2024). *Sphero*. https://sphero.com/

Spitale, M., Stower, R., Yadollahi, E., Parreira, M. T., Abbasi, N. I., Leite, I., & Gunes, H. (2023, September 4). *A Systematic Review on Reproducibility in Child-Robot Interaction*. arXiv.Org. https://arxiv.org/abs/2309.01822v1

Stower, R., Calvo-Barajas, N., Castellano, G., & Kappas, A. (2021). A Meta-analysis on children's trust in social robots. *International Journal of Social Robotics, 13*(8), 1979–2001. https://doi.org/10.1007/s12369-020-00736-8

Strait, M., Lier, F., Bernotat, J., Wachsmuth, S., Eyssel, F., Goldstone, R., & Šabanović, S. (2020). A Three-Site reproduction of the joint Simon effect with the NAO Robot. In *HRI '20: Proceedings of the 2020 ACM/IEEE International Conference on Human-Robot Interaction*, 103–111. https://doi.org/10.1145/3319502.3374783.

Strohkorb, S., Fukuto, E., Warren, N., Taylor, C., Berry, B., & Scassellati, B. (2016). Improving human-human collaboration between children with a social robot. In *25th IEEE International Symposium on Robot and Human Interactive Communication (RO-MAN)*, 551–556. https://doi.org/10.1109/roman.2016.7745172.

Su, J., Yang, W., & Zhong, Y. (2023). Influences of gender and socioeconomic status on children's use of robotics in early childhood education: A systematic review. *Early Education and Development, 34*(4), 910–926. https://doi.org/10.1080/10409289.2022.2078617

Tarrés-Puertas, M. I., Costa, V., Pedreira Alvarez, M., Lemkow-Tovias, G., Rossell, J. M., & Dorado, A. D. (2023). Child–Robot interactions using educational robots: An ethical and inclusive perspective. *Sensors, 23*(3), 3. https://doi.org/10.3390/s23031675.

The National Science Foundation, & Institute of Education Science. (2018). *Companion Guidelines on Replication & Reproducibility in Education Research: A Supplement to the Common Guidelines for Education Research and Development.* https://www.nsf.gov/pubs/2019/nsf19022/nsf19022.pdf.

The Nuremberg Code. (1996). The Nuremberg Code (1947). *British Medical Journal, 313*, 1448. https://media.tghn.org/medialibrary/2011/04/BMJ_No_7070_Volume_313_The_Nuremberg_Code.pdf

Thymio. (2024). *Thymio*. https://www.thymio.org/

Tolksdorf, N. F., Siebert, S., Zorn, I., Horwath, I., & Rohlfing, K. J. (2021). Ethical considerations of applying robots in Kindergarten settings: Towards an approach from a Macroperspective. *International Journal of Social Robotics, 13*(2), 129–140. https://doi.org/10.1007/s12369-020-00622-3

Torras, C. (2024). Ethics of social robotics: Individual and societal concerns and opportunities. *Annual Review of Control, Robotics, and Autonomous Systems, 7*(1), 1–18. https://doi.org/10.1146/annurev-control-062023-082238

Tractenberg, R. E. (2024). What is "ethical AI"? Leading or participating on an ethical team and/or working in statistics, data science, and artificial intelligence. In H. Doosti (Ed.), Ethics in Statistics. Ethics International Press. https://osf.io/preprints/socarxiv/8e6pv/

Ullman, D., Aladia, S., & Malle, B. F. (2021). Challenges and opportunities for replication science in HRI: A case study in Human-Robot trust. In *HRI '21: Proceedings of the 2021 ACM/IEEE International Conference on Human-Robot Interaction*, 110–118. https://doi.org/10.1145/3434073.3444652.

UNICEF. (2007). *Implementation Handbook for the Convention on the Rights of the Child* (3rd edn.). UNICEF. https://www.unicef.org/reports/implementation-handbook-convention-rights-child

United Nations. General Assembly. (1949). *Universal declaration of human rights*. Department of State, United States of America. https://www.un.org/en/about-us/universal-declaration-of-human-rights

U.S. House of Representatives. (2022). *H.R.8152—American Data Privacy and Protection Act*. https://www.congress.gov/bill/117th-congress/house-bill/8152/text

Vaajakallio, K., Lee, J.-J., & Mattelmäki, T. (2009). "It has to be a group work!": Co-design with children. In *IDC '09: Proceedings of the 8th International Conference on Interaction Design and Children*, 246–249. https://doi.org/10.1145/1551788.1551843.

Vaajakallio, K., Mattelmäki, T., & Lee, J.-J. (2010). Co-design lessons with children. *Interactions, 17*(4), 26–29. https://doi.org/10.1145/1806491.1806498

van den Berghe, R., Verhagen, J., Oudgenoeg-Paz, O., van der Ven, S., & Leseman, P. (2019). Social robots for language learning: A review. *Review of Educational Research, 89*(2), 259–295. https://doi.org/10.3102/0034654318821286

van Straten, C. L., Peter, J., & Kühne, R. (2023). Transparent robots: How children perceive and relate to a social robot that acknowledges its lack of human psychological capacities and machine status. *International Journal of Human-Computer Studies, 177*, 103063. https://doi.org/10.1016/j.ijhcs.2023.103063

van Straten, C. L., Peter, J., Kühne, R., & Barco, A. (2022). The wizard and I: How transparent teleoperation and self-description (do not) affect children's robot perceptions and child-robot relationship formation. *AI & Society, 37*(1), 383–399. https://doi.org/10.1007/s00146-021-01202-3

VEX Robotics. (2024). *VEX Robotics*. https://www.vexrobotics.com/

Walsh, G., Druin, A., Guha, M. L., Foss, E., Golub, E., Hatley, L., Bonsignore, E., & Franckel, S. (2010). Layered elaboration: A new technique for co-design with children. In *CHI '10: Proceedings of the SIGCHI Conference on Human Factors in Computing Systems*, 1237–1240. https://doi.org/10.1145/1753326.1753512.

Wang, F., & Cheung, A. C. K. (2024). Robots' social behaviors for language learning: A systematic review and Meta-Analysis. *Review of Educational Research*. https://doi.org/10.3102/00346543231216437

Wang, K., Sang, G.-Y., Huang, L.-Z., Li, S.-H., & Guo, J.-W. (2023). The effectiveness of educational robots in improving learning outcomes: A Meta-Analysis. *Sustainability, 15*(5). https://doi.org/10.3390/su15054637.

Westlund, J. M. K., & Breazeal, C. (2016). Transparency, teleoperation, and children's understanding of social robots. In *2016 11th ACM/IEEE International Conference on Human-Robot Interaction (HRI)*, 625–626. https://doi.org/10.1109/HRI.2016.7451888.

Whitebread, D., Basilio, M., O'Sullivan, L., & Zachariou, A. (2019). The importance of play, oral language and self-regulation in children's development and learning: implications for quality in early childhood education. *The SAGE Handbook of Developmental Psychology and Early Childhood Education*, 554–569. https://doi.org/10.4135/9781526470393.n32.

WHO. (2017). *Code of Conduct for Responsible Research*. https://www.who.int/about/ethics/code-of-conduct-for-responsible-research

Wilkinson, M. D., Dumontier, M., Aalbersberg, Ij. J., Appleton, G., Axton, M., Baak, A., Blomberg, N., Boiten, J.-W., da Silva Santos, L. B., Bourne, P. E., Bouwman, J., Brookes, A. J., Clark, T., Crosas, M., Dillo, I., Dumon, O., Edmunds, S., Evelo, C. T., Finkers, R., Mons, B. (2016). The FAIR guiding principles for scientific data management and stewardship. *Scientific Data, 3*(1), 1. https://doi.org/10.1038/sdata.2016.18.

Williams, R., Kaputsos, S. P., & Breazeal, C. (2021). Teacher perspectives on how to train your robot: a middle school AI and Ethics curriculum. *Proceedings of the AAAI Conference on Artificial Intelligence, 35*(17), 15678–15686. https://doi.org/10.1609/aaai.v35i17.17847

Wonder Workshop. (2024). *Dash and Dot*. https://www.makewonder.com/en/dash/

Woodward, J., McFadden, Z., Shiver, N., Ben-hayon, A., Yip, J. C., & Anthony, L. (2018). Using Co-Design to examine how children conceptualize intelligent interfaces. In *CHI '18: Proceedings of the 2018 CHI Conference on Human Factors in Computing Systems, Paper No. 575*, 1–14. https://doi.org/10.1145/3173574.3174149.

World Medical Association. (2013). *WMA Declaration of Helsinki: Ethical Principles for Medical Research Involving Human Subjects*. https://www.wma.net/policies-post/wma-declaration-of-helsinki-ethical-principles-for-medical-research-involving-human-subjects/

Zhang, Y., & Zhu, Y. (2022). Effects of educational robotics on the creativity and problem-solving skills of K-12 students: A meta-analysis. *Educational Studies*, 1–19. https://doi.org/10.1080/03055698.2022.2107873

The Use of Social Robots in Educational Settings: Acceptance and Usability

Daniela Conti, Carla Cirasa, Helene Høgsdal, and Santo F. Di Nuovo

Abstract Robots have increasingly permeated multiple dimensions of human life, and the field of robotics has transitioned from its traditional association with industrial manufacturing to a broader integration into social environments, where robots are increasingly seen as companions to humans. Moreover, advancements in anthropomorphism and artificial intelligence have enhanced robots' ability to engage in meaningful interactions with humans. These developments enable the creation of more sophisticated robots that can learn and adapt their behaviors based on feedback from human users. Such capabilities have paved the way for social robots to be utilized in providing assistance to vulnerable populations, including the elderly and children. However, while extensive research exists on the application of robotics in elderly care, studies exploring human-robot interaction with children remain limited. Specifically, there is still considerable uncertainty surrounding the effective integration of robotics into educational contexts. Several factors contribute to this challenge. First, the adoption of robotics in schools faces resistance from teachers, who may fear being replaced by robotic systems. Additionally, assessing children's acceptance of robotics in the classroom poses difficulties, as current evaluation tools primarily focus on the direct observation of interactions. The acceptance and usability of robotics in educational settings are influenced by numerous factors, many of which are the subject of ongoing research. It is crucial to further investigate how both children and teachers perceive robots in educational environments, and to examine whether either

D. Conti · C. Cirasa
Department of Humanities, University of Catania, Catania, Italy
e-mail: daniela.conti@unict.it

C. Cirasa
e-mail: carla.cirasa@phd.unict.it

H. Høgsdal
Regional Centre for Child and Youth Mental Health and Child Welfare – North, Faculty of Health Sciences, UiT The Arctic University of Norway, Tromsø, Norway
e-mail: helene.hogsdal@uit.no

S. F. Di Nuovo (✉)
Department of Educational Sciences, University of Catania, Catania, Italy
e-mail: s.dinuovo@unict.it

G. Lampropoulos and S. Papadakis (eds.), *Social Robots in Education*, Studies in Computational Intelligence 1194, https://doi.org/10.1007/978-3-031-82915-4_9

teachers or students experiences the Uncanny Valley phenomenon—a concept that suggests robots should neither appear too dissimilar from nor too similar to humans.

Keywords Social robots · Educational settings · Acceptance · Usability

1 Introduction

The term "robotics", in its modern sense, was used by scientist and writer Isaac Asimov in 1940 (Asimov, 1940). In his novel *I, Robot*, Asimov described a robot as a programmable electromechanical system equipped with various abilities, capable of performing specific tasks. Since then, the field of robotics has experienced remarkable growth, expanding its applications across numerous areas of human life. From the 1970s onwards, robotics became a fundamental component of industrial production, where machines were employed to replace humans in various tasks along the production chain. This shift led to a significant reduction in both production costs and time spent on important tasks. Shortly thereafter, the first service robots were introduced. These robots, still in use today, serve in space exploration, submarine missions, surgical procedures, meal preparation, and other tasks that assist humans. Advances in technology, particularly the miniaturization of circuits, have allowed the development of increasingly smaller robots, often designed to resemble humans more closely (i.e., social robots). A social robot is defined as an "*autonomous or semi-autonomous robot that interacts and communicates with humans, following the behavioral norms expected by the individuals with whom it interacts*" (Bartneck et al., 2004, p. 2). Social robots are capable of engaging in social interactions (Breazeal, 2003) and eliciting social responses from humans (de Graaf, 2016), and at the same time they are better to meet safety requirements and to integrate seamlessly into everyday human contexts. For a comprehensive review of the definition of social robots, see Sarrica et al. (2020). Due to their human-like features, such as eyes and mouth, social robots provide a natural mode of interaction that can stimulate children's learning. For example, humanoid social robots, with their embodied systems, have been shown to promote dialogue and language development more effectively than non-humanoid devices (van den Berghe et al., 2019; Randall, 2019). Moreover, with the contributions of the social sciences in the programming and implementation phases, robots have evolved into social devices capable of functioning in social settings such as in schools and elderly care facilities.

Currently, social robotics plays a significant role in education, following two main trajectories. Firstly, educational robotics (ER) is a key element within STEAM (Science, Technology, Engineering, Arts, and Mathematics) education. Numerous studies have underscored the pivotal role robotics education plays in fostering skills such as programming, computational thinking, inhibitory control, and working memory (Zhang, 2024). During robotics courses, students encounter various challenges in building and programming robots, which helps them develop important cognitive skills. This hands-on learning, combined with collaboration among peers, facilitates the acquisition of knowledge and abilities (Chookaew & Panjaburee, 2022). Research by Coskun & Filiz (2023) further emphasizes the contribution of robotics training to students' abilities in collaboration, flexibility, autonomy, innovation, and cognitive skills. Simultaneously, social robots are more often utilized as direct support for educators, such as teachers. These robots can assist both teachers and students, contributing to the educational process through the implementation of ER. In educational settings, social robots have demonstrated advantages over tablets in terms of performance, engagement, and enjoyment (Konijn et al., 2022). Similarly, Belpaeme et al. (2018) found that social robots offer greater benefits in engagement and learning outcomes compared to traditional tablets or computers.

However, the effectiveness of social robots, as well as students' engagement and learning outcomes, can vary depending on the role assigned to the robot—whether it acts as a tutor or a companion (Chen et al., 2020). Additionally, different teaching styles seem to influence the learning experience with robots (Okita et al., 2009).

Over the past decade, numerous studies have examined the performance of social robots in various educational contexts, evaluating their use in both teaching and learning. The market now offers a wide range of social robots, each with distinct features suited to different roles and functions. For instance, robots used in language learning are classified into four categories: anthropomorphic, zoomorphic, mechanomorphic, and cartoon-like (Randall, 2019). Depending on their design, certain robots may be better suited for interaction with specific user groups. The robot most effective in interacting with adults may not necessarily be the best choice for interacting with children. These robots' performance has been evaluated not only in classroom settings but also in group or one-on-one interactions. Research has assessed their impact on learning as well as on the acquisition of transversal skills such as anxiety management (Hsu et al., 2022; Kanero et al., 2022), motivation (Song et al., 2023), and memorization (van den Berghe et al., 2021). In classroom environments, social robots have proven effective not only in teaching curricular subjects but also in facilitating extracurricular workshops, such as those focused on food education (Balzotti et al., 2023) and environmental education (Castellano et al., 2021).

Despite the growing presence of robotics in various sectors, several research gaps remain in the context of educational settings, particularly with children. First, while there is a substantial body of work examining robotics in elderly care, relatively limited research has been conducted on human-robot interaction (HRI) with children, especially within educational environments. This highlights a clear need for more focused investigation in this area. Second, uncertainty persists regarding the effective integration of robotics into teaching and learning processes. Few studies have addressed how best to implement these technologies in classrooms, leaving this a largely unexplored challenge (Conti et al., 2017a).

Additionally, resistance from educators presents another barrier. Teachers may be hesitant to embrace robotics due to concerns about potential job displacement or doubts about the pedagogical value of these technologies. Furthermore, the measurement of children's acceptance of robots in educational settings remains an underdeveloped area. Existing methodologies primarily rely on direct observation, and there is a lack of robust tools for assessing children's acceptance of social robots.

The role that robots assume in educational contexts, whether as tutors, companions, or other roles, may also influence their effectiveness, depending on the teaching styles employed. However, this aspect has not been thoroughly explored in current research. Moreover, the design of robots—whether anthropomorphic, zoomorphic, or otherwise—could have significant implications for learning outcomes and user interactions. There is a particular need to understand how different robot forms may affect children and adults differently, as this remains insufficiently studied.

The study outlined here seeks to address these gaps, with three primary aims. First, it will explore the acceptability of social robots within educational contexts, focusing on how both children and teachers perceive these technologies. It will also consider how acceptance is influenced by cultural and religious factors, as well as the anthropomorphic characteristics of the robots. Second, the study aims to develop or refine tools to measure the acceptability of social robots in classrooms, with an emphasis on creating reliable and comprehensive methods of assessment. Finally, the research will evaluate the usability of social robots as educational support tools, identifying their strengths and weaknesses in real-world classroom applications.

The contributions of this research are multifaceted. It will address the scarcity of literature on child-robot interaction within educational settings, providing insights into how social robots can be successfully integrated into teaching and learning environments. By focusing on both acceptability and usability, the study will offer a nuanced understanding of the factors that influence the implementation of robotics in education, including cultural context, robot design, and the specific roles that robots take on in classrooms. In addition, the research will propose a framework for measuring robot acceptance, addressing the current limitations in evaluation methods. Importantly, the study will also emphasize the potential of social robots to enhance both cognitive and transversal skills, such as problem-solving, language development, anxiety management, motivation, and collaboration, thus contributing broadly to educational practices.

Through these contributions, this chapter aims to advance knowledge in these areas and provide practical insights for educators and developers seeking to integrate social robotics into educational environments. The chapter will also (2) define the concept of acceptability, emphasizing the importance of designing and implementing robots that achieve high levels of acceptance, which is influenced by both anthropomorphic features and the cultural or religious context of the end user. Furthermore, it will (3) discuss how to measure the acceptability of social robots in classrooms before addressing (4) the usability of social robots to support education.

2 What is Acceptability—The "Uncanny Valley" Phenomenon: Acceptability and Limitations

Social robots, defined as physically embodied artificial intelligence (AI) systems, are designed to interact and communicate with humans (Nazir et al., 2023). These machines, capable of performing various tasks, have increasingly integrated into numerous sectors of everyday life. The number and variety of social robots have expanded dramatically in the past decade, alongside significant growth in research exploring their use in natural environments. However, this proliferation does not guarantee automatic acceptance by potential users. A recent review by David et al. (2022), examining 43 studies conducted between 2010 and 2020, revealed that while 67.4% of participants generally exhibit positive attitudes toward social robots, a notable 11.6% express ambivalence. Common concerns include the appropriateness of social robots in assisting vulnerable populations, such as children, the elderly, and individuals with disabilities, as well as fears that robots might replace human workers. Therefore, it is critical to investigate the factors influencing people's perceived acceptability of robots, and how the perceived acceptability of robotic systems can be enhanced.

Ethology emphasizes that humans, as a social species, are naturally inclined to respond to interaction signals. As Nazir et al. (2023) points out, the design of social robots should consider that certain morphological and behavioral cues can evoke primary social responses in humans. For instance, physical features like a high forehead, large eyes, and a small mouth activate the "child schema" in humans (Lorenz, 1971), triggering nurturing behaviors in adults. Another key aspect of human social behavior is anthropomorphism, the tendency to attribute human-like qualities to non-human agents (Epley et al., 2007). This mechanism, shaped by biological and cultural evolution, allows humans to understand others—whether animate or inanimate—by projecting mental states onto them, aiding in both social interaction and the anticipation of potential dangers. When interacting with non-human entities such as robots, individuals often attribute emotions, intentions, and cognition to them.

Spatola et al. (2022) identified three distinct interpretations of anthropomorphism: (1) mentalization—the interpretation of behaviors in terms of mental states, (2) humanization—treating non-human entities as though they were human, and (3) spiritualism—attributing a spiritual nature to inanimate objects. Although anthropomorphism appears to be a universal phenomenon, some researchers have questioned whether cultural differences influence this tendency (Jensen & Blok, 2013). According to Jensen & Blok, several factors may shape the anthropomorphization of robots, including familiarity with robots, personal experiences, media exposure, and the technological development of a given society. Others have emphasized the importance of historical and religious contexts. For example, Japan, influenced by animist beliefs rooted in Shintoism, demonstrates a strong propensity for anthropomorphizing robots, a tendency further reinforced by the nation's high technological literacy. These cultural characteristics contribute to higher levels of acceptability for social robots compared to societies with different cultural norms.

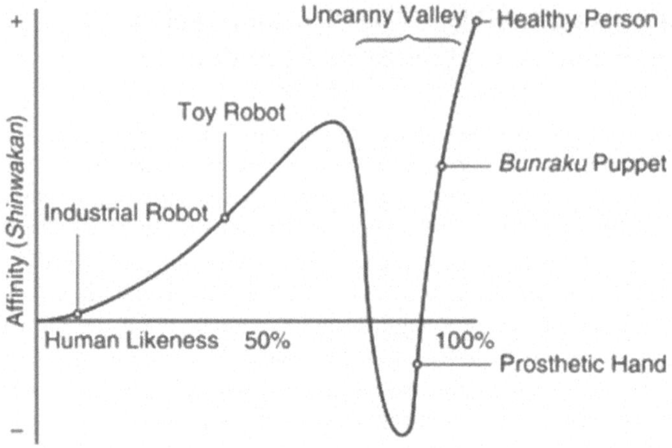

Fig. 1 The graph illustrates the uncanny valley, depicting the relationship between a robot's human likeness and the perceiver's affinity for it (Mori et al., 2012)

The role of media representations is also significant. While in many Eastern cultures robots are portrayed as "friends," in Western contexts—especially in the United States—robots are often depicted as "enemies" or "threats," a portrayal with a long cinematic tradition. Despite these varying portrayals, it is universally recognized that technological advancements have enabled robots to evolve from service machines with specific functions (e.g., mechanical arms in factories) to increasingly human-like entities, extending beyond their original functional scope.

These developments have led researchers to question whether a robot's resemblance to humans is beneficial or detrimental in terms of user acceptance. Mori et al. (2012) were among the first to theorize the phenomenon of the Uncanny Valley, observing that people's emotional responses to robots shift from empathy to repulsion when robots become too similar to humans (Fig. 1).

As noted by Wang et al. (2015), several explanations have been proposed for the Uncanny Valley effect. One hypothesis suggests that it is a perceptual phenomenon, linked to perceptual realism, occurring during the initial processing of visual stimuli. Another explanation is the expectation violation hypothesis (Saygin et al., 2012), which posits that the discomfort arises when a humanoid robot fails to meet the expectations humans project onto it. This phenomenon has been studied not only in relation to a robot's physical appearance but also in terms of its movement (whether natural or unnatural) and its language (whether human-like or mechanical).

In recent years, the Uncanny Valley has become one of the most extensively studied topics in human-robot interaction research. The goal is to mitigate its effects in order to design and implement humanoid robots that are widely accepted. This is particularly crucial in sensitive contexts such as education, where the acceptability of robots by both children and educators plays a critical role in their successful integration.

3 Acceptability of Socially Assistive Robots (SAR) in School Contexts

3.1 Measuring Acceptability in the Classroom

Introducing children to robots in an appropriate and safe manner is essential. According to David et al. (2022), it is not enough for interactions between humans and robots to be physically safe; they must also be perceived as psychologically safe and socially acceptable. This need for safety is particularly crucial when introducing children, who are considered a vulnerable group in research settings, to robots. Therefore, understanding how children of various ages evaluate robot acceptability and how this can be measured in educational contexts is vital.

While children often attribute human characteristics to robots (e.g., Beran et al., 2011; Manzi et al., 2020), research indicates that certain factors are assessed differently in robots than in humans. In educational contexts, children often express enthusiasm and interest in learning through technological advancements. However, studies show they do not trust smart devices as much as they trust other humans (Wojcik et al., 2022). Interestingly, when focusing on children's interactions with social robots, research suggests that children may find social robots more reliable than humans (Stower et al., 2024). A study by Flanagan et al. (2024) highlights the importance of robots providing accurate and credible information, as school-aged children tend to be more skeptical if a robot cannot deliver correct information. However, children appear more forgiving and regain trust if the robot apologizes for its mistakes, more so than when a human makes a similar apology. This difference may be due to human errors being perceived as more intentional than robot errors (Stower et al., 2024). Thus, robots introduced in classrooms should not only provide accurate information but also have mechanisms to correct errors to build and maintain trust among students.

The acceptability of social robots is often assessed either before or after interactions. Models such as the Almere Model or the Unified Theory of Acceptance and Use of Technology (UTAUT) are frequently used to evaluate acceptability (David et al., 2022). However, these models are typically applied to older populations and are not adapted to educational settings. Some studies have tested the general acceptability of social robots among school-aged children. For example, Al-Taee et al. (2016) measured the acceptance of specific services offered by a NAO robot among children aged 6 to 16, using a 10-item questionnaire administered after interaction. The results indicated high acceptability, particularly among younger children. However, this study focused on a clinical sample of children with Type 1 diabetes, and, like the Almere and UTAUT models, the questionnaire was not specifically tailored to educational contexts.

More tailored tools, such as the Technology-Specific Expectation Scale (TSES) and the Technology-Specific Satisfaction Scale (TSSS), have been developed to measure expectations and satisfaction with a social tutor robot in educational settings. These scales were tested with children aged 14 to 16, who completed the TSES before interacting with the robot and the TSSS afterward. The small gap between the TSES and TSSS scores suggests that the children's prior expectations largely aligned with their post-interaction satisfaction. The authors argue that both expectations and satisfaction are critical when assessing the effectiveness of robots in educational contexts, as these factors can influence learning outcomes (Alves-Oliveira et al., 2015). These findings indicate that the TSES and TSSS scales could be valuable tools for assessing human-robot interactions in school settings.

3.2 *Teachers' Acceptability of Robots*

While the development of educational technologies, including robots, has progressed rapidly, teachers' acceptance of robots in the classroom has lagged behind. Many teachers remain hesitant to accept robots in their educational environments, likely due to concerns that robots could replace their roles. Teachers' acceptability of robots seems to be closely tied to their beliefs about the real educational value these technologies can offer. Several scholars (Krägeloh et al., 2019; Nam, 2019) have emphasized the importance of distinguishing between "acceptability"—the evaluation that occurs before use or implementation—and "acceptance," which occurs after practical experience with the technology. David et al. (2022) described acceptability as a process that evolves through three phases: a priori acceptability, actual acceptability, and appropriation. The first phase occurs before direct experience, when users form subjective perceptions of the technology. After initial experiences, users begin to form judgments, which may involve both objective measures and subjective satisfaction. The final stage involves integrating the technology into daily life. Understanding the factors that drive this process is key to predicting whether a technology will be adopted in practice.

Research has shown that hands-on experience with a robot can significantly alter perceptions of acceptability. De Graaf et al. (2018) noted a discrepancy between pre-use and post-use beliefs, with many studies agreeing that direct interaction with robots increases users' acceptability levels. Teacher training appears to be a crucial factor in fostering the acceptability of social robots. Early exposure to robotics during teacher education may help ease teachers' concerns and enhance their willingness to incorporate robots into their pedagogy.

However, studies on pre-service teachers have not been encouraging. Istenic et al. (2021) found that a sample of 121 pre-service teachers exhibited resistance to the idea of using robots in classrooms. Similar results were reported by Istenic and colleagues in 2024 (Istenič et al., 2024), where concerns centered around robots' inability to develop social skills and form meaningful connections with students. Dutch teachers also expressed reservations about robots' ability to support children's emotional development (van Ewijk et al., 2020). Similarly, Slovenian and Russian pre-service teachers believed that the socio-emotional development of children should be handled by human teachers, not robots. A study conducted in Sweden, Portugal, and the UK revealed that one of the primary concerns is the potential dehumanization of children through increased interaction with robots (Serholt et al., 2017).

Currently, only 4.62% of studies on the acceptability of robotic systems focus on education, and the available data are often inconsistent (David et al., 2022). For example, a study by Conti et al. (2017b) found skepticism among professionals working with intellectual disabilities, whereas university students demonstrated more positive perceptions. These mixed results underscore the need for further research to understand how teachers might integrate robots into their educational practices and what factors could increase their acceptability. Future research should aim to identify the conditions under which teachers would be willing to engage with robots in their classrooms and how to foster greater acceptance of this emerging technology.

4 Usability of Social Robots to Support Education

4.1 Strengths and Weaknesses in Educational Contexts

The use of social robots in educational settings presents both significant opportunities and challenges. While robotics has found its way into classrooms worldwide, partly due to the rise of STEM education, its actual integration into daily teaching practices remains limited. Many teachers are hesitant to adopt these technologies, yet a growing body of research highlights the educational benefits that social robots can offer across various age groups.

One notable area of success is the use of social robots in educational interventions, particularly with techniques like fast mapping (FM), which helps children learn new words by comparing them with known ones. Esfandbod et al. (2023) demonstrated that social robots enhance children's engagement, attention, and retention of new words in such interventions. Similarly, research has shown positive results in fields like reading (Caruana et al., 2023; Liu et al., 2024) and storytelling (Conti et al., 2017b; Kori & Breazeal, 2014; Augello et al., 2023). For instance, Caruana et al. (2023) found that social robots create a non-judgmental environment that encourages children to improve their reading skills. Children perceived the robots as intelligent enough to read, listen, and comprehend stories. Moreover, robots have proven more engaging than human tutors in storytelling, thanks to their ability to exhibit social behaviors and gestures, which captivate children's attention (Conti et al., 2017b).

Another growing application is Robot-Assisted Language Learning (RALL), where social robots play an important role in foreign language acquisition. Effective language learning relies heavily on practice and interaction, and studies show that interacting with physical robots leads to better outcomes than virtual interactions (Al Hakim et al., 2022) or the use of tablets (Koniji et al., 2022). This is particularly effective in dialogue exercises, where robots help reduce anxiety and frustration (Alemi et al., 2014; Cumbal et al., 2020). RALL has shown promising results not only among children and adults but also in populations with severe learning difficulties (Vogt et al., 2017) and children with autism spectrum disorder (Fuglerud & Solheim, 2018).

Additionally, a key area of current research focuses on the ability of social robots to convey emotions during interactions. Konijn and Hoorn (2017) highlighted that users can form emotional bonds with robots, which could be leveraged to enhance engagement and learning. Emotional feedback from robots in response to student performance appears to play a crucial role in motivating students to learn (Błażejowska et al., 2023). Developing robots that can accurately express and respond to emotions may result in deeper engagement and better learning outcomes. Ongoing research seeks to understand individual differences in how well users interpret robot emotional expressions (Cirasa et al., 2024).

Despite these clear strengths, several limitations still hinder the widespread use of social robots in classrooms. One major challenge is the difficulty in synchronizing robot behavior with human interactions. Timing mismatches between a robot's responses and a student's actions can affect the robot's perceived trustworthiness, potentially undermining its effectiveness (Maehigashi & Yamada, 2024). Another significant issue is the learning curve associated with operating and programming these robots. While many have simple interfaces, teachers need time and training to effectively incorporate them into their teaching practices, and not all are willing to make that investment.

Cultural factors also play a role in limiting the usability of social robots in diverse educational contexts. Studies have shown that accents and dialects influence how students perceive robots, with local or native accents fostering greater trust (Foster & Stuart-Smith, 2023; Kühne et al., 2024; Sievers & Russwinkel, 2023). This could pose challenges in multicultural classrooms, where students from different backgrounds might respond differently to the robot's language and communication style, potentially disadvantaging certain groups. Similarly, cultural differences have been observed regarding the robot's voice gender, with studies indicating that male and female voices can elicit different perceptions (Makenova et al., 2018), potentially creating biases that could affect students' learning experiences.

Addressing these limitations is crucial for social robots to become effective educational tools. Ongoing research continues to explore how to mitigate these challenges, aiming to improve robot synchronization with human behaviors, make programming interfaces more accessible, and adapt robots to meet the diverse cultural needs of classroom settings. By overcoming these hurdles, social robots could become a valuable asset for enhancing learning outcomes across a range of educational contexts.

5 Discussion

Social robots are designed to engage with humans across various settings, including education, where they offer potential benefits for enhancing learning and engagement. Technological advancements in recent years have enabled these robots to develop human-like behaviors, making them more capable of fostering social interactions. However, despite these developments, the use of social robots in educational contexts remains limited, and their full potential has yet to be fully realized.

One of the key challenges in implementing social robots in schools is the discrepancy in acceptance between teachers and students. Research shows that children generally exhibit higher levels of acceptability toward interacting with social robots. They view these robots as reliable and engaging learning tools, and their interaction with them can improve skills in areas like reading, language learning, and social-emotional development. Studies suggest that robots can help create non-judgmental learning environments, facilitate foreign language learning, and reduce social anxieties in students. Moreover, robots' ability to engage students emotionally and respond to their performance positively impacts motivation and learning outcomes.

In contrast, teachers—across various geographic locations—remain hesitant to embrace social robots in their classrooms. A major concern among educators is the fear of being replaced by these technologies. Teachers also question the social capabilities of robots and whether they can truly support student learning and emotional development. This skepticism persists despite evidence that robots can complement teachers' work by engaging students in interactive and supportive ways, particularly when dealing with large or diverse classrooms.

Further research over the past decade has sought to identify the factors influencing teachers' acceptance of robots. Key findings suggest that teachers' concerns primarily stem from a perceived threat to their professional roles and skepticism about robots' ability to socially interact with students at a meaningful level. To address these concerns, efforts have been made to improve the design of social robots, both visually and behaviorally, to ensure that they foster positive social interactions without triggering the "Uncanny Valley" effect—a phenomenon where robots appear almost, but not quite, human-like, causing discomfort.

Additionally, the voices, accents, gestures, and overall communication patterns of robots are being fine-tuned to better align with the diverse needs of different user groups. This personalization helps improve robot acceptance, not only among students but also among teachers, by making the robots appear more relatable and capable of meaningful interaction.

6 Conclusions

After discussing the use of social robots in educational settings, it is crucial to conclude that social robots are not intended to replace teachers or educators. Rather, they are meant to serve as supportive tools that can enhance teaching and learning, both in the classroom and in home-based educational settings. By taking on repetitive tasks, offering individualized support, or providing additional engagement, social robots can free up teachers to focus on more complex and creative aspects of education. However, for this integration to be successful, more attention needs to be paid to teacher training and professional development so that educators feel confident in using robots as part of their pedagogical practice.

While the integration of social robots into educational settings holds promise, their widespread use is still in its infancy. Future research must continue to address both the technological development of these robots and the human factors that influence their acceptance, particularly from teachers. By overcoming barriers to acceptance and addressing the concerns of educators, social robots can become a valuable addition to modern classrooms, offering enhanced learning experiences and complementing the essential role of human teachers.

References

Al Hakim, V. G., Yang, S. H., Liyanawatta, M., Wang, J. H., & Chen, G. D. (2022). Robots in situated learning classrooms with immediate feedback mechanisms to improve students' learning performance. *Computers & Education, 182*, 104483. https://doi.org/10.1016/j.compedu.2022.104483

Al-Taee, M. A., Kapoor, R., Garrett, C., & Choudhary, P. (2016). Acceptability of robot assistant in management of type 1 diabetes in children. *Diabetes Technology & Therapeutics, 18*(9), 551–554. https://doi.org/10.1089/dia.2015.0428

Alemi, M., Meghdari, A., & Ghazisaedy, M. (2014). Employing humanoid robots for teaching English language in Iranian junior high-schools. *International Journal of Humanoid Robotics, 11*(03), 1450022. https://doi.org/10.1142/S0219843614500224

Alves-Oliveira, P., Ribeiro, T., Petisca, S., Di Tullio, E., Melo, F. S., & Paiva, A. (2015). An empathic robotic tutor for school classrooms: Considering expectation and satisfaction of children as end-users. In *Social Robotics: 7th International Conference, ICSR 2015*, Paris, France, October 26–30, 2015, Proceedings 7 (pp. 21–30). Springer International Publishing. https://doi.org/10.1007/978-3-319-25554-5_3.

Augello, A., Città, G., Gentile, M., & Lieto, A. (2023). A storytelling robot managing persuasive and ethical stances via act-r: An exploratory study. *International Journal of Social Robotics, 15*(12), 2115–2131. https://doi.org/10.1007/s12369-021-00847-w

Asimov, I. (1940). *I. robot.* Narkaling Productions.

Balzotti, A., De Carolis, B., Massaro, S., Perla, L., & Rossano, V. (2023). Healthy Pepper: Nutritional Education Through Social Robotics and Storytelling. In *2023 IEEE International Conference on Advanced Learning Technologies (ICALT)* (pp. 121–123). IEEE. https://doi.org/10.1109/ICALT58122.2023.00041.

Bartneck, C., & Forlizzi, J. (2004, September). A design-centred framework for social human-robot interaction. In *RO-MAN 2004. 13th IEEE International Workshop on Robot and Human*

Interactive Communication (IEEE Catalog No. 04TH8759) (pp. 591–594). IEEE. https://doi.org/10.1109/ROMAN.2004.1374827.

Belpaeme, T., Kennedy, J., Ramachandran, A., Scassellati, B., & Tanaka, F. (2018). Social robots for education: A review. *Science robotics*, 3(21), eaat5954.

Beran, T. N., Ramirez-Serrano, A., Kuzyk, R., Fior, M., & Nugent, S. (2011). Understanding how children understand robots: Perceived animism in child–robot interaction. *International Journal of Human-Computer Studies, 69*(7–8), 539–550. https://doi.org/10.1016/j.ijhcs.2011.04.003

Błażejowska, G., Gruba, Ł, Indurkhya, B., & Gunia, A. (2023). A study on the role of affective feedback in robot-assisted learning. *Sensors, 23*(3), 1181. https://doi.org/10.3390/s23031181

Breazeal, C. (2003). Toward sociable robots. *Robotics and Autonomous Systems, 42*(3–4), 167–175. https://doi.org/10.1016/S0921-8890(02)00373-1

Caruana, N., Moffat, R., Miguel-Blanco, A., & Cross, E. S. (2023). Perceptions of intelligence & sentience shape children's interactions with robot reading companions. *Scientific Reports, 13*(1), 7341. https://doi.org/10.1038/s41598-023-32104-7

Castellano, G., De Carolis, B., D'Errico, F., Macchiarulo, N., & Rossano, V. (2021). PeppeRecycle: Improving children's attitude toward recycling by playing with a social robot. *International Journal of Social Robotics, 13*(1), 97–111. https://doi.org/10.1007/s12369-021-00754-0

Cirasa, C., Høgsdal, H., & Conti, D. (2024). "I see what you feel": An exploratory study to investigate the understanding of robot emotions in deaf children. *Applied Sciences, 14*(4), 1446. https://doi.org/10.3390/app14041446

Chen, H., Park, H. W., & Breazeal, C. (2020). Teaching and learning with children: Impact of reciprocal peer learning with a social robot on children's learning and emotive engagement. *Computers & Education, 150*, 103836.

Chookaew, S., & Panjaburee, P. (2022). Implementation of a robotic-transformed five-phase inquiry learning to foster students' computational thinking and engagement: A mobile learning perspective. *International Journal of Mobile Learning and Organisation, 16*(2), 198–220. https://doi.org/10.1504/IJMLO.2022.121888

Conti, D., Di Nuovo, A., Cirasa, C., & Di Nuovo, S. (2017a). A comparison of kindergarten storytelling by human and humanoid robot with different social behavior. In *Proceedings of the companion of the 2017 ACM/IEEE international conference on human-robot interaction* (pp. 97–98). https://doi.org/10.1145/3029798.3038359.

Conti, D., Di Nuovo, S., Buono, S., & Di Nuovo, A. (2017b). Robots in education and care of children with developmental disabilities: A study on acceptance by experienced and future professionals. *International Journal of Social Robotics, 9*, 51–62. https://doi.org/10.1007/s12369-016-0359-6

Cumbal, R., Lopes, J., & Engwall, O. (2020). Detection of listener uncertainty in robot-led second language conversation practice. In *Proceedings of the 2020 International Conference on Multimodal Interaction* (pp. 625–629). https://doi.org/10.1145/3382507.3418873.

Coşkun, T. K., & Filiz, O. (2023). The impact of twenty-first century skills on university students' robotic achievements. *Education and Information Technologies, 28*(12), 16255–16283.

David, D., Thérouanne, P., & Milhabet, I. (2022). The acceptability of social robots: A scoping review of the recent literature. *Computers in Human Behavior, 137*, 107419. https://doi.org/10.1016/j.chb.2022.107419

de Graaf, M. M. (2016). An ethical evaluation of human–robot relationships. *International Journal of Social Robotics, 8*, 589–598. https://doi.org/10.1007/s12369-016-0368-5

de Graaf, M. M., Ben Allouch, S., & van Dijk, J. A. (2018). A phased framework for long-term user acceptance of interactive technology in domestic environments. *New Media & Society,* 20(7), 2582–2603. https://doi.org/10.1177/1461444817727264.

Epley, N., Waytz, A., & Cacioppo, J. T. (2007). On seeing human: A three-factor theory of anthropomorphism. *Psychological Review, 114*(4), 864. https://doi.org/10.1037/0033-295X.114.4.864

Esfandbod, A., Rokhi, Z., Meghdari, A. F., Taheri, A., Soleymani, Z., Alemi, M., & Karimi, M. (2023). Fast mapping in word-learning: A case study on the humanoid social robots' impacts on Children's performance. *International Journal of Child-Computer Interaction, 38*, 100614. https://doi.org/10.1016/j.ijcci.2023.100614

Flanagan, T., Georgiou, N. C., Scassellati, B., & Kushnir, T. (2024). School-age children are more skeptical of inaccurate robots than adults. *Cognition, 249*, 105814. https://doi.org/10.1016/j.cognition.2024.105814

Foster, M. E., & Stuart-Smith, J. (2023, March). Social robotics meets sociolinguistics: investigating accent bias and social context in HRI. In *Companion of the 2023 ACM/IEEE International Conference on Human-Robot Interaction* (pp. 156–160). https://doi.org/10.1145/3568294.3580063.

Fuglerud, K. S., & Solheim, I. (2018). The use of social robots for supporting language training of children. In *Transforming Our World Through Design, Diversity And Education* (pp. 401–408). IOS Press. https://doi.org/10.3233/978-1-61499-923-2-401.

Hsu, T. C., Chang, C., Wu, L. K., & Looi, C. K. (2022). Effects of a pair programming educational robot-based approach on students' interdisciplinary learning of computational thinking and language learning. *Frontiers in Psychology, 13*, 888215. https://doi.org/10.3389/fpsyg.2022.888215

Istenic, A., Bratko, I., & Rosanda, V. (2021). Are pre-service teachers disinclined to utilise embodied humanoid social robots in the classroom? *British Journal of Educational Technology, 52*(6), 2340–2358. https://doi.org/10.1111/bjet.13144

Istenič, A., Latypova, L., Rosanda, V., Turk, Ž, Valeeva, R., & Zhai, X. (2024). Reluctance to authenticity-imbued social robots as Child-Interaction partners. *Education Sciences, 14*(4), 390. https://doi.org/10.3390/educsci14040390

Jensen, C. B., & Blok, A. (2013). Techno-animism in Japan: Shinto cosmograms, actor-network theory, and the enabling powers of non-human agencies. *Theory, Culture & Society, 30*(2), 84–115. https://doi.org/10.1177/0263276412456564

Kanero, J., Oranç, C., Koşkulu, S., Kumkale, G. T., Göksun, T., & Küntay, A. C. (2022). Are tutor robots for everyone? The influence of attitudes, anxiety, and personality on robot-led language learning. *International Journal of Social Robotics, 14*(2), 297–312. https://doi.org/10.1007/s12369-021-00789-3

Konijn, E. A., & Hoorn, J. F. (2017). Parasocial interaction and beyond: Media personae and affective bonding. *The international encyclopedia of media effects*, 1–15. https://doi.org/10.1002/9781118783764.wbieme0071.

Konijn, E. A., Jansen, B., Mondaca Bustos, V., Hobbelink, V. L., & Preciado Vanegas, D. (2022). Social robots for (second) language learning in (migrant) primary school children. *International Journal of Social Robotics, 14*(3), 827–843. https://doi.org/10.1007/s12369-021-00824-3

Kory, J., & Breazeal, C. (2014). Storytelling with robots: Learning companions for preschool children's language development. In *The 23rd IEEE international symposium on robot and human interactive communication* (pp. 643–648). IEEE. https://doi.org/10.1109/ROMAN.2014.6926325.

Krägeloh, C. U., Bharatharaj, J., Sasthan Kutty, S. K., Nirmala, P. R., & Huang, L. (2019). Questionnaires to measure acceptability of social robots: A critical review. *Robotics, 8*(4), 88. https://doi.org/10.3390/robotics8040088

Kühne, K., Herbold, E., Bendel, O., Zhou, Y., & Fischer, M. H. (2024). "Ick bin een Berlina": Dialect proficiency impacts a robot's trustworthiness and competence evaluation. *Frontiers in Robotics and AI, 10*, 1241519. https://doi.org/10.3389/frobt.2023.1241519

Liu, X., Ma, J., & Wang, Q. (2024). A social robot as your reading companion: Exploring the relationships between gaze patterns and knowledge gains. *Journal on Multimodal User Interfaces, 18*(1), 21–41. https://doi.org/10.1007/s12193-023-00418-5

Lorenz, K. (1971). *Studies on animal and human behavior*. Harvard University Press.

Maehigashi, A., & Yamada, S. (2024, March). Effect of wait time on trust and reliance in Human-Robot Interaction. In *Companion of the 2024 ACM/IEEE International Conference on Human-Robot Interaction* (pp. 722–726). https://doi.org/10.1145/3610978.3640611.

Makenova, R., Karsybayeva, R., & Sandygulova, A. (2018). Exploring cross-cultural differences in persuasive robotics. In *Companion of the 2018 ACM/IEEE International Conference on Human-Robot Interaction* (pp. 185–186). https://doi.org/10.1145/3173386.31770799.

Manzi, F., Peretti, G., Di Dio, C., Cangelosi, A., Itakura, S., Kanda, T., & Marchetti, A. (2020). A robot is not worth another: Exploring children's mental state attribution to different humanoid robots. *Frontiers in Psychology, 11*, 2011. https://doi.org/10.3389/fpsyg.2020.02011

Mori, M., MacDorman, K. F., & Kageki, N. (2012). The uncanny valley [from the field]. *IEEE Robotics & Automation Magazine, 19*(2), 98–100. https://doi.org/10.1109/MRA.2012.2192811

Nam, T. (2019). Citizen attitudes about job replacement by robotic automation. *Futures, 109*, 39–49. https://doi.org/10.1016/j.futures.2019.04.005

Nazir, T. A., Lebrun, B., & Li, B. (2023). Improving the acceptability of social robots: Make them look different from humans. *PLoS ONE, 18*(11), e0287507. https://doi.org/10.1371/journal.pone.0287507

Okita, S. Y., Ng-Thow-Hing, V., & Sarvadevabhatla, R. (2009). Learning together: ASIMO developing an interactive learning partnership with children. In *RO-MAN 2009-The 18th IEEE International Symposium on Robot and Human Interactive Communication* (pp. 1125–1130). IEEE. https://doi.org/10.1109/ROMAN.2009.5326135.

Randall, N. (2019). A survey of robot-assisted language learning (RALL). *ACM Transactions on Human-Robot Interaction (THRI), 9*(1), 1–36. https://doi.org/10.1145/3345506

Sarrica, M., Brondi, S., & Fortunati, L. (2020). How many facets does a "social robot" have? A review of scientific and popular definitions online. *Information Technology & People, 33*(1), 1–21. https://doi.org/10.1108/ITP-04-2018-0203

Saygin, A. P., Chaminade, T., Ishiguro, H., Driver, J., & Frith, C. (2012). The thing that should not be: Predictive coding and the uncanny valley in perceiving human and humanoid robot actions. *Social Cognitive and Affective Neuroscience, 7*(4), 413–422. https://doi.org/10.1093/scan/nsr025

Serholt, S., Barendregt, W., Vasalou, A., Alves-Oliveira, P., Jones, A., Petisca, S., & Paiva, A. (2017). The case of classroom robots: Teachers' deliberations on the ethical tensions. *Ai & Society, 32*, 613–631. https://doi.org/10.1007/s00146-016-0667-2

Sievers, T., & Russwinkel, N. (2023, December). Talking like one of us: Effects of using regional language in a humanoid social robot. In *International Conference on Social Robotics* (pp. 71–80). Singapore: Springer Nature Singapore. https://doi.org/10.1007/978-981-99-8718-4_7.

Song, H., Huang, S., Barakova, E., Ham, J., & Markopoulos, P. (2023, July). How Social Robots can Influence Motivation as Motivators in Learning: A Scoping Review. In *Proceedings of the 16th International Conference on Pervasive Technologies Related to Assistive Environments* (pp. 313–320). https://doi.org/10.1145/3594806.3604591.

Spatola, N., Marchesi, S., & Wykowska, A. (2022). Different models of anthropomorphism across cultures and ontological limits in current frameworks the integrative framework of anthropomorphism. *Frontiers in Robotics and AI, 9*, 863319. https://doi.org/10.3389/frobt.2022.863319

Stower, R., Kappas, A., & Sommer, K. (2024). When is it right for a robot to be wrong? Children trust a robot over a human in a selective trust task. *Computers in Human Behavior, 157*, 108229. https://doi.org/10.1016/j.chb.2024.108229

Van den Berghe, R., Oudgenoeg-Paz, O., Verhagen, J., Brouwer, S., De Haas, M., De Wit, J., & Leseman, P. (2021). Individual differences in children's (language) learning skills moderate effects of robot-assisted second language learning. *Frontiers in Robotics and AI, 8*, 676248. https://doi.org/10.3389/frobt.2021.676248

Van den Berghe, R., Verhagen, J., Oudgenoeg-Paz, O., Van der Ven, S., & Leseman, P. (2019). Social robots for language learning: A review. *Review of Educational Research, 89*(2), 259–295. https://doi.org/10.3102/0034654318821286

Van Ewijk, G., Smakman, M., & Konijn, E. A. (2020). Teachers' perspectives on social robots in education: an exploratory case study. In *Proceedings of the interaction design and children conference* (pp. 273–280). https://doi.org/10.1145/3392063.3394397.

Vogt, P., De Haas, M., De Jong, C., Baxter, P., & Krahmer, E. (2017). Child-robot interactions for second language tutoring to preschool children. *Frontiers in Human Neuroscience, 11*, 73. https://doi.org/10.3389/fnhum.2017.00073

Wang, S., Lilienfeld, S. O., & Rochat, P. (2015). The uncanny valley: Existence and explanations. *Review of General Psychology, 19*(4), 393–407. https://doi.org/10.1037/gpr0000056

Wojcik, E. H., Prasad, A., Hutchinson, S. P., & Shen, K. (2022). Children prefer to learn from smart devices, but do not trust them more than humans. *International Journal of Child-Computer Interaction, 32*, 100406. https://doi.org/10.1016/j.ijcci.2021.100406

Zhang, X., Chen, Y., Li, D., Hu, L., Hwang, G.-J., & Tu, Y.-F. (2024). Engaging young students in effective robotics education: An embodied Learning-Based computer programming approach. *Journal of Educational Computing Research, 62*(2), 532–558. https://doi.org/10.1177/07356331231213548

“To Teach is to Learn Twice”: Using Social Robots as Tutees in Inclusive Education

Rita Cersosimo

Abstract The field of education is always looking for new ways to improve student learning, and one promising methodology is learning by teaching. This approach, well represented by the saying “to teach is to learn twice”, involves students teaching what they have learned to their peers. One possible realization is peer tutoring, where a tutor and a tutee learn in pairs; this helps students understand the material better, remember it longer, and boosts their motivation. But what happens if the tutee is a social robot? The learning by teaching approach has shown great potential in various subjects. Studies have also highlighted the promising results of this approach for students with Special Educational Needs (SEN). This chapter aims to provide a pedagogical framework for using robots as tutees, with students as tutors, especially for students with SEN. Interaction with robots can help these students develop important skills that can be used in human interactions. We suggest that teaching robots can serve as a preliminary step before engaging in actual peer tutoring for several reasons: it helps students develop language skills, model effective learner behaviors, address the need for explicit and flexible instruction for students with SEN, and enhance social and emotional competencies. The chapter also discusses ethical considerations and emphasizes the need for clear rules and roles to prevent inappropriate attachment to robots.

Keywords Social robots · Special educational needs · Learning by teaching · Peer tutoring

1 Introduction

Anyone who has ever taught something must have experienced the sensation of learning at the same time. This is not only because the interaction with students is enriching, but also because, when preparing to explain a topic, we need to organize our knowledge and ensure that everything is clear to us.

R. Cersosimo (✉)
University of Genoa, Genoa, Italy
e-mail: rita.cersosimo@unige.it

G. Lampropoulos and S. Papadakis (eds.), *Social Robots in Education*, Studies in Computational Intelligence 1194, https://doi.org/10.1007/978-3-031-82915-4_10

Several pedagogical approaches that place students at the center of the learning process involve peer mediation. In these models, knowledge building occurs through negotiation among individuals who share the same language and experiences, including frustrations and difficulties with the subject matter, and doubts are more directly addressed. We can distinguish between situations where there is a tutor and a tutee, who have different levels of knowledge, those where peers cooperate in a group with heterogeneous skills, benefiting from mutual exchange, and those where they collaborate, having similar abilities (Damon & Phelps, 1989).

The first scenario can be referred to as peer tutoring, that involves two students working together in a structured manner, with one taking the role as tutor, and the other taking the role as tutee. In peer tutoring, each student assumes a different role depending on whether they are the tutor or the tutee; while mutuality can vary depending on the receptivity of the tutee, it generally tends to be limited, as the interaction is dominated by the tutor. Tutors' perceptions of their role motivate them to learn the material, and their learning is supported by discussion and explanation, revisiting fundamentals, making links between conceptual areas, testing and clarifying their understanding, reorganizing and building ideas. Moreover, mental rehearsal of peer tutoring episodes helps them appreciate weaknesses in their own subject knowledge (Galbraith & Winterbottom, 2010).

The asymmetry between the roles of tutor and tutee can present significant challenges in classroom instruction, and research indicates that tutees may not learn as effectively as tutors (King et al., 1998; Walker et al., 2014; Thurston et al., 2021). One proposed solution to this issue is to alternate the roles of tutor and tutee among students. However, this approach requires additional time, effectively doubling the duration compared to traditional instructional methods where all students complete the same exercise simultaneously. Another promising solution emerges from research on educational technologies, specifically the development of virtual peers or teachable agents. These technologies ensure that each student consistently assumes the role of tutor, thereby maximizing the learning benefits associated with tutoring (Biswas et al., 2005; Schwartz et al., 2007). Virtual peers can provide immediate and continuous opportunities for students to engage in teaching, thereby enhancing their understanding and retention of the material. In this context, integrating virtual peers into classroom activities can address the challenges posed by the tutortutee asymmetry. By ensuring that every student has the opportunity to be a tutor, educational technologies can facilitate more effective and inclusive learning environments.

Interestingly, peer tutoring has been recognized as an effective strategy for promoting the inclusion of students with Special Educational Needs (SEN). This approach fosters interactions and relationships among peers, facilitating a supportive learning environment (Toulia et al., 2023). According to a metaanalytic review by Bowman-Perrott et al. (2013), there is no significant difference in learning gains between students with disabilities and their peers; however, the effect size is larger for students with disabilities (0.76) compared to those without (0.65). This indicates that while both groups benefit from peer tutoring, students with SEN may experience greater learning gains.

In this chapter, we will explore peer tutoring in relation to SEN and propose how learning by teaching can be enhanced by technology. Specifically, we suggest the integration of social robots into classroom activities as tutees in peer learning scenarios. Social robots and robotic technologies hold significant potential for advancing inclusive education by providing tailored support to diverse learners (see Sect. 3.3). Recent research highlights how social robots can create interactive, engaging learning environments that adapt to individual needs, thereby enhancing student participation and motivation. For instance, several studies found that social robots effectively supported children with autism by facilitating communication and social skills through structured interactions (e.g., Huijnen et al., 2019; Pennazio et al., 2020). A review by Belpaeme et al. (2018) demonstrated that robots could serve as adaptable learning companions, offering personalized feedback that aligns with the unique learning pace of each student. Other recent reviews also emphasized the role of social robots in providing emotional support, helping to create a more inclusive and empathetic classroom environment (Woo et al., 2021). These findings suggest that integrating social robots into educational settings can not only enhance accessibility but also create opportunities for more personalized and effective teaching, making them valuable tools in fostering inclusive education.

In this chapter, we propose that teaching robots enables every student to assume an active role as a tutor. This approach allows interaction with a "simpler" tutee, which can be programmed to align with the individual needs and can explicitly display its reasoning processes. Crucially, the integration of technology in these instructional designs is not intended to replace peer interactions—whose benefits are well-documented (see Sect. 2)—but rather to augment the tutor's teaching abilities, offering an effective model for question-asking and explicit learning. Teaching robots can thus serve as a preparatory phase to traditional peer tutoring.

Before presenting our main ideas, we will outline the pedagogical framework of learning by teaching (Sect. 2). Subsequently, we will review previous studies that have incorporated technology, particularly virtual peers, in this process (Sect. 3). We will then concentrate on the use of social robots as tutees, discussing the limited yet insightful studies that have used them within the learning by teaching framework (Sect. 3.3). In our discussion (Sect. 4), we will examine why social robots should be utilized to maximize the benefits of tutoring for students with SEN. Also, some ethical considerations will be provided in the conclusions (Sect. 5).

2 Theoretical Framework

Considering that learning by teaching is central to many widely-used and interconnected educational practices and methodologies, such as peer tutoring, but also cooperative learning (Johnson & Johnson, 2012), reciprocal teaching (Palincsar & Brown, 1984), flipped classroom (Akçayır & Akçayır, 2018) or small group discussions (Webb, 1982), it is essential for research to identify the pedagogical framework underlying the chosen methodology.

2.1 How Does a Student Learn by Teaching?

According to the classical Piagetian theory, learning is realized through the processes of assimilation and accommodation (Piaget, 1978). Assimilation involves integrating new information into existing cognitive structures, while accommodation requires altering these structures to fit the new information. When learners encounter information that doesn't align with their current understanding, they experience cognitive conflict, prompting either assimilation or accommodation to resolve the discrepancy.

Peer tutoring can enhance learning by balancing cognitive challenges within social interactions (Palinscar, 1998). During peer tutoring, both tutors and tutees encounter cognitive conflict, which drives deeper understanding and helps assimilate new information into existing cognitive structures or accommodate their structures to the new information (De Lisi & Goldbeck, 1999). This process leads to a deeper comprehension and better integration of new concepts. By dealing with cognitive conflict, learners develop critical thinking and reflection skills, essential for meaningful learning and long-term retention.

In peer tutoring, tutors often face questions or viewpoints that differ from their own understanding, something that prompts them to reassess and refine their knowledge. Similarly, tutees may experience cognitive conflict when the tutor presents information that challenges their preconceptions. This dynamic interaction encourages both parties to engage deeply with the material, promoting mutual growth and understanding.

Learning for oneself and learning with the aim of teaching appear to engage distinct cognitive processes. A seminal experiment by Bargh and Schul (1980) compared the learning outcomes of students who studied for their own understanding with those who were meant to teach the material to others; the study found that students who expected to teach performed better. This condition seemed to modify the learning process, encouraging students to invest more effort in selecting relevant information and organizing it into a coherent structure. The authors concluded that the expectancy to teach enhances the learning experience by promoting more effective cognitive strategies. Annis (1983) supported this hypothesis by involving 130 students in five different learning scenarios. Some students were taught the content directly, others read it, some both read and were taught the content, others learned the material with the expectation of teaching it but did not actually teach, and the final group learned the content and then taught it. By controlling variables such as group equivalence and activities, students were evaluated on their understanding of the topic learnt. The results showed that students who acted as teachers performed better, particularly those who had the chance to actually teach the material. Similarly, a relatively more recent study by Fiorella and Mayer (2013) compared two groups of students: those who expected to teach the material and those who actually taught it. Both groups performed better than a control group that only studied the material. However, the group that actually taught showed a deeper and more lasting understanding of the content when tested one week later.

Subsequent research has yielded similar results, showing that students who expect to teach exhibit better retention, improved organization (Nestojko et al., 2014), and enhanced comprehension and metacognitive processes (Muis et al., 2016). Therefore, the expectation to teach seems to alter cognitive processes, prompting students to revise the content and identify its fundamental structure (Duran & Topping, 2017). This expectation also appears to boost motivation (Fiorella & Mayer, 2013, 2014), which, according to Benware and Deci (1984), might also be attributed to the desire to avoid being unprepared in front of an audience.

Interestingly, some studies have found no significant learning benefits from expectancy alone (Kobayashi, 2021; Renkl, 1995). In order for the exchange to be successful, tutors must deliver high-quality explanations, identify areas for their own improvement, and use these insights to reorganize their knowledge and correct mistakes, engaging in meaningful metacognitive activities.

2.2 *How is the Knowledge Built?*

Delivering content to an audience do not only involve the simple sharing of information, known as knowledge telling, but also engaging in metacognitive processes to evaluate learning, comprehension, and the coherence of explanations, referred to as knowledge building (Roscoe & Chi, 2007, 2008). These processes of knowledge building are linked to a deeper understanding in learning by teaching contexts (Roscoe, 2014).

Vygotsky's (1985) concept of the Zone of Proximal Development (ZPD) is highly relevant in this context: it represents the area between what is already known and what it is possible to learn with the help of a mediator. It highlights the potential for cognitive development when learners are supported in performing tasks they cannot yet do independently. In peer tutoring, the tutor provides scaffolding to help the tutee bridge the gap between what they can do alone and what they can achieve with guidance. Scaffolding can take many forms, including questioning, modeling, and providing feedback. These strategies help to advance the tutee's skills and understanding, making the learning process more effective and fostering greater cognitive development within the ZPD. Thus, in the process of knowledge building, tutors not only improve their own understanding but also support their peers in reaching higher levels of comprehension through effective scaffolding.

Learning by teaching do not occur spontaneously (Matsuda et al., 2013): when students lack sufficient prior knowledge of the task and do not know how to teach effectively, they may become confused and learn less than expected. The teacher's role involves understanding what students already know (their level of actual development, LAD) and what they can learn with assistance (their level of potential development, LPD); by providing scaffolding and support within the ZPD, teachers can help students progress, and this support is gradually reduced as students' LAD increases.

Matsuda et al. (2023) identified two types of scaffolding commonly studied in the context of learning by teaching: scaffolding on how to teach (to encourage appropriate teaching activities) and scaffolding on how to solve problems (to ensure the correctness of the solutions taught). They concluded that only scaffolding on how to teach facilitated tutor learning; students who received this type of scaffolding were more likely to alternate between teaching and assessing their peers' competency compared to those who received scaffolding on how to solve problems. These findings suggest the importance of implementing adaptive scaffolding on how to teach to better support learning by teaching.

Students can be trained to be more effective tutors, which enhances the learning benefits for tutees and tutors themselves (Fuchs et al., 1997; King et al., 1998). They should be guided to provide high-quality explanations, recognize areas for improvement, reorganize their knowledge, and make inferences to correct errors. Using examples or different representations in their explanations can further deepen their understanding. However, research shows that tutors do not always fully utilize the learning opportunities presented by their role; often, instead of constructing knowledge, many tutors focus on simply stating information (knowledge telling), tending to respond to questions by summarizing information or describing procedures with minimal preparation or critical thinking. While this knowledge telling can positively impact the tutor's memory, it is not as effective as more engaged and reflective teaching strategies for deeper learning.

2.3 How Do We Learn Together?

In student interactions, the most significant processes occur during the collaboration phase, where students work together to construct knowledge (Graesser et al., 1997), rather than merely discussing it. During this phase, tutors should recognize the ZPD and provide appropriate levels of scaffolding to support their peers' learning.

The success of both tutors and tutees is interconnected, with mutual goals and reliance on each other enhancing learning outcomes, and effective peer tutoring depends on both roles being performed well, creating a positive interdependence that promotes cognitive growth. This motivational component has been emphasized in relation to cooperative learning (Johnson & Johnson, 2012), where individuals work together towards common goals, and the success of each member is linked to the success of the group as a whole. Positive interdependence is achieved through defined interaction patterns for both tutor and tutee, ensuring that each participant understands their role and responsibilities. Individual accountability is essential, requiring both participants to reflect on their performance and that of their partner, fostering a sense of personal responsibility and commitment to the learning process. Interaction patterns must be structured to encourage promotive communication and the development of social skills; this not only enhances academic learning but also helps students develop essential interpersonal skills.

Regarding communication, it is worth noting that the traditional classroom setup typically limits learning interactions to those between the teacher and students; this practice, still prevalent in many classrooms where the teacher dominates instruction, has at least two significant consequences (Duran, 2017): first, it confines the sources of learning assistance to just the teacher, thereby limiting the amount of individual help each student can receive; second, teachers often feel frustrated because they are unable to tailor their teaching to meet the individual educational needs of each student. Today, peer interactions are increasingly recognized as valuable mechanisms for learning. While not all conversations between students promote learning, just as not all interactions between teachers and students do, it is widely acknowledged that students can learn from each other through well-organized interactions. Interactive dialogue plays a crucial role in negotiating meaning, fosters shared self-regulation and scaffolds support within a framework of mutual understanding (Topping, 2005).

Various dialogic behaviors can be taught to students involved in peer tutoring to enhance their interactions and learning outcomes (Duran, 2017): hinting involves giving subtle clues or suggestions rather than direct answers, thereby sharing the cognitive load and responsibility for solving a task; prompting encourages the tutee to complete their knowledge by providing relevant information; splicing refers to the process of weaving the conversation together, requiring high cooperation as tutors carefully insert correct information and make necessary corrections; pumping involves extracting information from the tutee, using positive or neutral feedback or explicit requests to gauge the tutee's reasoning and understanding; finally, summarizing entails having the tutee independently construct the correct answer or solution after collaborative effort, demonstrating their mastery of the content. All these processes occur within the dynamic interactions that characterize peer learning, facilitating the construction of knowledge by the tutor through engagement with the tutee. The enriched interaction between tutor and tutee fosters an environment where both participants can enhance their understanding. This collaborative learning process benefits not only the learner but also the tutor, as it provides increased opportunities for both to develop and refine their knowledge and communicative competence.

2.4 *Is Learning by Teaching Inclusive?*

Inclusive education focuses on addressing the unique needs of students with disabilities and learning differences, providing tailored instruction and support to help them succeed academically and socially. It is aimed at fostering a sense of belonging and promoting equality by ensuring that all students, regardless of their abilities, have access to the same educational opportunities. Inclusion not only benefits students with disabilities by exposing them to the general curriculum and social interactions with their peers, but it also enriches the learning environment for all students by promoting diversity, empathy, and collaboration.

In a whole-class instructional setting, it can be challenging to tailor content to the individual needs of each student. Peer tutoring offers more opportunities for personalized assistance, practice, and clarification for students with SEN as well (Okilwa & Shelby, 2010). Metaanalytic reviews (Bowman-Perrott et al., 2013; Talbott et al., 2017) demonstrated moderate to significant academic improvements in vocabulary, mathematics, reading, and spelling for students with SEN. These reviews also highlighted positive effects on social skills, irrespective of the type of disability. Additionally, Carter et al. (2015) found that peer tutoring can foster friendships even among students with severe disabilities.

One effective model of peer support is reciprocal peer tutoring (Okilwa & Shelby, 2010). In this approach, students with disabilities also take on the role of tutor or coach, which helps to maintain equal status among peers in the classroom. Topping and Ehly (1998) described reciprocal peer tutoring as a robust, consistent, and more equitable form of peer tutoring compared to other models.

Students without disabilities also benefit from participating in peer tutoring. They show improvements in academic performance, behavior, and attitudes towards people with disabilities. Additionally, in a study by Cushing and Kennedy (1997), the teacher selected peer tutors from students without disabilities who had difficulties following directions, staying on task, and completing assignments. Results showed increased engagement and achievement among these students when they served as peer supports. These positive outcomes persisted over time, as indicated by follow-up assessments.

Long-term benefits of peer tutoring were observed in a study by Brooks (2014). Students who had served as peer tutors were more likely to pursue careers in teaching, health sciences, and human services. As adults, they demonstrated higher levels of comfort with hiring, working with, living near, and assisting people with disabilities. These former peer tutors were also more accepting and inclusive of individuals with disabilities.

Overall, peer tutoring can significantly enhance the academic achievement and social skills of students in diverse, inclusive, and self-contained classrooms. This approach seems not only to support students with disabilities but also to foster a more inclusive attitude for all students.

3 The Use of Technologies in Learning by Teaching

Several research studies have indicated that teaching content to others is effective even when the audience is not physically present (for a review see Lachner et al., 2022), because the expectation of teaching itself can lead to positive learning outcomes. However, there are also studies on learning through non-interactive teaching, where explanations are generated for imaginary or absent individuals, which indicate that the act of explaining to a real person may be needed (Hoogerheide et al., 2016). Moreover, the presence of an interacting peer has been shown to be useful for effective knowledge construction, particularly when one is in the role of the tutor.

The integration of technology in learning by teaching paradigms addresses these points effectively. First, technology ensures that every student can take on an active role as a tutor. Second, it allows for interaction with a "simpler" tutee, which can be programmed according to the tutor's needs and whose reasoning processes can be explicitly displayed. Note that the goal of incorporating technology in these designs is not to eliminate peer interactions, which are highly beneficial, as previously discussed. Instead, technology aims to enhance the tutor's teaching capabilities and to provide tutees with an effective model for asking questions and learning explicitly. Consequently, teaching to virtual peers can serve as a preliminary step before engaging in actual peer tutoring; this approach can be particularly useful, for instance, for homework or in situations where face-to-face interaction is limited.

To create educational technological tools that effectively incorporate the concept of learning by teaching in real-world settings, researchers have developed and studied various learning environments that will be briefly described in the following section. They show a trend towards more sophisticated, personalized, and context-aware systems that leverage emerging technologies to create more effective and engaging learning experiences.

3.1 Teachable Agents

A teachable agent is a computer-simulated peer designed for students to teach the content and concepts they have learned. A notable feature of teachable agents is that students can observe the agent's reasoning processes, which provide valuable feedback to reinforce the students' own understanding (Chin et al., 2013). Research on teachable agents has demonstrated that students who engage in teaching these virtual peers achieve greater learning gains compared to those who study independently.

One foundational study by Biswas et al. (2005) introduced the concept of teachable agents through the "Betty's Brain" system, where students taught a virtual character named Betty by constructing concept maps. This approach effectively enhanced students' understanding of scientific topics by requiring them to articulate their knowledge and receive immediate feedback on their instruction. Subsequent studies have reinforced these findings, highlighting the potential of virtual agents to promote active learning and self-regulation skills. For instance, Chase et al. (2009) showed that middle school students who engaged with the "Betty's Brain" system improved their metacognitive skills, as they had to monitor and evaluate both their own and the agent's understanding. In the same study, the authors suggest that the so-called protegé effect extends to teachable agents: "students make greater effort to learn for their teaching agents than they do for themselves" (p. 334).

Schwartz et al. (2007) further explored the role of teachable agents in fostering deep learning by comparing the outcomes of students who taught a virtual agent with those who engaged in traditional study methods. Their research found that students who taught the agent demonstrated superior problem-solving abilities and retained information longer, underscoring the efficacy of the learning by teaching model implemented through teachable agents.

Leelawong and Biswas (2008) and Matsuda et al., (2011, 2013, 2014, 2020) expanded on these works by investigating how different scaffolding techniques within the teachable agents affected student learning. They discovered that students benefited most from adaptive scaffolding, which dynamically adjusted to their needs, thus supporting both the acquisition of domain knowledge and the development of teaching strategies.

Specifically, the body of research by Matsuda and his team developed a teachable agent, often referred to as a metatutor, which provides adaptive scaffolding to students as they teach a synthetic peer. Their analysis revealed that students frequently taught the teachable agent incorrectly without realizing their errors. To address this, the researchers first introduced metacognitive scaffolding into the APLUS learning environment by embedding a metatutor agent. This agent offered adaptive assistance on how to teach effectively while students engaged with the teachable agent. A classroom study demonstrated that incorporating metacognitive scaffolding significantly enhanced the learning by teaching effect compared to a baseline environment without such support (Matsuda et al., 2014). Subsequently, the team implemented cognitive scaffolding as an additional functionality of the metatutor. This scaffolding, driven by the model-tracing technique commonly used in cognitive tutors (Anderson & Pelletier, 1991), provided adaptive help on solving equations. However, the results indicated that while metacognitive scaffolding enhanced the learning by teaching process, cognitive scaffolding did not produce the same effect (Matsuda et al., 2020). This suggests that the type of scaffolding plays a crucial role in the efficacy of learning through teaching.

Recent investigations are starting to use large language models (LLMs) such as Chat GPT as teachable agents in learning by teaching paradigms to simulate tutees. Advancements in LLMs have demonstrated their abilities in creating contextual dialogues, mimicking roles, and learning from demonstrations (Markel et al., 2023; Ouyang et al., 2022). Teachable agents that incorporate LLMs capabilities can engage in more realistic and natural tutoring interactions compared to earlier systems that relied on prescripted and limited interaction channels. This enhanced flexibility in interaction seems to enable learners to ask free-form questions and experiment with various teaching methods, thereby improving their knowledge construction and metacognitive skills (Aflalo, 2021). For instance, a study by Jin et al. (2024) involved students who wanted to learn how to code; in a preliminary phase, 15 of them were let free to interact with ChatGPT, but the outcomes were poor because this LLM could proficiently write code without needing to be taught and did not pose questions that could drive deeper elaboration and knowledge-building. Several needs emerged, such as limiting the knowledge level of LLM agents, incorporating agent-initiated "why" and "how" questions, and providing in-conversation feedback on tutors' teaching methods. To address these issues, the authors developed AlgoBo, a teachable agent that scaffolds learners' knowledge-building through the use of "why" and "how" questions, and TeachYou, a web-based algorithm learning system that integrates learning by teaching with AlgoBo, offering metacognitive feedback on teaching through real-time conversation analysis. This analysis indicated that posing questions significantly increased the density of knowledge-building messages in the conversations with a notable effect size (Cohen's $d = 0.71$). Additionally, the features of TeachYou facilitated participants' reflection on their teaching methods and encouraged strategic sequencing of their questions; however, no significant improvement in participants' metacognitive skills was observed.

In summary, teachable agents powered by large language models (LLMs) offer significant advantages for tutor-tutee interactions and activities in learning by teaching contexts. Unlike traditional non-LLM teachable agents, which limit learners to predefined assessment methods like multiple-choice questions, LLM-based agents enable learners to pose free-form questions and explore diverse teaching strategies. This flexibility can enhance the learning experience by allowing more personalized and dynamic interactions. However, this field is still in its early stages, and several challenges remain. Key issues include the need for LLM-based agents to effectively simulate tutoring scenarios and actively support learners' knowledge-building and metacognitive skills. Addressing these challenges is crucial for fully realizing the potential of LLM-based teachable agents in educational settings.

3.2 *Immersive Virtual Reality*

Recently, immersive virtual reality (IVR) has garnered significant attention for its potential impact on learning. IVR can simulate realistic and interactive educational environments using head-mounted visors, creating a fully immersive sensory experience that makes users feel present in a digital environment (Slater, 2009). Unlike low-immersion virtual reality, which relies on external displays and user interaction via a mouse or keyboard, IVR completely surrounds the user and incorporates head and hand motion tracking for interactivity.

Adapting learning by teaching to IVR is promising because it allows the simulation of social experiences without another person and offers greater immersion and spatial presence compared to traditional teachable agents. The effectiveness of learning by teaching in IVR may depend on whether students find the virtual scenario plausible, especially since the recipient is not real.

A study by Chiquet et al. (2023) compared learning outcomes after participants taught an agent in IVR and on a computer screen (a "traditional" teachable agent); some were instructed to imagine helping a peer pass a test (student instruction), while others were instructed to teach a computer-based agent (agent instruction). Results showed that individuals who taught an agent in IVR had better learning outcomes if they had a high absorption tendency. Absorption is defined as the ability to focus attention on sensory experiences and ignore distractions (Tellegen & Atkinson, 1974). High-absorption individuals perceive virtual experiences as more real: in the study, these participants were likely to be more motivated and able to accept the computer-based agent as a real recipient, enabling them to focus on task-relevant cognitive processing, such as ensuring their explanations were clear and accurate.

Thus, further research is needed to establish whether learning by teaching in IVR environments is genuinely effective in terms of learning gains. Current findings suggest that students with higher absorption tendencies benefit more from IVR, but this outcome does not necessarily promote inclusivity for all learners.

3.3 Social Robots

Compared to the technologies presented so far, social robots add an important component: physical presence. This is particularly true for humanoid robots, which are widely used in education and therapeutic settings as interactive partners. They have proven effective in terms of concentration, motivation, and initiating positive attitudes towards conversations (Tapus et al., 2007). A review by Jamet et al. (2018) explored studies that used robots in learning by teaching settings, highlighting their potential in various educational domains such as reading (Yadollahi et al., 2018), writing (Chandra et al., 2017, 2018; Hood et al., 2015; Johal et al., 2016), language learning (Tanaka & Matsuzoe, 2012; Tanaka et al., 2015), reasoning (Masson et al., 2016), and metacognitive abilities (Jacq et al., 2016; Rohrbeck et al., 2003).

For instance, Hood et al. (2015) utilized a tablet where the robot NAO (https://corporate-internal-prod.aldebaran.com/en/nao) was instructed to draw letters with its finger, with children teaching and correcting it. While specific outcomes on writing skills were not reported, the study did observe positive effects on children's engagement and willingness to teach. The authors noted that children engaged in a cognitive relationship with the robot, attempting to understand and correct its mistakes. Here, the robot's physical presence and embodiment stimulated a sense of agency and anthropomorphism, deeply involving the child in the learning process.

Chandra et al. (2017) examined children's perceptions of a social robot's learning abilities and behavior in an educational setting. The study involved a robot asking for help to correct the shapes of letters it wrote, across two conditions: one where it learns and one where it does not improve. The results indicated that children's social relationship with the robot was unaffected by the robot's learning abilities, showing empathy and patience regardless. Some other studies have shown that children notice the robot's progress, enjoy teaching more as the robot improves, and feel like better teachers after each interaction (Chandra et al., 2018; Johal et al., 2016). Situations where the robot needs help appear particularly useful for developing not only academic skills but also caretaking behaviors. Tanaka et al. (2015) provided interesting results in a study on foreign language teaching. They found that children significantly improved their performance in learning English action verbs when teaching the robot compared to being taught by an experimenter. Additionally, the fact that participants had siblings or pets influenced learning performance, suggesting that habitual interactions in a caregiving context may be projected to robots.

The concept of the robot being ignorant and knowing less than the child can also lead to positive learning outcomes. Masson et al., (2016) conducted a study on developing the concepts of hyponyms and hypernyms (e.g., class "fruit" and sub-class "apple") in children. In this experiment, children were instructed to teach new things to NAO, who was explicitly described as knowing less than them. NAO's introduction and interaction, where it made deliberate mistakes, prompted children to correct and teach it repeatedly until it understood the categories correctly. Qualitative analysis showed that children engaged in a conversational manner, starting patiently and becoming more assertive as they corrected NAO's mistakes. Therefore, social robots seem to elicit a protégé effect, similar to traditional teachable agents, but with enhanced engagement due to their physical presence and their ability to communicate both verbally and nonverbally (Kristoffersson et al., 2013; Takayama & Pantofaru, 2009). Studies indicate that children are more interested in interacting with social robots, finding them more engaging than non-physical agents (Lindberg et al., 2017).

However, compared to interactions with human peers, robots still require significant improvements before they can engage in autonomous and unsupervised interactions with children. Recent studies by Serholt et al. (2022) and Pareto et al. (2022) involved students acting as tutors in a collaborative mathematics game, where they taught both a robot tutee and a younger student. The results indicated that the two scenarios were similar in terms of enjoyment, willingness to play again, and perceived learning gains. Both conditions offered valuable learning opportunities for a tutor, but in distinct ways: interacting with the child tutee offered more opportunities for explaining concepts, experiencing smooth collaboration, and gaining motivation through social responsibility for the younger peer; on the other hand, the robot tutee condition provided opportunities to answer challenging questions, receive ample feedback, and communicate using mathematical language. In fact, significant differences were found in communication and collaboration, which participants found easier with a human tutee. They also felt less autonomous in their tutoring roles with the robot tutee, often needing their teacher's help, and perceived the activity as clearer and more effective when working with the human tutee. These findings suggest that while children can enjoy tutoring a robot, having a human peer as a tutee offers more advantages and should remain the ultimate goal of such activities involving robots.

4 Why Using Social Robots as Tutees in Inclusive Education

Social robots have proven particularly useful in inclusive education due to their unique features, which cater to the needs of students with SEN. Robots offer flexibility and can help enhance social skills, making them well-suited for inclusion. Modularity and adaptability are essential features that a robot should have (Cabibihan et al., 2013): the former allows the robot to perform various functions tailored to different individuals, the latter means progressively increasing the complexity of interactions (Giullian et al., 2010). Some scholars view the use of robotic platforms as a way to bridge the gap between the stable, predictable environment of a simple toy and the complex, unpredictable world of human communication (Cunha Costa, 2014). This is especially relevant for conditions affecting communication, such as autism (Pennazio & Cersosimo, 2025): social behavior can be unpredictable and may seem threatening to individuals with autism, but using a robot can provide a simplified environment (e.g., in language or gestures) that gradually increases in complexity based on the individual's needs and abilities.

Research on the use of robots in learning by teaching is still limited, but it shows promising potential. One of the earliest studies in this field, conducted by Jimenez et al. (2017a), involved students with autism. The study used the robot Ifbot (http://www.theoldrobots.com/ifbot.html) with an 11-year-old child with autism spectrum disorder and a teacher. The session was divided into two stages: a 10-min familiarization period where the child, teacher, and robot played a game together, followed by 40 min of collaborative work. During this work session, the teacher alternated between asking the child and the robot to read aloud a page from a comic book. The robot asked the student some questions about the story and made guesses; when it made a mistake, the child corrected it and even moved closer to the robot to do so. However, if the robot made too many consecutive mistakes, the child became more reluctant to correct it. The same authors also analyzed the outcomes of a collaborative work involved building a story using Lego blocks, with the robot asking clarification questions. The results showed that the two children with autism involved were engaged in the task and explained the story to the robot.

Students with developmental disabilities were involved in a study (Jimenez et al., 2017b) in which a learning by teaching group, a group where the child learns alone, and a group where children learn together were compared to measure the time that each of them dedicated to study the material; interestingly, students focused more in the learning by teaching group, and appreciated the activity of teaching to a robot.

A study by Jacq et al. (2016) involved students with different difficulties: one exhibiting significant inhibition and low motivation for writing, another experiencing visual impairments and difficulties in maintaining focus, and a group of eight children facing varying degrees of challenges in learning cursive writing. The experimental setup involved a child and NAO robot, each equipped with a tablet featuring a customized program tailored to address the child's specific drawing errors. The child begins by drawing a letter on the tablet and then presents it to NAO, which reproduces it; the child then corrects the drawing multiple times until satisfied with the result. Gains were reported in terms of enjoyment in the activity and focus, and parents noticed an improvement in writing abilities.

Based on previous studies and on the theoretical foundation of peer tutoring that was delineated in this chapter, we will conclude with some reasons why we believe that, among various modalities where social robots can be integrated into classroom activities, the robot acting as tutee appears the most promising.

4.1 Language Competence Development

Students with SEN exhibit highly diverse language abilities. For instance, in the case of autism, 25–30% of individuals remain non-verbal or have minimal verbal abilities (Tager-Flusberg & Kasari, 2013). Even those who develop language skills often face persistent challenges in social communication (Wetherby et al., 1997). Similar language difficulties are seen in other disabilities, such as Down syndrome (Gunn & Crombie, 1996) and Williams syndrome (Kozel et al., 2021), as well as in learning differences like dyslexia (Cappelli & Noccetti, 2022). Generally, these students may struggle with comprehension and achieving speech fluency, often complicated by functional and pragmatic impairments.

The flexibility of social robots, particularly their programmable nature, offers potential solutions to these language difficulties. Robots can be programmed to use simple language, avoid figurative speech, and eliminate complex implicatures. Additionally, they can be set to use the student's first language if the student is learning in a second language. Customizing teachable agents to align with students' cultural backgrounds can significantly improve engagement and learning outcomes, modifying gestures, proxemics, and tone of voice to suit each student's cultural context. Furthermore, the integration of robots with LLMs enhances the ability to address specific user needs, enabling more natural and effective conversations, which gradually become more challenging.

As previously discussed, peer tutoring relies on effective dialogic behaviors to facilitate knowledge co-construction, and students with SEN may need additional support to become effective tutors. One area for initial engagement might be questioning, which involves both the ability of asking and responding to questions. Engaging in deeper questioning that integrates prior and new knowledge, re-organizes mental models, and fosters inferences and metacognitive self-regulation (King, 1998) is particularly beneficial. Tutors generate questions that prompt tutees to think deeply about the material, exploring contrasting concepts, applications, and causal relationships. Thus, with the help of a robot, tailored training can develop this competence, benefiting learning by teaching and broader life skills. Robots can also model good questioning techniques, helping tutors learn how to pose effective inquires.

4.2 Modelling Productive Learner Behaviors

A robot can serve as an exemplary learner, demonstrating attitudes and strategies that students can emulate to enhance their learning techniques. Peer tutoring provides a unique opportunity for students to step into the teacher's role, gaining insight into the complexities of explaining concepts and reflecting on their responsibilities as learners. This metacognitive process can be especially beneficial for students with learning difficulties, who often find it challenging to establish a constructive learner-teacher relationship.

A robot can be programmed to mimic a challenging student, one who has trouble focusing, requiring the tutor to devise effective engagement strategies. Conversely, it can also act as an exemplary student, exhibiting positive dialogic behaviors that encourage the tutor to broaden their knowledge and engage in deeper reflections. It is crucial, however, that the robot strikes a balance: it should be sufficiently ignorant to maintain a meaningful tutor-tutee relationship, yet not so error-prone that it discourages the tutor.

The integration of LLMs into robot tutees presents both opportunities and challenges in this regard. If not properly calibrated, LLMs might acquire knowledge too rapidly, disrupting the natural knowledge-building process between tutor and tutee. For instance, in the aforementioned study involving ChatGPT and AlgoBo (Jin et al., 2024), participants appreciated AlgoBo-Basic's conversational abilities and its role in simulating a learner by asking specific questions. However, they also reported awkwardness due to its competence. Initially, AlgoBo-Basic sought help, but it quickly began providing highly competent responses, which did not align with the typical learning curve of a novice learner.

Therefore, careful programming is essential to ensure that the robot maintains an appropriate level of ignorance and challenge, facilitating a productive learning environment. By doing so, the robot can effectively model good learning behaviors, help tutors develop their teaching strategies, and foster a more profound understanding and reflection on their own learning processes.

4.3 Need for Explicitness

In peer tutoring, it is crucial for the tutor to monitor the tutee's performance and be attentive to any new ideas introduced by the tutee. Understanding the tutee's thoughts is a fundamental aspect of the learning-by-teaching process. Students with SEN often struggle with monitoring their own and others' thinking, so this skill requires special attention.

Attributing mental states and beliefs to others is closely linked to Theory of Mind abilities, which play a significant role in communication by allowing individuals to infer the thoughts and emotions of others during interactions (Bosco et al., 2018). Research indicates that Theory of Mind is often impaired in various SEN conditions, such as mild intellectual disability, hearing impairment (Smogorzewska, 2018), Down syndrome (Giaouri et al., 2010), and particularly autism (Rosello et al., 2020). A recent literature review by Alghamdi et al. (2023) suggests that social robots could aid in developing Theory of Mind and its precursors, including joint attention, symbol use, imitation, and other non-verbal abilities like gestures and eye contact.

The impact of learning by teaching on mental state attribution has also been examined in relation to technologies. Studies reveal that children view teachable agents as social entities, attributing mental states and responsibilities to them (Chase et al., 2009). Lindberg et al. (2017) found that mental state attribution to social robots is similar to that of teachable agents. Notably, many teachable agents are designed to make the tutee's thinking visible through structured visual representations and intuitive reasoning mechanisms (see the section on teachable agents).

One key feature of peer tutoring is recursive feedback, which is the information tutors receive by observing their tutees' post-instruction performance. Making this feedback explicit enhances the interaction and can be effectively facilitated through technology. This approach can significantly help in developing Theory of Mind abilities of students with SEN, while also clarifying the teaching and learning process.

The need for explicit instruction is essential for all students, particularly those with SEN, who often struggle with implicit aspects of learning. Previous research has shown that clear, structured, and systematic instruction can help address the diverse challenges these students face by making the knowledge construction process explicit (Roitsch et al., 2021; Schnepel & Aunio, 2022). This breaks down complex tasks into manageable steps, making it easier for students with SEN to understand and retain new information.

If a robot provides the student-tutor with recursive feedback that clarifies what the tutee has understood, the tutor can more effectively address comprehension issues. This process enhances the co-construction of knowledge between tutor and tutee, making the learning experience more beneficial for both.

4.4 Flexibility and Openness of Scenarios

Learning by tutoring a robot aligns well with the principles of Universal Design for Learning (UDL), an inclusive framework designed to optimize teaching and learning for all students (Hall et al., 2012). UDL emphasizes the importance of providing choices to sustain interest and motivation, offering varied activities, and making content accessible in multiple formats such as visual, auditory, and tactile. Additionally, UDL supports enabling learners to express their knowledge in diverse ways.

Robots are particularly effective in supporting UDL principles because they can be customized to meet the needs of different learners and allow for various modes of interaction. For instance, robots like Pepper (https://corporate-internal-prod.aldebaran.com/en/pepper) are equipped with a tablet that can display pictures. This feature is beneficial for students with verbal difficulties, who can use images to communicate with the robot, or even Augmentative and Alternative Communication (AAC) methods (see Loncke et al., 2020, for a review of AAC). Students with hearing impairments can read what the robot is saying on the tablet, while those with sight impairments can rely on auditory feedback. Moreover, the pictures and fonts on the tablet can be tailored to individual students' needs, enhancing accessibility.

Another significant advantage of using robots in this domain is the openness of scenarios they offer. Unlike human peers, whose interaction topics might be limited, robots can be programmed to address a wide range of subjects, tailored to the specific interests and abilities of each student. This customization means that learning by teaching can be simplified for students with more severe disabilities or made more challenging for those with greater potential. Additionally, robots can facilitate the inclusion of topics that might not be part of the standard curriculum but are of particular interest to students with SEN. The relevance of the learning content to students' lives and interests is a key aspect emphasized by UDL. Robots can help address this by allowing educators to select topics that resonate with individual students, making learning more engaging and meaningful. This adaptability ensures that educational experiences are inclusive and cater to the diverse needs and preferences of all students, ultimately fostering a more supportive and effective learning environment.

4.5 Social and Emotional Skills

Previous studies have demonstrated that social robots can enhance empathy and the willingness to care for others. Specifically, they can be used as care-receiving robots, a framework experimented with in a long-term field trial in US schools (Tanaka et al., 2012). This trial involved integrating a small humanoid robot into a preschool classroom. Observations from this study revealed significant evidence of socialization between the children and the robot. The behavioral data obtained from these children provided important insights for the care-receiving robot. It was found that the robot encouraged the children's caretaking behaviors more than other toys did in the classroom. This increased the children's motivation in their teaching tasks, as they perceived the robot as fragile and in need of care. Robots can be seen as toys with additional features such as speech, gestures, and movement.

Students with SEN, particularly those with severe disabilities, often do not have the opportunity to fully benefit from the richness of play experiences. This "play deprivation" might be caused by functional limitations, such as motor, cognitive, or sensory deficits, and environmental barriers, such as the lack of suitable contexts and accessible play materials (Pennazio, 2015). As we have seen, robots can be adapted to meet the needs of students and can be accessible through flexible and interactive activities, thus allowing for play experimentation.

Some students with SEN, especially those with disabilities that limit social skills, might benefit from interacting with a robot, thereby extending this social competence to interactions with peers. For instance, the use of robots has proven beneficial in acquiring collaborative behaviors and applying them in relationships with human partners (Charron et al., 2017). Soares et al. (2019) explored how a humanoid robot impacts the development of socioemotional skills in children with autism; they assessed the children's performance in in-game scenarios designed to improve their ability to recognize facial expressions. The study's findings suggest that employing a humanoid robot effectively enhances socioemotional skills in interventions for children with autism, as evidenced by increased engagement and positive learning outcomes.

In the tutor-tutee relationship, a sense of care typically arises, as noted in the theoretical introduction. However, SEN students who struggle with social skills may find it difficult to initiate such activities with a peer. Introducing a robot as an intermediary—a less emotionally complex entity than humans—can reduce anxiety and facilitate the development of social and cooperative skills, which can then be transferred to interactions with others.

5 Ethical Considerations

When designing educational activities with robots, especially for students with SEN, it is crucial to consider ethical implications carefully. This is particularly important when the activities involve aspects of care, social skills, and emotional engagement. Critics might argue that these activities pose a risk of inappropriate attachment to the robots. For example, in a case study by Lemaignan et al. (2016), a child teaching a robot to write became emotionally distressed during the final session and repeatedly asked to write letters to the robot after the sessions ended. Such scenarios highlight the potential for emotional attachment, which can be problematic, as it may blur the lines between human and robotic relationships, potentially affecting the child's understanding of social interactions.

In addition to concerns about emotional attachment, there are broader ethical considerations related to privacy and data security. Robots often collect data about students' interactions, learning progress, and even emotional responses. It is essential to ensure that this data is handled in accordance with privacy regulations such as GDPR, protecting students' sensitive information from unauthorized access. Consent from parents or guardians should be obtained, clearly explaining how the data will be used, stored, and shared. This transparency is crucial to maintaining trust and safeguarding students' rights. Furthermore, schools and educators must take steps to ensure that data collected through interactions with robots is not used for commercial purposes or shared with third parties without explicit consent, to avoid exploitation of students' data.

Another important ethical consideration is the potential reinforcement of biases through robot interactions. Robots are often programmed with algorithms that, if not carefully designed, could reflect and perpetuate existing social biases or stereotypes. For example, if a robot's interactions are based on a limited dataset that does not account for the diversity of student backgrounds and experiences, it may unintentionally reinforce stereotypes or exclude certain students from meaningful engagement. It is essential to ensure that the design and programming of educational robots consider inclusivity and are tested for fairness, avoiding the risk of reinforcing harmful biases that could negatively impact students' learning experiences.

Additionally, the introduction of robots into the classroom raises questions about the displacement of human roles, particularly the role of teachers. While robots can serve as useful tools to support certain aspects of learning, it is important to ensure that they do not undermine the role of teachers or create a reliance on technology that could diminish human-centered education. Teachers play a crucial role in adapting lessons, offering empathy, and understanding the nuances of student behavior in ways that robots cannot replicate. There is a risk that over-reliance on robots might devalue these essential aspects of teaching, which could lead to a more standardized and less personalized learning experience for students. The use of robots should be positioned as an enhancement to human teaching, not as a substitute, preserving the human touch that is critical in educational settings.

The ethical considerations also extend to ensuring accessibility and fairness in providing access to these technologies. Schools with fewer resources may find it difficult to acquire or maintain robotic technologies, potentially creating disparities between well-funded institutions and those in underprivileged areas. This raises the question of whether the benefits of robotic learning are accessible to all students, or if they inadvertently widen the gap between students who have access to advanced learning tools and those who do not. Policymakers and educators must consider how to equitably integrate robots into classrooms, ensuring that all students, regardless of their socioeconomic background, have the opportunity to benefit from these innovations.

To mitigate these risks, several measures should be implemented. First, teachers should clearly present the activity as a role-play exercise, explaining that the student is playing the role of a tutor and the robot is the tutee, and outlining the goals and expectations of the activity. This helps set clear boundaries and manage students' understanding of the interaction, reducing the likelihood of forming attachments. Second, the robot should be introduced gradually to observe the child's reactions and adjust accordingly. If a child shows signs of excessive attachment or emotional distress, sessions should be paused or stopped to prevent further distress. Additionally, limiting the number and duration of sessions can help manage and prevent the development of strong emotional bonds.

Moreover, ongoing monitoring and assessment are essential. Teachers should observe interactions closely, ready to intervene if signs of undue attachment appear. Open communication with students about the nature of the robot and the purpose of the activities can help maintain a healthy perspective, emphasizing the educational purpose of the interaction rather than fostering emotional dependency. It is also beneficial to integrate debriefing sessions, where students can discuss their feelings and experiences with teachers, helping them process their interactions with the robot and maintain a balanced perspective. Additionally, ongoing training for teachers on the ethical use of robots is critical. Educators should be equipped with the skills and understanding needed to manage robot-student inter-actions responsibly, ensuring that they can address any ethical challenges that arise.

In summary, while integrating robots into educational activities for students with SEN offers numerous benefits, it is essential to approach these initiatives with careful ethical consideration. By setting clear expectations, gradually introducing the technology, monitoring student reactions, limiting session duration, and ensuring transparency regarding data use, educators can create a safe and effective learning environment that maximizes the benefits while minimizing potential emotional risks. It is equally important to ensure that the use of robots is inclusive, free of bias, and supportive of the essential role of human educators. This approach not only safeguards the well-being of students but also promotes the ethical and responsible use of technology in inclusive education, ensuring that all students have the opportunity to benefit from these advancements.

6 Conclusions

The integration of social robots as tutees in inclusive education holds significant promise, offering a transformative approach to enhancing educational experiences for students with SEN. This chapter has explored the multifaceted benefits of employing social robots in educational settings, particularly through the learning-by-teaching paradigm. By acting as tutees, robots provide students with a unique opportunity to engage in reciprocal teaching and learning processes that are both motivating and enriching.

Key findings indicate that social robots can effectively foster empathy, care-taking behaviors, and social skills among students. For instance, the introduction of robots in classroom environments has been shown to encourage children's caretaking behaviors, thereby enhancing their willingness to engage in collaborative activities. This aligns with the principles of UDL, which emphasize the importance of providing multiple means of engagement, representation, and expression to cater to diverse learning needs. Moreover, the adaptability of robots to various students' needs—be it through customizable content, multimodal interaction capabilities, or assistive features—makes them invaluable tools for inclusive education.

However, the use of social robots in educational settings also necessitates careful ethical considerations. Ensuring informed consent and assent, safeguarding privacy and data security, and monitoring emotional well-being are critical to the responsible use of robots. Additionally, addressing potential risks such as inappropriate emotional attachment and ensuring cultural sensitivity are paramount to fostering a safe and supportive learning environment.

In conclusion, among various modalities where social robots can be integrated into classroom activities, such as robots acting as teachers, the role of the robot as a tutee appears most promising and is well-supported by pedagogical theories. The use of social robots as tutees in inclusive education offers a compelling strategy for enhancing educational outcomes by leveraging the unique capabilities of robots to support diverse learning needs. Future research should continue to explore the long-term impacts of these interventions, ensuring that the benefits are sustainable and integrated seamlessly into broader educational practices. Through thoughtful implementation and continuous ethical vigilance, social robots can become powerful allies in the quest for educational equity.

References

Aflalo, E. (2021). Students generating questions as a way of learning. *Active Learning in Higher Education, 22*(1), 63–75. https://doi.org/10.1177/1469787418769120

Akçayır, G., & Akçayır, M. (2018). The flipped classroom: A review of its advantages and challenges. *Computers & Education, 126*, 334–345. https://doi.org/10.1016/j.compedu.2018.07.021

Alghamdi, M., Alhakbani, N., & Al-Nafjan, A. (2023). Assessing the potential of robotics technology for enhancing educational for children with Autism spectrum disorder. *Behavioral Science, 13*(7), 598–598. https://doi.org/10.3390/bs13070598

Annis, L. F. (1983). The processes and effects of peer tutoring. *Human Learning, 2*, 39–47.

Bargh, J., & Schul, Y. (1980). On the cognitive benefits of teaching. *Journal of Educational Psychology, 72*(5), 593–604. https://doi.org/10.1037/0022-0663.72.5.593

Benware, C., & Deci, E. (1984). Quality of learning with an active versus passive motivational set. *American Educational Research Journal, 21*(4), 755–765. https://doi.org/10.3102/00028312021004755

Biswas, G., Leelawong, K., Schwartz, D., Vye, N., & The Teachable Agents Group at Vanderbilt. (2005). Learning by teaching: A new agent paradigm for educational software. *Applied Artificial Intelligence, 19*(3–4), 363–392. https://doi.org/10.1080/08839510590910200

Bowman-Perrott, L., Davis, H., Vannest, K., Williams, L., Greenwood, C., & Parker, R. (2013). Academic benefits of peer tutoring: A Meta-Analytic review of Single-Case research. *School Psychology Review, 42*(1), 39–55. https://doi.org/10.1080/02796015.2013.12087490

Bosco, F. M., Tirassa, M., & Gabbatore, I. (2018). Why Pragmatics and Theory of Mind Do Not (Completely) Overlap. *Frontiers in Psychology, 9*. https://doi.org/10.3389/fpsyg.2018.01453

Brooks, R. J. (2014). Exploring the long-term benefits from peer tutoring high school students with significant disabilities. Doctoral dissertation, The Claremont Graduate University.

Cabibihan, J. J., Javed, H., Ang, M., & Aljunied, S. M. (2013). Why robots? A survey on the roles and benefits of social robots in the therapy of children with autism. *International Journal of Social Robotics, 5*, 593–618. https://doi.org/10.1007/s12369-013-0202-2

Cappelli, G., & Noccetti, S. (Eds.). (2022). A Linguistic Approach to the Study of Dyslexia. Multilingual Matters. https://doi.org/10.21832/cappel5966

Carter, E. W., Moss, C. K., Asmus, J., Fesperman, E., Cooney, M., Brock, M. E., & Vincent, L. B. (2015). Promoting inclusion, social connections, and learning through peer support arrangements. *Teaching Exceptional Children, 48*(1), 9–18. https://doi.org/10.1177/0040059915594784

Chandra, S., Paradeda, R., Yin, H., Dillenbourg, P., Prada, R., & Paiva, A. (2017). Affect of robot's competencies on children's perception. In *AAMAS'17: Proceedings of the 16th Conference on Autonomous Agents and MultiAgent Systems* (pp. 1490–1492).

Chandra, S., Paradeda, R., Yin, H., Dillenbourg, P., Prada, R., & Paiva, A. (2018). Do children perceive whether a robotic peer is learning or not?. In *Proceedings of the 2018 ACM/IEEE International Conference on Human-Robot Interaction* (pp. 41–49). https://doi.org/10.1145/3171221.3171274

Charron, N., Lewis, L., & Craig, M. (2017). A robotic therapy case study: Developing joint attention skills with a student on the Autism spectrum. *Journal of Educational Technology Systems, 46*(1), 137–148. https://doi.org/10.1177/0047239516687721

Chase, C. C., Chin, D. B., Oppezzo, M. A., & Schwartz, D. L. (2009). Teachable agents and the protégé effect: Increasing the effort towards learning. *Journal of Science Education and Technology, 18*, 334–352. https://doi.org/10.1007/s10956-009-9180-4

Chin, D. B., Dohmen, I. M., & Schwartz, D. L. (2013). Young children can learn scientific reasoning with teachable agents. *IEEE Transactions on Learning Technologies, 6*(3), 248–257. https://doi.org/10.1109/tlt.2013.24

Chiquet, S., Martarelli, C. S., Weibel, D., & Mast, F. W. (2023). Learning by teaching in immersive virtual reality—Absorption tendency increases learning outcomes. *Learning and Instruction, 84*, 101716. https://doi.org/10.1016/j.learninstruc.2022.101716

Costa, S. C. C. (2014). Affective robotics for socio-emotional development in children with Autism spectrum disorders. Doctoral dissertation, Universidade do Minho, Portugal.

Cushing, L. S., & Kennedy, C. H. (1997). Academic effects of providing peer support in general education classrooms on students without disabilities. *Journal of Applied Behavior Analysis, 30*(1), 139–151. https://doi.org/10.1901/jaba.1997.30-139

De Lisi, R., & Golbeck, S. L. (1999). Implications of Piaget's theory for peer-learning. In A. M. O'Donnell & A. King (Eds.), *Cognitive perspectives on peer-learning* (pp. 3–37). Erlbaum. https://doi.org/10.4324/9781410603715

Duran, D., & Topping, K. J. (2017). Learning by teaching: Evidence-based strategies to enhance learning in the classroom. *Routledge*. https://doi.org/10.4324/9781315649047-1

Fiorella, L., & Mayer, R. (2013). The relative benefits of learning by teaching expectancy. *Contemporary Educational Psychology, 38*(4), 281–288. https://doi.org/10.1016/j.cedpsych.2013.06.001

Fiorella, L., & Mayer, R. E. (2014). Role of expectations and explanations in learning by teaching. *Contemporary Educational Psychology*, 39(2), 75–85. https://doi.org/ https://doi.org/10.1016/j.cedpsych.2014.01.001

Fuchs, L. S., Fuchs, D., Hamlett, C. L., Phillips, N. B., Karns, K., & Dutka, S. (1997). Enhancing students' helping behavior during peer-mediated instruction with conceptual mathematical explanations. *Elementary School Journal, 97*(3), 223–249. https://doi.org/10.1086/461863

Galbraith, J., & Winterbottom, M. (2010). Peer-tutoring: What's in it for the tutor? *Educational Studies, 37*(3), 321–332.

Giaouri, S., Alevriadou, A., & Tsakiridou, E. (2010). Theory of mind abilities in children with down syndrome and non-specific intellectual disabilities: An empirical study with some educational implications. *Procedia-Social and Behavioral Sciences, 2*(2), 3883–3887. https://doi.org/10.1016/j.sbspro.2010.03.609

Giullian, N., Ricks, D., Atherton, A., Colton, M., Goodrich, M., & Brinton, B. (2010). Detailed requirements for robots in autism therapy. In *2010 IEEE International Conference on Systems, Man and Cybernetics* (pp. 2595–2602). IEEE. https://doi.org/10.1109/icsmc.2010.5641908

Graesser, A. C., Bowers, C. A., Hacker, D. J., & Person, N. K. (1997). An anatomy of naturalistic tutoring. Scaffolding student learning: Instructional approaches and issues, 145–184. In K. Hogan & M. Pressley (Eds.). *Scaffolding student learning* (pp. 145–184). New York: Brookline Books.

Gunn, P., & Crombie, M. (1996). Language and speech. In B. Stratford & P. Gunn (Eds.), *New approaches to Down syndrome* (pp. 249–267). Cassell.

Hall, T. E., Meyer, A., & Rose, D. H. (Eds.). (2012). *Universal design for learning in the classroom: Practical applications*. Guilford press.

Hood, D., Lemaignan, S., & Dillenbourg, P. (2015). When children teach a robot to write: An autonomous teachable humanoid which uses simulated handwriting. In *Proceedings of the tenth annual ACM/IEEE international conference on human-robot interaction* (pp. 83–90). https://doi.org/10.1145/2696454.2696479

Hoogerheide, V., Deijkers, L., Loyens, S. M., Heijltjes, A., & van Gog, T. (2016). Gaining from explaining: Learning improves from explaining to fictitious others on video, not from writing to them. *Contemporary Educational Psychology, 44*, 95–106. https://doi.org/10.1016/j.cedpsych.2016.02.005

Huijnen, C. A., Lexis, M. A., Jansens, R., & de Witte, L. P. (2019). Roles, strengths and challenges of using robots in interventions for children with autism spectrum disorder (ASD). *Journal of Autism and Developmental Disorders, 49*, 11–21.

Jacq, A., Lemaignan, S., Garcia, F., Dillenbourg, P., & Paiva, A. (2016). Building successful long child-robot interactions in a learning context. In *2016 11th ACM/IEEE International Conference on Human-Robot Interaction (HRI)* (pp. 239–246). IEEE. https://doi.org/10.1109/hri.2016.7451758

Jamet, F., Masson, O., Jacquet, B., Stilgenbauer, J.-L., & Baratgin, J. (2018). Learning by teaching with humanoid robot: A new powerful experimental tool to improve children's learning ability. *Journal of Robotics, 2018*, 1–11. https://doi.org/10.1155/2018/4578762

Jimenez, F., Yoshikawa, T., Furuhashi, T., Kanoh, M., & Nakamura, T. (2017a). Feasibility of collaborative learning and work between robots and children with autism spectrum disorders. In *New Frontiers in Artificial Intelligence: JSAI-isAI 2015 Workshops, LENLS, JURISIN, AAA, HAT-MASH, TSDAA, ASD-HR, and SKL, Kanagawa, Japan, November 16–18, 2015, Revised*

Selected Papers (pp. 454–461). Springer International Publishing. https://doi.org/10.1007/978-3-319-50953-2_32

Jimenez, F., Yoshikawa, T., Furuhashi, T., Kanoh, M., & Nakamura, T. (2017b). Collaborative learning between robots and children with potential symptoms of a developmental disability. In *2017 IEEE Symposium Series on Computational Intelligence (SSCI)* (pp. 1–5). IEEE. https://doi.org/10.1109/ssci.2017.8280917

Jin, H., Lee, S., Shin, H., & Kim, J. (2024). Teach AI how to code: Using large language models as teachable agents for programming education. In *Proceedings of the CHI Conference on Human Factors in Computing Systems*, 1–28. https://doi.org/10.1145/3613904.3642349

Johal, W., Jacq, A., Paiva, A., & Dillenbourg, P. (2016). Child-robot spatial arrangement in a learning by teaching activity. In *2016 25th IEEE International Symposium on Robot and Human Interactive Communication (RO-MAN)* (pp. 533–538). https://doi.org/10.1109/roman.2016.7745169

Johnson, D. W., & Johnson, R. T. (2012). Restorative justice in the classroom: Necessary roles of cooperative context, constructive conflict, and civic values. *Negotiation and Conflict Management Research, 5*(1), 4–28. https://doi.org/10.1111/j.1750-4716.2011.00091.x

King, A., Staffieri, A., & Adelgais, A. (1998). Mutual peer tutoring: Effects of structuring tutorial interaction to scaffold peer learning. *Journal of Educational Psychology, 90*(1), 134–152. https://doi.org/10.1037/0022-0663.90.1.134

Kobayashi, K. (2021). Effects of collaborative versus individual preparation on learning by teaching. *Instructional Science, 49*, 811–829. https://doi.org/10.1007/s11251-021-09561-6

Kozel, B. A., Barak, B., Kim, C. A., Mervis, C. B., Osborne, L. R., Porter, M., & Pober, B. R. (2021). Williams syndrome. *Nature Reviews Disease Primers, 7*(1), 42. https://doi.org/10.1038/s41572-021-00283-0

Kristoffersson, A., Severinson Eklundh, K., & Loutfi, A. (2013). Measuring the quality of interaction in mobile robotic telepresence: A pilot's perspective. *International Journal of Social Robotics, 5*, 89–101. https://doi.org/10.1007/s12369-012-0166-7

Lachner, A., Hoogerheide, V., Van Gog, T., & Renkl, A. (2022). Learning by teaching without audience presence or interaction: When and why does it work? *Educational Psychology Review, 34*(2), 575–607. https://doi.org/10.1007/s10648-021-09643-4

Leelawong, K., & Biswas, G. (2008). Designing learning by teaching agents: The Betty's Brain system. *International Journal of Artificial Intelligence in Education, 18*(3), 181–208. https://doi.org/10.1007/s40593-015-0057-9

Lemaignan, S., Jacq, A., Hood, D., Garcia, F., Paiva, A., & Dillenbourg, P. (2016). Learning by teaching a robot: The case of handwriting. *IEEE Robotics & Automation Magazine, 23*(2), 56–66. https://doi.org/10.1109/mra.2016.2546700

Lindberg, M., Månsson, K., Johansson, B., Gulz, A., & Balkenius, C. (2017). Does a robot tutee increase children's engagement in a learning by teaching situation?. In *Intelligent Virtual Agents: 17th International Conference, IVA 2017, Stockholm, Sweden, August 27–30, 2017, Proceedings 17* (pp. 243–246). Springer International Publishing. https://doi.org/10.1007/978-3-319-67401-8_31

Loncke, F. (2020). Augmentative and alternative communication: Models and applications (Vol. 1). Plural publishing.

Markel, J. M., Opferman, S. G., Landay, J. A., & Piech, C. (2023). Gpteach: Interactive ta training with Gpt-based students. In *Proceedings of the tenth acm conference on learning@ scale* (pp. 226–236). https://doi.org/10.35542/osf.io/r23bu

Masson, O., Baratgin, J., Jamet, F., Ruggieri, F., & Filatova, D. (2016). Use a robot to serve experimental psychology: Some examples of methods with children and adults. In *2016 International Conference on Information and Digital Technologies (IDT)* (pp. 190–197). IEEE. https://doi.org/10.1109/dt.2016.7557172

Matsuda, N., Yarzebinski, E., Keiser, V., Raizada, R., Stylianides, G. J., Cohen, W. W., & Koedinger, K. R. (2011). Learning by teaching SimStudent—An initial classroom baseline study comparing with cognitive tutor. In G. Biswas & S. Bull (Eds.), *Proceedings of the international conference on artificial intelligence in education* (pp. 213–221). Berlin, Heidelberg: Springer. https://doi.org/10.1007/978-3-642-21869-9_29

Matsuda, N., Yarzebinski, E., Keiser, V., Raizada, R., William, W. C., Stylianides, G. J., & Koedinge, K. R. (2013). Cognitive anatomy of tutor learning: Lessons learned with SimStudent. *Journal of Educational Psychology, 105*(4), 1152–1163. https://doi.org/10.1037/a0031955

Matsuda, N., Griger, C. L., Barbalios, N., Stylianides, G., Cohen, W. W., & Koedinger, K. R. (2014). Investigating the effect of meta-cognitive scaffolding for learning by teaching. In S. Trausen-Matu, K. Boyer, M. Crosby, & K. Panourgia (Eds.), *Proceedings of the international conference on intelligent tutoring systems* (pp. 104–113). Switzerland: Springer. https://doi.org/10.1007/978-3-319-07221-0_13

Matsuda, N., Weng, W., & Wall, N. (2020). The effect of metacognitive scaffolding for learning by teaching a teachable agent. *International Journal of Artificial Intelligence in Education, 30*(1), 1–37. https://doi.org/10.1007/s40593-019-00190-2

Matsuda, N., Lv, D., & Zheng, G. (2023). Teaching How to teach promotes learning by teaching. *International Journal of Artificial Intelligence in Education, 33*(3), 720–751. https://doi.org/10.1007/s40593-022-00306-1

Muis, K. R., Psaradellis, C., Chevrier, M., Di Leo, I., & Lajoie, S. P. (2016). Learning by preparing to teach: Fostering self-regulatory processes and achievement during complex mathematics problem solving. *Journal of Educational Psychology, 108*(4), 474–492. https://doi.org/10.1037/edu0000071

Nestojko, J. F., Bui, D. C., & Kornell, N. (2014). Expecting to teach enhances learning and organization of knowledge in free recall of text passages. *Memory & Cognition, 42*(7), 1038–1048. https://doi.org/10.3758/s13421-014-0416-z

Okilwa, N. S. A., & Shelby, L. (2010). The effects of peer tutoring on academic performance of students with disabilities in grades 6 through 12: A synthesis of the literature. *Remedial and Special Education, 31*(450), 463. https://doi.org/10.1177/0741932509355991

Ouyang, L., Wu, J., Jiang, X., Almeida, D., Wainwright, C., Mishkin, P., & Lowe, R. (2022). Training language models to follow instructions with human feedback. *Advances in Neural Information Processing Systems, 35*, 27730–27744.

Palinscar, A. S. (1998). Social constructivist perspectives on teaching and learning. *Annual Review of Psychology, 49*, 345–375. https://doi.org/10.4324/9780203022214-20

Palincsar, A., & Brown, A. (1984). Reciprocal teaching of comprehension-fostering and metacognitive strategies. *Cognition and Instruction, 1*, 117–175.

Pareto, L., Ekström, S., & Serholt, S. (2022). Children's learning by teaching with a social robot versus a younger child: Comparing interactions and tutoring styles. *Frontiers in Robotics and AI, 9*, 875704. https://doi.org/10.3389/frobt.2022.875704

Pennazio, V. (2015). Disabilità, gioco e robotica nella scuola dell'infanzia. *Tecnologie Didattiche, 23*(3), 155–163.

Pennazio, V., & Cersosimo, R. (2025). Promoting social communication skills in ASD through robotics and virtual words. In S. Papadakis, G. Lampropoulos (Eds.), *Intelligent educational robots in maker education. Toward personalized problem-based learning environments*, De Gruyter.

Pennazio, V., Fedeli, L., Datteri, E., & Crifaci, G. (2020). Robotica e mondi virtuali per losviluppo delle abilità sociali nei bambini con autismo: una riflessione metodologica. *Sistemi intelligenti, 32*(1), 139–154.

Piaget, J. (1978). The development of thought: Equilibration of cognitive structures. *Basil Blackwell*. https://doi.org/10.2307/1175382

Renkl, A. (1995). Learning for later teaching: An exploration of meditational links between teaching expectancy and learning results. *Learning and Instruction, 5*, 21–36. https://doi.org/10.1016/0959-4752(94)00015-H

Rohrbeck, C. A., Ginsburg-Block, M. D., Fantuzzo, J. W., & Miller, T. R. (2003). Peer-assisted learning interventions with elementary school students: A meta-analytic review. *Journal of Educational Psychology, 95*(2), 240. https://doi.org/10.1037/0022-0663.95.2.240

Roitsch, J., Gumpert, M., Springle, A., & Raymer, A. M. (2021). Writing instruction for students with learning disabilities: Quality appraisal of systematic reviews and meta-analyses. *Reading & Writing Quarterly, 37*(1), 32–44. https://doi.org/10.1080/10573569.2019.1708221

Roscoe, R. (2014). Self-monitoring and knowledge building in learning by teaching. *Instructional Science, 42*(3), 327–351. https://doi.org/10.1007/s11251-013-92834

Roscoe, R., & Chi, M. (2007). Understanding tutor learning: Knowledge-building and knowledge-telling in peer tutors' explanations and questions. *Review of Educational Research, 77*(4), 534–574. https://doi.org/10.3102/0034654307309920

Roscoe, R., & Chi, M. (2008). Tutor learning: The role of explaining and responding to questions. *Instructional Science, 36*, 321–350. https://doi.org/10.1007/s11251007-9034-5

Rosello, B., Berenguer, C., Baixauli, I., García, R., & Miranda, A. (2020). Theory of mind profiles in children with autism spectrum disorder: Adaptive/social skills and pragmatic competence. *Frontiers in Psychology, 11*, 567401. https://doi.org/10.3389/fpsyg.2020.567401

Schnepel, S., & Aunio, P. (2022). A systematic review of mathematics interventions for primary school students with intellectual disabilities. *European Journal of Special Needs Education, 37*(4), 663–678. https://doi.org/10.1080/08856257.2021.1943268

Schwartz, D. L., Blair, K. P., Biswas, G., Leelawong, K., & Davis, J. (2007). Animations of thought: Interactivity in the teachable agent paradigm. *Learning with animation: Research and implications for design*, 114–140.

Serholt, S., Ekström, S., Küster, D., Ljungblad, S., & Pareto, L. (2022). Comparing a robot tutee to a human tutee in a learning by teaching scenario with children. *Frontiers in Robotics and AI, 9*, 836462. https://doi.org/10.3389/frobt.2022.836462

Slater, M. (2009). Place illusion and plausibility can lead to realistic behaviour in immersive virtual environments. *Philosophical Transactions of the Royal Society b: Biological Sciences, 364*(1535), 3549–3557. https://doi.org/10.1098/rstb.2009.0138

Smogorzewska, J., Szumski, G., & Grygiel, P. (2018). Same or different? Theory of mind among children with and without disabilities. *PLoS ONE, 13*(10), e0202553. https://doi.org/10.1371/journal.pone.0202553

Soares, F. O., Costa, S. C., Santos, C. P., Pereira, A. P. S., Hiolle, A. R., & Silva, V. (2019). Socio-emotional development in high functioning children with Autism Spectrum Disorders using a humanoid robot. *Interaction Studies, 20*(2), 205–233. https://doi.org/10.1075/is.15003.cos

Takayama, L., & Pantofaru, C. (2009, October). Influences on Proxemic behaviors in human-robot interaction. In *2009 IEEE/RSJ international conference on intelligent robots and systems* (pp. 5495–5502). IEEE. https://doi.org/10.1109/iros.2009.5354145

Talbott, E., Trzaska, A., & Zurheide, J. L. (2017). A Systematic Review of Peer Tutoring Interventions for Students with Disabilities. In M. T. Hughes & E. Talbott (A c. Di), *The Wiley Handbook of Diversity in Special Education* (1ª ed., pp. 319–356). Wiley. https://doi.org/10.1002/9781118768778.ch16

Tager-Flusberg, H., & Kasari, C. (2013). Minimally verbal school-aged children with Autism Spectrum disorder: The neglected end of the spectrum. *Autism Research : Official Journal of the International Society for Autism Research, 6*(6), 468–478. https://doi.org/10.1002/aur.1329

Tanaka, F., & Matsuzoe, S. (2012). Children teach a care-receiving robot to promote their learning: Field experiments in a classroom for vocabulary learning. *Journal of Human-Robot Interaction, 1*(1), 78–95. https://doi.org/10.5898/jhri.1.1.tanaka

Tanaka, F., Isshiki, K., Takahashi, F., Uekusa, M., Sei, R., & Hayashi, K. (2015). Pepper learns together with children: Development of an educational application. In *2015 IEEE-RAS 15th International Conference on Humanoid Robots (Humanoids)* (pp. 270–275). IEEE. https://doi.org/10.1109/humanoids.2015.7363546

Tapus, A., Maja, M., & Scassellatti, B. (2007). The grand challenges in socially assistive robotics. *IEEE Robotics and Automation Magazine, 14*(1).

Tellegen, A., & Atkinson, G. (1974). Openness to absorbing and self-altering experiences ("absorption"), a trait related to hypnotic susceptibility. *Journal of Abnormal Psychology, 83*(3), 268–277. https://doi.org/10.1037/h0036681

Thurston, A., Cockerill, M., & Chiang, T.-H. (2021). Assessing the differential effects of peer tutoring for tutors and tutees. *Education Sciences, 11*(3), 97. https://doi.org/10.3390/educsci11030097

Topping, K. J. (2005). Trends in peer learning. *Educational Psychology, 25*(6), 631–645. https://doi.org/10.1080/01443410500345172

Topping, K. J., & Ehly, S. W. (1998). *Peer assisted learning*. Routledge. https://doi.org/10.4324/9781410603678

Vygotsky, L. (1985). *Mind in society*. Continuum.

Walker, E., Rummel, N., & Koedinger, K. R. (2014). Adaptive intelligent support to improve peer tutoring in algebra. *International Journal of Artificial Intelligence in Education, 24*(1), 33–61.

Woo, H., LeTendre, G. K., Pham-Shouse, T., & Xiong, Y. (2021). The use of social robots in classrooms: A review of field-based studies. *Educational Research Review, 33*, 100388.

Yadollahi, E., Johal, W., Paiva, A., & Dillenbourg, P. (2018, June). When deictic gestures in a robot can harm child-robot collaboration. In *Proceedings of the 17th ACM conference on interaction design and children* (pp. 195–206). https://doi.org/10.1145/3202185.3202743

Pedagogical Guidelines for Telepresence Robots Users: A Pilot

Tiina Kasuk, Janika Leoste, Kristof Fenyvesi, Branko Anđić, and Mati Heidmets

Abstract Telepresence robots (TPRs) represent an innovative shift in educational technology, offering significant potential to enhance student engagement, facilitate communication, and promote inclusivity through improved social connectedness. Our proposed chapter will explore the pedagogical integration of TPRs in educational settings, aiming to create equitable learning opportunities for all students, irrespective of physical constraints. We will delve into practical implementations of TPRs to support students unable to attend classes due to health constraints and educators teaching from remote locations, thus expanding the educational reach and personalized learning experiences. The chapter will outline strategic approaches to overcome common barriers to TPR integration, such as connectivity issues, maneuverability limitations, and user adaptation challenges. We will discuss the necessity for multimodal planning to maximize the benefits of TPRs within pedagogical designs, emphasising the development of a conducive learning environment that offers high-quality learning and teaching experience for all participants; acknowledge, clarify, and overcome potential cognitive and physical limitations; responsive

T. Kasuk (✉) · J. Leoste
Tallinn University of Technology, Tallinn, Estonia
e-mail: tiina.kasuk@taltech.ee

J. Leoste
e-mail: janika.leoste@taltech.ee

J. Leoste · K. Fenyvesi
School of Educational Sciences, Tallinn University, Tallinn, Estonia
e-mail: kristof.fenyvesi@jyu.fi

K. Fenyvesi
Finnish Institute for Educational Research, University of Jyväskylä, Jyväskylä, Finland

B. Anđić
University of Vienna, Vienna, Austria

J. Leoste · M. Heidmets
Tallinn University, Tallinn, Estonia
e-mail: mati.heidmets@tlu.ee

G. Lampropoulos and S. Papadakis (eds.), *Social Robots in Education*, Studies in Computational Intelligence 1194, https://doi.org/10.1007/978-3-031-82915-4_11

to ethical considerations and ensure privacy and cybersecurity protections. Additionally, the chapter will address the training needs for educators and students to effectively utilize TPR technology, enhancing both learning outcomes and user experience. Our analysis will contribute to understanding digital pedagogy and human–robot interaction by providing evidence-based strategies and innovative lesson designs. This will support the educational community in implementing adaptive learning activities that leverage the capabilities of social robots to foster creativity, critical thinking, and problem-solving skills among students.

Keywords Telepresence robots · Education · Inclusive learning · Digital pedagogy · Personalized learning · Human–robot interaction

1 Introduction

In education, telepresence robots (TPRs) offer promising tools to extend social interaction and inclusion. However, there is a notable research gap concerning comprehensive pedagogical guidelines for implementing TPRs in education, especially based on best practices for teachers and students using them effectively. This chapter aims to contribute to filling this gap by presenting structured, research-based guidelines that go beyond the technical setup, focusing on the educational integration of TPRs by putting into practice pedagogical principles of inclusivity and engagement. The study addresses the practical challenges (technical, pedagogical and social-emotional) faced by educators and provides evidence-based strategies to maximize the benefits of TPRs, particularly in supporting students unable to attend classes physically. Through this exploration, the chapter aims to advance the discussion on digital pedagogy and human–robot interaction in hybrid learning environments, offering innovative solutions for personalized and inclusive learning.

TPRs are advanced technological systems designed to facilitate virtual presence, enabling individuals to be socially and interactively present in locations they cannot physically access. The TPR is like a video conferencing system on wheels (Tsui et al., 2011). It has speakers, microphones, a display, and cameras. TPRs' operator controls what they can see, moves around in a remote space independently and chooses their own conversational partner (Tuli et al., 2021). These robots enhance our ability to engage in social interactions and sensory experiences remotely, making it possible to participate in conversations, meetings, and activities with others as if we were physically there (Almeida et al., 2022). In the context of hybrid learning environments, TPRs represent a significant advancement over traditional two-dimensional audio and video communication methods. They create a more immersive, three-dimensional experience where participants, regardless of their physical location, can interact as if they are all in the same room. This technology not only bridges geographical distances but also fosters a sense of inclusion and presence, enriching the overall learning and social experience for all involved (Buchem et al., 2024).

TPRs are used in different fields like business, medicine, elder care, and education, and their use has been rising over the years (Virkus et al., 2023). Implementing TPRs in educational environments could significantly enhance accessibility and inclusion, particularly for students who cannot attend school physically due to health conditions, such as chronic illnesses, geographical distance, or other barriers. TPRs help maintain educational continuity by enabling real-time class participation and interaction with teachers and classmates. This helps mitigate academic setbacks and reduces the likelihood of grade repetition, ensuring that students stay on track with their peers (Powell et al., 2021). Besides having direct academic benefits, TPRs are instrumental in maintaining social connections, which are vital for the psychological well-being of students. They allow students to engage in social activities and maintain friendships while reducing feelings of loneliness and isolation. Social interactions facilitated by TPRs can significantly improve the students' mental health and help reduce anxiety and depression that often accompany prolonged absences from school (Choquette et al., 2016). The integration of TPRs in hybrid educational environments has also been linked to enhanced well-being and a sense of agency among students. By allowing them to keep up with their studies and participate in school activities, TPRs provide a sense of normalcy and control, which is particularly important during treatment for chronic illnesses. This sense of control can significantly improve psychological outcomes and adherence to treatment, empowering students and reducing feelings of helplessness (Bellizzi et al., 2012). The post humanist, new materialist, and intersectionalist perspectives further underscore the potential of TPRs to not only bridge physical distances but also challenge traditional educational boundaries, promoting a more inclusive and equitable learning environment. This comprehensive view of TPRs' impact on accessibility and inclusion has been explored in depth by some of the authors of this contribution in a book chapter published in 2024, offering critical insights into the transformative possibilities of this technology (Leoste et al., 2024).

Although using TPRs in the learning process seems promising and creates opportunities, implementing TPRs in learning and teaching is problematic today. This is mostly because teachers, instructional designers, and other stakeholders do not have accessible, well-structured classroom usage guidelines for implementing TPRs into learning. Manufacturers of TPRs usually provide potential users with technical guides on how to set up a TPR, drive, generate links and users, etc. However, they do not guide how teachers use TPRs in an educational setting. The manufacturers also see the need for detailed pedagogical guidelines to support the implementation of TPRs in education (Marmor et al., 2024). Therefore, in this chapter, we present the results of our pilot study, aimed at proposing an initial version of Pedagogical Guidelines for Telepresence Robot Users. These Pedagogical Guidelines would assist teachers in using TPRs not only from a technical standpoint but also as a set of guidelines in a pedagogical sense.

The Chapter is organized as follows: Sect. 2 offers an overview of the related literature reviews. Section 3 outlines the study materials and methods. Section 4 presents the pedagogical guidelines for telepresence robot users. Section 5 provides the discussion, and Sect. 6 concludes the paper.

2 Literature Review

Recent studies have explored various aspects of implementing TPRs in educational settings, highlighting their impact on both academic and social engagement. Zhang and Zhu (2023) introduced a framework for integrating institutional functions to support TPRs usage, emphasising the role of instructional guidance and technological support. LeTendre and Gray (2024) provided insights on teacher preparation and student engagement, underscoring the importance of handling technical limitations for effective TPRs implementation. These studies collectively underscore the transformative potential of the TPRs in creating inclusive and supportive educational environments.

The prerequisite for using TPRs is that the learner or teacher has a stable and quality internet connection. Confusion and frustration will arise during the lesson if a good internet connection cannot be ensured (Gallon et al., 2019). Using the mobile Wi-Fi hotspot helps to ensure a stable Wi-Fi connection, and those should move with the robot (Ahumada-Newhart & Olson, 2019). To ensure the stability and speed of the internet connection, the new 5G primary internet connection, which is rapidly entering the market, is suitable (Mordor Intelligence, 2023). TPRs require a classroom with internet coverage to ensure independent movement. Otherwise, situations may arise where TPRs cannot move because it has no internet connection, experiencing Wi-Fi dead spots (Ahumada-Newhart & Olson, 2019). If TPRs' operator loses connection with TPRs, communicating with classmates and participating in discussions is lost (Page et al., 2023).

When using TPRs in education, it's vital to ensure that TPR-mediated learners can actively participate in the lesson and have an equal learning experience. It can be challenging for TPR-mediated learners to understand what's shown on the screen, so using larger text sizes and more contrasting colours in the slideshows and other learning materials is recommended (Elmimouni et al., 2024). Keep the room evenly lit to ensure better video quality for TPRs. Text readiness is better from still pictures than TPRs' live video (Talisainen et al., 2023).

Understanding the learning material is essential for achieving academic learning outcomes, but TPR-mediated students also should understand what is happening in the room and achieve better social interactions. This can be achieved by skillfully manoeuvring TPRs. The person in a TPR should be able to turn or look in the right direction and hear what others are saying (Elmimouni et al., 2024; Virkus et al., 2023). Hearing others can be difficult when using TPRs, especially when several people speak simultaneously. To ensure that TPR-mediated learners can hear their peers during discussions, the teacher should repeat the ideas, comments, and answers from other students in the classroom. Additional microphones could help TPR-mediated students hear everything discussed in the classroom, or TP-mediated students have the option of which microphones are on and off for reduced noise (Elmimouni et al., 2024). The sound is important for TPR-mediated students to communicate clearly with other participants (Wernbacher et al., 2022a, 2022b).

It is worthwhile to share the necessary learning content in an online repository to ensure the availability of learning materials (Elmimouni et al., 2024). Sharing learning materials in an open repository or learning management system increases their accessibility to all students (including the TPR-mediated) so they can recall learning content later (Kasuk & Virkus, 2024). In emerging situations (pandemics, weather disasters and the like), the role of accessible materials is even more crucial (García-Morales et al., 2021).

The advantage of using a robot is that TPRs enable the operator to move around the room. Moving independently around the classroom and participating in discussions and activities increases learners' inclusion (Fitter et al, 2020). Possible obstacles should be removed (Elmimouni et al., 2024; Leoste et al., 2022), and movement corridors should be created (Leoste et al., 2022). To reduce the cognitive load of maneuvering with TPRs, automated solutions (pre-define maps, moving paths) could be used to move between points in the classroom (Elmimouni et al., 2024). In addition, an element for driving with TPRs is the battery life, which may vary; potential charging points in the classroom where TPRs can recharge during the lesson must also be considered (Leoste et al., 2022).

It is vital to support the practical skill development of students and teachers (LeTendre & Gray, 2024). The lesson's effectiveness depends on how familiar and efficient the teachers and students manage and operate TPRs. Competent training and practice can achieve the necessary experience (Talisainen et al., 2024). The school administration and parents must also be informed and involved in this preparation process (Gallon et al., 2019).

When problems arise, the TPR operator needs additional communication channels to get help from technical support or the teachers (Ahumada-Newhart & Olson, 2019). Well-planned, effective communication that removes extra cognitive load is vital (Echeverria et al., 2019).

Different authors have described various problems and provided recommendations on how to use TPRs effectively in the learning process. However, systematic guidelines that support the implementation of TPRs in the learning process have not yet been proposed. Based on the previous overview, implementing TPRs in the educational process requires overcoming various technical limitations and re-designing or adapting the existing learning process. To support teachers, written pedagogical guidelines would help facilitate the adoption of new technology in the classroom. A pedagogical guideline with a clear structure and layout can provide a positive initial experience for teachers and students. Practical guides will aid teachers and others in shaping the educational environment in creating an inclusive learning process that considers all learners.

3 Materials and Methods

Our version of the Pedagogical guidelines relies on three pilot studies. Figure 1 describes the process of development of pedagogical guidelines. The first one was conducted from October to November 2023 and was a semi-structured interview with six (four male and two female) Estonian higher education teachers who had previous experience using TPRs in the learning process. They had experience teaching with a TRP or teaching in lessons where students were TPR-mediated. With semi-structured interviews, we mapped out relevant best practices and recommendations when implementing a telepresence robot in a learning process.

The second study was conducted from February to May 2024. It was an elective online course with six students that lasted 13 weeks. Students participated in lessons via TPRs. The course aimed to test the first draft of the Pedagogical Guidelines.

The third study was conducted from May to July 2024. It included 13 practical hands-on workshops with 56 participants (16 male and 40 female) from different educational levels (mainly from higher education) and different countries (mainly from Estonia but also from Cyprus, Austria, Finland, Belgium, and Serbia). In one workshop, there were 4–6 participants. All workshops had the same design and structure. The workshop's design was for participants to get basic knowledge and skills to use the TPR in learning via participating mediated via the TPR. The feedback session by focus group interviews was at the end of the workshops. The structure

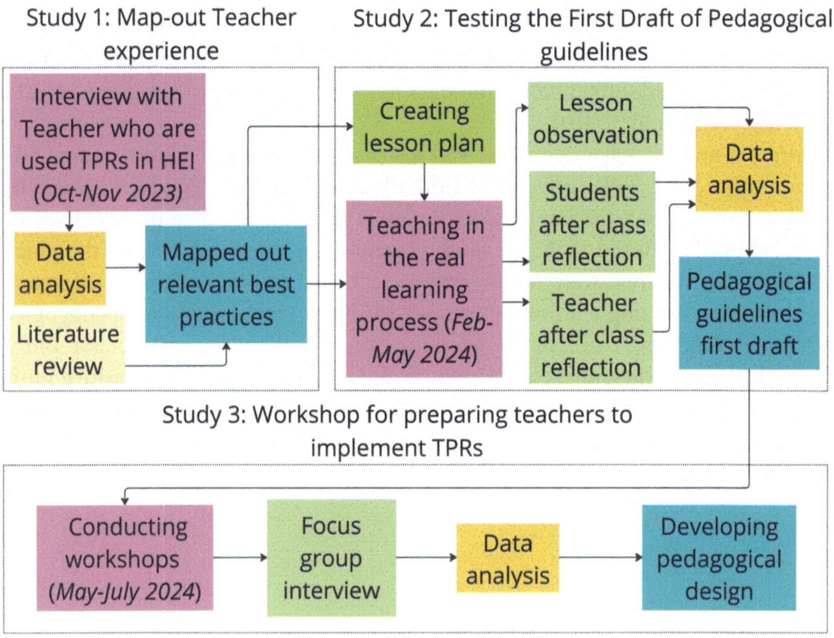

Fig. 1 Diagram of the process of development of pedagogical guidelines

of the Pedagogical guidelines for a TPR user was formulated based on the data analysis of focus group interviews. The analysis included categorizing the results of group interviews, formulating them in the form of guidelines/recommendations, and thematic structuring. The analysis was conducted by four independent experts, whose results were coordinated and are presented in the next section of the chapter.

This chapter explores how TPRs can ensure educational continuity, maintain social connections, allow for customization, engage stakeholders, and provide a sense of agency, thereby significantly enhancing these students' academic and emotional well-being. The scope of the chapter includes a detailed analysis of the various factors involved in the successful implementation of TPRs, such as addressing technical and privacy challenges, ensuring stakeholder engagement, and creating a supportive learning environment.

By examining these topics, the chapter aims to highlight TPRs' transformative potential in creating more inclusive and accessible educational environments.

Our version of the Pedagogical guidelines includes the following topics:

- The suitable environment for operating TPRs, including recommendations for both remote and classroom settings;
- Technical solutions for operating TPRs, including sound quality and camera configuration;
- Strategies for ensuring good audio and visual communication, and recommendations for the surrounding environment;
- Addressing cognitive limitations, including presenting content on the TPR screens and reducing cognitive load;
- Overcoming physical limitations, detailing the role of assistants and preparation for the TPR usage;
- Ethical and cybersecurity aspects of using TPRs in the learning process;
- Recommendations for instructional methods and practical implementation in classrooms.

4 Pedagogical Guidelines for Telepresence Robots Users

In the following guidelines, we have deliberately used lists to simplify the guidelines application for potential users. The lists are intended to be used as possible checklists.

4.1 Suitable Environment for Operating TPRs

This section provides guidelines for creating an optimal environment for operating TPRs remotely. It outlines recommendations for selecting a suitable room that minimises background noise and distractions and details the technical setup necessary for effective TPRs communication, including sound quality and camera configuration. Key aspects include the importance of good audio equipment, multiple screens in classrooms, and proper camera positioning and lighting to ensure clear and effective interactions between the TPR operator and the classroom.

4.1.1 Recommendation for the TPRs Educational Environment

On the remote side, where the person operating a TPR is located, we must create an environment that supports mediating a TPR operator into another room. When choosing the room or place to be in the session with a TPR, consider the following recommendations:

- Avoid loud noise from the background;
- Avoid bright light from the background;
- Use a room where there are no other distractions;
- Avoid operating TPRs in public places (e.g. cafeteria) or while travelling by car, train, or bus.

It's necessary to create an environment where you, as a TPR operator, can focus all your attention on manipulating a TPRs and interacting with people in the classroom.

4.1.2 The Technical Solution for Operating TPRs

Good sound quality is essential when working with TPRs. Therefore, if possible, use headphones with a noise-cancelling system. The question, "What type of communication through a TPR will you participate in?" needs to be addressed. The minimum requirements for a TPR for one-to-one communication include a computer with an integrated microphone, speakers, and a web camera. The recommended setup should feature a high-quality web camera with a built-in microphone and external speakers or headphones.

When using a TPR in a smaller classroom and sharing learning content with students, it is crucial to use two screens. One screen should have interactive capability (e.g., an iPad) to enable annotations if needed. In larger classrooms, video conference software can be used to share audio with all students through the classroom AV system. Additional headphones may be useful for listening to a TPR audio if needed.

4.1.3 Recommendation for the Configuring Camera

Usually, only the mediated person's face is presented on a TPR screen. Proper video framing enhances communication quality and access to mediated human interaction. Bittner (2021) describes the basics of webcam cinematography as necessary in every video-based online communication. Key aspects include composition, framing, camera angle, and correct lighting.

Composition: When mediating through a TPR, ensure your face occupies as much of the screen as possible. Position yourself in the centre of the frame, with your eyes two-thirds of the way up. Leave minimal space above your head, at most two finger widths, to avoid distracting dead space. Pay attention to composition to prevent parts of your face from cropping off due to a TPR screen's framing

Face Size in the Frame: Make your face as large as possible on a TPR screen for optimal representation. Use a neutral but contrasting background. Adjust the zoom of your webcam to change the size of your face. Use the webcam manufacturer's software, a manually zooming camera or software that creates a virtual camera (e.g. OBS Studio). Be careful not to zoom in too much so that pictures of your chin or head are not cut off.

Camera Angle: Position your web camera at eye level to be an equal communication partner while mediating through a TPR. Elevate your computer with a stand or suitable materials if using a laptop-integrated web camera.

Light Direction: Correct lighting ensures that communication partners can clearly see the mediated person on a TPR screen. The light source should be in front of the person and behind the web camera. A light source behind the mediated person can interfere with camera sensors, making the video dark or unclear. When using natural light (e.g., from a window), ensure it is directly in front or at an angle; light from the side can cast half the face in shadow.

4.2 Suitable Educational Environment for TPRs

Creating an optimal classroom environment for TPRs is crucial for their effective participation in the learning process. Different TPRs models have varying capabilities in terms of movement, vision, and audio; however, they generally have limitations compared to human abilities. Ensuring that TPRs can navigate the classroom independently, have a stable Wi-Fi connection, and are positioned to see and hear adequately is essential. Additionally, addressing potential obstacles, maintaining good lighting, and ensuring high-quality audio communication are critical for supporting both academic and social interactions. This section provides detailed recommendations on how to prepare the classroom environment to maximise the effectiveness of TPRs in educational settings.

4.2.1 Ensuring that TPRs Can Move Independently

Before a TPR enters the room, it is essential to verify that the space is suitable for the robot to navigate independently. Ensuring that TPRs can move freely without encountering obstacles is crucial (Leoste et al., 2022). Common barriers include:

- Chairs and tables;

- Cables connecting devices (e.g., external TVs with HDMI, laptop charging cables, power cords, extension cords);
- Narrow gaps between furnishings (e.g., tables and chairs) that TPRs cannot pass through;
- Other belongings obstructing the movement path;
- Elevation changes in the room (e.g., thresholds) that TPRs cannot traverse without assistance.

Classrooms with movable furniture are most conducive to creating ideal learning environments where TPRs can operate independently. In cases where furniture is fixed (e.g., computer labs), testing is necessary to determine whether there is adequate space for TPRs to manoeuvre.

TPRs can only move independently if their batteries are charged. Therefore, a charging station must be added to the classroom for sessions longer than the battery's lifespan. While charging, TPRs can be used for telepresence functions (hearing, talking, and seeing) but not for navigating the classroom. When planning lessons with TPRs, it is essential to select activities that can be conducted while the robots are charging. Leoste et al. (2022) acknowledge the importance of thinking about where to locate charging stations. To optimise the placement of charging stations, consider the following questions:

- Where are the power sockets located?;
- Will TPRs be able to see the screen and easily hear the teacher?;
- Does an additional power cord enable the charging station to be placed in a location that enhances visibility and audibility for the TPR-mediated person?.
- Stable lighting conditions are crucial in the classroom. When driving around the room, TPRs' operator may need to adjust camera settings frequently to maintain good visibility. The following recommendations can help ensure good visibility:
    - Keep window coverings drawn forward;
    - Switch on room lights to ensure even light distribution, especially in rooms without windows, as this prevents the room from being too dark.

The quality of the Wi-Fi connection is critical to the success of a TPR session. A TPR operator will experience a smooth drive and high-quality video and audio if the Wi-Fi connection is stable. Also, a good connection is vital to be part of discussions and other learning activities (Page et al., 2023). Both the classroom and the remote locations must have good internet quality. Public Wi-Fi is generally unsuitable for connecting TPRs from the classroom for two reasons: it is not secure and often lacks the necessary upload and download speeds. Using a dedicated Wi-Fi network exclusively for TPRs is preferable. Ahumada-Newhart and Olson (2019), if we do not ensure that Wi-Fi covers all classrooms, there are dead spots where the TPR operator will lose the connection to the TPR.

4.2.2 Ensuring Seeing Shared Content and Good Audio Communication

When a mediated student enters the lesson, the robot is often positioned at the end of the room, which can lead to issues with hearing audio and seeing shared content. The primary problem with hearing is that TPRs microphone is unsuitable for picking up sound from a long distance. While increasing the microphone volume is possible, it also amplifies the noise level. TPRs usually have built-in noise reduction and echo-cancelling systems, but these systems further process the audio as the volume increases, resulting in decreased audio quality. Additionally, the mediated student may find it challenging to follow conversations if multiple people are talking simultaneously, further complicating the audio experience.

Another audio issue is that a mediated person in a TPR can only listen to one person at a time. The TPRs microphone's directional pattern picks up sound from all around TPRs, meaning the operator can hear people from both the front and the back. In a room full of people, TPRs microphones capture all sounds, making it difficult for the operator to discern who is speaking and what is being said. This can be particularly problematic during group discussions or when multiple students are asking questions simultaneously.

When TPRs are placed at the far side of the room, the operator cannot see the screen or whiteboard effectively due to the following reasons:

- The distance from the screen or whiteboard is too great for clear viewing.
- Other students may obstruct the view.

Jakonen and Jauni (2021) recommended in their study that TPR should be located in the front of the classroom. In cases where the presenter (teacher or students) is at the front of the room, the location of the TPRs must meet the following criteria:

- A short distance from the presenting person;
- A clear view of the screen or whiteboard;
- Only one person should speak at a time.

To enhance the effectiveness of TPRs, it is essential to ensure that the robot is positioned in a way that maximises both audio and visual access, allowing the mediated student to fully participate in the lesson.

4.2.3 Recommendations for the Environment Preparation

Recommendations for the Surrounding Environment in using TPRs in the classroom:

- Ensure minimal gaps where TPRs drive through are at least 50 cm wide. The gaps between chairs should be wider, as chairs often change position during lessons;
- Place cables near the walls and out of the areas where TPRs drive to prevent any entanglements;

- Analyses possible moving trajectories. If any thresholds or other floor drops or raisings exist, test to determine whether TPRs need assistance to navigate these obstacles;
- Provide charging stations in locations where TPRs can see and hear presentations and participate in active discussions or group work that does not involve driving. This ensures that TPRs remain functional throughout more extended class periods;
- Maintain even lighting in the classroom to support TPRs' orientation and ensure the mediated person can see clearly. Avoid shadows and excessively bright spots that could affect visibility;
- Ensure a stable Wi-Fi connection throughout the classroom so that the connection to TPRs is not interrupted. Public Wi-Fi is generally not suitable due to security and performance issues; a separate, dedicated Wi-Fi network for TPRs is preferable. The same recommendations were given by Ahumada-Newhart and Olson (2019) in their study.

These recommendations will help create a supportive environment for TPRs, allowing them to function effectively and ensuring a seamless integration into the classroom setting.

4.3 Cognitive Limitations

This section addresses the challenges of cognitive load associated with using TPRs in educational settings and provides strategies to mitigate these limitations. Presenting content through TPRs often requires the operator to manage multiple tasks simultaneously, which can be cognitively demanding. The section explores effective methods for sharing content in one-to-one and one-to-many communication scenarios, ensuring optimal visibility and interaction. Recommendations are offered for minimising cognitive load, including practical steps for preparing the learning environment and using technology efficiently. By following these guidelines, educators can create a more effective and engaging learning experience for students using TPRs.

4.3.1 Presenting Content to the Screen While Using TPR

New learning content is regularly presented on a projector or TV screen during the learning process. This content typically includes presentations (e.g., PowerPoint, Keynote, Google Slides), text documents, or videos. Sharing content mediated by a TPR imposes a significant cognitive load on a TPR operator, who must manage both a TPR interface and the content-sharing process simultaneously.

Content sharing can be categorised based on communication types: (1) In one-to-one communication, content can be presented directly on a TPRs screen; (2) In one-to-many communication, content is shared via external channels, often requiring assistance from classroom support personnel.

In one-to-one communication, sharing content via a TPR screen is easy. Different TPRs models offer different options for sharing content, including sharing hosted computer screens, on-screen text, web pages and sharing satellite cameras via web page links. Some models also feature touch-activated screens that enhance user interaction and control. This versatility makes TPRs' screens a valuable tool for seamless information sharing. In addition, the intuitive user interface ensures that even those with limited technical knowledge can use them effectively. In one-to-many communication, external content sharing is often necessary. Assistance from support personnel, such as a support worker or student, is required to start content sharing from the teacher's computer and activate the local projector or TV screens.

The following external classroom content-sharing channels can be used while presenting learning content with TPRs:

- Videoconference-based solutions (e.g., Zoom, MS Teams);
- Live presentation environments (e.g., Live MS PowerPoint, Miro).

The presenter initiates the presentation and shares it remotely in videoconference meetings. The shared presentation is then displayed on a classroom projector or other multimedia screens. Table 1 provides a checklist of essential activities to ensure effective content delivery.

The videoconference solution offers the capability to use the built-in whiteboard for adding additional annotations to slides during live sessions. This functionality enhances interactivity and allows for real-time feedback and updates during presentations. Additionally, we can utilise various web services that enable live presentations, providing flexibility and accessibility for different types of content delivery. Suitable online tools include Online MS PowerPoint 365, which allows for live presentations (available to Business or Education users), and Miro board (free for educational users). These tools support collaborative engagement and facilitate dynamic presentations with interactive features.

The presenter must initiate the live presentation on the remote side, ensuring that all necessary materials are prepared and accessible. The same live presentation must be projected onto a projector or other multimedia screen in the classroom, ensuring that all students can view the content clearly. This setup requires proper coordination and testing to ensure seamless integration between remote and in-class technologies.

Table 1 Checklist for video conference solutions

Activities on the remote side	Activities in the classroom
1. Start a video conference meeting 2. Share the access link with the assistant (student or co-worker) 3. Mute microphone and webcam 4. Share content (e.g. slides) in meetings	1. Join a videoconference meeting from the on-site computer (e.g., the stationary computer at the teacher's table) 2. Start the local projector/TV screens so all people in the classroom can see the slides

In some applications, there may also be a need to mute the microphone and webcam to avoid distractions or feedback issues during the presentation. This step is crucial for maintaining audio clarity and ensuring that the focus remains on the content being shared. Additionally, presenters should be familiar with the technical aspects of these tools to troubleshoot any issues that may arise during the session. By effectively utilising these technologies, educators can create a more engaging and interactive learning environment that benefits both remote and in-class students.

4.3.2 Recommendation for Remote Computer Screen Layout

Using an external channel to deliver a presentation to the classroom introduces the risk that a TPR operator may lose sight of a TPR interface and, consequently, classroom awareness. To mitigate this risk, it is advisable to use two monitors or a single large monitor (such as a 4 K display). This setup helps maintain the visibility of all necessary application windows simultaneously, ensuring the operator remains situationally aware.

A TPR operator must arrange the layout of their screens to ensure that all critical application windows are visible at once. This arrangement is crucial for maintaining awareness and effective classroom management. In a typical lesson where the teacher is mediated through a TPR and uses an external channel to share learning content, the following application windows should always be visible:

- A TPR interface;
- Communication channel with the assistant (technical support);
- Shared content (presentation);
- External video conference system.

Maintaining situational awareness in this manner helps the operator perform better by reducing cognitive load and ensuring smooth interaction between the remote and classroom environments. By carefully managing these application windows, the operator can respond promptly to any issues, maintain effective communication with the classroom, and deliver a seamless educational experience.

Additionally, the operator should conduct regular checks and practice sessions to familiarise themselves with the setup, ensuring that they can quickly and efficiently navigate between the necessary applications. This preparation will further enhance their ability to manage the classroom effectively and maintain a high level of engagement with the students.

4.3.3 Recommendation for Lowering the Cognitive Load

To reduce cognitive load in the learning process, a TPR operator can take several steps to ensure effective operation and communication:

- **Familiarise with a TPR User Interface**: Spend adequate time learning the TPR's user interface to ensure smooth navigation and operation during sessions;
- **Practice Driving the TPR**: Regular practice in driving a TPR enhances proficiency and reduces the likelihood of operational errors;
- **Select a Quiet Operating Location**: Choose a quiet spot for operating a TPR to minimise distractions and improve focus;
- **Plan Note-Taking Strategies**: Think through effective methods for taking notes during the session, ensuring important information is captured without disrupting the operation;
- **Arrange Necessary Application Windows**: Simultaneously display all necessary application windows, including a TPR user interface, slide-sharing platform, and any other relevant applications, to maintain awareness and control;
- **Choose an Appropriate TPR Model**: If possible, select a TPR model that best suits the specific session's needs, considering factors such as audio and video quality, mobility, and interaction capabilities;
- **Practice Presentation and Manipulation**: Before giving a presentation with a TPR, find time to practice both presenting and manipulating a TPR to ensure smooth delivery.

In preparation for integrating a TPR into the learning process, making relevant changes in the classroom environment is crucial to further reduce cognitive load. The following recommendations can help create an optimal learning environment:

- **Prepare Low-Noise Learning Environments**: Ensure that the learning environment has low noise levels. High background noise can make it difficult for the speaker to be heard clearly, affecting the quality of communication;
- **Provide Clear Instructions to Students**: Give explicit instructions to students using the TPR, outlining their roles and responsibilities to avoid confusion;
- **Emphasise One-at-a-Time Speaking**: Explain to all participants the importance of speaking one at a time. This helps a TPR operator distinguish between different speakers and follow the conversation more effectively;
- **Offer Instant Audio Feedback**: Provide immediate feedback to a TPR operator about audio levels, making adjustments as necessary to ensure clear communication;
- **Optimise TPR Sound Levels**: Adjust a TPR sound levels in the classroom to ensure optimal audio quality, as a TPR operators may not be aware if the audio level is too low or too high;
- **Assist in Optimal TPR Placement**: Help a TPR operators position a TPR in the best spot within the classroom to ensure they can hear and speak effectively, facilitating better interaction and understanding;
- **Consider Robot Height in Seating Arrangements:** When arranging classroom chairs and tables, consider the robot's height. For example, place the chairs in an arc so that everyone can see the shorter robot's face, enhancing engagement;
- **Maintain Adequate Lighting:** Keep the classroom lights on during the lesson; avoid dimming the lights as it may hinder visibility for a TPR operator.

By following these comprehensive guidelines, TPRs' operators can effectively manage their cognitive load, leading to a more productive, engaging, and seamless learning experience for all participants.

4.4 Physical Limitations

This section addresses how to overcome physical limitations when integrating TPRs in the classroom. While Sect. 4.2 covered general physical limitations, this section delves into specific challenges encountered during the learning process. For instance, TPRs that are not stationary require assistance to navigate thresholds and elevators. Additionally, TPRs' operators cannot manipulate classroom computers, AV systems, or handle physical objects due to lack of hands. To mitigate these issues, a support person, referred to as an assistant, is essential. This section outlines the role and preparation of the assistant, offers recommendations for conducting lessons with TPRs, and emphasises the importance of pretesting to ensure successful implementation.

Additional shortcomings in the learning process include:

- Non-stationary TPRs cannot independently navigate to the classroom; they need assistance to cross thresholds and use elevators;
- TPRs' operators are unable to manipulate classroom computers or start sharing learning content on projectors or TV screens;
- They cannot adjust classroom light, audio, and video systems;
- The learning environment must be prepared to make it suitable for TPRs participation;
- Most TPRs lack hands and, therefore, cannot handle objects or demonstrate physical materials.

To overcome these shortcomings, a support person, referred to as an assistant, is needed. The assistant can be one of the students or a support person from the university.

4.4.1 The Assistant Role Description

To enable sessions to run smoothly and efficiently, it is sometimes necessary to have the support of a team member or assistant in the classroom. Their role can be crucial in ensuring that everything runs smoothly from start to finish when administering TPRs. The assistant's help is regularly needed for the following tasks:

- **TPRs Logistics:** Transport the TPRs into the classroom before the session and return them to storage afterward;
- **Session Preparation**: Ensure stable Wi-Fi connectivity and place the charging station appropriately;

- **Connection Control**: Confirm that audio and video functions are working properly and that the TPR operator's audio volume is suitable for the session;
- **Classroom Layout Preparation**: Assist in setting up the classroom layout, as the teacher in the TPR cannot do this;
- **Audio Setup**: If needed, select and set up appropriate microphones and speakers;
- **AV System Operation**: If required, start the classroom computer and AV system, including room controls, projectors, or TV screens, for sharing learning content;
- **Instant Feedback**: Provide immediate feedback during lessons about audio volume levels, connection quality, etc.;
- **Hands-on Assistance**: Act as the TPR's hands when needed, such as showing teaching tools to the students;
- **Technical Support**: Support the TPR operator during the lesson by solving any issues that arise, such as a TPR getting stuck, changing Wi-Fi settings, low battery, or needing a reboot.

4.4.2 The Assistant Preparation

Assistant training is necessary before one can effectively assist with TPR operations. The assistant must be knowledgeable in the following areas:

- **Basics of a TPR Function**: Understanding the fundamental operations of TPRs;
- **Essential Environment**: Knowing what kind of environment is required for the TPR to move, listen, speak, and see effectively;
- **Technical Skills**: Learning how to manoeuvre the TPR, change Wi-Fi settings, reboot the system, and place it into charging stations.

During the learning process, it is crucial for the TPR operator and the assistant to maintain a separate communication channel (e.g., chat in Teams, Discord, Messenger) for instant messaging.

When recruiting a student to serve as an assistant, consider the following recommendations:

- **Classroom Access**: Ensure that the student can access the necessary classroom.
- **Convenience and Motivation**: Arrange a schedule that is convenient for the student and motivates them to assist during the agreed period.
- **Extended Periods**: If the assistant is needed for extended periods, the consultant recruits more than one student to share the responsibilities.

4.4.3 Recommendations for Conducting the Lesson with the Help of the Assistant

To ensure the success of lessons involving TPRs, it is crucial to meticulously plan and test the lesson beforehand. If the assistant has prior experience with TPRs in educational settings, they can contribute significantly to the lesson design. Keep the following considerations in mind:

- **Clear Expectations**: Ensure that assistants are fully aware of what is expected from them;
- **Plan Testing**: Collaboratively test the lesson plan with the assistant. This step is vital to identify and address any potential issues before the actual lesson;
- **Involving the Assistant**: Including the assistant in the testing phase can uncover previously unnoticed aspects that might need adjustment.

During the planning phase, consider the following elements:

- **Classroom Layout**: Determine what changes in table and chair arrangements are necessary;
- **Location of TPRs**: Decide where the TPR will be positioned during the lesson;
- **Planned Activities**: Identify the learning activities that will take place;
- **Content Presentation**: Determine if the TPRs will present or share content, and if so, what technical solutions will be employed;
- **Classroom Size**: Assess the size of the classroom to ensure adequate space for TPRs' movement and functionality;
- **Audio System**: Consider whether a classroom audio system is needed to support teachers and presenters effectively.

4.4.4 Recommendation for Pretest

If a plan for integrating TPRs into the learning process is developed, conducting a pretest is essential to verify that everything functions correctly and to identify any areas needing improvement. Pretesting should be supported and include checks on the following aspects:

Access and Setup:

- Does the assistant have access to the classroom and can they open doors before the lesson?
- Does the assistant have access to the necessary software?
- Are external microphones working so the TPR operator can hear other participants?
- Is the TPR battery fully charged?
- Are TPRs' access links available and accessible?
- Is the classroom suitable for TPRs participation? Consider a test drive to ensure the environment suits TPRs' movement.

Content Sharing:

- Which computer will be used to share content on the classroom projector or TV screens?
- Can the assistant log in to the computer to share content with the classroom?
- Is the content on the slides visible (e.g., correct background, font colour, and type)?
- Are the projector or TV screens functioning properly?

Video Conference Sessions:

- Are the video conference links created and working?
- If audio is shared through the video conference, do the additional microphones, speakers, and cameras work?
- If using a whiteboard, is it functioning correctly?
- If using the video conference to transmit the TPR operator's voice, what kind of speakers are used to amplify the voice, and are they working?

Live Presentation Environment:

- Is the live presentation session working correctly?
- Is the audio of the live presentation session muted when necessary?

Ensuring these aspects are tested and functional will help guarantee a smooth integration of TPRs into the learning process, minimising disruptions and technical issues during actual lessons.

4.5 Inclusive, Ethical and Cybersecurity Aspects of Using the TPRs in Learning

This section addresses the critical considerations for ensuring inclusiveness, ethical practices, and cybersecurity when integrating TPRs into the learning process. Emphasising these aspects is crucial to creating an equitable and secure educational environment for all students, including those mediated through TPRs. This section discusses strategies to provide equal learning opportunities, maintain high ethical standards, and safeguard against cybersecurity threats, ensuring that TPRs are used responsibly and effectively in educational settings.

4.5.1 Inclusive Learning with the TPRs

"Inclusive" means that all students have equal learning opportunities, whether attending in person or mediated through TPRs. Both groups must be able to participate in all learning activities to achieve the desired learning outcomes. When a teacher is mediated via a TPR, the quality of instruction must remain equivalent to that provided by an in-person teacher.

To ensure inclusive learning with the implementation TPRs, the following aspects must be addressed:

- Hearing and speaking capabilities;
- Design of learning content;
- Freedom of movement;
- Participation of all students in every learning activity;
- Socio-economic factors include access to a suitable and stable internet connection.

Table 2 Main audio issues that can appear in TPRs and class communication and possible solutions to solve them

Situation	Possible problems	Solution
The teacher is TPR-mediated, and students do not hear or understand what the teacher is saying	A TPR speakers are too weak	Share audio via videoconference to the classroom AV system
TPR-mediated teacher must hear the students' comments, questions, or answers	A TPR microphone range does not cover	A teacher must move closer to the student. The downside is that moving to a new location takes time and energy Use a microphone system that allows quick microphone transfer (e.g. CatchBox, handheld microphone)
The student is TPR-mediated and wants to hear fellow students' comments, questions, or answers	A TPR microphone range does not cover all classroom	Use a microphone system that allows quick microphone transfer (e.g. CatchBox, handheld microphone)

Aspects related to movement and internet accessibility are discussed in Suitable Learning Environment for TPRs.

A critical element of inclusion is ensuring everyone has the opportunity to hear and be heard. All participants must be able to listen to others, whether they are in-person attendees or mediated by TPRs. Agreeing on common ground rules or norms, such as when to speak, when to ask questions, etc., can ensure better engagement of the TPR-mediated student (Elmimouni et al., 2024). Depending on the classroom size, there may be instances where TPRs cannot hear in-person participants and vice versa. Table 2 outlines potential problems and their solutions.

4.5.2 Ethical Aspects

When integrating TPRs into the classroom, it is crucial to address several ethical considerations to ensure their responsible use. Before using TPRs in the learning process, participants and other stakeholders (in the case of minors their parents) must consider how to create safe communication for all parties involved. It is essential to inspect what TPR enables remote students to see and what the remote student is showing with a camera from their environment (Newhart & Olson, 2017). This section discusses key ethical issues, such as explaining the purpose of TPRs to all participants, maintaining appropriate distances, using zoom and recording functions responsibly, and ensuring user authentication. By adhering to these ethical guidelines, educators and students can foster a respectful and secure learning environment that protects the privacy and comfort of all participants while leveraging the benefits of

TPRa technology. The first and most crucial step is to explain the purpose of the TPR in the lesson to all participants. This includes explaining to students:

- Why the teacher or student is using the TPR;
- The goal of using the TPR.

If students are unfamiliar with TPRs, a brief introduction is appropriate to clarify any misconceptions and alleviate any fears or discomfort related to robots (robomorphism).

Several ethical aspects must be addressed while operating TPRs. This section outlines the most relevant considerations:

- **Maintaining Appropriate Distance**: When driving the TPR, it can be challenging to estimate the distance between the robot and other objects or individuals. The TPR operator should avoid getting too close, which might cause discomfort. Asking for feedback on the appropriateness of the distance is advisable;

- **Ethical Use of Zoom**: The zooming function on TPRs should be used responsibly. Some models allow for multiple magnifications, providing detailed views that can exceed human capability. Avoid using zoom to capture inappropriate images;
- **Camera and Microphone Usage**: Muting the microphone when not speaking is recommended. There are situations where turning off the camera is necessary, such as during emergencies or to protect privacy. Generally, the TPR operator's face should remain visible, but temporary camera shutdowns are acceptable when justified;
- **Photography Limitations**: Some TPRa models allow for taking screenshots secretly. Ethical use requires obtaining permission from participants before taking any screenshots, ensuring they are not used for malicious purposes;
- **Background Visibility**: The TPR operator should be mindful of what is visible in the background while operating remotely. Following camera configuration recommendations (see Sect. 4.3) helps maintain appropriate visuals. Horizontal screens on TPRs may display more background than intended, so careful setup is essential;
- **User Authentication**: Verify the identity of the person using the TPR. With advancements in technology, there are risks of impersonation through deep fake audio and video. Ensure that the classroom participants' privacy is protected by using secure methods, such as operating from a separate room and using headphones;
- **Appropriate Usage Times**: TPRs should only be used for their intended educational purposes and during agreed-upon times. Using TPRs as surveillance tools to observe others without their knowledge is unethical.

By addressing these ethical considerations, the integration of TPRs in the learning environment can be conducted responsibly, maintaining trust and respect among all participants.

4.5.3 Cybersecurity Aspects

Ensuring cybersecurity is paramount when integrating TPRs into the learning environment. This section outlines potential cybersecurity challenges and offers strategies to mitigate risks, safeguarding both data and privacy. Using TPRs on public Wi-Fi networks is discouraged due to low cybersecurity standards. Instead, it is recommended to use private, password-protected Wi-Fi connections to minimise vulnerabilities and reduce the risk of unauthorised access and data breaches.

When TPRs are used outside the classroom, it is essential to consider the environment. Teachers and assistants must ensure that no sensitive information is inadvertently captured by the TPR. Adjusting the room setup can help maintain privacy and prevent the disclosure of confidential information.

Each TPR's session should generate unique access links that are valid for a limited time to prevent unauthorised access. Sharing these links through one-to-one communication minimises the number of people who can access them. Creating weekly or monthly access links ensures controlled access over time for recurring lessons. Unauthorised use of access links can lead to sensitive information being exposed or the TPR being moved without permission.

If a connection is made outside of planned sessions, there is a risk that the TPR operator could overhear sensitive information. While some TPRs models send audio notifications when someone connects, others allow these notifications to be turned off, enabling secret connections. It is crucial to manage these settings to protect privacy.

Lastly, unauthorised movement of the TPR can occur, allowing it to navigate and observe areas without permission. Ensuring strict access controls and monitoring usage can prevent such unauthorised activities.

4.5.4 Recommendation to Inclusive, Ensure Ethical and Cybersecurity Use of TPRs in the Learning Process

To effectively integrate TPRs in educational settings, adopting practices that promote inclusivity, uphold ethical standards, and ensure cybersecurity is essential. The following recommendations provide a framework for planning and conducting lessons with TPRs, ensuring that all students can participate fully and safely.

When planning a learning process, ensure that all students, including those mediated by TPRs, can engage in the activity. Agree on the methods of participation for TPR-mediated students and prefer digital tools such as web-based interactive boards and collaborative writing platforms. For paper worksheets, use a larger font size (18 pt) or provide digital copies to accommodate TPR-mediated students. Adjust activities to ensure that TPR-mediated students can achieve the same learning outcomes as their in-person peers.

During lessons involving TPRs students or teachers, it is crucial to maintain ethical behaviour. Avoid actions that intentionally block the TPR's movement or vision, and always seek permission before moving the TPR to another location, even when assistance is required.

Guidelines for teachers to ensure the ethical and secure use of TPRs include:

- **Explain the Purpose**: Clearly explain to students what a TPR is, why it is being used, and the goals of its use;
- **Introduce Ethical Practices**: Educate students on ethical practices related to TPR usage, such as respecting privacy and maintaining appropriate behaviour during lessons involving TPRs;
- **Control Access**: Generate access links for TPR sessions for a limited period. This task can be handled by an assistant or the person responsible for managing access;
- **Limit Link Distribution**: Share access links only with those who need them to prevent unauthorised access.

By following these guidelines, educators can create an inclusive, ethical, and secure learning environment that enhances the educational experience for all students, whether they are physically present or participating via TPRs.

4.6 Instructional Methods

This section provides recommendations for instructional methods to implement TPRs in the classroom effectively. It explores various teaching strategies to enhance the learning experience when integrating TPRs, particularly in hybrid learning environments where online students participate in class activities (Kalya et al., 2022). Practical recommendations include using active learning methods, breaking down lessons into smaller chunks, and utilising mini-lectures to introduce new material. Additionally, the section covers the preparation of the learning environment, the design of suitable content for TPRs users, and strategies for effective group work and interactive discussions.

Teachers can employ a variety of instructional methods in the learning process. Using TPRs in traditional lectures, which last 90 min or more and involve little interaction with students, may not be as effective and can be replaced by video conferencing or lecture capture. The TPR should only be involved in the classroom if it can enrich the learning process. The TPR's ability to move independently helps eliminate obstacles in hybrid learning, where online students cannot participate in class activities.

The most useful is TPR in the learning process, where it is used as an active learning method. While teaching via telepresence, robots start lessons with mini-lectures. Interactive mini-lectures are an excellent way to introduce new material in the learning process. These lectures should be concise, typically lasting 10–20 min, and should focus on a specific topic that defines or sets the context for the lesson. After the mini-lecture, plan an engaging activity (e.g., group work, pair work, experiments) that allows students to explore the topic more deeply and apply their knowledge. Conclude the topic by summarising the activities and highlighting the key takeaways.

For longer lessons, incorporating multiple mini-lectures can maintain engagement and provide a structured learning experience.

To enhance interaction during mini-lectures, incorporate discussions and interactive questions. Utilising online surveys or interactive question tools (e.g., Slido, MS Forms, Poll Everywhere) can add questions directly to the presentation, encouraging student participation and engagement. This approach is particularly beneficial when teaching via TPRs, as it helps to reduce the cognitive load on the TPRs' operator by fostering a dynamic and interactive learning environment.

Additionally, consider the following tips to maximise the effectiveness of mini-lectures and interactive teaching:

- **Use Visual Aids**: Integrate slides, videos, and other visual aids to make the mini-lecture more engaging and easier to follow;
- **Frequent Check-ins**: Regularly check in with students to ensure they understand the material and feel comfortable participating in activities;
- **Active Learning Techniques**: Encourage students to ask questions, participate in discussions, and engage in problem-solving activities;
- **Feedback and Reflection**: Provide immediate feedback on activities and encourage students to reflect on what they have learned, promoting deeper understanding and retention of the material;
- **Adaptability**: Be flexible and ready to adjust the pace or content of the mini-lecture based on student responses and engagement levels.

By incorporating these strategies, mini-lectures can become a powerful tool in creating an interactive and effective learning environment, especially when using telepresence technology.

4.6.1 Group Work

In group or pair work, where listening and speaking with others is essential, it is crucial to create an environment that effectively engages the TPR. Ensuring that the TPR can hear and interact with other group members is vital for meaningful participation. Explore the recommendations below to improve the participation and effectiveness of group work when implementing TPRs:

- **Utilise Separate Physical Rooms**: If possible, assign different groups to separate physical rooms. This setup minimises background noise and allows the TPR to focus on its group's discussion without interference from other groups;
- **Spread Groups within One Room**: If separate rooms are not feasible, spread the groups across a single room, leaving significant gaps between them. This spacing helps TPRs operators hear and participate in their group discussions more effectively;
- **Position the TPR Optimally**: In smaller rooms or when there are many groups, move the TPR as close as possible to its group members. This proximity helps the TPR operator listen more effectively and engage in the conversation;

- **Use Audio Enhancement Tools**: Consider using microphones or other audio enhancement tools to ensure clear communication between group members and the TPR. These tools can help overcome any limitations in the TPR's audio capabilities;
- **Minimise Background Noise**: Encourage all participants to minimise background noise and speak clearly. This practice is particularly important in ensuring that the TPR can hear and contribute effectively to group discussions.

By following these recommendations, you can support development of an inclusive and engaging environment for TPRs during group or pair work, facilitating better communication and collaboration among all participants.

4.6.2 Preparation of the Learning Process

Effective lesson planning is crucial for integrating TPRs into the classroom. Perifanou (2023) provides a pedagogical model with four steps to support teachers in planning lessons with TPR-mediated participation. The following recommendations will help create a conducive learning environment and prepare engaging content for TPRs' users:

1. Prepare the Learning Environment:
   (a) Ensure the classroom setup accommodates both in-person and TPR-mediated participation (refer to Sect. 4.2 for detailed guidelines).
2. Develop and Design Learning Content:
   (a) Create slides, materials, and worksheets, including digital formats, to facilitate seamless learning (refer to Sect. 4.5 for specific guidelines);
   (b) Divide learning content into smaller, manageable chunks to maintain student engagement;
   (c) Use a mini-lecture format to introduce new material, keeping each segment between 10 and 20 min;
   (d) Plan interactive activities, such as group work or experiments, to allow students to explore topics more deeply;
   (e) Conclude each lesson with a summarising activity to reinforce key takeaways (e.g., discussions or addressing the muddiest point).
3. Utilise Cooperative Tools and Active Learning Methods:
   (a) Use cooperative tools to manage group work effectively, ensuring active participation from all students, including those mediated by TPRs;
   (b) Incorporate active learning methods to engage students and enhance their understanding of the material.
4. Design Content Suitable for TPRs Users:

(a) Use font sizes of 24 or 28 pt for slides and 18 pt for printed materials to ensure readability;
(b) Choose high-contrast colour schemes for slides, such as black and white, to improve visibility;
(c) Use broader lines for whiteboard sketches to ensure clarity;
(d) Ensure the TPR has an optimal view of learning materials by finding the best angle and spot in the classroom;
(e) Utilize good lighting conditions, and if necessary, take photos with the TPR for better clarity, as photo quality is often superior to live camera feeds.

By following these comprehensive recommendations, educators can create an inclusive and engaging learning experience that effectively integrates TPRs technology, enhancing the educational experience for all students.

5 Discussion

The aim of this pilot study is to develop comprehensive pedagogical guidelines for the effective integration of telepresence robots (TPR) in educational settings, thereby filling a significant gap in the literature. The results of the three pilot studies involving educators and students from different countries indicating that technical preparation, adaptation to the environment, cognitive and physical limitations, ethical considerations, cybersecurity and teaching methods are the most important factors for the successful use of TPR in education.

The essence of our research shows that it is necessary to create a suitable environment for working with TPR, both remotely and within the classroom. This is in line with the recommendations of Ahumada-Newhart and Olson (2019), who point out the importance of stable internet connections and avoiding Wi-Fi signal 'blind spots', which are crucial for the continuous operation of TPR. Based on the results of our study and the study mentioned above, we extend these guidelines to it, pointing out the importance of clear paths of at least 50 cm for TPR navigation and maintaining consistent lighting to improve the visibility of TPR operators. The results of our study also suggest that correct control of TPR requires the simultaneous performance of multiple tasks, which may increase cognitive load. This finding in our study is supported by the results of Talisainen et al. (2023), who suggest that the readability of text is better with static images, indicating the need for clear and accessible presentation of content. With this in mind, we propose practical solutions such as using two monitors to manage different application windows and organising the screen layout to keep all necessary interfaces visible, thus reducing the cognitive effort of the operator. Physical limitations in the use of TPR were also identified in our study. For example, the inability to manipulate objects or navigate independently on certain terrains. To successfully overcome this obstacle in practise, we propose the inclusion of assistants based on the results of our study. In addition to the results of our study, this recommendation is based on previous studies by Leoste et al. (2022) and Elmimouni et al.

(2024), which emphasise the importance of assistants in facilitating the movement and work of TPR. Ethical and cybersecurity considerations are emphasised as critical components for the integration of TPR. While previous studies such as Gallon et al. (2019) focused on technical readiness, our guidelines delve deeply into ethical practises, such as respecting privacy by obtaining consent before screen recording and ensuring that TPR is only used at agreed times to prevent unauthorised surveillance. These considerations are in line with Echeverria et al. (2019), who emphasise the importance of effective communication to reduce cognitive load, including ethical communication practises. The results of our research show the importance of integrative teaching practises when using TPR. Our findings are consistent with those of LeTendre and Grey (2024), who emphasise teacher preparation and student engagement. Active learning methods such as mini-lectures and group work with special considerations for students delivered through TPR can potentially contribute to the development of an inclusive learning environment. Future research should help to confirm or reject this hypothesis.

6 Conclusion

Our guidelines recommend and guide how to redesign or adjust the teaching and learning process compared to the manufacturers' technical user manuals. It also recommends adjusting learning environments and ensuring good audio and video quality. Furthermore, our guidelines highlight ethical and cybersecurity aspects of using TPRs with various people. Compared with articles that also give recommendations and guidelines on using TPRs effectively in learning, our guidelines combine the summaries and critical aspects that need to be known to ensure good quality teaching and learning with TPRs. Our guidelines are systematically structured. The pedagogical guidelines we offer are not the final version. It needs further development to be used in different fields (e.g. workplace-based learning, task-based, office workers, healthcare, care homes).

The integration of TPRs in hybrid educational environments significantly enhances accessibility and inclusion for students who cannot attend school physically due to chronic illnesses or other conditions. Key benefits include ensuring educational continuity, maintaining social connections, allowing customization, engaging stakeholders, and providing a sense of agency, all of which improve the academic and emotional well-being of these students.

The limitations of the pedagogical guidelines presented in this chapter is that most participants were females from higher educational institutions in Estonia. The workshops were conducted in a hybrid format, with all participants participating online. The sample was not selected randomly, and participants joined voluntarily. None of the participants had the pressure to start implementing TPRs in their classrooms.

Further work would be to improve the guidelines by validating it with teachers who need to teach through a TPR or teach TPR-mediated students. In addition, longitudinal studies are needed to better evaluate and gather ideas for improvements.

Aspects that are important at the beginning of implementation may not be so crucial in longer use cases. Future research should also continue exploring the long-term impacts of TPRs on students' academic performance, emotional well-being, and social development. Additionally, investigating ways to improve the technical aspects and user experience of TPRs will further enhance their effectiveness and acceptance in educational settings. Overall, TPRs hold great promise in creating a more inclusive and accessible education system, benefiting students, educators, and the broader community.

Successful implementation of TPRs requires addressing technical and privacy challenges through proper training, support, and planning. Comprehensive stakeholder engagement is essential to creating a supportive environment and ensuring the technology effectively supports students' educational and social goals. By leveraging TPRs, educational institutions can provide a more inclusive and supportive learning environment, ensuring that all students have equal opportunities to succeed academically and socially.

References

Ahumada-Newhart, V., & Olson, J. S. (2019). Going to school on a robot. *ACM Transactions on Computer-Human Interaction, 26*(4), 1–28. https://doi.org/10.1145/3325210

Almeida, L., Menezes, P., & Dias, J. (2022). Telepresence social robotics towards co-presence: A review. *Applied Sciences, 12*(11), Article 11. https://doi.org/10.3390/app12115557

Bellizzi, K. M., Smith, A., Schmidt, S., Keegan, T. H. M., Zebrack, B., Lynch, C. F., Deapen, D., Shnorhavorian, M., Tompkins, B. J., & Simon, M. (2012). Positive and negative psychosocial impact of being diagnosed with cancer as an adolescent or young adult. *Cancer, 118*(20), 5155–5162. https://doi.org/10.1002/cncr.27512

Bittner, B. (2021, December 14). Webcam Cinematography—Brett Bittner—Medium. *Medium.* https://medium.com/@brettbittner/webcam-cinematography-d759bfd4ae44

Buchem, I., Sostak, S., & Christiansen, L. (2024). Human–robot co-facilitation in collaborative learning: A comparative study of the effects of human and robot facilitation on learning experience and learning outcomes. *Multidisciplinary Scientific Journal, 7*(3), 236–263. https://doi.org/10.3390/j7030014

Choquette, A., Rennick, J. E., & Lee, V. (2016). Back to school after cancer treatment. *Cancer Nursing, 39*(5), 393–401. https://doi.org/10.1097/ncc.0000000000000301

Echeverria, V., Martinez-Maldonado, R., & Buckingham Shum, S. (2019). Towards collaboration translucence: Giving meaning to multimodal group data. In *Proceedings of the 2019 CHI conference on human factors in computing systems* (pp. 1–16). https://doi.org/10.1145/3290605.3300269

Elmimouni, H., Rode, J. A., & Šabanović, S. (2024). Articulation work for supporting the values of students attending class via telepresence robots. *International Journal of Human-Computer Studies, 190*, 103318. https://doi.org/10.1016/j.ijhcs.2024.103318

Fitter, N. T., Raghunath, N., Cha, E., Sanchez, C. A., Takayama, L., & Matarić, M. J. (2020). Are we there yet? Comparing remote learning technologies in the university classroom. *IEEE Robotics and Automation Letters, 5*(2), 2706–2713. https://doi.org/10.1109/LRA.2020.2970939

Gallon, L., Charrier, J. L., Dejean, C., Jannin, P., Couronne, R., & Rigaud, A. S. (2019, July). Using a telepresence robot in an educational context. In *Proceedings of the 15th international conference on frontiers in education: Computer science and computer engineering* (FECS).

García-Morales, V. J., Garrido-Moreno, A., & Martín-Rojas, R. (2021). The transformation of higher education after the COVID disruption: Emerging challenges in an online learning scenario. *Frontiers in Psychology*, 12. https://doi.org/10.3389/fpsyg.2021.616059

Jakonen, T., & Jauni, H. (2021). Mediated learning materials: Visibility checks in telepresence robot mediated classroom interaction. *Classroom Discourse, 12*(1–2), 121–145. https://doi.org/10.1080/19463014.2020.1808496

Kalya, S., Lee, C. H., & Teo, T. H. (2022). Enhanced learning experience for remote students in hybrid class model using 360° view camera and telepresence robot. In 2022 *IEEE international conference on teaching, assessment and learning for engineering* (TALE) (pp. 744–748). https://doi.org/10.1109/TALE54877.2022.00133

Kasuk, T., & Virkus, S. (2024). Exploring the power of telepresence: Enhancing education through telepresence robots. *Information and Learning Science, 125*(1–2), 109–137. Scopus. https://doi.org/10.1108/ILS-07-2023-0093

Leoste, J., Burnard, P., Harris, D., Virkus, S., Blakeborough, J., Fenyvesi, K., Lavicza, Z., & Marmor, K. (2024). A critical retelling of the implementation of telepresence robots to enhance learners' collaboration: Through a posthumanist new materialist and intersectionalist lens. In S. Papadakis, & M. Kalogiannakis (Eds.), *Education, development, and intervention* (Integrated Science, Vol. 23). Springer. https://doi.org/10.1007/978-3-031-60713-4_8

Leoste, J., Virkus, S., Kasuk, T., Talisainen, A., Kangur, K., & Tolmos, P. (2022). Aspects of using telepresence robots in a higher education STEAM workshop. *Lecture Notes in Computer Science* (pp. 18–28). https://doi.org/10.1007/978-3-031-21047-1_2

LeTendre, G. K., & Gray, R. (2024). Social robots in a project-based learning environment: Adolescent understanding of robot–human interactions. *Journal of Computer Assisted Learning, 40*(1), 192–204. https://doi.org/10.1111/jcal.12872

Marmor, K., Leoste, J., & Heidmets, M. (2024). Evaluating users' self-efficacy in using social service robots in public areas. *International Journal of Social Robotics*. In press.

Mordor Intelligence. (2023). *Telepresence robots market—Growth, trends, COVID-19 impact, and forecasts* (2023–2028). https://www.mordorintelligence.com/industry-reports/telepresence-robots-market/market-trends

Newhart, V. A., & Olson, J. S. (2017). My student is a robot: How schools manage telepresence experiences for students. In *Conference on human factors in computing systems—Proceedings* (pp. 342–347). https://doi.org/10.1145/3025453.3025809

Page, A., Charteris, J., & Berman, J. (2023). Telepresence robot use for children with chronic illness in Australian schools: A scoping review and thematic analysis. *International Journal of Social Robotics, 13*(6), 1281–1293. https://doi.org/10.1007/s12369-020-00714-0

Perifanou, M. (2023). "TRINE 4D Pedagogical Model": Designing teaching scenarios for using telepresence robots in education. In *Lecture notes in computer science (including subseries lecture notes in artificial intelligence and lecture notes in bioinformatics)*, 14200 LNCS (pp. 605–610). Scopus. https://doi.org/10.1007/978-3-031-42682-7_49

Powell, T., Cohen, J., & Patterson, P. (2021). Keeping connected with school: Implementing telepresence robots to improve the wellbeing of adolescent cancer patients. *Frontiers in Psychology, 12*, 749957. https://doi.org/10.3389/fpsyg.2021.749957

Talisainen, A., Leoste, J., & Virkus, S. (2023). Comparative analysis of telepresence robots' video performance: Evaluating camera capabilities for remote teaching and learning. *Applied Sciences, 14*, 233. https://doi.org/10.3390/app14010233

Talisainen, A., Leoste, J., & Virkus, S. (2024). Exploring challenges in remote communication: Comparing teleconferencing software and telepresence robot usage in higher education. In Future technologies conference (FTC). Accepted for Publication.

Tuli, T. B., Terefe, T. O., & Rashid, M. M. U. (2021). Telepresence mobile robots design and control for social interaction. *International Journal of Social Robotics, 13*(5), 877–886.

Tsui, K. M., Desai, M., Yanco, H. A., & Uhlik, C. (2011). Exploring use cases for telepresence robots. In *Proceedings of the 6th international conference on human-robot interaction* (pp. 11–18). https://doi.org/10.1145/1957656.1957664

Virkus, S., Leoste, J., Marmor, K., & Kasuk, T. (2023). Telepresence robots from the perspective of psychology and educational sciences. *Information and Learning Sciences, 124*(1/2), 48–69. https://doi.org/10.1108/ils-09-2022-0106

Wernbacher, T., Pfeiffer, A., Häfner, P., Buchar, A., Denk, N., König, N., DeRaffaele, N. C., Attard, A., Economides, A. A., Perifanou, M. (2022a). TRINE: Telepresence robots in education. In *INTED proceedings*. https://doi.org/10.21125/inted.2022.1653

Wernbacher, T., Pfeiffer, A., Häfner, P., Buchar, A., Denk, N., König, N., DeRaffaele, C., Attard, A., Economides, A. A., & Perifanou, M. (2022b). Trine: Telepresence robots in education. In *INTED proceedings* (pp. 6514–6522). https://doi.org/10.21125/inted.2022.1653

Zhang, W., & Zhu, C. (2023). Institutional presence: Toward a further developed community of inquiry model integrating institutional functions in online and blended learning environments. *Frontiers in Psychology, 14*, 1132204. https://doi.org/10.3389/fpsyg.2023.1132204

Exploring Nao's Role in Enhancing Motor Activity for Young Children with ASD in Inclusive Classrooms

Loredana Perla, Berardina De Carolis, Ilenia Amati, Stefania Massaro, and Giuseppe Palestra

Abstract Ground-breaking research into the use of social robots with children with Autism Spectrum Disorder (ASD) is paving the way for innovative applications in early childhood education. This experimental study carried out a social robot-based intervention in a preschool setting to investigate the potential of robotic technology to transform inclusive pedagogical approaches with young children with ASD. The study examined the relationships between children and educational figures and between children and the Nao robot during motor activities in two comparative preschool settings. 20 children aged between 24 and 36 months were involved, divided into two groups of ten subjects. Each group included a child with autism. The research aims to provide evidence for the creation of specific robotic educational mediation devices, to promote communicative-relational skills of educators in the preschool context, and to evaluate the effectiveness of the NAO robot during motor activity sessions for children with autism, examining its impact on children's involvement, attention, and on motor skills and coordination. The research tools and phases, carried out within the micro-pedagogical and phenomenological theoretical-methodological approach, are described. The interactions and meanings are pedagogically analyzed in the framework of the didactics of the implicit. The results show that children were engaged with the robot and children with ASD had an interest in

L. Perla (✉) · I. Amati · S. Massaro
Department of Education Sciences, Psychology, Communication, University of Bari, Bari, Italy
e-mail: loredana.perla@uniba.it

I. Amati
e-mail: ilenia.amati@uniba.it

S. Massaro
e-mail: stefania.massaro@uniba.it

B. De Carolis · G. Palestra
Department of Computer Science, University of Bari, Bari, Italy
e-mail: berardina.decarolis@uniba.it

G. Palestra
e-mail: giuseppe.palestra@uniba.it

G. Lampropoulos and S. Papadakis (eds.), *Social Robots in Education*, Studies in Computational Intelligence 1194, https://doi.org/10.1007/978-3-031-82915-4_12

the robot sessions. In one case, this involvement was prompted by the child's spontaneous curiosity, while in the other case, it was facilitated by the educator's support in fostering the child's relation with the robot alongside his peers.

Keywords Inclusive education · Social robots in education · Social robotics for autism · Early childhood education · Child–robot interaction

1 Introduction

Social robots are robotic agents that can interact socially with humans (Malerba et al., 2019). Recent studies show that if integrated into class activities as tutors or peers they can transform lessons into interactive and engaging learning environments (Belpaeme et al., 2018; Perla et al., 2023a, 2023b). The advent of social robots has brought with it the potential for these devices to become innovative tools that promote participation and collaborative learning within the principles of social constructivism (Díaz-Boladeras et al., 2023). They have the potential to act as facilitators, not only improving learning outcomes, but also fostering a sense of community among students of different abilities in inclusive classrooms.

Social robots are the subject of particular attention on intervention practices with children with Autism Spectrum Disorder (ASD). ASD is a developmental disability that generally affects the development of proper social communication and behavior skills (Bennett & Goodall, 2016). The use of social robots has been found to have positive effects on the education of children and young people on the autism spectrum, with benefits observed across a range of areas, including attention, learning, behavioral regulation and the management of restricted and repetitive patterns of behavior (Palestra et al., 2017). Studies on robot-assisted interventions for autism indicate that robots can foster enthusiasm, with children with autism often engaging comfortably with robotic technologies (Dubois-Sage et al., 2024; Lytridis at al., 2018).

Richardson et al. (2018) suggest that robots as therapeutic tools for children with autism are based on findings that individuals with ASD: (a) exhibit strengths in understanding the physical (object-related) world and relative weaknesses in understanding the social world; (b) are more responsive to feedback, even social feedback, when administered via technology rather than a human being; (c) are more intrinsically interested in treatment when it involves electronic or robotic components.

These robots allow for the promotion of a communication channel that passes through the attraction and channeling of the attention (eye contact), continues with the triggering of motivation, and lead to the imitation and enactment of new social behaviors and joint attention (Pennazio, 2017).

This experimental study examines the innovative aspect of robot-based interventions in preschool settings, namely the potential of robotic technology to transform pedagogical approaches with children with ASD through integration into early childhood inclusive education. This approach is aligned with the framework that views inclusion as a complex concept to be addressed in the context of class-group relations rather than through a deficit paradigm and it emphasizes equity as the right of students to universally designed learning opportunities (Field et al., 2007; Perla, 2013).

The main aim of this study is to explore the use of NAO, a social assistive robot, as a coach for preschool children's motor activities, specifically targeting those with ASD, in a Montessori school setting. Positioned at the intersection of Socially Intelligent Robots and Assistive Robots, Socially Assistive Robots (SAR) are designed to help students reduce social impairments through social interaction (Fang et al., 2022; Palestra et al., 2016). The study intends to investigate how robotic interventions can enhance motor skills and, in turn, improve social interaction and communication among young children.

An increasing number of studies have identified motor difficulties in children with ASD that are likely to negatively affect social-communicative development. Teacher's sense of inefficacy and bullism often take place. An overview of the benefits of motor activity for individuals with ASD reveals physical benefits such as improvement of gross and fine motor skills, enhancement of control and coordination (Ji et al., 2023). Additionally, there are benefits related to cognitive and emotional growth. Finally, motor activity through peer interaction can improve skills related to social interaction and communication.

Despite the growing body of literature on robot-based interventions for children with ASD, there remains a significant research gap regarding the systematic exploration of how these interventions can specifically enhance motor skills. Most studies focus on improving emotional and social competencies, often overlooking the critical role that motor skills play in overall development. This is particularly concerning given the established correlation between motor proficiency and the severity of communication skills in children with ASD (Jouaiti & Hénaff, 2019). A child's motor ability remains a significant predictor of social behavior (Cummins et al., 2005). Therefore, there is an urgent need to investigate how robotic technology can be leveraged to promote motor activities and, consequently, enhance social interaction and communication among these children. Nevertheless, authors identify a significant minority of research focusing on robotic application for autism in school settings (Yang et al., 2023). More research is needed to understand the actual added value in therapy and education.

The contribution of this study lies in addressing the underexplored area of motor skills in children with ASD within the context of robot-based interventions. By integrating these robotic technologies into early childhood education, this study offers significant insights that contribute to foster inclusivity and promote holistic development for children with ASD, thereby bridging the gap between traditional educational approaches and modern technological solutions.

Key components of the chapter include the rationale behind specific decisions on hardware and software selection, as well as the design of motor activity exercises tailored to the study's objectives. The chapter outlines the experimental protocol, describing each phase of the research, and details the analysis process, which relies on observation grids and qualitative data collected through observer-completed sheets.

Finally, the findings are discussed in depth, providing insights into the effectiveness and implications of the intervention.

2 Related Studies

The results of studies on robot-led interventions for the development of motor skills in children with ASD indicate improvements in motor performance, imitation skills, social engagement and coordination. Huijnen et al. (2016) conducted a study aimed at identifying therapeutic and educational goals relevant to children with ASD and highlighting state-of-the-art robots suited to these goals. Their findings describe motor experiences and skills as one of the nine main domains, as noted by practitioners in focus group sessions. They specifically identify the use of an imitator robot to teach children coordinated behaviors, encompassing a range of motor skills, such as fine and gross motor skills, motor coordination, imitation, and sensorimotor skills.

In a pilot study, Moorthy and Pugazhenthi (2016) explored the feasibility of using a robot to train imitation skills in children with autism. They proposed that enhanced imitation skills in ASD may help improve both psychomotor and communication abilities. The study's scope focused on improving specific psychomotor skills, including grasping, directional movement, and backward walking. Their findings suggest that robotic training can effectively teach imitation skills, resulting in positive study outcomes. Despite limitations in sample size and lack of follow-up, the study offers preliminary evidence that robotic interventions could enhance motor and imitation skills in children with ASD.

Zheng et al. (2016) presented the design and development of a novel robot-mediated imitation skill training system, potentially addressing core deficits in young children with ASD. This system was able to detect imitation performance in real time and provide feedback to an administrator for adaptive intervention. In proof-of-concept testing, they found that young children with ASD displayed greater attention and performed better in robot-assisted imitation tasks compared to human-administered sessions under the same protocol. Notable strengths of the system include its non-invasive nature, avoiding wearable sensors, and adaptability, which eliminates the need for pre-collected training data and enhances scalability. Limitations of the study include its small sample size and short interaction duration, which may restrict the generalizability of findings. Additionally, the brief duration raises concerns as to whether children's preference for robotic interaction was due to the novelty effect rather than sustainable engagement.

The study by Srinivasan et al. (2015) examined the effects of rhythm-based, robotic-assisted, and standard-of-care interventions on motor and social skills in children with ASD. Key findings indicate that both rhythm and robotic interventions significantly improved body coordination. Notably, all groups made gains in imitation and praxis skills, although only the rhythm and robotic groups showed marked improvements in interpersonal synchrony over time. Critically, the study provides evidence that socially engaging, movement-based activities like rhythm and robotics can uniquely enhance motor coordination and social synchrony in ASD. This suggests that incorporating such activities into routine care could enrich standard therapies by addressing specific motor and social needs. Further research could examine long-term effects of these interventions and determine if improvements in interpersonal synchrony are generalizable to broader social contexts.

Ground-breaking research in this field has paved the way for social robots to be used in preschool and kindergarten settings (Gómez-Espinosa et al., 2024). Social robots have shown promise as effective mediators in early interactions, with evidence indicating that effects are stronger in clinical settings than at home or in schools, especially for younger children, suggesting a better developmental match (Landa, 2018). Meta-analyses and systematic reviews conclude that early interventions produce positive effects on development for young children with ASD (Kouroupa et al., 2022). The Handbook of Autism and Pervasive Developmental Disorders (Volkmar et al., 2014) argues that children with ASD benefit significantly from early and continuous intervention tailored to their specific needs.

3 System Architecture

Motor activity intervention for children with ASD is a specific intervention modality focused on body movement-centered approaches conducted by a physical therapist. The proposed system consists of multiple components, hardware and software, to provide a comprehensive motor activity training system based on social robotics:

- NAO V6 robot;
- Windows 10 Workstation;
- Motor activity software module;
- NAOqi APIs module;
- Wi-Fi router.

The system architecture is depicted in Fig. 1. The system utilizes the NAO V6 humanoid social robot and a workstation equipped with an Intel i7 8th generation CPU and 16 GB of RAM. The NAO robot was chosen because it has already been used in motor imitation tasks with children with ASD (Grossard et al., 2018). The NAO robot, created by Aldebaran Robotics, is a small humanoid robot standing 57 cm tall and weighing 4 kg. It features a total of 25 Degrees of Freedom (DoF), with 11 DoF dedicated to the lower limbs, including the legs and pelvis, and 14 DoF for the upper limbs, including the trunk, arms, and head. Additionally, the manufacturer provides several software tools for use with the robot, such as Choregraphe, a Windows application that allows users to create robot behaviors and test them on a simulated or real robot. To facilitate the robot's movements, a direct connection via an Ethernet cable was avoided; instead, the robot was connected to the workstation through a Wi-Fi connection.

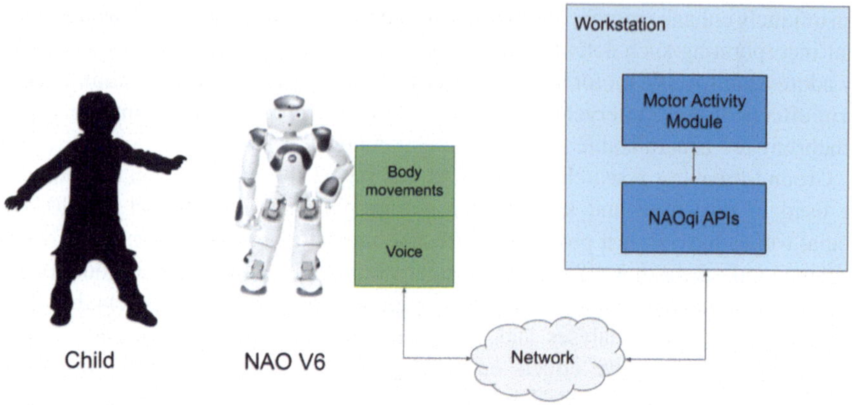

Fig. 1 System architecture. On the workstation are installed the motor activity software module, the NAOqi APIs. The workstation is connected to the robot through a Wi-Fi router

3.1 Motor Activity Software Module

The motor exercises have been selected from specific studies related to the motor impairments of autistic individuals (Gargot at al., 2022). The motor activity intervention exercises were developed using Choregraphe, a rapid programming tool provided by the robot's manufacturer, along with Python. The developed software was installed on the workstation and utilized NAOqi APIs to perform the five exercises of the intervention, incorporating both speech and body movements.

The motor activity software module includes a set of exercises for motor intervention. In this study, motor activity exercises are defined as a collection of educational and therapeutic approaches that address the individual's body movement expression related to social, emotional, and cognitive functioning, as facilitated by a psychal therapist.

In our study, a subset of five motor activity exercises has been implemented on the robot, including:

1. Raise both arms up and lower them down along the sides.
2. Raise the arms up and lower them down along the sides, alternating. First the left, then the right.
3. Knee bends, up and down (like a full sack and an empty sack).
4. Open both arms and then bring them back down along the sides (like an airplane).
5. Open the arms and then bring them back down along the sides, alternating. First the left, then the right.

The system can provide a stimulus and give the child positive reinforcement. The reinforcement consists of a randomly selected word from the following: great job!, well done!, very good!, good!, great!, super!, yay!, fantastic!, you're doing great!, how nice!, perfect!, very well!

4 The Experimental Protocol

This section provides a comprehensive overview of the experimental protocol. The protocol was designed to ensure consistency in data collection. Each phase of the experiment is described, including preparatory steps and data collection procedures. We outline research question and research tools through a structured approach that facilitates the analysis of the experimental outcomes.

4.1 Description

The study examined the relationship established during motor activities in two compared settings, specifically between children and educational figures and children and the Nao robot. The research took place in 'Le Matite Colorate Baby' nursery school in Matera, involving 20 children aged between 24 and 36 months, divided into two groups of ten subjects. Each group included a child with autism. The research had three main objectives:

1. to gather evidence for the creation of specific devices for robot co-piloted educational mediation;
2. to develop communicative-relational skills of educators in the context of nursery school, in the presence of children with and without disabilities;
3. to evaluate the efficacy of the NAO robot as a support tool during motor activity sessions for children with autism, examining the impact of interaction with the robot on children's engagement and attention and exploring whether interaction with the NAO robot can improve the motor skills and coordination of children with autism.

In the study, particular attention was given to the process of relationship and trust that exists between adult and educational figures and children. The interactions and attributions of the meaning given by the protagonists in the process were pedagogically analysed, also taking into account the didactics of the implicit (Perla, 2014). The study employed a micro-pedagogical and phenomenological theoretical-methodological approach (Mortari, 2007). The phenomenological approach represents a key epistemic orientation within the ecological paradigm, standing in contrast to the positivist paradigm, which is insufficient for investigating educational phenomenon, characterized by uniqueness and unrepeatability. The hermeneutic phenomenological orientation shifts focus from describing the phenomenon to uncovering the meaning that the experience holds for the subjects involved, developing interpretative descriptions through a process of continuous researcher's self-reflection. A micro-pedagogical orientation aligns with the principle that 'the observer includes themselves in the observation', employing tools such as interviews, case studies, participant observation, and the collection of life histories. This orientation emphasizes attention to the subjective, and the specific living environments of individuals.

4.2 Research Phases I and II

The study was conducted through a systematic framework comprising five distinct operational phases, each designed to structure and guide the research process. These phases provided a coherent, step-by-step approach, ensuring that all necessary elements of the study were addressed comprehensively and in alignment with the research objectives. Each operational phase was developed to build upon the previous one, enabling the collection, analysis, and interpretation of data in a structured manner (Fig. 2).

In the preparatory phase, the study received the approval of the Ethics Committee of the Department of Education, Psychology and Communication Sciences at the University of Bari, as it complied with the ethical requirements for research involving human participants. The parents of the children signed an Informed Consent form to participate in the study, allowing them to join the research in a responsible

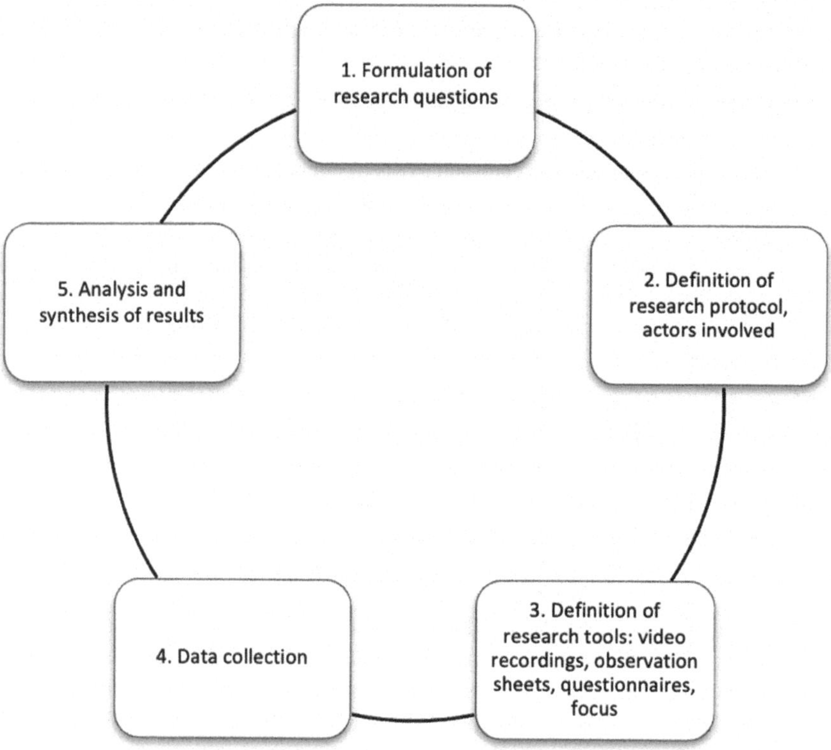

Fig. 2 Research phases

manner. They also signed a data protection agreement for the processing of sensitive personal data, ensuring the protection of children's data. The children's identities were protected by ensuring that they were filmed only from behind and above, safeguarding their anonymity throughout the study.

The first phase saw the definition of research questions (RQ) by the research team:

(RQ1)—Do children with ASD show an improvement in motor skills after participating in Nao-led activities compared to those led by the educator?
(RQ2)—What differences in engagement exist between children with ASD in Nao's group and those in the control group?

The researchers, therefore, asked how the quality of the execution of motor exercises varies between robot- and educator-led activities, how the child's ability to adapt and follow instructions varies between robot- and educator-led activities, and whether there is a difference in the child's relationship, attention and concentration time.

4.3 Phase III: Research Tools

In phase III of the research, the research tools to be used were defined: video recordings of the motor activity sessions with the robot and the educator, observation sheets to be filled in by three female observers, questionnaires administered to the three female observers, parents and educators, and focus groups with parents and educators. In this paper, for space saving, the qualitative data relating to observations from video recordings are provided.

Video research in learning contexts (Perla & Schiavone, 2017) and in our specific case in the nursery school is a useful process for assessing children's development, social, communicative and motor skills, as well as for identifying any specific difficulties or challenges that may need to be addressed. The observations from the video recordings reported here are further confirmed by the analysis of the video recordings of the motor activity sessions that followed the phases shown in the diagram in Fig. 3.

The video research process followed the following five phases (Fig. 4):

1. Video recording. The recording made it possible to capture the children's interactions at different moments of the motor activity and to observe situations, such as play, interaction with others, interaction with the robot and interaction with the educator. It was important to ensure that the children were comfortable and that the situations were representative of their everyday environment. The children had become familiar with the Nao robot in the pre-research phase the previous day.
2. Behavioral observation. The researchers observed the videos carefully, noting the behavior of the children in the two groups. They focused on aspects such as verbal and non-verbal communication, socialization, response to stimuli, and repetitive

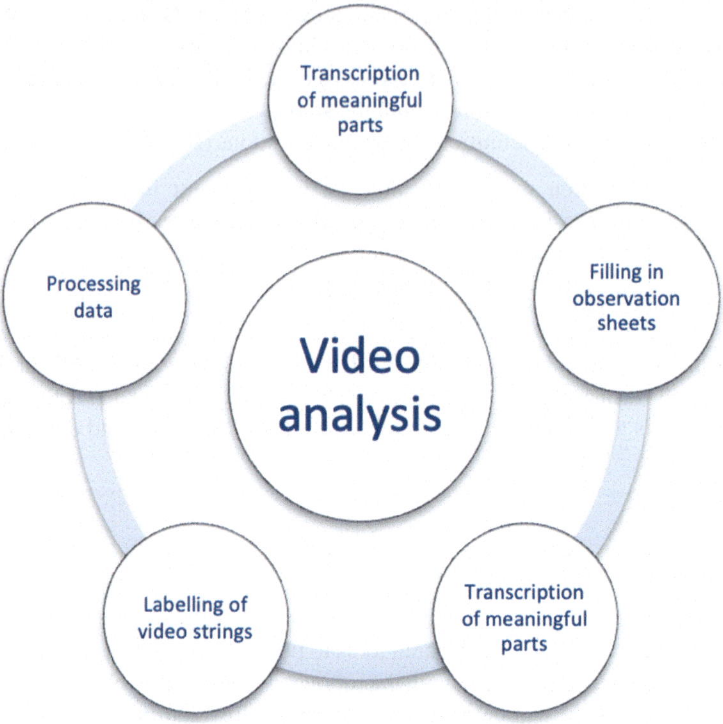

Fig. 3 Video analysis process

Fig. 4 Video research phases

behaviors. To guide the analysis, the structured observation forms constructed by the research team were used, inspired by the Autism in Children Scale, the Checklist for Autism in Children.

3. Identification of challenges and skills. During the analysis, areas were identified in which the groups and the specific child with autism (child 1 in the first group and child 2 in the second group, respectively a boy and a girl) show specific difficulties or challenges, such as lack of eye contact, poor verbal communication, limited social interaction and gestural repetitiveness. At the same time, skills and strengths of the group and individual children were also identified, such as motor skills, selective attention, and understanding of instructions.
4. Assessment of progress. The researchers used the videos as a benchmark to assess the progress of the group and the children in the different sessions. Subsequently, the videos recorded at different times were compared to identify any improvements or changes in the child's behavior and skills. This helped to monitor the effectiveness of the ongoing educational and motor interventions.
5. Involvement of observers. Three observers were involved for an in-depth evaluation of the videos and to obtain specific recommendations for the intervention. The observers then filled out the forms.

The analysis of the videos was carried out from a progression perspective, according to a 'telescope-like' progression (Perla et al., 2022) starting from an overview and arriving at a focused view on specific aspects (Fig. 5). The progression view in video analysis is an approach that focuses on the identification and tracking of moving objects within a temporal sequence. This approach is particularly useful for understanding the changes and dynamics in video. The analysis of the videos then involved further five phases described in Fig. 6.

The fourth phase of the research involved data collection consisting of two steps: quantitative and qualitative. To obtain qualitative data, observation sheets were used. Regarding quantitative data, four questionnaires were administered: session 1—robot, session 1—physical therapist, session 2—robot, session 2—physical therapist. These questionnaires will not be returned in this study.

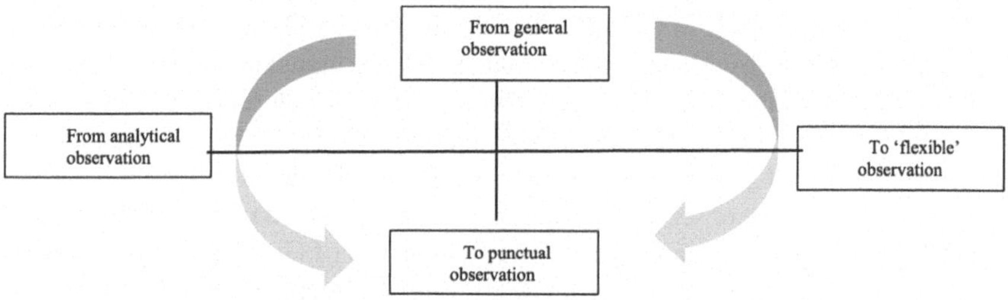

Fig. 5 Progression perspective in video analysis

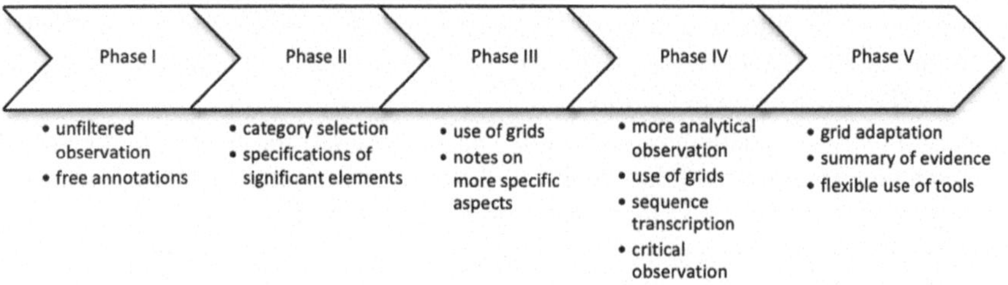

Fig. 6 Video analysis phases

4.4 Phase V: Analysis and Synthesis of the Results

The results were collected using observation grids. Each grid consists of a personal data section relating to the observer and 11 items to be observed:

(1) Response to the robot/educator's pronunciation of the name
(2) Eye contact
(3) Shared attention
(4) Response to auditory stimuli
(5) Exploration of environment and objects
(6) Shared play
(7) Sensory behaviors
(8) Facial expressions during activity or interaction
(9) Imitation
(10) Imitation of expressions or sounds
(11) Reaction to changes in teaching and educational activity.

Results are presented as follows:

- Table 1 presents the observation grid for items 1–6, which relate to Group 1 and to Child 1 with autism in Group 1. The frequencies of Group 1 and frequencies referring to Child 1 with autism in Group 1 are also included.
- Table 2 presents the observation grid for items 7–11, which relate to Group 1 and to Child 1 with autism in Group 1. The frequencies of Group 1 and frequencies referring to Child 1 with autism in Group 1 are also included.
- Table 3 presents the observation grid for items 1–6, which relate to Group 2 and to Child 2 with autism in Group 2. The frequencies of Group 2 and frequencies referring to Child 2 with autism in Group 2 are also included.
- Table 4 presents the observation grid for items 7–11, which relate to Group 2 and to Child 2 with autism in Group 2. The frequencies of Group 2 and frequencies referring to Child 2 with autism in Group 2 are also included.

From the qualitative data collected, it can be deduced that the groups of children participated actively in the sessions coordinated by the Nao robot (Fig. 7), especially in the first session, while they responded better with the mediation of

Table 1 Observation and frequencies of group 1 and child 1 from item 1 to 6

Item	Group 1 observation	Child 1 observation (group 1)	Group 1 frequency	Child 1 frequency (group 1)
(1) Response to the robot/educator's pronunciation of the name	The group responds promptly to the robot and educator pronouncing their names. When the robot calls a member of the group, the children react enthusiastically with nods of assent. Likewise, when the educator calls out their names, the children respond attentively and actively participate. Personalized interaction, both through the robot and the educator, contributes positively to the involvement of the children	The child is involved in the situation with the educator and the robot only initially, then he leaves the activity in both situations. He still responds to the pronunciation of his name by the educator, because of the educator's ability to involve him, and by the robot because he is taken by surprise	The response rate is high for the educator and the robot, due to the previous relationship with the educator in the first case and the unexpected attention by the robot in the second case	The child always answers the pronunciation of his name from both the educator and the robot initially, but he withdraws from the activity in both contexts as the session progresses. Despite this, he consistently responds to the educator's calls by name, likely due to the educator's skill in maintaining engagement, and to the robot's interactions, which surprise him
(2) Eye contact	The group makes eye contact with the robot, showing particular interest, and with the educator	The child maintains eye contact with the robot even if he leaves the activity. Child's eye contact with the robot fades after the initial curiosity and thanks to the educator's intervention the child does not leave the activity. With the educator he stops eye contact to end the activity but continues to watch from aside	The children regularly watch both the educator and the robot, maintaining eye contact especially with Nao	The educator's increased guidance makes the child frequently watch Nao. He regularly stares also at the educator and keep eye contact with except for some moments of complaints in both cases

(continued)

Table 1 (continued)

Item	Group 1 observation	Child 1 observation (group 1)	Group 1 frequency	Child 1 frequency (group 1)
(3) Shared attention	In both cases, the children share attention with the educator and the robot. By the end of the activity with Nao a greater number of children appear to be attentive	The child is attentive to the educator but leaves the activity. Tiredness due to prolonged activities and the heat of the day, can be factors triggering a loss of attention. He also leaves with robot but continues sharing attention due to the intervention of the educator that sits with him	The group is more attentive to Nao with high frequency	During the activity with the robot, the child's initial shared attention declined progressively showing a decline in the frequency of responses, to the point of abandoning the activity, until the educator makes him watching sitting with him and attention arises again. Attention in the activity with the educator soon fades
(4) Response to auditory stimuli	The group responds positively to auditory stimuli, showing participation when given instructions. The children react actively to stimuli by both the educator and the robot, demonstrating a good level of involvement	At the beginning of the activity, the child responds to the robot's auditory stimuli and reacts appropriately, performing the required movements. Before long, the response to the robot's auditory stimuli begins to diminish, suggesting that fatigue may be affecting its ability to concentrate. He doesn't respond to the educators' stimuli and sits on a small sofa in a corner	The group responds positively to auditory stimuli from both the educator and robot with high frequency but at the end the frequency decrease with the educator	The child responds more often to auditory stimuli given by the educator than by the robot

(continued)

Table 1 (continued)

Item	Group 1 observation	Child 1 observation (group 1)	Group 1 frequency	Child 1 frequency (group 1)
(5) Exploration of environment and objects	The group doesn't appear distracted exploring the environment or objects	The child, during the activity with both the robot and the educator, does not explore his surroundings and objects showing signs of disinterest. Initially involved, he progressively abandons the activity, preferring neither to participate nor to interact further in both cases	The group demonstrates a low frequency of exploration of the environment or objects in both activities	During the activity with both the robot and the educator, the child shows a low frequency of exploration of his surroundings and objects, indicating signs of disinterest
(6) Shared play	The children actively participate in the shared play activity in both situations, showing initial enthusiasm and involvement. However, signs of distraction are also observed as the activity continues. At the end of the activity with the robot more children are still participating	The child does not participate in the shared playful activity with the robot: despite the educator's efforts to involve him, he prefers not to take part in the game but to observe	The children exhibit high frequency of shared play activities in both scenarios. At the end of the robot's activity a higher frequency of engagement in shared play is provided	The child frequently abstains from participating in the shared playful activity with the robot and educator, except at the beginning where he seemed to be involved in shared play with the robot

Table 2 Observation and frequencies of group 1 and child 1 from item 7 to 11

Item	Group 1 observation	Child 1 observation (group 1)	Group 1 frequency	Child 1 frequency (group 1)
(7) Sensory behaviors	The children do not show sensory behaviors	During the activity with the educator, the child seems to show signs of sensory under-stimulation, which affects his participation in shared play: the child does not seem to receive enough stimulation from the environment to maintain his interest, leading him to withdraw from the activity and complains. With the robot he shows less sensory behaviors while watching from aside		The child more frequently complains of tiredness or takes refuge in a secluded corner during activities with the educator
(8) Facial expressions during activity or interaction	The children display facial expressions of attention in both cases for approximately half of the participants, and show satisfaction when the robot gives them compliments	While observing the robot, the child makes facial grimaces at certain moments and complains more frequently when working with the educator	The children exhibit facial expressions of attention with high frequency in both cases, affecting approximately half of the participants. They show heightened satisfaction with increased frequency when the robot gives them compliments	The child sometimes makes facial grimaces while observing the robot and exhibits an increased frequency of complaints during activities with the educator

(continued)

Table 2 (continued)

Item	Group 1 observation	Child 1 observation (group 1)	Group 1 frequency	Child 1 frequency (group 1)
(9) Imitation	Half the group imitates the proposed activities in both cases until the end. At the end about one more child with the robot than the educator remains to imitate the activity	In the initial phase of the activity, the child imitates the robot by repeating certain exercises with enthusiasm and precision. He observes Nao's movements and tries to replicate them. However, he continues the imitation activity in the arms of the educator who supports him. Imitation fails with the educator as he leaves the activity crying	Half of the group mimics the robot and the educator with high frequency, replicating the movements proposed during the activity	In the initial phase of the activity, the child imitates the robot with a constant frequency, repeating some exercises several times. He observes Nao's movements and tries to replicate them, showing interest and involvement. After a few repetitions, the frequency with which he participates gradually decreases, and finally, he abandons the activity
(10) Imitation or expressions of sounds	The children do not make particular sounds but seem quieter with the robot as providing attention to the novelty	The child makes complaining sounds with the educator, but is quieter with the robot as he remains in observation	The children often seem quieter with the robot, likely due to their heightened attention to its novelty	The child frequently makes complaining sounds with the educator, but is often quieter with the robot as he remains in observation
(11) Reacting to changes in teaching and educational activity	The group reacted positively to the change in teaching activity showing more curiosity to the robot	The child initially shows interest in the change in activity represented by Nao, and react less with the educator	The group frequently showed signs of curiosity towards the robot. This engagement was particularly noticeable at the start	The child's lack of engagement both within the educator's and the robot's activity becomes more evident over time. However, with constant support from the educator, the child shows more stabilized reactions

Table 3 Observation and frequencies of group 2 and child 2 from item 1 to 6

Item	Group 2 observation	Child 2 observation (group 2)	Group 2 frequency	Child 2 frequency (group 1)
(1) Response to the robot/educator's pronunciation of the name	The group responded to the pronunciation of the name by both the robot and the educator. The activities of the second group with the robot were followed with some difficulties, probably due to the fatigue accumulated during the last session	The child responds to the robot's pronunciation of her name by the educator. The interaction with the robot is more stimulating: she greets it often and tries to reach out to touch it during the activity	The group responds to the pronunciation of the name by both the robot and the educator. with a regular and positive frequency	The child responds promptly to the robot's and the educator's pronunciation of the name, showing a high frequency of reactions
(2) Eye contact	The group exhibits eye contact with the robot, indicative of curiosity. Additionally, the children also establish eye contact with the educator, likely due to their well-established relationship with them	The child initiates eye contact with the robot, and doesn't seek eye contact with the educator and gives mostly brief glances	Eye contact frequency level is higher with the robot and lower with the educator	The child seeks eye contact with the robot with high frequency, less with the educator
(3) Shared attention	The group showed attention towards the robot but about half of the children dispersed their attention and became distracted. Even with the educator, despite the initial interest, the children showed signs of distraction and fatigue, which affected their ability to actively participate and maintain attention during the proposed activities. At the end of the activity, children are more attentive with the educator	The child is initially attentive to the robot and disperses shared attention about halfway through the activity, showing overall attention for the novel experience. She diverts her initial shared attention with the educator by seeking out another friend to play with and by looking around	The frequency of responses and interaction decreased over the course of both the activities, especially with the robot	The child maintains a good, shared attention span with the robot during the activity, less so with the educator with whom she is more distracted

(continued)

Table 3 (continued)

Item	Group 2 observation	Child 2 observation (group 2)	Group 2 frequency	Child 2 frequency (group 1)
(4) Response to auditory stimuli	The group responds to the auditory stimuli in both cases but with the educator showed signs of being livelier. With the robot although the children initially react to the auditory stimuli and instructions, their participation becomes less lively and slower as the activity continues	The child responds to auditory stimuli from the robot. She responds also to the auditory prompts from the educator and engages in the motor activity but does not complete the task and ends her efforts	There is a reduction in the frequency of active participation of children with the robot	The child responds to the robot's auditory stimuli with high frequency, reacting to sounds and instructions. She regularly responds to the auditory cues from the educator and begins the motor activity, but frequently fails to complete the task and stops her efforts prematurely

(continued)

Table 3 (continued)

Item	Group 2 observation	Child 2 observation (group 2)	Group 2 frequency	Child 2 frequency (group 1)
(5) Exploration of environment and objects	The group does not explore environments and objects during both activities, probably because they are more familiar to them than the proposed motor activity	The child directs her attention to the surrounding environment ceasing the act of imitation approximately halfway through the activity with the educator. She instead explores Nao very curiously during its activity, and she often directs towards it, not exploring the environment	The group explores its surroundings minimally during both activities	The child explores her surroundings often with the educator and explores the object Nao frequently during its activity
(6) Shared play	Participation in shared play with the robot by the group seems to be limited by the robot's inability to use reinforcement tools. The inclusion of musical elements could have provided more enjoyment. The group seemed to share play activities with the educator	The child shares play activities with the robot, actively participating and showing interest in much of the activity, especially at the beginning. The child engages in less shared play with the educator and is more focused on seeking external stimuli in the environment	The group exhibits limited engagement in share playful interactions with the robot and more shared play with the educator	The child shares play activities with the robot with good frequency, actively participating and showing interest in much of the activity. The frequency soon decreases with the educator, resulting in the child not completing the activity it halfway through

Table 4 Observation and frequencies of group 2 and child 2 from item 7 to 11

Item	Group 2 observation	Child 2 observation (group 2)	Group 2 frequency	Child 2 frequency (group 1)
(7) Sensory behaviors	The children do not show sensory behaviors	The child with the educator moves around the room for sensory stimulation and snuggles on the ground perhaps for self-regulation. The little girl with the robot swings her arms but shows no particular behavior		The child, when with the educator, frequently moves around, especially in the second part. In contrast, the little girl, while interacting with the robot, intermittently swings her arms but exhibits no distinct or frequent behaviors
(8) Facial expressions during activity or interaction	The group participates in both activities, but maintains a serious attitude during interaction with the robot	During the activity with the robot, the child shows joyful expressions and emits giggles of enthusiasm. These signs indicate that she is engaged and amused by the interaction with the robot, indicating a high level of participation	The group participates in the activity with moderate frequency, but they consistently maintain a serious attitude throughout the interaction with the robot	Expressions indicate that the child is constantly engaged and amused by the interaction with the robot. The child does not display notable facial expressions when interacting with the educator

(continued)

Table 4 (continued)

Item	Group 2 observation	Child 2 observation (group 2)	Group 2 frequency	Child 2 frequency (group 1)
(9) Imitation	The children imitate Nao during the initial phase of the activity, but by the midpoint, half of them cease participating due to fatigue, although they continue to observe the robot. The group engages more in imitation with the educator predominantly during the early part of the session, with some children maintaining focus and participation as the activity progresses	The child enthusiastically imitates Nao in the first part of the activity, actively participating and replicating the robot's movements. Subsequently, her attention shifts to a more passive observation but remains focused on Nao. The child initially imitates the educator during the first part of the activity, but subsequently becomes distracted and turns her back to the educator	The children frequently imitate Nao during the initial phase of the activity, less in the second part. The group more frequently imitate the educator throughout the session	The child energetically mimics Nao with high frequency during the initial phase of the activity, actively engaging and replicating the robot's movements. Later, her attention shifts to a more passive mode of observation, though she continues to pay moderate attention to Nao. In contrast, the child initially engages in high-frequency imitation of the educator but then becomes distracted and frequently turns her back to the educator

(continued)

Table 4 (continued)

Item	Group 2 observation	Child 2 observation (group 2)	Group 2 frequency	Child 2 frequency (group 1)
(10) Imitation or expressions of sounds		The little girl gives light screams while interacting with the robot		The little girl sometimes gives light screams interacting with the robot
(11) Reacting to changes in teaching and educational activity	The group responded positively to changes in the teaching and educational activities, demonstrating a keen interest in both the activities led by the educator and those facilitated by the robot. Maybe familiarity with the educator made them more responsive	The child reacts positively to Nao's proposed activity changes. In contrast, the child's engagement with the educator diminishes, as she increasingly shows less involvement in activities led by the educator	The group frequently responded positively to changes in the teaching and educational activities, showing significant interest in both the educator-led and robot-facilitated activities	The child reacts positively to Nao's proposed activity changes with high frequency. In contrast, the child's engagement with the educator decreases over time, as she shows progressively less involvement in activities led by the educator

the educator in the second sessions. Regarding analyses of the behavior of children with autism, the child in the first group (with Level II autism, impairment in the area of communication and social interaction and limited interests and repetitive behaviors) (Kandola, 2019) showed signs of fatigue and distraction during the activity, receiving support from the educator several times. The child in the second group (child 2) showed more autonomous involvement and curiosity throughout the activities. This may be since, even if their level is the same, they are different in the spectrum, and this outline the need for personalized robot intervention. Moreover, the relationship between children with and without autism and educators is of fundamental importance in the development and learning of children (Perla & Riva, 2016). Educators provide emotional, cognitive and social support that helps children grow, learn and develop in a healthy and balanced way. The child-educator relationship was characterized by several key elements as empathy, trust, mutual understanding and the ability to adapt to individual needs. Indeed, the educators have interpreted and responded to the nuances of emotions, non-verbal signals and children's needs, creating a safe and stimulating educational environment. On the other hand, the relationship between children and robots is still in development and cannot completely replace the relationship with educators. Robots can offer complementary educational support, but, at present, they cannot provide the same range of interactions and personalized responses that educators can offer. However, robots can play a significant role in supporting education, especially when used as assistive tools or to support the learning of certain skills. Robots can provide immediate feedback, personalized exercises, and represent interactive characters that engage and motivate children in educational activities.

Fig. 7 Children from group 1 and group 2 interacting with Nao

5 Discussion

The findings of this study underscore the promising role of robotic interventions in fostering motor and social skill development in children with ASD. Using observation grids with specific items such as response to auditory stimuli, eye contact, shared attention, and imitation, this study analyzed behaviors across two mixed groups of children, with focused observation on a child with ASD in each group. Results showed that all children actively engaged with the Nao robot in initial sessions, displaying high participation, especially during the robot-led activities. However, in subsequent sessions, children responded more favorably to the educator-led interventions, suggesting that while robotic activities effectively capture initial interest, the sustained engagement of children, especially those with ASD, may benefit from human mediation. This indicates that robotic tools are most effective when used as complementary supports to educator-facilitated interventions.

Comparative observations of the two children with ASD revealed meaningful differences, emphasizing the need for personalized intervention strategies. For example, child 1, who has Level II autism with notable communication and social impairments, showed signs of fatigue and required repeated support from the educator to maintain engagement, whereas child 2 demonstrated autonomous curiosity and active participation throughout.

The findings from this study resonate strongly with Huijnen et al. (2016), who explored the integration of robotic technologies into therapeutic and educational interventions for children with ASD. Their identification of therapy and educational objectives, particularly those related to motor skills such as coordination and imitation, aligns with our observations of how robotic interventions can effectively promote these skills. As both studies emphasize the adaptability of robotic platforms to meet diverse motor skill needs, they collectively call for further empirical research to validate the long-term effectiveness of these interventions in supporting motor skill development among children with ASD.

The findings from this study also align with Moorthy and Pugazhenthi's review (2016), which underscores the potential of robots to enhance therapeutic outcomes in ASD, particularly by supporting motor and social skills. This study corroborates their insights, showing how robot-assisted interventions promote engagement and facilitate motor and social skill acquisition, which the authors attribute to the structured and predictable nature of robotic interactions.

The study's results align with Zheng et al. (2016), who observed higher engagement and task performance during robot-assisted tasks. In our study, robot-led activities initially captured children's attention, but longer-term involvement appeared to necessitate educator interaction, perhaps due to the complex social support educators provide. Moreover, Srinivasan et al. (2015) found that rhythm- and robot-based interventions led to improved motor coordination and interpersonal synchrony. This reinforces our findings, as interactions with both robots and educators enhanced social behaviors among children, especially when paired with peer support, suggesting that integrating robots within a socially rich context optimizes their effectiveness.

While these findings are valuable, they highlight the limitations of our study, including its small sample size and short duration, which restrict the generalizability of the results. Expanding this research with larger, longitudinal studies will be essential to explore the sustained effects of robotic interventions on both motor and social skills in children with ASD. Additionally, future studies should examine individual responses to robotic interventions across the autism spectrum, potentially guiding the development of highly adaptive, personalized robotic systems.

In summary, this study contributes with valuable insights to the growing field of robot-based interventions for children with ASD. By exploring the impact of robot-assisted activities on motor and social development, this research highlights the benefits of combining robotic and human-led approaches to promote inclusive and holistic educational practices for children with ASD.

6 Conclusions

It is important to consider that the child-educator relationship and the child-robot relationship can coexist and complement each other. Educators can use robots as additional resources to enrich the educational experience, facilitate collaborative learning and encourage creativity and innovation. The relationship between children with autism and robots can offer several significant benefits. Here are some of the main advantages that emerged:

1. Communication and social interaction: robots can be programmed to interact with children with autism in specific and predictable ways. This can facilitate communication and social interaction, providing a comfortable and non-threatening environment for children. Robots can encourage children to express their emotions, develop conversational skills and improve social skills.
2. Structured learning: children with autism tend to benefit from a structured and predictable environment. Robots can provide structured learning activities and programmes, adapted to the individual needs of children. This can help children develop cognitive, language and motor skills through interaction with the robot.
3. Reducing anxiety and stress. For many children with moderate or severe autism, social interactions can be a source of anxiety and stress. Robots can provide a reassuring and less intimidating alternative to interactions with humans. This can help children feel more comfortable and reduce anxiety and stress during learning and socializing activities.
4. Neutral feedback: robots can provide neutral and consistent feedback to children with autism. This can be especially helpful for children who may be sensitive or react negatively to human feedback. The robot's neutral feedback can be used to encourage and reinforce positive behaviors, providing a non-judgmental learning environment.

5. Motivation and involvement: robots can be designed to be engaging and interesting for children with autism. The interactive and playful features of robots can stimulate children's motivation and attention, encouraging interest in and commitment to learning activities.

It is important to emphasize that the use of robots as a support tool for children with autism should be necessarily integrated with human intervention and customized to the needs of each child. Robots do not replace human relationships, but they can be a valuable complement in helping children with autism to develop social skills, communicative and cognitive and can be considered a reassuring alternative for children with autism compared to interactions with humans for several reasons that emerge from the observations of the video recordings. We summarize them below:

1. Predictability: robots can be programmed to follow predictable routines and behaviors. This provides children with autism a structured and stable environment, in which interactions are consistent and not surprising. The predictability of the robot's actions can help children feel more confident and reduce anxiety.
2. Non-judgmental: robots do not make judgements about children's mistakes or performance. No matter how many mistakes they make or how long it takes them to learn something, robots respond in a neutral way and encourage children to keep trying without fear of being judged or criticized. This can help create a positive and pressure-free learning environment.
3. Patience: robots do not get tired or bored. They can take time to interact with children, repeat instructions or practice skills without showing signs of frustration. This can be especially helpful for children with autism, who may take longer to process the information or perform certain activities.
4. Consistent feedback: robots can provide consistent and objective feedback. While educators may vary in the way they provide feedback or expression of emotions, robots respond in a consistent and predictable manner. This allows children to better understand feedback and develop a greater awareness of their actions.
5. Reduction of complex social interaction: for many children with autism, complex social interactions can be a source of anxiety and stress. Robots offer simplified social interaction, where children can focus on specific skills without having to interpret complex social signals that may be difficult to understand. This reduces social pressure and can help children develop confidence in their own social skills.
6. Interest and attractiveness: robots can be designed to be interesting and attractive for children. They can have looks and features that catch the children's attention, such as bright colors, funny movements or funny voices. This can motivate children to interact with the robot and engage in the proposed activities.

In conclusion, the relationship between children and educators remains of fundamental importance for children's development. Although robots may play a complementary role in education, the presence and interaction with educators remain essential to provide emotional, cognitive and social support that promote the learning and integral development of children. Although robots can offer these advantages, they do not replace human and educational relationships. Educators can provide empathy, understanding and social connection that robots cannot currently replicate.

Future research will incorporate focus groups with both parents and educators to deepen our understanding of the impacts and potential improvements in robotic interventions for children with ASD. These insights could provide valuable qualitative data on practical challenges, observed benefits and areas for adjustment in the educational setting. Parents' perspective would be essential to understand how robotic interventions influence children's motor skills and social interactions outside structured sessions, providing a holistic view of the intervention's effects in daily life. Educators could offer perspectives on how these interventions integrate with existing curricula and suggest strategies for classroom-based application to foster inclusivity.

Moreover, focus groups could explore parental and teacher perceptions regarding the usability and accessibility of robotic technologies, helping to identify any barriers that may affect the implementation of such tools. Feedback from these stakeholders would also be invaluable in ensuring that future iterations are user-friendly and contextually adaptable. Overall, involving parents and educators in future research would not only contribute to a more comprehensive evaluation of robotic interventions but would also support the design of practical and sustainable applications in both educational and home settings.

References

Belpaeme, T., Kennedy, J., Ramachandran, A., Scassellati, B., & Tanaka, F. (2018). Social robots for education: A review. *Science Robotics, 3*(21), eaat5954. https://doi.org/10.1126/scirobotics.aat5954

Bennett, M., & Goodall, E. (2016). A meta-analysis of DSM-5 autism diagnoses in relation to DSM-IV and DSM-IV-TR. *Review Journal of Autism and Developmental Disorders, 3*, 119–124. https://doi.org/10.1007/s40489-016-0070-4

Cummins, A., Piek, J. P., & Dyck, M. J. (2005). Motor coordination, empathy, and social behaviour in school-aged children. *Developmental Medicine & Child Neurology, 47*(7), 437–442. https://doi.org/10.1017/s001216220500085x

Díaz-Boladeras, M., Claver i Díaz, A., & García-Sánchez, M. (2023). Robots for inclusive classrooms: A scoping review. *Universal Access in the Information Society*. https://doi.org/10.1007/s10209-023-01065-z

Dubois-Sage, M., Jacquet, B., Jamet, F., & Baratgin, J. (2024). People with autism spectrum disorder could interact more easily with a robot than with a human: reasons and limits. *Behavioral sciences, 14*(2), 131. https://doi.org/10.3390/bs14020131

Fang, N., Zhang, C., Sankaran, S., & Ren, S. (2022). Role of socially assistive robots in reducing anxiety and preserving autonomy in children. In *HRI 2022—Proceedings of the 2022 ACM/IEEE international conference on human-robot interaction* (pp. 754–759). https://doi.org/10.1109/HRI53351.2022.9889561

Field, S., Kuczera, M., & Pont, B. (2007). No more failures: Ten steps to equity in education. *OECD Publishing*. https://doi.org/10.1787/9789264032606-en.Accessed30october2024

Gargot, T., Archambault, D., Chetouani, M., Cohen, D., Johal, W., & Anzalone, S. M. (2022). Automatic assessment of motor impairments in autism spectrum disorders: A systematic review. *Cognitive Computation, 14*(2), 624–659. https://doi.org/10.1007/s12559-021-09940-8

Gómez-Espinosa, A., Moreno, J. C., & Pérez-de la Cruz, S. (2024). Assisted robots in therapies for children with autism in early childhood. *Sensors, 24*(5), 1503. https://doi.org/10.3390/s24051503

Grossard, C., Palestra, G., Xavier, J., Chetouani, M., Grynszpan, O., & Cohen, D. (2018). ICT and autism care: State of the art. *Current Opinion in Psychiatry, 31*(6), 474–483. https://doi.org/10.1097/YCO.0000000000000455

Huijnen, C. A. G. J., Lexis, M. A. S., Jansens, R., & de Witte, L. P. (2016). Mapping robots to therapy and educational objectives for children with autism spectrum disorder. *Journal of Autism and Developmental Disorders, 46*(6), 2100–2114. https://doi.org/10.1007/s10803-016-2740-6

Ji, Y. Q., Tian, H., Zheng, Z. Y., Ye, Z. Y., & Ye, Q. (2023). Effectiveness of exercise intervention on improving fundamental motor skills in children with autism spectrum disorder: A systematic review and meta-analysis. *Frontiers in Psychiatry, 14*, 1132074. https://doi.org/10.3389/fpsyt.2023.1132074

Jouaiti, M., & Hénaff, P. (2019). Robot-based motor rehabilitation in autism: A systematic review. *International Journal of Social Robotics, 11*(5), 753–764. https://doi.org/10.1007/s12369-019-00598-9

Kandola, A. (2019). *Levels of autism: Everything you need to know*. Retrieved October 30, 2024, from https://www.medicalnewstoday.com/articles/325106.php

Kouroupa, A., Laws, K. R., Irvine, K., Mengoni, S. E., Baird, A., & Sharma, S. (2022). The use of social robots with children and young people on the autism spectrum: A systematic review and meta-analysis. *PLoS ONE, 17*(6), e0269800. https://doi.org/10.1371/journal.pone.0269800

Landa, R. J. (2018). Efficacy of early interventions for infants and young children with, and at risk for, autism spectrum disorders. *International Review of Psychiatry, 30*(1), 25–39. https://doi.org/10.1080/09540261.2018.1432574

Lytridis, C., Vrochidou, E., Chatzistamatis, S., & Kaburlasos, V. G. (2018). Social engagement interaction games between children with autism and humanoid robot NAO. In *International conference on European transnational education* (Vol. 771, pp. 562–570). https://doi.org/10.1007/978-3-319-94120-2_55

Malerba, D., Appice, A., Buono, P., Castellano, G., De Carolis, B., de Gemmis, M., Polignano, M., Rossano, V., & Rudd, L. M. (2019). Advanced programming of intelligent social robots. *Journal of E-Learning and Knowledge Society, 15*(2). https://doi.org/10.20368/1971-8829/1611

Moorthy, R. S., & Pugazhenthi, S. (2016). Imitation based training to enhance psychomotor skills in autistic children using a snatcher robot. In: *2016 International conference on robotics and automation for humanitarian applications (RAHA)* (pp. 1–6). https://doi.org/10.1109/RAHA.2016.7931903

Mortari, L. (2007). Cultura della ricerca e pedagogia. Prospettive epistemologiche, Carocci, Roma

Palestra, G., De Carolis, B., & Esposito, F. (2017). Artificial intelligence for robot-assisted treatment of autism. In *WAIAH@AI*IA* (pp. 17–24).

Palestra, G., Varni, G., Chetouani, M., & Esposito, F. (2016). A multimodal and multilevel system for robotics treatment of autism in children. In *Proceedings of the international workshop on social learning and multimodal interaction for designing artificial agents* (pp. 1–6). https://doi.org/10.1145/3005338.3005341

Pennazio, V. (2017). Social robotics to help children with autism in their interactions through imitation. *Research on Education and Media, 9*(1), 10–16. https://doi.org/10.1515/rem-2017-0003

Perla, L. (2013). Per una didattica dell'inclusione. In *Prove di formalizzazione*. Pensa Multimedia.

Perla, L. (2014). Didattica dell'implicito. In *Ciò che l'insegnante non sa*. La Scuola.

Perla, L., De Carolis, B., Massaro, S., & Vinci, V. (2023a). Promoting health and wellbeing: Harnessing the potential of social robots in English L2 for elderly cognitive decline prevention. In *2023 HELMeTO, 5th international conference on higher education learning methodologies and technologies online* (pp. 146–148).

Perla, L., Garofoli, F. J., Amati, I., Santacroce, M. T. (2022). La forza mite dell'educazione. In *Ricerca e strategie di intervento per il contrasto del bullismo a scuola*. FrancoAngeli.

Perla, L., Massaro, S., De Carolis, B., Vinci, V., & Balzotti, A. (2023b). Social robot-based nutritional education for children's engagement on healthy lifestyles. In L. Perla, S. L. Agrati, V. Vinci, & A. Scarinci (Eds.), *Living and leading in the next era: Connecting teaching, research, citizenship and equity* (pp. 153–154). Pensa MultiMedia.

Perla, L., & Riva, M. G. (Eds.). (2016). *L'agire educativo: Manuale per educatori e operatori socio-assistenziali.* La Scuola.

Perla, L., & Schiavone, N. (2017). Video-analysis and self-assessment in teaching work. In P. G. Rossi, & L. Fedeli (Eds.), *Integrating video into pre-service and in-service teacher training* (pp. 191–209). IGI Global. https://doi.org/10.4018/978-1-5225-0711-6.ch010

Richardson, K., Coeckelbergh, M., Wakunuma, K., Billing, E., Ziemke, T., Gomez, P., Vanderborght, B., & Belpaeme, T. (2018). Robot enhanced therapy for children with autism (DREAM): A social model of autism. *IEEE Technology and Society Magazine, 37*(1), 30–39. https://doi.org/10.1109/mts.2018.2795096

Srinivasan, S. M., Kaur, M., Park, I. K., Gifford, T. D., Marsh, K. L., & Bhat, A. N. (2015). The effects of rhythm and robotic interventions on the imitation/praxis, interpersonal synchrony, and motor performance of children with autism spectrum disorder (ASD): A pilot randomized controlled trial. *Autism Research and Treatment, 736516*(2015), 736516. https://doi.org/10.1155/2015/736516

Volkmar, F., Rogers, S., Paul, R., & Pelphrey, K. A. (Eds.). (2014). *Handbook of autism and pervasive developmental disorders* (4th ed.). John Wiley & Sons Inc.

Yang, Q., Feng, H., & Gong, S. (2023). Analyzing the potential of using social robots in autism classroom settings. In *SHS web of conferences*. https://doi.org/10.1051/shsconf/202317403023

Zheng, Z., Young, E. M., Swanson, A. R., Weitlauf, A. S., Warren, Z. E., & Sarkar, N. (2016). Robot-mediated imitation skill training for children with autism. *IEEE Transactions on Neural Systems and Rehabilitation Engineering: A Publication of the IEEE Engineering in Medicine and Biology Society, 24*(6), 682–691. https://doi.org/10.1109/TNSRE.2015.2475724

The Application of Emerging Technologies in Autistic Students During the Period 1996–2023: A Conceptual, Social and Intellectual Review

Gonzalo Lorenzo-Lledó, Santiago Meliá Beigbeder, and Diana Gadzhimusieva

Abstract In contemporary societies, technology plays a crucial role in the development of daily tasks. One of the areas where it is having a greater expansion is education. Technology, therefore, enables the adaptation to the diverse learning needs of students by facilitating the creation of personalized learning plans. In recent years, students with autistic disorder are increasingly present in the classroom. Consequently, various emerging technologies such as virtual reality, robotics, augmented reality, mixed reality, and artificial intelligence has been developed to cater to the visual learning and systematic tasks required by these students by creating personalized learning systems. Thus, the objective is to perform a bibliometric analysis of trends in the use of emerging technologies during the period 1996–2023 for students with autism spectrum disorder. For the development of the research, the Web of Science database was chosen, where a series of search criteria were applied, and the documents were filtered using the PRISMA methodology. A sample of 888 documents was obtained. The results showed that during the period 2015–2023 it was the most productive period with the highest number of appointments. Furthermore, virtual reality and robotics were the emerging technologies most used in areas such as emotions, imitation or joint attention. The findings underscore the necessity for an increased application of AI to generate personalized learning plans for autistic students, in conjunction with other tools.

Keywords Artificial intelligence · Emerging technologies · Autism · Augmented reality · Mixed reality

---

G. Lorenzo-Lledó (✉)
Department of Developmental Psychology and Didactics, University of Alicante, Alicante, Spain
e-mail: glledo@ua.es

S. M. Beigbeder · D. Gadzhimusieva
Department of Computer Language and Systems, University of Alicante, Alicante, Spain
e-mail: santi@ua.es

D. Gadzhimusieva
e-mail: diana.gadzhimusi@gmail.com

G. Lampropoulos and S. Papadakis (eds.), *Social Robots in Education*, Studies in Computational Intelligence 1194, https://doi.org/10.1007/978-3-031-82915-4_13

1 Introduction

In contemporary societies, technology is one of the most widely used tools in everyday life activities. Consequently, technology is employed in a wide range of fields of knowledge. In the educational environment, technology has the potential to generate learning situations adapted to the needs of learners. In this sense, an increasing number of students with specific educational support needs (SEN) have been observed in school classrooms. Within this group, in recent years there has been an increase in the detection of students with autism spectrum disorder (ASD). Globally, 0.6% of the population manifests ASD (Salari et al., 2022). According to Durkin et al. (2017) there are several reasons for this increase. Firstly, the change in diagnostic criteria due to the numerous changes in the DSM-V Secondly, the existence of more sensitive screening tools (Durkin et al., 2017).

ASD is defined as a neurological disorder that affects two areas: communication and social interaction, and restrictive and stereotyped behaviors (APA, 2013). Thus, autistic students have different perception, communication and interaction skills when interacting with the environment (Newbutt et al., 2020). This results in difficulties in responding to different stimuli in a social situation (Schmidt et al., 2021). Therefore, these students show problems in identifying changing social information in social contexts (Moon & Ke, 2021). In this line, it can be stated that due to problems in information processing (Horwitz et al., 2020), autistic students have difficulties in understanding and interpreting the emotions of their interlocutor (Talaat, 2023). As a consequence, they have issues anticipating actions associated with moods. Similarly, autistic students show deficits in the ability to create new ideas and responses (Verté et al., 2006). Furthermore, they have troubles in the process of monitoring, re-evaluating and updating a planned sequence of actions, as well as in flexibility of thought and action (Semrud-Clikeman et al., 2014). Despite these challenges, autistic learners possess two fundamental strengths in their learning process. First, learning through visual information and highly specific reasoning (Lorenzo et al., 2016). Secondly, systematized task-based learning.

Learning through visual information for autistic students is the main reason for the use of emerging technologies (Lorenzo et al., 2016). These technologies include virtual reality (VR), robotics, augmented reality (AR) and mixed reality (MR). VR is defined as a simulation of a three-dimensional environment in which the user can interact with the environment with strong similarities to reality (Mosher et al., 2022). This tool allows autistic learners to practice social interactions in a context where the overstimulation of real environments is avoided (Bradley & Newbutt, 2018). There are two types of virtual reality depending on the level of immersion (Radianti et al., 2020). Firstly, immersive virtual reality (IVR) where the user is immersed in a virtual environment where they can interact with stimuli in a natural way (Luo & Du, 2022). Secondly, non-immersive virtual reality where desktop virtual reality stands out. In this case, interaction occurs through the computer screen using external peripherals such as a mouse or keyboard (Lee & Wong, 2014).

The next step in terms of the level of interaction with the real world is augmented reality (AR). AR is defined as the projection of virtual images onto real-world entities (Verhey et al., 2020). The connection between VR and AR is through mixed reality (MR). This technology allows the user to manipulate real objects and have their movements reflected in virtual environments (Hu et al., 2019). AR, VR and MR can be grouped into a new term called extended reality (Morimoto et al., 2022). Extended reality tools can be combined with artificial intelligence (AI) to generate tasks tailored to the needs of autistic learners (Xia et al., 2022). The latest emerging technology used to work with autistic learners is robotics. Its combination with artificial intelligence will enable the acquisition of social behaviors (Maroto-Gómez et al., 2023) in interactions with autistic learners. In other words, robots can act as supportive elements in social interactions (Singh et al., 2023).

The use of these emerging tools in the learning of autistic students is justified for several reasons. Firstly, robots improve attention, enhance emotion recognition, and provide controlled and predictable environments for autistic students (Robins et al., 2005; Sartorato et al., 2017). Secondly, AR allows for more realistic interactions with the physical world, enabling autistic learners to develop manipulative activities that facilitate the transfer of learning to real environments (Quintero et al., 2019). Thirdly, VR enables the creation of virtual environments where autistic learners do not experience the negative consequences of real environments (Lorenzo et al., 2023). Additionally, it is possible to create environments with a reduced number of stimuli, which facilitates learning for autistic learners. In fourth place, AI can make real-time changes to activities in a manner similar to human adjustments (Kaplan et al., 2023). In this way, it could adapt to the learning that autistic students have.

In recent years, more and more reviews of the scientific literature have been produced. For Noyons et al. (1999) there are two types of bibliometric tools that can be used. The first type, called performance analysis, defines the field of knowledge using bibliometric indicators such as scientific production, most productive authors, etc. The second type, called scientific mapping, consists of the creation of scientific maps of the field based on various units of analysis (Martínez et al., 2015). In this study, a combination of the two bibliometric techniques will be performed. Based on the bibliometric techniques indicated by Noyons et al. (1999), several previous investigations have been undertaken using scientific mapping or performance analysis.

As for scientific mapping, Aria et al. (2021), examined trends in the publication of articles over the last 34 years on the cognitive processes of dogs. In medicine, Fortuna et al. (2020) studied the complex oral sensitivity disorder using scientific mapping. Rodríguez-Soler et al. (2020) developed a study on global trends in rural depopulation using bibliometrics. Lorenzo et al. (2022) explored trends in the use of VR applications for autistic students. In this study, despite not being the primary objective, the close thematic relationship between VR and augmented reality was confirmed. With respect to performance analysis, Toma et al. (2024) delved into the use of VR, AR, and mixed reality over a period of 10 years, analyzing only 28 publications. The results showed that AR still lacked a related numerical evaluation system, and its peak production period was 2015–2019. Similarly, Mosher et al.

(2022) analyzes the use of VR, AR, MR and XR in autistic students, examining improvements, acquisitions, and maintenance of social skills. The results showed the maintenance of social skills where autistic students had difficulties, such as emotion recognition or executive function.

Therefore, this review offers several advantages compared to previous ones. Firstly, it covers the period from 1996 to 2023, allowing for examination of the origins of technologies such as VR and AR. Secondly, a detailed study of the scientific mapping is carried out, enabling an in-depth analysis of the relationships among authors of the field and emerging thematic trends. This way, information can be obtained about the current implementation status of emerging technologies. Thirdly, a combination of both methodologies is used, providing comprehensive insights.

Based on the knowledge gaps that exist in this field, the aim of the study is to perform a bibliometric analysis of trends in the use of emerging technologies during the period 1996–2023 for students with autism spectrum disorder. From the general objective, a series of research questions were established.

RQ1. What is the social and intellectual structure of the authors in the application of emerging technologies to autistic students.

RQ2. What is the intellectual structure of the documents analyzed in the application of emerging technologies in autistic students?

RQ3. What is the conceptual structure of the documents analyzed in the application of emerging technologies in autistic students?

RQ4. What is the conceptual structure of the sources analyzed in the application of emerging technologies in autistic students?

The present research has been organized as follows. Section 2 describes the method implemented for the analysis of the documents. This section has several subsections such as: the participants where the characteristics of the documents are explained, the instrument used, i.e. the reasons that justify the use of the Web of Science, the third point is the design and procedure for the search of the documents. The last subsection is the data analysis. Section 3 focuses on the description of the results in terms of the research questions. Section 4 provides a discussion of the results obtained by analyzing the reasons for the findings. Section 5 highlights the conclusions of the study.

2 Method

To examine the existing trends in the field of application of emerging technologies and autism, a bibliometric methodology has been chosen because it allows us to quantify, evaluate and estimate the quality and productivity of a field in addition to its thematic evolution (Hinojo-Lucena et al., 2019). Specifically, the objective of the work was to delve into the situation of publications, citations and sources of information. Similarly, within bibliometric techniques, scientific mapping has been selected. The reason is due to allowing a representation of the structure of the field in the conceptual, social and intellectual spheres (Montero-Díaz et al., 2018). In this sense, Donthu et al. (2021) defines conceptual structure as the co-occurrence of keywords established by authors or databases in different documents. In addition, these authors define the intellectual structure as the citation patterns existing between sources, authors or documents. While the social structure is defined as the collaboration networks between authors, countries and institutions (Glänzel, 2002). In this way, the application of co-citation, co-words and citation analysis techniques is encouraged. The use of them allows us to answer the research questions. Furthermore, scientific mapping makes it possible to analyze data in different periods of time (Zupic & Cater, 2015). Likewise, this methodology can collect the influence of other publications outside the field (White & McCain, 1998). With the aim of deepening the use of bibliometrix and achieving its greater dissemination, Donthu et al. (2021) try to establish a series of steps to implement an analysis using scientific mapping combined with bibliometric techniques. According to Aria and Cuccurullo (2017), this tool is applied in the following way:

- In the study design phase, the research questions must be established and the bibliometric techniques to work with the documents must be chosen. According to Aria and Cuccurullo (2017) the questions in this phase can be of three types: The first type focuses on identifying the knowledge base of the field. The second type where the intellectual structure and the evaluation of the conceptual structure are delved into. While the third type works on the production of a structure of a scientific community.
- The second phase consists of data collection or also called the procedure. In this phase, researchers generate their own database by applying a series of filtering criteria. In the case of the authors, the PRISMA methodology was applied (Page & Moher, 2017)
- The third and fourth phase consists of data analysis and visualization. To analyze the data, the statistical analysis library called Bibliometrix, is used. Subsequently, different bibliometric techniques are chosen to visualize the data, such as co-citation, co-words or collaboration networks. Finally, the fifth phase consists of interpretation where the description of the results and their interpretation are carried out. An overview is given in Fig. 1, which shows all the phases of the process.

2.1 Participants

The final sample used for the study was 888 documents. The average number of citations per document was 13.72, while the average age of the documents was 6.24 years. In this sense, the total number of references is 24594 being the Keyword Plus of 863 and the Keyword Authors is 1956. There were 2800 authors in the field, 51 of whom had single-authored documents. In addition, the average number of

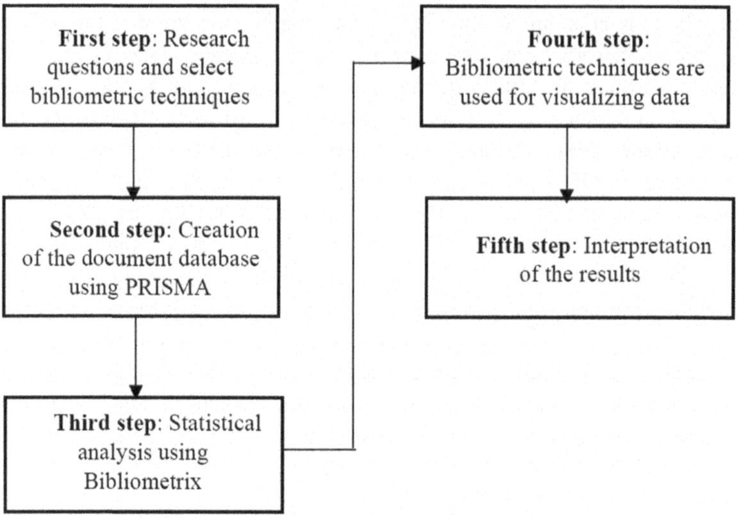

Fig. 1 Methodology process

authors per document was 4.45. While 22.52% of the papers had international co-authorships outside their country of origin. As for the documents, there were 315 articles, 502 documents that were papers of congresses. There were also 37 reviews of the field of work.

2.2 *Instrument*

For the development of the research, the Web of Science has been chosen as a database, which is composed of the Science Citation Index, Social Science Citation Index, Conference Proceeding Citation Index-Science, Conference Proceedings Index Social Science & Humanities, Book Citation Index-Science, Book Citation Index Social Science & Humanities and Emerging Source Citation Index.

The reasons why WOS was chosen for the document search were as follows. Firstly, WOS allows a document to be assigned to several categories (De Moya-Anegón et al., 2006). In this way, a classification is achieved that is much more appropriate to the subject they are working on. Secondly, the WOS includes information not available in other databases such as the institutional addresses of the authors, the bibliographic references of each article, the authors' bibliometric indices (Repiso & Torres-Salinas, 2016). Thirdly, WOS articles are mostly published in North America and Western Europe (Meho & Rogers, 2008), which is where the most important centres of scientific production are located. Fourthly, WOS has internal databases referring to various types of documents such as the Social Science Citation Index

(Journals), Book Citation Index (Articles) or Conference Proceeding Citation Index-Science (Conference Proceedings). These features are not available in databases such as Scopus. Fifthly, the journals with the highest scientific recognition and citations are included in WOS. Sixth, WOS publications can cover documents that were published in the year 1900 unlike Scopus whose coverage starts in the year 1996 (Delgado et al., 2019). In addition, the number of journals indexed in WOS versus those only in Scopus is higher in WOS. WOS has more than 22,000 while Scopus has 16,257. Therefore, the use of this database for searching documents is justified.

2.3 Design and Procedure

After making the database selection, the filtering of the documents was carried out using the criteria established in PRISMA (Page & Moher, 2017). This methodology is applied in systematic reviews (García-González & Ramírez-Montoya, 2019). In this way, PRISMA methodology is divided into different phases. In the first phase, the search words chosen were all those terms that have been used throughout the different DSM to refer to autism spectrum disorder. However, following Haas et al. (2020), a screening was made by searching the UNESCO and ERIC thesauri. In this way, the terminology that was most commonly included in the articles and recognized by the scientific community was selected (Haas et al., 2020). Emerging technologies were treated in the same way. In addition, acronyms were avoided for two reasons. Firstly, they were not included in the UNESCO and ERIC thesauri. Secondly, articles from other fields of knowledge might appear.

> ("Low functioning autism" or "autism" or "autistic students" or "asperger" or "autism spectrum disoders" or "high functioning autism") and ("virtual reality" or "immersive virtual reality" or "artificial intelligence" or "augmented reality" or "robotics" or "extended reality" or "mixed reality").

The application of the equation gave a total of 2488 WOS documents. In the second Screening phase, inclusion criteria 1, 2, 3 (Table 1) and exclusion criteria 1 and 2 were applied (Fig. 1). During this phase, a card was prepared so that the various authors could extract information from the publications. The first of the inclusion criteria is the time 1996–2023. The second filter focuses on those WOS categories related to informatics and education. These categories are Computer Science Artificial Intelligence, Robotics, Computer Science Theory Methods, Engineering Electrical Electronic, Computer Science Cybernetics, Computer Science Information Systems, Computer Science Software Engineering, Automation Control Systems, Computer Science Hardware Architecture, Engineering Manufacturing, Education Educational Research, Education Science Disciplines, Education Special, Social Science Interdisciplinary, Information Science Library Science. The third criterion refers to the language being English. Inclusion criteria 4, 5, 6 have a thematic focus since interventions or reviews focused on VR, AR, MR, XR, robotics or AI applied

Table 1 Inclusion and exclusion criteria

Inclusion criteria	Exclusion criteria
IC1. Period 1996–2023	IEX1. Presence of duplicate publications
IC2. Categories of the WOS	IEX2. No publication access
IC3. Language: English	IEX3. Study area
IC4. Type of study: interventions and reviews IC5. Thematic AR, VR, MR, robotics, AI and ASD	IEX4. Type of technology
IC6. Educational topics	IEX5. Disability

to autistic students with an educational approach should be chosen. As for the exclusion criteria, the first two refer to duplicate and inaccessible documents. The third exclusion criterion refers to documents that are not related to the educational field. The last two focus on the type of technology and type of disability.

The third phase of the PRISMA document filtering methodology consists of a content analysis where the inclusion criteria 4, 5, 6 and exclusion criteria 3, 4, 5 are applied. The fourth phase defines a final sample size of 888. Figure 2 shows the various steps that have been carried out.

2.4 Data Analysis

The data analysis was developed with the bibliometrix tool of Aria and Cuccurullo (2017). This tool works in the same way as scientific mapping. That is, through data collection, analysis and visualization. To develop this analysis, it is necessary to establish the different units and levels of analysis, their metrics, bibliometric techniques, statistical techniques and their structure as explained by Rodríguez-Soler et al. (2020).

Three levels of analysis were established in order to be able to deepen the importance of each topic in each of the of the knowledge components (Rodríguez-Soler et al., 2020). To this end, authors, sources and documents were chosen. Similarly, a series of metrics were established to study the prevalence of each of the topics at different levels. These metrics were based on performance analysis. For example, for documents, the frequency of occurrence of keywords was chosen. Under authors, the metric of most productive and most cited authors was selected. In this way, the authors who publish the most documents and those who receive the most citations are analyzed. Meanwhile, for sources, the most productive sources were chosen as the metric. This metric means which journals publish the most articles. The units of analysis are the fundamental components on which the analysis is undertaken. For the study, journals were chosen as the most important source. From the documents, words were selected as the unit element from which to obtain thematic trends. Whereas for the authors, the unit of analysis was the authors themselves.

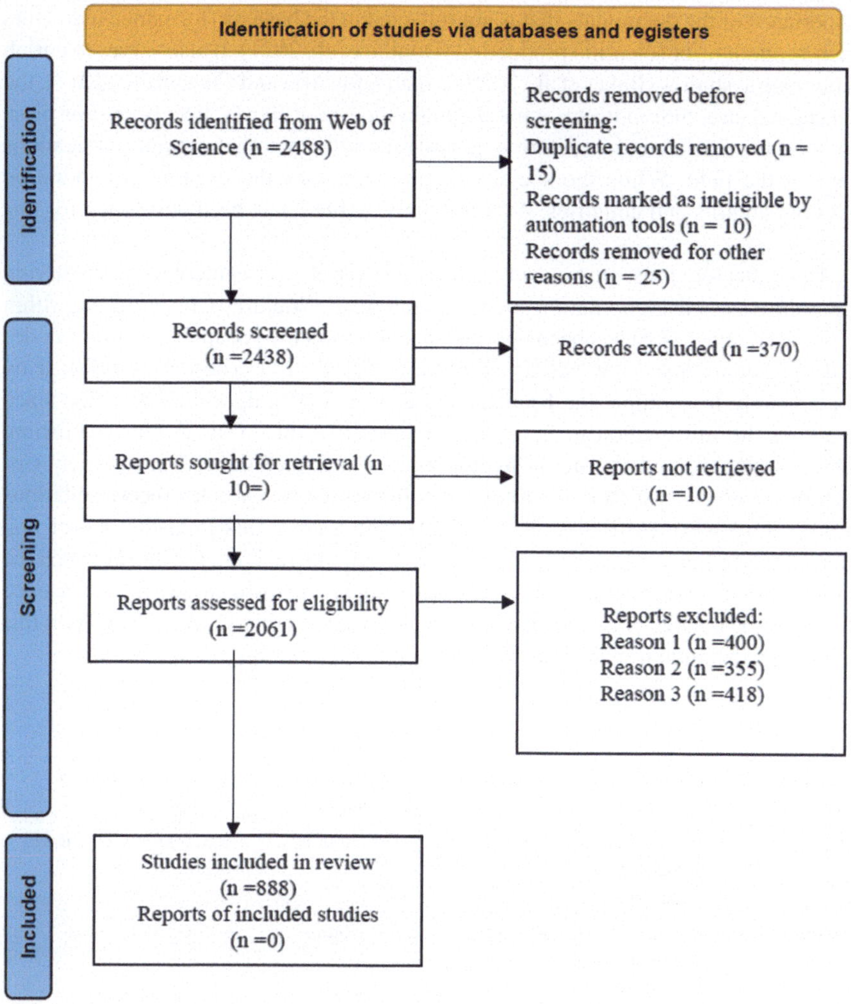

Fig. 2 Flow using PRISMA

As for bibliometric techniques, which allow researchers to show the relationships between the different components of research, co-words, co-citation are highlighted (Donthu et al., 2021). Co-words is a scientific mapping technique where there is a co-occurrence of words in different publications (Aria & Cuccurulo, 2017). That is, those words that are most repeated in different publications show a thematic similarity. The words extracted for the analysis come from the title, abstract, keywords or body of the document. Co-citation is a scientific mapping technique where two publications are thematically connected when they appear frequently in the reference list of the other publication (Donthu et al., 2021). Following the presentation of the analysis parameters, the statistical technique using networks allows a representation of the importance of the documents that is not reflected in the basic performance indicators such as citations or scientific production (Donthu et al., 2021). It also serves to enrich bibliometric studies (Baker et al., 2020). Finally, for Aria and Cuccurullo (2017), the conceptual structure indicates what the main themes or trends have been. For these authors, the intellectual structure expresses the influence that certain publications have in the field. While the social structure expresses the existing collaboration between authors and countries. As a summary, Table 2 has been developed for this purpose.

The indicators presented must be adjusted in the R environment using clustering algorithms as well as co-citation nodes. Therefore, the Walktrap Clustering algorithm (Pons & Latapi, 2006) has been chosen since it allows the grouping of author nodes and documents that show certain similarities based on a criterion established by the research. In addition, the Kamada and Kawai (1989) algorithm was used since it allows the information to be classified into different levels. After establishing the clustering algorithms, the following bibliometric techniques were chosen. Co-citation networks, which is the frequency with which two articles receive citations from a third party (Mustafee et al., 2020). The greater this link, the denser and more solid the networks will be. In this way, the intellectual relationship between the authors is achieved. Another of the techniques used is co-words where the occurrence or joint appearances of two or more terms is measured (Wen et al., 2021). With this technique, thematic trends are analyzed.

Table 2 Variables for the analysis

Level of analysis	Metrics	Unit of analysis	Bibliometric techniques	Statistical technique/ structure
Authors	Most productive authors, most cited authors	Authors	Co-citation	Network/intellectual and social
Sources	More productive sources	Journal	Co-citation	Network/conceptual
Documents	Most cited documents, frequency of "keyword authors"	References "Keyword Authors"	Co-words	Network-thematic mapping-thematic evolution/ intellectual-conceptual

3 Results

3.1 Social and Intellectual Structure of the Authors

The top three most productive authors are N. Sarkar with 27 papers followed by Z. Warren with 22 papers and the third position is occupied by A. Swanson with 18 papers. While the most cited authors are Scassellati with 92 citations, Sarkar with 88 citations and Dautenhahn with 85 citations. Similarly, the most productive research centers are "State University System of Florida" with 31 publications, the second is the "Vanderbilt University" with 28 publications while the third position is occupied by the "University of Alicante" with 22 papers. In reference to the collaboration networks between authors, there are a total of 27 clusters, with Tony Belpaeme obtaining the highest value of Betweenness with 263 points followed by Baxter with 242 and Yussof with 162. In the same way, cluster 4 is the one that accumulates the greatest number of authors with the highest "Betweenness "which are Yussof, Hashim, Hanapiah and Ismail with 162, 140, 134, 72 respectively. The most numerous clusters are 1 and 5 with 11 and 12 authors correspondingly. Each of the clusters is represented with a different color while the most important author has the circle that represents it highlighted. There are a total of three clusters, with APA obtaining the highest value of Betweenness with 343 points followed by Sarah Parsons with 230 and Brian Scassellati with 130. In the same way, clusters 1, 2 are that accumulates the greatest number of authors with the highest "Betweenness". The most numerous cluster is three with 43 authors.

3.2 Intellectual Structure of the Documents

Based on the results obtained, the period of highest scientific productivity starts in 2015 and ends in 2023. During this interval, the average production is 81 articles, while from 1996 to 2014, the average production is 9 articles. In 1996–2014, the most productive year, which was 2014, 45 papers were published. Meanwhile, for 2015–2023, the most productive year was 2022 with 100 papers. Furthermore, the findings of the study show that from 1996 to 2014, the average number of citations received by the documents is 55 citations. The year with the most citations was 2014 with 271. Also, the most cited article was published in this period with 317 citations, which corresponds to Cabibihan et al. (2013). During this same period, the second and third most cited papers were also written by Kozima et al. (2009) and Bainbridge et al. (2011) with 289 and 252 citations, respectfully. With respect to the time interval from 2015 to 2023, the average number of citations was 1139 with the highest year being 2023 with 2102 citations. In this period, the article by Scassellati et al. (2018) obtained a total of 167 citations. Additionally, when analyzing the papers based on the citation they received, three main clusters arose. The first cluster is made up of 53 documents. The most important papers due to high betweenness (btw) values are

Parsons and Mitchell (2002) with 226,490 and Kandalaft et al. (2013) with 162,232. This parameter measures the frequency with which a node is chosen by others to make a path within a co-citation network (Pradhan & Pal, 2020) As for the second cluster, it is formed by 13 papers with the work of Lorenzo et al. (2019) having the highest btw value with 165,929. The third cluster is composed of 69 papers being the most important papers with the highest btw value those of Scassellati et al. (2012) with 504 points and Diehl et al. (2012) with 132 points. Finally, the fourth cluster is composed of twelve elements being the most important APA (2022) with btw of 5,036 points followed by Boucenna et al. (2014) with 784 points and the third position is for Lord et al. (2000) with 745,181. This is reflected in the fact that the node that represents them is represented in a larger size.

In terms of global or meso citation topics, 602 documents are grouped around autism and developmental disorders, followed by robotics with 87 documents, neuro-exploration with 57 documents and human–robot interaction with 32 documents in the fourth place. Regarding citation microtopics, within developmental disorders and autism, the topic of autism stands out with 503 documents. As for robotics, the topic of human–robot interaction can be included. The third macrotopic which was "Neuroscaning" includes aspects such as emotions, action recognition, facial recognition, functional connectivity and deep learning. The fourth macrotopic which was human–computer interaction has as an internal topic augmented reality with a network of 31 papers.

3.3 Conceptual Structure of the Documents

For the conceptual structure of the field of knowledge, an analysis of the frequency of occurrence of the words established by the authors called "Keyword authors" will be developed. Therefore, this information establishes which topics are being researched in the field of knowledge. The word autism is the most used by the authors with 221 occurrences followed by autism spectrum disorder with 154 occurrences and virtual reality with 141 occurrences. The fourth and fifth places are occupied by human–robot interaction and augmented reality with 64 and 53 occurrences respectively. Artificial intelligence and machine learning stand out in the seventh and tenth position with 35 and 30 occurrences. To complete the analysis, Fig. 3 shows the co-citation network of keywords used by the authors. In this way, it can be seen in detail how the knowledge topics are grouped.

There are a total of 9 clusters, the smallest being clusters 7, 8, 9 which contain only one word, Autism Spectrum Disorders, Human Computer Interaction and Facial Expressions respectively. Cluster 6 has 7 words related to artificial intelligence and emotion recognition with words like classification, machine learning, deep learning, emotion recognition. Cluster 5 is formed by only two words which are autism spectrum condition and child-robot interaction. Continuing with the study, cluster 4 is the most numerous with 59 words. There are different topics within the cluster being virtual reality, autism, autism spectrum disorder as the most important ones due to the

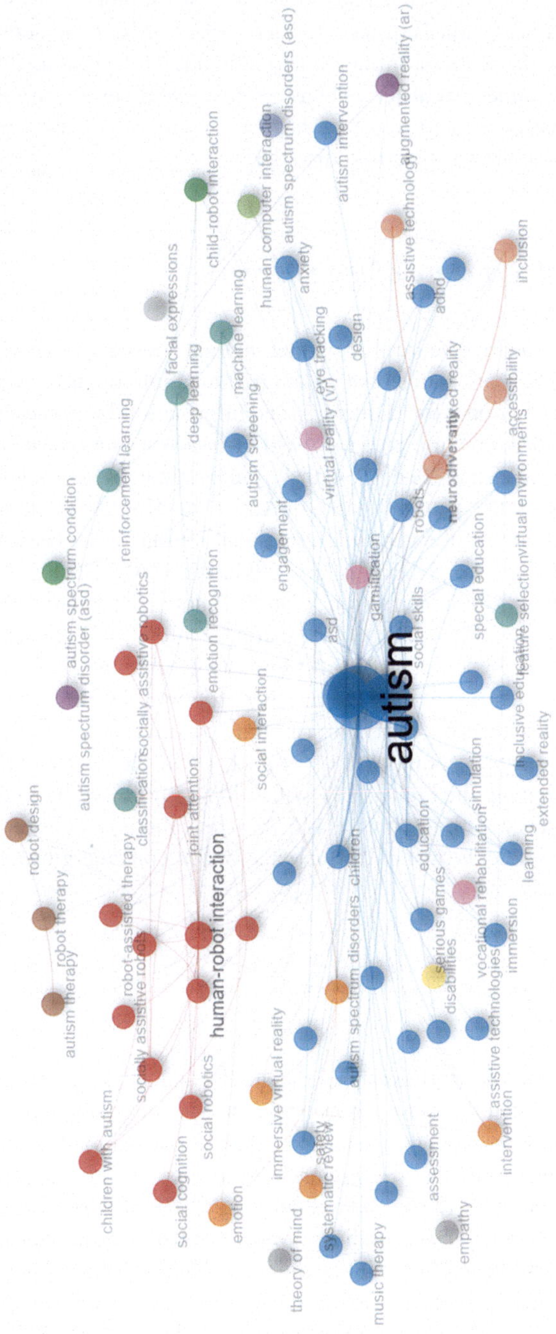

Fig. 3 Conceptual structure

value of btw. There is also research on robotics, facial expressions, theory of mind, empathy, mindfulness and inclusive education. Also, we should not forget research on mixed reality. While the third cluster is formed by 14 words. In this sense, the most important topics are the work on imitation and joint attention combined with social robotics tasks. In reference to cluster 2, it has 3 words that focus on gamification and vocational rehabilitation. Finally, cluster 1 has three words which are virtual reality, augmented reality and autism spectrum disorder.

3.4 Conceptual Structure of the Sources

Among the most productive sources are "International Journal of Social Robotics" with 56 documents, "Frontiers in Robotics and Artificial Intelligence" with 21 documents, and the third position is for "Education and Information Technologies" with 12 publications. The fourth position goes to "ACM Transactions on accessible computing" and "Frontiers in Virtual Reality" with 11 publications. Regarding the number of citations received by the source, Journal of Autism and Developmental Disorders has received 1816 citations followed by International Journal of Social Robotics with 607 citations, while the third position is for a conference "ACM/IEEE International Conference on Human–Robot Interaction" with 510 citations. From the point of view of co-citation of sources, Fig. 4 shows that there are a total of 2 clusters. The first cluster is composed of 38 sources, the most important being the Journal of Autism and Developmental Disorders with btw = 335.56, followed by Research in Autism Spectrum Disorders with btw = 103.74. The third position is for Autism with 64 points. The second cluster is composed of 37 papers with the International Journal of Social Robotics having the highest btw with 190.845 followed by Lecture Notes in Artificial Intelligence with the btw of 145.74. Thus, there is a cluster of Autism and another cluster where the rest of the journals working on emerging technologies are located.

4 Discussions

The first research question (RQ1) analyses the intellectual and social structure of the authors. It was found that the most productive author is among the three most cited authors. However, he does not have the largest collaborative team. This situation may be a consequence of the fact that the authors who receive the most citations are not the most reputable authors in the field (Pradhan & Pal, 2020). In this way, work teams are generated on new topics. These findings coincide with those obtained by Lorenzo et al. (2022). It has also been observed that there are many working teams composed of few authors. This indicates that we are in a field where the great diversity of groups ensures the existence of a wide array of research lines (Deep-Arora & Chakraborty, 2021). Continuing with the analysis, the most productive centers are in the United

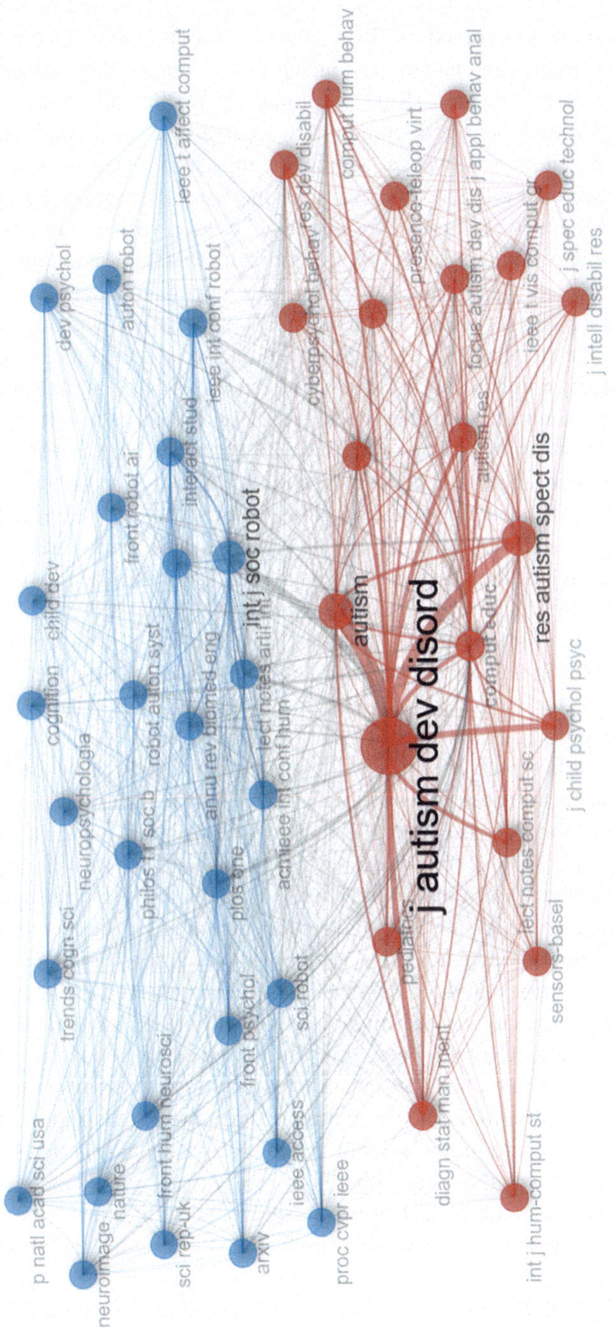

Fig. 4 Co-citation sources

States. The reason is that the vast majority of emerging devices and technologies have been developed in this country (Kavanagh et al., 2017). It is significant to learn that the third position is occupied by Spain, which without having specific centers on emerging technologies and autism has outstanding authors on the subject.

The second research question (RQ2) explores the intellectual structure of the documents. It was found that the document that received the most citations did not have the widest co-citation networks. Moreover, they took place in times of low scientific production and citation. For example, the work of Parsons and Mitchell (2002) is a review that delves into the work done on desktop virtual reality in autistic students. According to Albert et al. (2000), this indicates that this document acts as an accumulator of knowledge and control over other documents. In other words, anyone who is going to start working with VR consults this publication to find out how it can be applied to autistic students. This first cluster of citation documents is the second most important in the number of documents. The cluster with the highest number of papers with 69 is the one with Scasellati et al. (2012) and Diehl et al. (2012) with the highest btw. The btw values indicate that these are the knowledge nodes (Brandes et al., 2016) to which all researchers in the field of robotics in autistic students turn to because they present those fundamental features that robots must fulfill to develop a successful intervention with autistic students. This predominant position would explain why it has the largest co-citation node. As for the document to which everyone turns to, and which is reflected in the high btw values is APA (2022). This is due to the acquisition of information about the characteristics of autistic students. In another of the smaller nodes, the document by Lord et al. (2000), which is another of the main documents for characterizing autistic students, acquires great importance.

Regarding the third research question (RQ3), related to the thematic structure of the documents, it has been observed that autism and virtual reality are the most worked topics as reflected in the Keyword Authors. This may be a consequence of the 2014 appearance of the Oculus Quest glasses that were the first low-cost, wearable RVI device (Parsons, 2016) unlike previous research that worked with traditional cave or L-shaped systems. Although it is a small but incipient cluster, the application of artificial intelligence, machine learning and deep learning for the recognition and classification of emotions appears. This may be a consequence of the increasing demand to find standardized tools to diagnose autism from the earliest ages and thus be able to generalize standardized learning plans (Shahamiri & Thabtah, 2020). With the application of artificial intelligence, this whole procedure, which also requires the work of a specialist, is reduced and the implementation of response measures is speeded up. Continuing with the analysis, robotics acquires great importance with a cluster of 14 words where imitation, joint attention and social robotics stand out. The diffusion of robotics in the social sphere is a consequence of two factors: first, it facilitates the transfer of skills in real situations due to the similarities it has with the human interlocutor (Sani-Bozkurt & Sani-Genc, 2023). Secondly, the growing use of this tool for working with autistic students (Pérez-Vázquez et al., 2020). Finally, areas such as joint attention and imitation are worked on due to the ease with which they can establish eye contact with these interlocutors and therefore the activities that the robots propose can be developed in a simpler way (Cao et al., 2019).

Finally, the fourth research question (RQ4) delves into the conceptual structure of the sources published in the field of autism and emerging technologies. Currently, the field is made up of two clusters. The first of them where all the magazines that are related to autism appear. Journal of Autism and Developmental Disorders and Research in Autism Spectrum Disorders stand out. This is because according to Brandes et al. (2016) these journals act as knowledge accumulators where new trends in research are published and act as control elements for other journals. This causes researchers to publish in these reviews and cite their work. In addition, these journals are classified in the WOS in medical, psychological or educational categories. In the second cluster, there are magazines that work on technology and the way in which it is used by autistic students. For example, the most important journal is "International Journal of Social Robotics". This is in the WOS robotics category but has publications on the use of robotics in autistic students. This may be due, as indicated by Lorenzo et al. (2022), to the fact that there is a constant increase in publications where technology is used with these students, such as those by Lorenzo et al. (2024), Zhao et al. (2024), or Ip et al. (2024)

5 Conclusions

The study confirms that scientific production regarding the use of emerging technologies such as VR, AR, XR, MR, AI and robotics in autistic students during the period 1996–2023 is continuously growing, with many unresolved aspects remaining for use in the educational environment. Based on the research developed, the following conclusions are drawn.

In reference to the intellectual and social structure of the authors, the author who produces the largest number of articles is not part of the collaborative network of authors with the largest number of authors. One of the most important authors in terms of number of contributing authors is Tony Belpaeme who is not the largest producer and does not receive the highest number of citations.

The intellectual structure of the documents shows that the document that received the most citations took place in the least productive period and in the one with the least citations, which is from 1996 to 2014. Also, the citation structure is focused on Scassellati et al. (2012) with its publication entitled "Robots for use in Autism Research", Lord et al. (2000) with its publication entitled "Autism Spectrum Disorders" and APA (2022). These documents achieve the highest btw values.

The conceptual structure of the documents shows that virtual reality and robotics are the emerging technologies most applied with autistic students where expressions, theory of mind and empathy are worked on. In addition, there is a line of work on AI for the recognition and classification of emotions. Finally, there is a specific line for robotics, imitation and joint attention.

The conceptual structure of the sources indicates that the International Journal of Social Robotics is the most productive journal, but it is the Journal of Autism and Developmental disorders that receives the most citations. Thus, two co-citation clusters are generated. The first is focused on autism and the second on emerging technologies.

From the findings of the study, it can be concluded that virtual reality and robotics are the emerging technologies that are having the greatest diffusion in their application with autistic students. Furthermore, in recent years, artificial intelligence has emerged as a powerful detection tool for autistic students. Similarly, the development of virtual reality has allowed the export of improvements to the rest of the emerging technologies. From the results of the study, we derive a series of practical implications that can be the starting point for future lines of work. Firstly, information has been obtained on the emerging technologies that are most developed. In this way, there is the possibility of providing training in the centers for their application with autistic students. Secondly, it has been possible to analyze which areas have developed the most emerging technology interventions with autistic students. Therefore, a range of activities that can be applied in classrooms with autistic students is available. In addition, lines of collaboration can be established between universities and centers in order to develop a broader catalog of activities that can be available according to the characteristics of autistic students. Thirdly, the use of AI in schools is a technology that can provide real-time activities for teachers that can be adapted to the diversity of characteristics of autistic students. Therefore, it is possible to have a tool where teachers can consult activities to develop in the classroom.

Limitations that have been found in the study are diverse. For example, the influence that may have been generated by personal biases in the inclusion and exclusion criteria. This is conditioned by the perception of each researcher during the application of the filtering criteria. Other limitations are centered on the use of various databases from other fields which, despite being of lower importance, can also contribute documents. In addition, more search terms could have been included that referred to the thematic areas worked on. Likewise, the representation of the thematic maps should be much clearer, since when many nodes are chosen, it is not clear which are the components. Finally, the use of much more intuitive parameters for measuring the nodes.

Acknowledgements This publication is part of the doctoral thesis of Diana Gadzhimusieva who has received a grant with reference UAFPU22-04 of the University of Alicante's own research program. This article was supported by the Programa Estatal de I+D+i Orientado a los Retos de la Sociedad del Ministerio de Ciencia e Innovación Español. PID2020-112611RB-I00/AEI/https://doi.org/10.13039/501100011033 and the Agencia Estatal de la Investigación. Title of the project "The application of virtual reality and robotics in communication and social interaction of students with autism spectrum disorder". The publication has also been carried out in collaboration with the research group Advanced deveLopment and empIrical research on Software (ALISoft).

References

Albert, R., Jeong, H., & Barabasi, A. (2000). Error and attack tolerance of complex networks. *Nature, 406*(6794), 378–382.

American Psychiatric Association. (2013). *Diagnostic and statistical manual of mental disorders* (5th ed.). American Psychiatric Publishing.

American Psychiatric Association. (2022). *Diagnostic and statistical manual of mental disorders* (7th ed.). American Psychiatric Publishing.

Aria, M., Alterisio, A., Scandurra, A., Pinelli, C., & D'Aniello, B. (2021). The scholar's best friend: Research trends in dog cognitive and behavioral studies. *Animal Cognition, 24*(1), 541–553. https://doi.org/10.1007/s10071-020-01448-2

Aria, M., & Cuccurullo, C. (2017). Bibliometrix: An R-tool for comprehensive science mapping analysis. *Journal of Informetrics, 11*(4), 959–975. https://doi.org/10.1016/j.joi.2017.08.007

Bainbridge, W.A., Hart, J.W., Kim, E. S., & Scassellati, B. (2011). The benefits of interactions with physically present robots over video-displayed agents. *International Journal of Social Robotics, 3*, 41–52. https://doi.org/10.1007/s12369-010-0082-7

Baker, H., Kumar, S., & Pattnaik, D. (2020). Twenty-five years of the Journal of Corporate Finance: A scientometric analysis. *Journal of Corporate Finance, 66*(1), 1–24. https://doi.org/10.1016/j.jcorpfin.2020.101572

Boucenna, S., Narzisi, A., Tilmont, E., Muratori, F., Pioggia, G., Cohen, D., & Chetouani, M. (2014). Interactive technologies for autistic children: A review. *Cognitive Computation, 6*, 722–740. https://doi.org/10.1007/s12559-014-9276-x

Bradley, R., & Newbutt, N. (2018). Autism and virtual reality head-mounted displays: A state-of-the-art systematic review. *Journal of Enabling Technologies, 12*(3), 101–113. https://doi.org/10.1108/JET-01-2018-0004

Brandes, U., Borgatti, S., & Freeman, L. (2016). Maintaining the duality of closeness and betweenness centrality. *Social Networks, 44*(1), 153–159. https://doi.org/10.1016/j.socnet.2015.08.003

Cabibihan, J., Javed, H., Ang, J. R. M., & Aljunied, S. (2013). Why robots? A survey on the roles and benefits of social robots in the therapy of children with autism. *International Journal of Social Robotics, 5*(1), 593–618. https://doi.org/10.1007/s12369-013-0202-2

Cao, W., Song, W., Li, X., Zheng, S., Zhang, G., Wu, Y., He, S., Zhu, H., & Chen, J. (2019). Interaction with social robots: Improving gaze toward face but not necessarily joint attention in children with autism spectrum disorder. *Frontiers in Psychology, 10*(1503), 1–10. https://doi.org/10.3389/fpsyg.2019.01503

De Moya-Anegón, F., Vargas-Quesada, B., Chinchilla-Rodríguez, Z., Herrero-Solana, V., & Corera-Álvarez, E. (2006). Análisis de la cobertura de la base de datos Scopus. *El Profesional De La Información, 15*(2), 142–144.

Deep-Arora, S., & Chakraborty, A. (2021). Intellectual structure of consumer complaining behavior (CCB) research: A bibliometric analysis. *Journal Business Research, 122*(1), 60–74. https://doi.org/10.1016/j.jbusres.2020.08.043

Delgado, A., Vázquez-Cano, E., Belando, M., & López-Meneses, E. (2019). Análisis bibliométrico del impacto de la investigación educativa en diversidad funcional y competencia digital: Web of Science y Scopus. *Aula abierta, 48*(2), 147–156. https://doi.org/10.17811/rifie.48.2.2019.147-156

Diehl, J., Schmitt, L., Villano, M., & Crowell, C. (2012). The clinical use of robots for individuals with autism spectrum disorders: A critical review. *Research in Autism Spectrum Disorders, 6*(1), 249–262. https://doi.org/10.1016/j.rasd.2011.05.006

Donthu, N., Kumar, S., Mukherjee, D., Pandey, N., & Lin, W. (2021). How to conduct a bibliometric analysis: An overview and guidelines. *Journal of Business Research, 133*(1), 285–296. https://doi.org/10.1016/j.jbusres.2021.04.070

Durkin, M., Maenner, M., Baio, J., Christensen, D., Daniel, J., Fitzgerald, R., Imm, P., Li-Ching, L., Schieve, L., Van-Naardem, K., Wingate, M., & Yeargin-Allsopp, M. (2017). Autism spectrum disorder among US children (2002–2010): Socioeconomic, racial, and ethnic disparities. *American Journal of Public Health, 107*(11), 1818–1826. https://doi.org/10.2105/AJPH.2017.304032

Fortuna, G., Aria, M., Iorio, C., Mignogna, M. D., & Kalsser, G. D. (2020). Global research trends in complex oral sensitivity disorder: A systematic bibliometric analysis of the framework. *Journal of Oral Pathology & Medicine, 49*(6), 555–564. https://doi.org/10.1111/jop.13076

García-González, A., & Ramírez-Montoya, M. (2019). Systematic mapping of scientific production on open innovation (2015–2018): Opportunities for sustainable training environments. *Sustainability, 11*(6), 1–15. https://doi.org/10.3390/su11061781

Glänzel, W. (2002). Coauthorship patterns and trends in the sciences (1980–1998): A bibliometric study with implications for database indexing and search strategies. *Library Trends, 50*(3), 461–473.

Haas, A., Vannest, K., Thompson, J., Fuller, M., & Wattanawongwan, S. (2020). Peer mediated instruction and academic outcomes for students with autism spectrum disorders: A comparison of quality indicators. *Mentoring & Tutoring: Partnership in Learning, 28*(5), 625–642. https://doi.org/10.1080/13611267.2020.1859330

Hinojo-Lucena, F. J., Aznar-Díaz, I., Cáceres-Reche, M. P., & Romero-Rodríguez, J. M. (2019). Artificial intelligence in higher education: A bibliometric study on its impact in scientific literature. *Education Sciences, 9*(1), 1–9. https://doi.org/10.3390/educsci9010051

Horwitz, E., Schoevers, R., Greaves-Lord, K., de Bildt, A. & Hartman, C. (2020). Adult manifestation of milder forms of autism spectrum disorder; autistic and non-autistic psychopathology. *Journal of Autism and Developmental Disorders, 50*(8). 2973–2986. https://doi.org/10.1007/s10803-020-04403-9

Hu, H., Feng, X., Shao, Z., Xie, M., Xu, S., Wu, S., Wu, X., & Ye. (2019). Application and prospect of mixed reality technology in medical field. *Current Medical Science, 39*(1), 1–6. https://doi.org/10.1007/s11596-019-1992-8

Ip, H., Wong, S., Chan, D., Li, C., Kon, L., Ma, P., Lau, K., & Bryne, J. (2024). Enhance affective expression and social reciprocity for children with autism spectrum disorder: Using virtual reality headsets at schools. *Interactive Learning Environments, 32*(3), 1012–1035. https://doi.org/10.1080/10494820.2022.2107681

Kamada, T., & Kawai, S. (1989). An algorithm for drawing general undirected graphs. *Information Processing Letters, 31*(1), 7–15. https://doi.org/10.1016/0020-0190(89)90102-6

Kandalaft, M., Didehbani, N., Krawczyk, D., Allen, T., & Chapman, S. (2013). Virtual reality social cognition training for young adults with high-functioning autism. *Journal of Autism and Developmental Disorders, 43*(1), 34–44. https://doi.org/10.1007/s10803-012-1544-6

Kaplan, A., Kessler, T., Brill, J. & Hancock, P. (2023). Trust in artificial intelligence: Meta-analytic findings. *Human Factors: The Journal of the Human Factors and Ergonomics Society, 65(*2), 337–359. https://doi.org/10.1177/00187208211013988

Kavanagh, S., Luxton-Reilly, A., Wuensche, B., & Plimmer, B. (2017). A systematic review of virtual reality in education. *Themes in Science and Technology Education, 10*(2), 85–119.

Kozima, H., Michalowski, M., & Nakagawa, C. (2009). Keepon. *International Journal of Social Robotics, 1*(1), 3–18. https://doi.org/10.1007/s12369-008-0009-8

Lee, E., & Wong, K. (2014). Learning with desktop virtual reality: Low spatial ability learners are more positively affected. *Computers & Education, 79*, 49–58. https://doi.org/10.1016/j.compedu.2014.07.010

Lord, C., Cook, E., Leventhal, B., & Amaral, D. (2000). Autism spectrum disorders. *Neuron, 28*(2), 355–363. https://doi.org/10.1016/S0896-6273(00)00115-X

Lorenzo, G., Lledó, A., Pomares, J., & Roig-Vila, R. (2016). Design and application of an immersive virtual reality system to enhance emotional skills for children with autism spectrum disorders. *Computers and Education, 98*(1), 192–205. https://doi.org/10.1016/j.compedu.2016.03.018

Lorenzo, G., Gómez-Puerta, M., Arráez-Vera, G., & Lorenzo-Lledó, A. (2019). Preliminary study of augmented reality as an instrument for improvement of social skills in children with autism spectrum disorder. *Education and Information Technologies, 24*, 181–204. https://doi.org/10.1007/s10639-018-9768-5

Lorenzo, G., Newbutt, N., & Lorenzo-Lledó, A. (2022). Global trends in the application of virtual reality for people with autism spectrum disorders: Conceptual, intellectual and the social structure of scientific production. *Journal of Computers in Education, 9*(1), 225–260. https://doi.org/10.1007/s40692-021-00202-y

Lorenzo, G., Newbutt, N., & Lorenzo-Lledó, A. (2023). Designing virtual reality tools for students with autism spectrum disorder: A systematic review. *Education and Information Technologies, 28*(1), 9557–9605. https://doi.org/10.1007/s10639-022-11545-z

Lorenzo, G., López-Núñez, J. A., Lorenzo-Lledó, A., & López-Belmonte, J. (2024) Creation of social interaction environments with robot for autistic children: Analysis of assessment instruments and type of activities. *Journal of Enabling Technologies, 18*(1), 13–27. https://doi.org/10.1108/JET-06-2023-0019

Luo, Y., & Du, H. (2022). Learning with desktop virtual reality: Changes and interrelationship of self-efficacy, goal orientation, technology acceptance and learning. *Smart Learning Environments, 9*(22), 1–22. https://doi.org/10.1186/s40561-022-00203-z

Maroto-Gómez, M., Alonso-Martín, F., Malfaz, M., Castro-González, A., Castillo, J. & Salich, M. (2023). A systematic literature review of decision-making and control systems for autonomous and social robots. *International Journal of Social Robotics, 15*(1), 745–789. https://doi.org/10.1007/s12369-023-00977-3

Martínez, M. A., Cobo, M., Herrera, M., & Herrera-Viedma, E. (2015). Analyzing the scientific evolution of social work using science mapping research on social work using science mapping. *Research on Social Work Practice, 25*(2), 257–277. https://doi.org/10.1177/104973151452210

Meho, L., & Rogers, Y. (2008). Citation counting, citation ranking, and h-index of human-computer interaction researchers: A comparison of Scopus and web of science. *Journal of the American Society for Information Science and Technology, 59*(11), 1711–1726. https://doi.org/10.1002/asi.20874

Montero-Díaz, J., Cobo, M., Gutiérrez-Salcedo, M., Segado-Boj, F., & Herrera-Viedma, E. (2018). Mapeo científico en la categoría "Comunicación" en WOS (1980–2013*). Comunicar, 55*(1), 81–91. https://doi.org/10.3916/C55-2018-08

Moon, J., & Ke, F. (2021). Exploring the treatment integrity of virtual reality-based social skills training for children with high functioning autism. *Interactive Learning Environments, 29*(6), 939–953. https://doi.org/10.1080/10494820.2019.1613665

Morimoto, T., Kobayashi, T., Hirata, H., Otani, K., Sugimoto, M., Tsukamoto, M., Yoshihara, T., Ueno, M., & Mawatari, M. (2022). XR (Extended reality: Virtual reality, augmented reality, mixed reality) technology in spine medicine: Status Quo and Quo Vadis. *Journal of Clinical Medicine, 11*(2), 470. https://doi.org/10.3390/jcm11020470

Mosher, M., Carreon, A., Graig, S., & Ruhter, L. (2022). Immersive technology to teach social skills to students with autism spectrum disorder: A literature review. *Review Journal of Autism Spectrum Disorder*, *9*(1), 334–350. https://doi.org/10.1007/s40489-021-00259-6

Mustafee, N., Bessis, N., Taylor, S., Hou, J., & Matthew, P. (2020). Co-citation analysis of literature in e-science and e-infrastructures. *Concurrency and Computation: Practice and Experience, 32*(9), 1–17. https://doi.org/10.1002/cpe.5620

Newbutt, N., Bradley, R., & Conley, I. (2020). Using virtual reality head-mounted displays in schools with autistic children: Views, experiences, and future directions. *Cyberpsychology, Behavior, and Social Networking, 23*(1), 23–33. https://doi.org/10.1089/cyber.2019.0206

Noyons, E., Moed, H., & Luwel, M. (1999). Combining mapping and citation analysis for evaluative bibliometric purposes: A bibliometric study. *Journal of the American Society for Information Science, 50*, 115–131. https://doi.org/10.1002/(SICI)1097-4571(1999)50:2

Page, M., & Moher, D. (2017). Evaluations of the uptake and impact of the preferred reporting items for systematic reviews and meta-analyses (PRISMA) statement and extensions: A scoping review. *Systematic Reviews, 6*(1), 263. https://doi.org/10.1186/s13643-017-0663-8

Parsons, S. (2016). Authenticity in virtual reality for assessment and intervention in autism: A conceptual review. *Educational Research Review, 19*(1), 138–157. https://doi.org/10.1016/j.edurev.2016.08.001

Parsons, S., & Mitchell, P. (2002). The potential of virtual reality in social skills training for people with autistic spectrum disorders. *Journal Intellectual Disability Research, 46*(5), 430–443. https://doi.org/10.1046/j.1365-2788.2002.00425.x

Pérez-Vázquez, E., Lorenzo, G., Lledó, A., & Lorenzo-Lledó, A. (2020). Evolution and identification from a bibliometric perspective of the use of robots in the intervention of children with ASD. *technology, Knowledge and Learning, 25*, 83–114. https://doi.org/10.1007/s10758-019-09415-8

Pons, P., & Latapi, M. (2006). Computing communities in large networks using random walks. *Journal of Graph Algorithms and Applications, 10*(2), 191–218. https://doi.org/10.1007/11569596_31

Pradhan, T., & Pal, S. (2020). A hybrid personalized scholarly venue recommender system integrating social network analysis and contextual similarity. *Future Generation Computer Systems, 110*(1), 1139–1166. https://doi.org/10.1016/j.future.2019.11.017

Quintero, J., Baldiris, S., Rubira, R., Cerón, J., & Velez, G. (2019). Augmented reality in educational inclusion. A systematic review on the last decade. *Frontiers in Psychology, 10*(1), 1–14. https://doi.org/10.3389/fpsyg.2019.01835

Radianti, J., Majchrzak, T. A., Fromm, J., & Wohlgenannt, I. (2020). A systematic review of immersive virtual reality applications for higher education: Design elements, lessons learned, and research agenda. *Computers & Education, 147*, 103778. https://doi.org/10.1016/j.compedu.2019.103778.robots

Repiso, R., & Torres-Salinas, D. (2016). Características e implicaciones de la base de datos emerging sources citation index (thomson reuters): Las revistas en estado transitorio. *Anuario Thinkepi, 10*(1), 234–236. https://doi.org/10.3145/thinkepi.2016.46

Robins, B., Dautenhahn, K., Boekhorst, R., & Billard, A. (2005). Robotic assistants in therapy and education of children with autism. Can a small humanoid robot help encourage social interaction skills? *Universal Access in the Information Society, 4*(1), 105–120. https://doi.org/10.1007/s10209-005-0116-3

Rodríguez-Soler, R., Uribe-Toril, J., & De Pablo-Valenciano, J. (2020). Worldwide trends in the scientific production on rural depopulation, a bibliometric analysis using bibliometrix R-tool. *Land Use Policy, 97*(1), 1–20. https://doi.org/10.1016/j.landusepol.2020.104787

Salari, N., Rasoulpoor, S., Rasoulpoor, S., Shohaimi, S., Jafarpour, S., Abdoli, N., Khaledi-Paveh, B., & Mohammadi. (2022). The global prevalence of autism spectrum disorder: A comprehensive systematic review and meta-analysis. *Italian Journal of Pediatrics, 48*(1), 1–16. https://doi.org/10.1186/s13052-022-01310-w

Sani-Bozkurt, S., & Sani-Genc, G. (2023). Social robots for joint attention development in autism spectrum disorder: A systematic review. *International Journal of Disability, Development and Education, 70*(5), 625–643. https://doi.org/10.1080/1034912X.2021.1905153

Sartorato, F., Przybylowski, L., & Sarko, D. (2017). Improving therapeutic outcomes in autism spectrum disorders: Enhancing social communication and sensory processing through the use of interactive robots. *Journal of Psychiatric Research, 90*(1), 1–11. https://doi.org/10.1016/j.jpsychires.2017.02.004

Scassellati, B., Admoni, H., & Mataric, M. (2012). Robots for use in autism research. *Annual Review of Biomedical Engineering, 14*(1), 275–294. https://doi.org/10.1146/annurev-bioeng-071811-150036

Scassellati, B., Boccanfuso, L., Huang, C., Mademtzi, M., Qin, M., Salomons, N., Ventola, P., & Shic, S. (2018). Improving social skills in children with ASD using a long-term, in-home social robot. *Science Robotics, 3*(21), 1–19. https://doi.org/10.1126/scirobotics.aat7544

Schmidt, M., Schmidt, C., Glaser, N., Beck, D., Lim, M., & Palmer, H. (2021). Evaluation of a spherical video based virtual learning intervention to teach adaptive skills for adults with autism: A preliminary report. *Interactive Learning Environments, 29*(3), 345–364. https://doi.org/10.1080/10494820.2019.1579236

Semrud-Clikeman, M., Fine, J., & Bledsoe, J. (2014). Comparison among children with autism spectrum disorder, nonverbal learning disorder and typically developing children on measures of executive functioning. *Journal of Autism and Developmental Disorders, 44*(2), 331–342. https://doi.org/10.1007/s10803-013-1871-2

Shahamiri, S., & Thabtah, F. (2020). Autism AI: A new autism screening system based on artificial intelligence. *Cognitive Computation, 12*(1), 766–777. https://doi.org/10.1007/s12559-020-09743-3

Singh, A., Raj, K., Kumar, T., Verma, S., & Roy, A. (2023). Deep learning-based cost-effective and responsive robot for autism treatment. *Drones, 7*(2), 81–99. https://doi.org/10.3390/drones7020081

Talaat, F. (2023). Real-time facial emotion recognition system among children with autism based on deep learning and IoT. *Neural Computing and Applications, 35*(1), 12717–12728. https://doi.org/10.1007/s00521-023-08372-9

Toma, V., Turcu, C., Turcu, C., Vlad, S., Tiliute, D., & Pascu, P. (2024). Extended reality-based mobile app solutions for the therapy of children with autism spectrum disorders: Systematic literature review. *JMIR Serious Games, 12*, e49906. https://doi.org/10.2196/49906

Verhey, J., Haglin, J., Verhey, E., & Hartigan, D. (2020). Virtual, augmented, and mixed reality applications in orthopedic surgery. *The International Journal of Medical Robotics and Computers Assisted Surgery, 16*(2), e2067. https://doi.org/10.1002/rcs.2067

Verté, S., Geurts, H., Roeyers, H., Oosterlaan, J., & Sergeant, J. (2006). Executive functioning in children with an autism spectrum disorder: Can we differentiate within the spectrum?. *Journal of Autism and Developmental Disorders, 36*(3), 351–372. https://doi.org/10.1007/s10803-006-0074-5

Wen, Q. J., Ren, Z. J., Lu, H., & Wu, J. F. (2021). The progress and trend of BIM research: A bibliometrics-based visualization analysis. *Automation in Construction, 124*(1), 1–15. https://doi.org/10.1016/j.autcon.2021.103558

White, H. D., & McCain, K. W. (1998). Visualizing a discipline: An author co-citation analysis of information science, 1972–1995. *Journal of the American Society for Information Science, 49*(4), 327–355. https://doi.org/10.1002/(SICI)

Xia, Q., Chiu, T. K. F, Lee, M., Temitayo I., Dai, Y., & Chai, C. S. (2022). A self-determination theory design approach for inclusive and diverse artificial intelligence (AI) K-12 education. *Computers & Education, 189*, 104582. https://doi.org/10.1016/j.compedu.2022.104582.

Zhao, W., Xu, S., Zhang, Y., Li, D., Zu, C., & Wang, K. (2024). The application of extended reality in treating children with autism spectrum disorder. *Neuroscience Bulletin*. https://doi.org/10.1007/s12264-024-01190-6

Zupic, I., & Cater, T. (2015). Bibliometric methods in management and organization. *Organizational Research Methods, 18*(3), 429–472. https://doi.org/10.1177/1094428114562629

Practical Applications of Social Robots

A Meta-Analysis of the Efficacy of Conversational AI Use in EFL Learning Contexts

Wenjie Gu

Abstract In recent years, technological advancements have spurred the emergence of conversational AI as a promising tool for enhancing language learning and teaching outcomes. Despite extensive exploration into the effects of chatbot technology on educational outcomes within language learning contexts, there remains a dearth of research synthesizing previous experimental findings to assess the efficacy of chatbot-supported learning in EFL education quantitatively. This study seeks to bridge this gap by examining the impact of chatbot-supported learning on various domains of English language acquisition and investigating the influence of different moderating variables on its effectiveness. A coding scheme was devised based on the theoretical framework of activity theory, and a meta-analysis was conducted, encompassing 5,031 participants from 55 experimental studies. The results indicated that chatbots moderately to substantially improved overall English language learning outcomes, with insignificant moderators, such as intervention duration, English proficiency, and educational levels. Furthermore, the analysis demonstrated that chatbots significantly enhance overall English language proficiency and most specific language skills while positively affecting affective outcomes such as learning motivation and interest. However, no significant differences were observed in the domains of English writing skills and learning anxiety. The results support the notion that conversational AI technology-mediated language instruction holds promise in revolutionizing English language learning, highlighting its positive effects in the realm of language education. This study also suggests the incorporation of a broader range of learning outcome domains in future research involving AI chatbot-based interventions in EFL education.

Keywords English education · Chatbot-assisted language learning · Activity theory · Meta-analysis

W. Gu (✉)
Faculty of International Studies, Beijing Language and Culture University, Beijing, China
e-mail: guwenjie021@163.com

G. Lampropoulos and S. Papadakis (eds.), *Social Robots in Education*, Studies in Computational Intelligence 1194, https://doi.org/10.1007/978-3-031-82915-4_14

1 Introduction

The booming Artificial Intelligence (AI) provides fertile ground for AI in language education. The integration of AI technology into English Language Learning (ELL) is revolutionizing education (Jiang, 2022; Kumar, 2023). Among its applications, conversational AI has captured the considerable attention of English as a Foreign Language (EFL) researchers in English education. Conversational AI is a computer program designed to simulate conversations with human users over the internet based on the Natural Language Processing (NLP) model from information collecting to query routing (Hawanti & Zubaydulloevna, 2023). It stimulates Human–Computer Interaction (HCI) by processing auditory or textual data. AI chatbots have been valued by EFL researchers as effective teaching and learning tools that support a range of English learning practices for increasing motivation, proficiency, and preference over traditional methods (Okyar, 2023; Younis et al., 2023).

AI-based chatbots are becoming immensely popular in EFL educational environments as they can perform various tasks based on human speech recognition, comprehension, and synthesis technology (Yang et al., 2022). Chatbots are beneficial enough to act as an interlocutor for English language learning. Chatbot-based language learning programs potentially serve as a supplementary language learning tool for learners who want to practice the target language in authentic contexts, fostering immersive and interactive learning environments conducive to effective two-way communication and enhancing EFL teaching methods (Kyung, 2019). The real-time interactive features of chatbots aided EFL students in enhancing their language proficiency by facilitating comprehension and production of English conversation, thereby diminishing speaking anxiety and refining pronunciation for greater authenticity and accuracy (Chien et al., 2022; Han, 2020). From this perspective, many studies have illustrated the capacity of AI-driven chatbots to improve EFL learners' English skills, making language acquisition more engaging and instructional delivery more interactive (Escalante et al., 2023; Hwang et al., 2023).

Despite its usefulness in providing rapid and accurate responses to a wide range of questions, many studies expressed concerns. The application of chatbots in language education might reinforce biases, misinformation, data privacy leakage, low transparency, and harmed ethical principles, hindering students' development of critical thinking and research skills and potentially (Kohnke et al., 2023; Levy et al., 2023). Therefore, numerous experimental studies sought to elucidate the effectiveness of chatbot-assisted language learning, yielding inconsistent results. Few studies have reviewed those experimental studies to evaluate the overall effect of chatbots, especially in the EFL educational context (Jeon et al., 2023; Okyar, 2023).

Recent review articles provided basic information on the types of chatbot applied in language education, chatbot roles, research trends, affordances, and challenges through systematic reviews, bibliometric, and content analyses (Table 1). Regarding the effectiveness of chatbot-assisted language learning, three meta-analyses synthesized the previous experimental studies, calculated the overall effects, and evaluated the influence of certain moderating variables, such as school level, chatbot type, and treatment period. Among them, only Je-Young and Yohan (2022) focused on EFL learning, but the research scope was limited to the Korean context. Therefore, this study aimed to bridge the research gap by synthesizing the results of previous experimental studies to investigate the effect of chatbot use in EFL learning on various components and potential variables moderating the effectiveness.

The subsequent chapters of this study will be structured as follows. Firstly, the authors will introduce the theoretical framework guiding the research and conduct an extensive literature review of prior studies examining the utilization of AI chatbots in EFL educational contexts to formulate 13 null hypotheses (Sect. 2). Secondly, it will outline the methodology for the meta-analysis, including the search and screening process for relevant experimental studies, coding procedures for the included studies, and examination of effect sizes across various outcome domains such as English language skills, learning motivation, interest, and anxiety, as well as moderating variables including intervention duration, educational levels, and English proficiency (Sect. 3). Finally, it will present the analysis of results (Sect. 4), discuss potential reasons for the findings (Sect. 5), outline the contributions of the study and provide insights for future research (Sect. 6). The outcomes of this quantitative meta-analysis

Table 1 Relevant review studies of chatbot-assisted language learning

No.	Relevant review studies	Themes	Types of review
1	Huang et al. (2022), Kohnke et al. (2023)	The technological, pedagogical, and social affordances and challenges of using chatbots for L1, L2, and foreign language learning	Systematic review
2	Ji et al. (2023)	Empirical evidence of human–computer collaboration in AI chatbots integrated into language learning environment	Systematic review
3	Jeon et al. (2023)	Research foci, locations of chatbot use, research designs, and chatbot roles in the reviewed empirical studies on speech-recognition chatbots for language learning	Systematic review
4	Kartal and Yesilyurt (2024), Traymbak et al. (2024), Zheltukhina et al. (2024)	Current trends, prominent themes, and influential authors, documents, and sources of chatbot use in L2 education	Bibliometric analysis
4	Meniado (2023), Eisenring et al. (2024), Jiang (2022), Okyar (2023)	Advantages and challenges of using chatbots in EFL education	Systematic review/ content analysis
5	Chen et al. (2020), Je-Young and Yohan (2022), Zhang et al. (2023b)	The overall effects of chatbot-assisted language learning and the influence of some moderating variables	Meta-analysis

aim to offer valuable references for educators, researchers, technology developers, policymakers, and individuals interested in integrating AI chatbots into EFL learning environments.

2 Literature Review

2.1 Activity Theory in Chatbot-Assisted Language Learning

In exploring the dynamics among learners, tools, and activities, researchers proposed Activity Theory (AT) as a suitable framework for evaluation (Jonassen & Rohrer-Murphy, 1999). The Activity theory describes actions through six related elements: object, subject, context, tools, division of labor, and rules (Engeström, 1999). The activity theory was originally derived from Vygotsky's (1978) concept of social mediation in human cognitive development, suggesting that human activities were not isolated but shaped by social, cultural, and historical contexts. Thus, Vygotsky formed the triad model comprising subject, object, and tools for psychological development. The subsequent generation of AT was refined by Leont'ev (1974) and Engeström (1999). Leont'ev incorporated community into the model to account for the interrelationships between subjects and their community (Leont'ev, 1974). Engeström further expanded the framework by integrating rules, community, and division of labor into the interaction process, emphasizing the dynamic and collective aspects of activity systems and the interplay between two or more activity systems (Engeström, 1999).

The Activity Theory provided an interpretive framework to explain how the chatbot platform facilitated EFL learning. This study adopted Engeström's (1999) version of AT. As shown in Fig. 1, the subjects in this study are participants in the previous experimental studies of chatbot-assisted language learning in EFL learning contexts. The objects are the motives or goals driving the subjects to engage in the activity, i.e., learning outcome domains consisting of English language skills and affective aspects. The tools in this study include both text- or voice-based AI-assisted chatbots mediating the relationship between the subjects and objects through communication. The rules are certain control measurements in the implementation of AI chatbots in the EFL learning process, such as intervention duration, English proficiency levels and educational levels. Labor division is the organization of English learning and teaching activity and distribution of responsibility among the teachers and chatbots during the experimenting process.

The concept of community refers to the social groups that engage in the chatbot-assisted learning process with the subjects. For example, when chatbots were utilized as virtual learning companions to support English language learning, they effectively became community members who aided students in their English language acquisition alongside other community members (Guo et al., 2024). Regarding the interplay among the six elements, subjects (EFL learners in the experimental experiments)

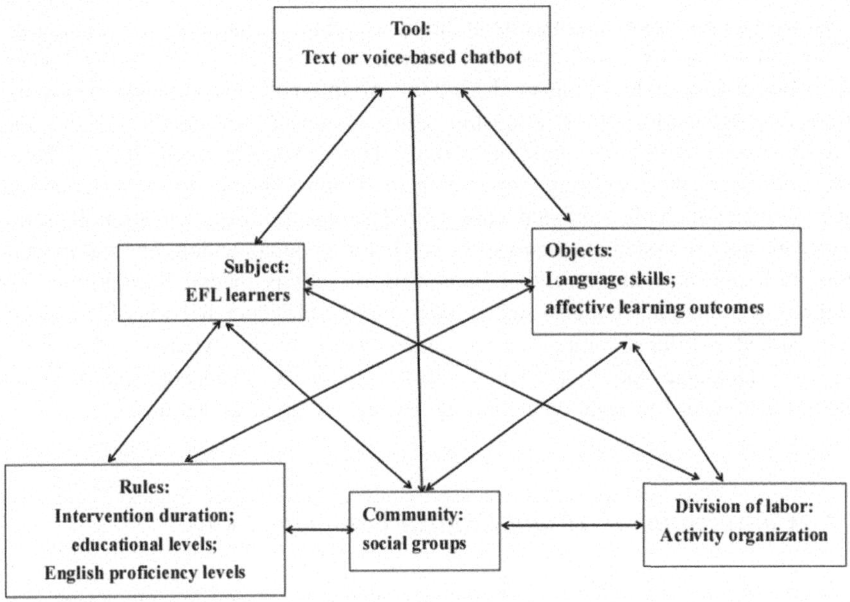

Fig. 1 The adapted framework of AT to chatbot-assisted language learning

engaging in actions aims at achieving a desired objective (e.g. improving certain English skills). Throughout this process, the subject may utilize various mediating tools (text or voice-based chatbots). Rules represent certain norms or conventions that may guide the subject's behavior. The subject can also be influenced by the community within a particular social context. Division of labor, which deals with the distribution of roles within an activity, may emerge as the subject pursues their objective when task responsibilities are allocated in some manner (Fig. 1).

The Activity Theory provided a valuable analytical framework applicable to a wide range of human–computer interaction contexts, both general and specific. Previous empirical studies and review studies utilized AT to examine various aspects of AI technology-supported language learning activities. Wu and Li (2024) conducted a systematic meta-analysis guided by AT theoretical principles to assess the impact of robot-assisted language learning on the development of EFL skills. From the perspective of AT, AI-supported technology as a mediating tool demonstrated some effectiveness in language learning outcomes but needed further improvement in the use of language for communication and collaborative design (Yang & Kyun, 2022). Students formed an efficient learning community during the interaction with chatbots in writing class, and the student–chatbot collaborative writing was shaped by rules associated with meeting task requirements and following argumentative writing conventions (Guo et al., 2024). In summary, activity theory served as a scientific theoretical framework offering valuable insights into how language learners utilized AI-based chatbots to enhance their learning experiences.

However, previous meta-analysis studies in the realm of chatbot-assisted language learning often lacked a well-grounded theoretical framework (Je-Young & Yohan, 2022; Lee & Lee, 2022; Zhang et al., 2023b). Additionally, few carried out systematic reviews to explore the relationship between students and chatbots within the activity system by drawing upon the activity theory (Yang & Kyun, 2022). Therefore, addressing these existing gaps, this meta-analysis intended to investigate the overall impact of chatbot-assisted learning in the context of EFL education. It aimed to provide a more systematic analysis of key learning outcome domains and moderators, utilizing the rich conceptual framework of activity theory. Furthermore, by integrating contextual elements such as community, rules, and division of labor, this study aimed to offer a comprehensive understanding of the dynamics of chatbot-supported language learning activities within EFL contexts. This contribution could add to the broader literature on technology-enhanced language learning.

2.2 Conversational AI and English Language Skills

AI chatbots fulfilled a pedagogical function by engaging in real-time interaction with users within virtual online environments, employing either text or speech communication (Agarwal et al., 2022). Numerous experimental studies demonstrated that chatbot-assisted language learning exerted a discernible impact on learners' cognitive outcomes. These outcomes encompassed language learning achievement (Jeon, 2023; Kim, 2018a; Wei, 2023; Yuan, 2023) and retention of knowledge (Kim, 2022). Additionally, chatbot-assisted language learning affected learners' affective outcomes, including their interest in learning (Han, 2020; Kim, 2018b), motivation (Ryong et al., 2023; Song & Song, 2023), anxiety levels (Çakmak, 2022; Lee et al., 2023a, 2023b), confidence (Wei et al., 2023), etc.

Language skills refer to a person's capability to use a certain language effectively in both spoken and written forms for various purposes. The four main language skills included listening, speaking, reading, and writing (LSRW). Listening and reading, as the basic skills of language learning, were known as receptive skills, while speaking and writing were called productive skills (Burns & Siegel, 2018). The goal of learning a language was to enhance the learners' four skills, with language vocabulary and grammar proficiency as the base (Dash, 2022). The importance of grammar knowledge and progressive improvement in vocabulary facilitated both productive and receptive skills and played a significant role in the development of language skills (Hu et al., 2022).

Various studies explored the effect of AI chatbots on language skills improvement for EFL learners. A majority of EFL students found chatbot-assisted learning engaging and enjoyable, making it easier to practice English outside the classroom (Kumar, 2023). Engaging in chat interactions with chatbots improved various language competencies, including listening proficiency, reading comprehension, and grammar skills (Kim, 2018b, 2019). AI chatbots were effective and useful for enhancing students' performance and engagement in learning English prepositions

(Nghi et al., 2019). Using AI-based chatbots also effectively improved speaking skills among EFL students (Hamuddin et al., 2020). Additionally, quantitative analysis revealed significant improvements in writing skills among students who received AI chatbot-assisted instruction compared with the control group (Wei et al., 2023). However, Vogt et al. (2020) conducted a large-scale study and found insignificant differences in reading competence between an experimental group that utilized chatbots and a control group that did not. The inconsistent findings could confound future research, pedagogical practice, and technological designs of AI chatbots. To clarify this issue, the authors proposed the following null hypotheses:

H1. The use of AI chatbots could not significantly improve overall language skills at the 0.05 level.

H2. The use of AI chatbots could not significantly improve listening competence at the 0.05 level.

H3. The use of AI chatbots could not significantly improve speaking competence at the 0.05 level.

H4. The use of AI chatbots could not significantly improve reading competence at the 0.05 level.

H5. The use of AI chatbots could not significantly improve writing competence at the 0.05 level.

H6. The use of AI chatbots could not significantly improve vocabulary learning at the 0.05 level.

H7. The use of AI chatbots could not significantly improve grammar skills at the 0.05 level.

2.3 English Learning Motivation, Interest, and Anxiety

As a critical psychological state, motivation refers to the process where students establish objectives and strive to attain them, which supports, encourages, and sustains learning (McInerney, 2019). Within the framework of the self-determination theory, motivation could be categorized into intrinsic and extrinsic motivation (Vallerand, 1997). Intrinsic motivation remains an important construct with inner satisfaction derived from the learning process serving as an internal driving force, which reflects the natural human propensity to learn and assimilate. However, extrinsic motivation is defined as behavior driven by external influence or rewards, varying considerably in its relative autonomy and thus can either reflect external control or true self-regulation (Ryan & Deci, 2000). The application of chatbots in English education could motivate learners more in L2 writing (Kim et al., 2020). The quantitative analysis conducted by Wei (2023) revealed that the experimental group with the instruction of an AI chatbot exhibited greater L2 motivation compared with the control group. Meanwhile, EFL learners' motivation was enhanced under an AI chatbot contextual learning environment, though these enhancements did not yield significant differences in extrinsic motivation (Chien et al., 2022). The authors then presented the following null hypothesis to address the inconsistent findings on the effect of AI chatbots on EFL learning motivation:

H8. The use of AI chatbots could not significantly improve English learning motivation at the 0.05 level.

Interest represents a learner's inclination or initiative to engage in assigned learning tasks (Krapp, 1999). Previous research demonstrated that AI voice-based chatbots increased learners' interest in reading English books (Lee et al., 2023a, 2023b) and enhanced college students' enthusiasm for learning English vocabulary (Kim, 2018a). Moreover, employing chatbot apps for conversation practice in authentic contexts stimulated learners' interest and encouraged them to engage in more meaningful English conversation (Hwang et al., 2022). A survey conducted by Han (2020) revealed a positive change over time in students' interest in AI chatbot-assisted English learning. However, contrasting findings were reported by Kim et al. (2020), where using AI chatbot applications did not lead to a significant improvement in students' interest in English learning. Facing the controversial results, the authors thus raised the following null hypothesis:

H9. The use of AI chatbots could not significantly improve English learning interest at the 0.05 level.

In this study, anxiety refers specifically to computer or technology anxiety, indicating negative attitudes or perceptions toward the use of chatbot technologies or other computer applications (Korukonda, 2007). Research showed that AI chatbot-based instruction with instant feedback alleviated students' anxiety about learning English writing (Hawanti & Zubaydulloevna, 2023), while the AI chatbot grammar learning approach reduced English learning anxiety (Chen & Lin, 2023). Chatbots lowered the barrier to learning natural English conversation, with EFL learners using chatbots significantly outperforming those in non-chatbot groups in terms of oral fluency, while experiencing less anxiety (Azizimajd, 2023). However, Çakmak (2022) found that students reported difficulties in being precisely understood, and actual interaction with chatbots might not reliably reduce their anxiety with second-language speaking. Therefore, the authors proposed the following hypothesis:

H10. The use of AI chatbots could not significantly reduce English learning anxiety at the 0.05 level.

2.4 *Intervention Duration, Proficiency, and Educational Levels*

Previous studies varied in intervention durations and yielded diverse outcomes. For instance, an intervention lasting less than an hour resulted in insignificant differences in speaking performance between chatbot and control groups (Goda et al., 2014). Conversely, a study implementing treatment within one week observed perceived improvement in English achievement among the experimental group using chatbot applications in classes (Pasang et al., 2024). In another study, fifteen instructional sessions conducted over three weeks enhanced students' comprehension of English adjectival constructions but did not lead to improvement in generating prepositional constructions (Kim, 2022). Furthermore, speaking test scores in the experimental group were significantly higher than those of the control group after a 16 week experiment (Hsu et al., 2023a, 2023b). Based on the diverse findings, the duration might be a critical moderator of the impact of AI chatbots on EFL learning outcomes. Therefore, the authors developed the following null hypothesis:

H11. Intervention duration could not influence the effect of chatbot-assisted English learning at the 0.05 level.

Language proficiency levels categorized a speaker's ability to communicate in a specific language. The Common European Framework of Reference (CEFR) organized language proficiency into six levels, A1–C2, which could be regrouped into three broad levels, i.e., beginner, intermediate, and advanced levels (North, 2014). Previous experimental studies on AI chatbots for EFL education were conducted with participants at different English proficiency levels and received different results. Some researchers suggested that learners at the intermediate level or higher derived greater benefits from chatbots in enhancing their language skills (Meskill, 2009; Satar & Özdener, 2008). However, Kim (2016) observed positive improvements in speaking ability among participants as a result of engaging in voice chat, irrespective of their proficiency levels. Thus, the authors developed the following null hypothesis to clarify the role of language proficiency in chatbot-assisted EFL learning:

H12. Proficiency levels could not influence the effect of chatbot-assisted English learning at the 0.05 level.

Numerous experimental studies on AI chatbots in EFL education investigated learning outcomes among participants at various educational levels. While many of these studies focused on participants at higher education levels, previous findings consistently demonstrated that the application of AI chatbots can enhance EFL learners' language skills at other pre-college levels (Han, 2020; Hwang et al., 2023; Wei, 2023). Meanwhile, some research results revealed that chatbots didn't exert a significant influence on learners' English learning achievement or affective domains neither at higher educational levels (Lee et al., 2023a, 2023b; Zhang et al., 2023a, 2023b) nor at the pre-college levels (Chien et al., 2022; Mageira et al., 2022; Vogt et al., 2020). The authors therefore raised the following null hypothesis to clarify the role of educational levels in chatbot-assisted EFL learning:

H13. Educational levels could not influence the effect of chatbot-assisted English learning at the 0.05 level.

Based on the literature review, this study raised 13 null hypotheses listed in Table 2, aiming to reach a quantitative evaluation of the overall effect and the specific effect of chatbot use on various outcome domains including language skills, English learning motivation, learning interest, and anxiety, and whether the effectiveness could be influenced by different moderators including intervention duration, educational and English proficiency levels.

Table 2 The list of 13 null hypotheses

No.	Null hypotheses
H1	The use of AI chatbots could not significantly improve overall language skills at the 0.05 level
H2	The use of AI chatbots could not significantly improve listening competence at the 0.05 level
H3	The use of AI chatbots could not significantly improve speaking competence at the 0.05 level
H4	The use of AI chatbots could not significantly improve reading competence at the 0.05 level
H5	The use of AI chatbots could not significantly improve writing competence at the 0.05 level
H6	The use of AI chatbots could not significantly improve vocabulary learning at the 0.05 level
H7	The use of AI chatbots could not significantly improve grammar skills at the 0.05 level
H8	The use of AI chatbots could not significantly improve English learning motivation at 0.05 level
H9	The use of AI chatbots could not significantly improve English learning interest at the 0.05 level
H10	The use of AI chatbots could not significantly reduce English learning anxiety at the 0.05 level
H11	Intervention duration could not influence the effect of chatbot-assisted English learning at the 0.05 level
H12	Proficiency levels could not influence the effect of chatbot-assisted English learning at the 0.05 level
H13	Educational levels could not influence the effect of chatbot-assisted English learning at the 0.05 level

3 Research Methods

To test the above null hypotheses, a meta-analysis was conducted aiming at synthesizing and assessing previous research findings. This process involved identifying relevant experimental studies, coding the included studies, and calculating effect sizes. The review was carried out following the Preferred Reporting Items for Systematic Reviews and Meta-Analyses (PRISMA) guidelines, which were proposed to enhance the reliability, transparency, and completeness of review research and reporting (Page et al., 2021). The PRISMA statement consisted of a 27-item checklist and a four-phase flow diagram designed to guide the authors in reporting systematic reviews and meta-analyses, thus ensuring readers' thorough understanding of the methods, results, and conclusions (Page et al., 2021). The subgroups of this study were categorized based on the effects of AI-based chatbots on various outcome domains, including language skills, learning interest, learning motivation, and learning anxiety.

3.1 *Study Searching and Screening*

3.1.1 Identification of Search Keywords

A visualization analysis was initially conducted to determine keywords for clustering relevant literature. Using the search themes ((TOPIC) ("chatbot*" OR "conversational agent*" OR "conversational AI") AND (TOPIC) ("language learn*" OR "language teach*" OR "language educat*")), 722 results were retrieved from the Web of Science Core Collection. The results were exported as plain text files and imported into VOSviewer (Van Eck & Waltman, 2010) for co-occurrence analysis, with all keywords as the unit of analysis. A minimum keyword occurrence of six was set, which refined the results to 127 items meeting the threshold and clustered them into 5 groups (Fig. 2). The top ten frequently occurring keywords were identified: chatbot (N = 325), artificial intelligence (N = 200), ChatGPT (N = 142), natural language processing (N = 126), conversational agent (N = 92), machine learning (N = 90), deep learning (N = 51), AI (N = 48), education (N = 40), and large language models (N = 35).

The most prominent clusters in the above visualization map included keywords like "chatbot," "artificial intelligence," "natural language processing," and "machine

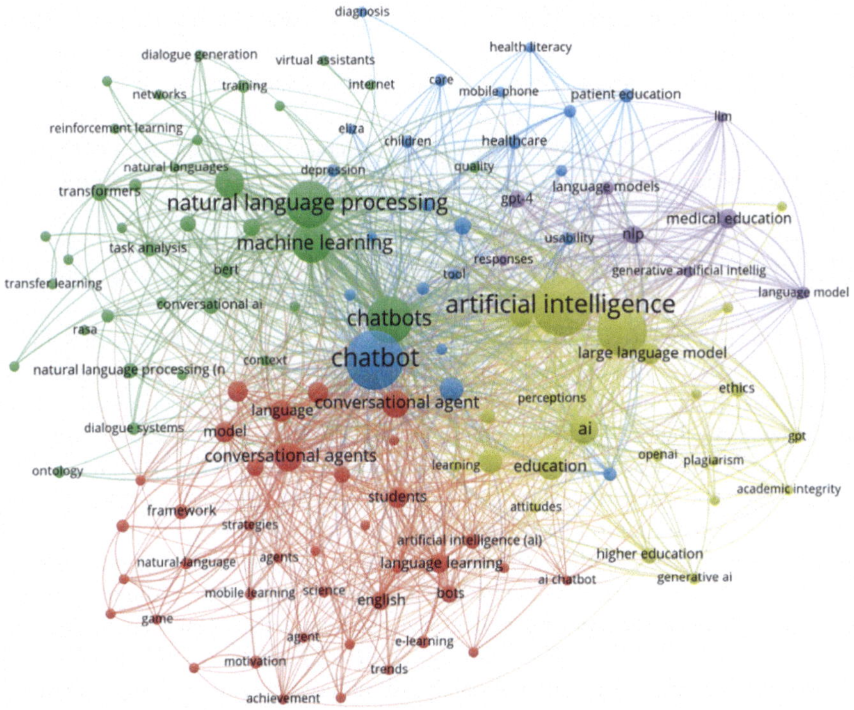

Fig. 2 Clustering words of chatbot-assisted language learning

learning", which depicted the development of chatbot technology and emphasized the increasing use of AI-based chatbots in language teaching and learning research and their potential to provide personalized feedback (Fig. 2). Another notable cluster represented by "conversational agent" highlighted a growing research interest in exploring the effectiveness of chatbot use on learning achievement and affective domains like learning motivation. It suggested a need for further research in this area to evaluate the potential of chatbots in the enhancement of specific language skills and other affective domains such as learning interest, anxiety, etc.

3.1.2 Literature Searching

The authors collected a total of 2,331 studies from online databases on April 2nd, 2024. The primary databases included Web of Science Core Collection, Elsevier, Springer Link, Wiley Online Library, Taylor & Francis Online, and Google Scholar. Specifically for this study's purpose, the authors retrieved 260 results from the Web of Science Core Collection by applying the sequent terms to "Topic": (chatbot OR "conversational agent") AND (English educat* OR English learn* OR English teach*) AND ("control group" OR experim* OR experient*). From Elsevier, 67 studies were retrieved by entering "chatbot OR conversational agent" in the title AND "English education OR learning OR teaching OR control OR experiment" in the title, abstract, and keywords. Springer yielded 54 records, and Google Scholar provided 1,867 results by entering "chatbot" in the title and "control group" in the exact phrase, AND "English education OR English learn OR English teach" in at least one of the words. Additionally, 38 results were obtained from Wiley and 45 results from Taylor & Francis by entering "chatbot OR conversational agent" AND "English education OR learning OR teaching OR control OR experiment" in the abstract.

3.1.3 Inclusion and Exclusion Criteria

The authors adhered to specific inclusion and exclusion criteria when selecting literature for the study. The identified studies were required to (1) clarify the effect of chatbot-based learning in EFL contexts; (2) include both an experimental group that utilized AI chatbots in English learning and a control group that utilized traditional learning approaches; (3) report sufficient and clear statistics, such as sample sizes, means, and standard deviations, to calculate effect sizes; and (4) ensure homogeneity between the control and experimental groups, with equivalent prior learning outcomes for students. Studies were excluded if they (1) were irrelevant to the educational use of chatbots in EFL contexts; (2) lacked control groups; (3) did not provide adequate statistical information for effect-size calculations; or (4) were not written in acceptable English language. Following these criteria, 39 studies were included in this meta-analysis after screening. Additionally, 16 studies were added from the previous version of the review using a snowballing approach to identify articles not extracted in the initial database search through citation checking (Wohlin, 2014). During the

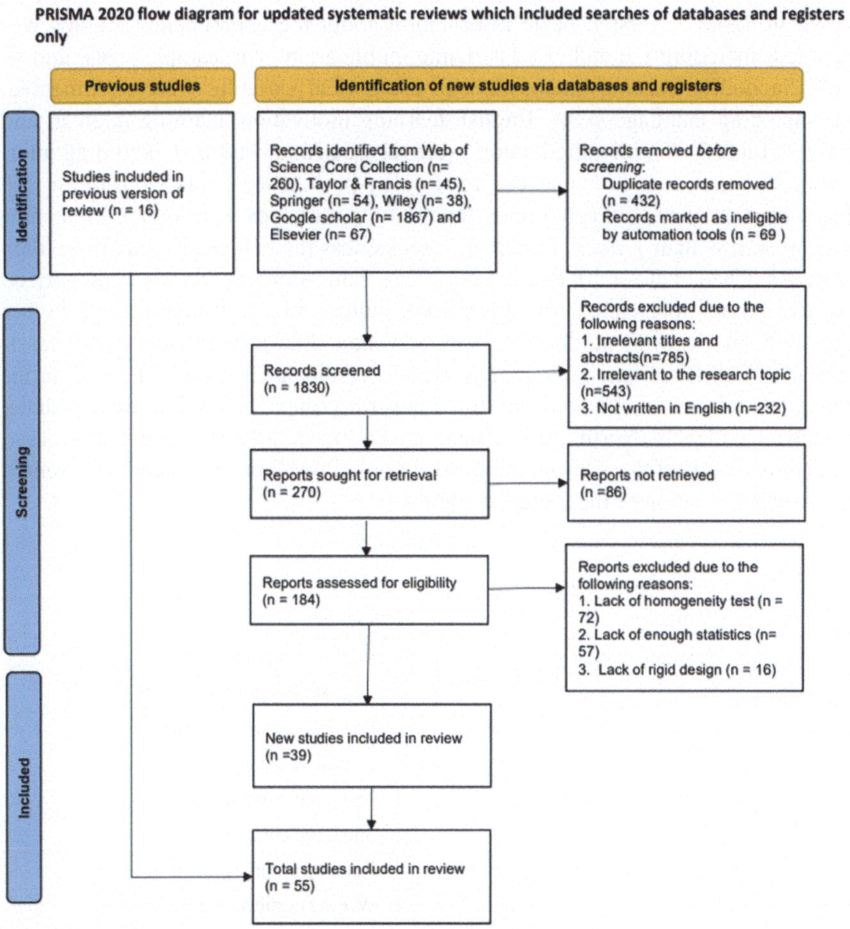

Fig. 3 A PRISMA-based flow diagram of study search and screening in this meta-analysis

snowballing stage, the website *Bing* was utilized to manually search for specific articles based on their titles. Consequently, a total of 55 studies were included in this meta-analysis. Figure 3 displays complete literature search and selection procedures in this study.

3.2 Coding Procedures of the Included Studies

To analyze the effects of AI chatbots in EFL contexts, a specific coding scheme was developed to facilitate a comprehensive analysis of the 55 selected articles, based on the elements of objects and rules in the activity theory framework, as shown in Table 3. Firstly, basic information about the experimental studies was recorded, including the author's last name, publication year, sample sizes, and so forth. Secondly, learning outcomes with categorical variables were documented, encompassing language skills, English learning motivation, learning interest, and anxiety (Table 3). Thirdly, moderators were recorded and classified, including intervention duration, educational levels, and English proficiency levels. Considering the characteristics of the collected data, the duration variables were divided into four categories: less than 1 week, less than 5 weeks, less than 10 weeks, and more than 10 weeks (Chen et al., 2020; Lee & Lee, 2022). Educational levels were categorized into five groups: preschool, primary school, junior and senior high school, higher education, and working adults (Je-Young & Yohan, 2022). According to the classification of CEFR and the homogeneity pre-test results in each study, English proficiency levels were categorized into three general groups: beginning, intermediate, and advanced levels (North, 2014; Zhang et al., 2023a, 2023b). Table 4 showcased the coding results of the relevant included studies. Table 9 in the appendix presented the basic information of the included studies.

Table 3 The coding scheme

Elements	Variables	Categories
Objects	Learning outcomes	1. English language skills
		2. English learning motivation
		3. English learning interest
		4. English learning anxiety
	Subgroup of language skills	1. Listening competence
		2. Speaking competence
		3. Reading competence
		4. Writing competence
		5. Vocabulary knowledge
		6. Grammar proficiency
Rules	Educational levels	1. Preschool
		2. Primary school
		3. Junior and senior high school
		4. Higher education
		5. Working adults
	English proficiency levels	1. Beginning level
		2. Intermediate level
		3. Advanced level
	Intervention periods	1. Less than 1 week
		2. Less than 5 weeks
		3. Less than 10 weeks
		4. More than 10 weeks

Table 4 The coding results of the relevant included studies

No.	Learning outcomes	Included studies
1	Listening competence	Behforouz and Ghaithi (2024), Chien et al. (2022), Dizon (2020), Han et al. (2008), Kim (2018b), Pasang et al. (2024)
2	Speaking competence	(Adipat (2023), Chien et al. (2022), Dizon (2020), Fathi et al. (2024), Goda et al. (2014), Hamuddin et al. (2020), Han (2020), Hsu et al. (2023a, 2023b), Hwang et al. (2022), Kim (2016), Kim et al. (2021), Lee et al. (2023a, 2023b), Lin and Mubarok (2021), Qiao and Zhao (2023), Shafiee Rad (2024), Yuan (2023)
3	Reading competence	(Adipat (2023), Chen and Lin (2023), Hong et al. (2016), Hwang and Nurtantyana (2022), Hyun et al. (2008), Kim (2018a, 2018b), Mageira et al. (2022), Vogt et al. (2020)
4	Writing competence	(Escalante et al. (2023), Hwang et al. (2022, 2023), Kim et al. (2020), Kwon et al. (2023), Liu et al. (2023), Rad et al. (2023), Song and Song (2023), Wei et al. (2023), Zhang et al. (2023a, 2023b)
5	Vocabulary knowledge	Alemi et al. (2014), Hsu et al. (2023a, 2023b), Jeon (2023), Kim (2018b), Levitskaia (2023), Polyzi and Moussiades (2023), Qasem et al. (2023), Zhang and Huang (2024)
6	Grammar proficiency	Abdollahzadeh et al. (2022), Ahn (2022), Chen and Lin (2023), Kim (2019, 2022), Lee et al. (2023a, 2023b), Nghi et al. (2019), Zou et al. (2023)
7	Learning motivation	Chien et al. (2022), Han (2020), Hong et al. (2016), Kim (2018b), Kim et al. (2020), Ryong et al. (2023), Wei (2023), Zhang et al. (2023a, 2023b), Zou et al. (2023)
8	Learning interest	Han (2020), Han et al. (2008), Kim (2018b), Kim et al. (2020), Lee et al. (2023a, 2023b)
9	Learning anxiety	Azizimajd (2023), Çakmak (2022), Chen and Lin (2023), Han (2020), Hawanti and Zubaydulloevna (2023), Kim et al. (2020), Lee et al. (2023a, 2023b)

3.3 *Data Analysis*

This study utilized Stata MP/14.0 to conduct the meta-analysis, which involved calculating Cohen's d as effect sizes, Q statistics, heterogeneity measures (I^2), p-values, weights of individual studies, 95% confidence intervals (CIs), and pooled effect sizes. Cohen's d was calculated by dividing the mean difference between the experimental and control groups by the standard deviation of the study results for both groups (Sedgwick & Marston, 2013). In total, 77 effect sizes were extracted from 55 identified studies, with a combined sample size of 5031. The calculation of the effect size involved three steps. Firstly, publication bias was assessed using the funnel plot, Begg's test, and Egger's test. The shape of the funnel plot was examined to preliminarily assess publication bias. Symmetrically distributed dots along the no-effect line indicate insignificant publication bias, while an asymmetric plot suggests the presence of significant publication bias.

Begg's and Egger's tests yield *p*-values to determine the presence or absence of publication bias. A *p*-value less than 0.05 suggests the presence of publication bias, while a *p*-value greater than 0.05 indicates its absence. In case of significant publication bias, the trim and fill technique in Stata should be applied, correcting the biased effect size to a more balanced one. Secondly, sensitivity analysis was conducted, and the heterogeneity of each effect size was calculated using the I^2 test. Heterogeneity, which is unavoidable in published literature, is categorized as follows: not important ($I^2 < 40\%$), moderate (I^2 between 30 and 60%), substantial (I^2 between 50 and 90%), and considerable (I^2 between 75 and 100%). Generally, heterogeneity is deemed significant when I^2 is larger than 50%, and the random effect model should be adopted (Higgins et al., 2003). Otherwise, heterogeneity is considered insignificant, and the fixed effect model should be used. Following this, effect sizes for each outcome domain were calculated and evaluated according to Cohen's effect-size criteria, which classified 0.2, 0.5, and 0.8 as small, medium, and large effect sizes respectively (Cohen, 1988).

4 Results

Section 4 presented and analyzed the results of the collected data of 55 studies, including the results of publication bias, results of sensitivity analysis, effect sizes calculation of each learning outcome domain and three moderator variables. By analyzing the *p* value, it testified the above 13 null hypotheses to estimate the overall effect of the use of AI chatbots on English learning.

4.1 Results of Publication Bias

As depicted in Table 4, the results of both tests suggested the absence of publication bias in the overall and specific outcome domains, except for the presence of publication bias in speaking competence according to Egger's test results ($t = 7.61, p = 0.042$). This result demonstrated the necessity of the effect size correction through the trim and fill technique. Meanwhile, the funnel plot for overall outcome domains displayed a relatively symmetrical distribution along both sides of the no-effect line, indicating that the results were relatively robust and less susceptible to publication bias (Fig. 4). Therefore, it was likely that there was no significant publication bias in the included studies (Table 5).

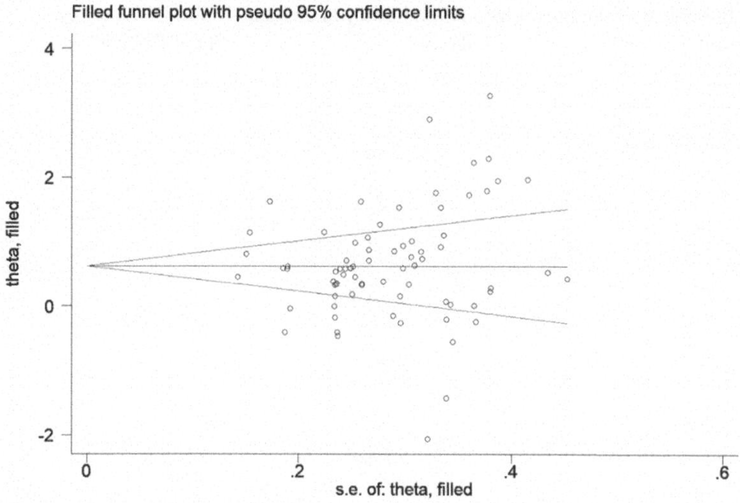

Fig. 4 Funnel plot of overall outcome domains

Table 5 Results of publication bias

Outcome	n	Egger's test		Begg's test				Continuity corrected	
		Bias	*p*	Score	sd	*z*	*p*	*z*	*p*
Listening competence	6	7.46	0.111	9	5.32	1.69	0.091	1.50	0.133
Speaking competence	16	7.61	0.042	26	22.21	1.17	0.242	1.13	0.260
Reading competence	8	2.66	0.559	−2	8.08	−0.25	0.805	0.12	0.902
Writing competence	10	4.77	0.091	−3	11.18	−0.27	0.788	0.18	0.858
Vocabulary learning	8	−0.18	0.933	12	8.08	1.48	0.138	1.36	0.174
Grammar learning	8	−0.90	0.615	−4	8.08	−0.49	0.621	0.37	0.711
Learning motivation	9	3.75	0.408	6	9.59	0.63	0.532	0.52	0.602
Learning interest	5	7.56	0.232	6	4.08	1.47	0.142	1.22	0.221
Learning anxiety	7	3.05	0.068	−5	6.66	−0.75	0.453	0.60	0.548
Overall outcomes	77	1.25	0.447	361	218.61	1.65	0.099	1.65	0.100

Note n: number of studies; sd: standard deviation

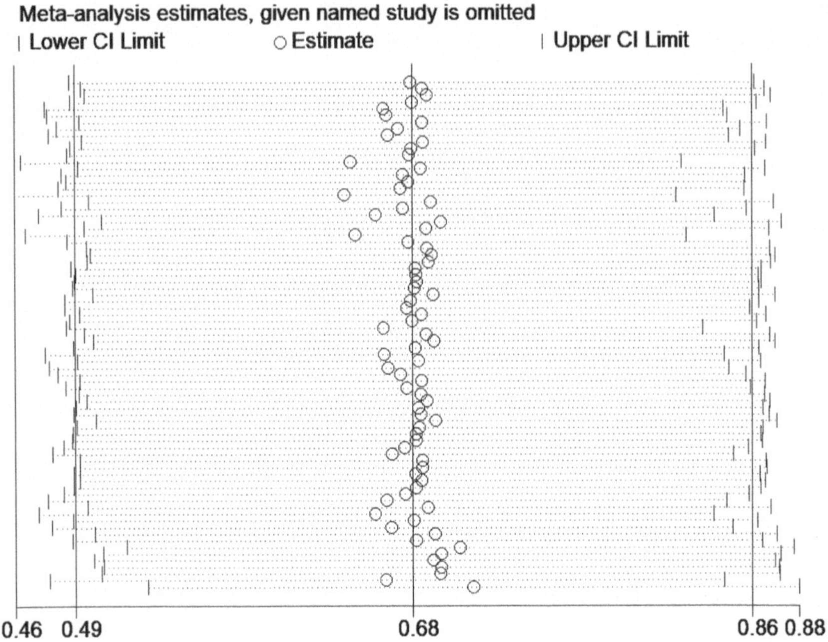

Fig. 5 Results of sensitivity analysis

4.2 *Results of Sensitivity Analysis*

The authors conducted a sensitivity analysis using Stata 14.0 to test the reliability of the pooled effect size. Through this analysis, it evaluated whether the exclusion of any single study could significantly impact the pooled result. As illustrated in Fig. 5, all estimates fell within the range from the lower confidence interval limit (2.5% CI = 0.49) to the upper confidence interval limit (97.5% CI = 0.86). This observation suggested that the exclusion of any single study did not substantially alter the pooled effect size. Therefore, it could be concluded that the meta-analytical results were reliable and not unduly influenced by any individual study.

4.3 *Results of Effect Sizes Calculation*

Table 6 presented the pooled effect sizes of each outcome domain and the overall results. Notably, the analysis revealed considerable heterogeneity in overall English language skills ($Q = 417.74$) with an I^2 value of 86.8% at $p < 0.001$. Therefore, the authors employed the random effect model for the meta-analysis of overall language skills. As indicated in Table 6, a significant difference was observed in the domain of English language skills between the experimental and control groups ($d = 0.786$,

95% CI = [0.588, 0.983], $z = 7.79, p < 0.001$). This finding suggested that English skills in the chatbot-assisted group were significantly higher than control group. Therefore, H1 was rejected.

Given the statistics indicating considerable heterogeneity in listening competence ($Q = 25.30$) with an I^2 value of 80.2% and speaking competence ($Q = 148.91$) with an I^2 value of 89.9% at $p < 0.001$, the authors adopted the random effect model for conducting the meta-analysis related to listening competence and speaking competence. Due to the presence of publication bias in speaking competence, the trim and fill method was applied, and the pooled effect size of speaking competence was corrected to $d = 1.118$, with 95% CI = [0.668, 1.568]. Despite the significant publication bias, the effect of chatbots on speaking competence was significantly positive after the correction. As depicted in Table 6, significant differences were observed in the domain of listening competence ($d = 0.906$, 95% CI = [0.338, 1.473], $z = 3.13, p < 0.001$) and speaking competence ($z = 4.87, p < 0.001$) between the experimental and control groups. This suggested that listening competence and speaking competence in the chatbot-assisted group were significantly higher than the control group. Consequently, hypotheses 2 and 3 were rejected.

Table 6 The effect sizes of learning outcome domains

Outcome domains	Effect size				Heterogeneity test			
	d	95% CI	*z*	*p*	*Q*-value	df	*p*	I^2 (%)
English skills	0.786	0.588, 0.983	7.79	<0.001	417.74	55	<0.001	86.8
Listening competence	0.906	0.338, 1.473	3.13	0.002	25.30	5	<0.001	80.2
Speaking competence	1.118	0.668, 1.568	4.87	<0.001	148.91	15	<0.001	89.9
Reading competence	0.631	0.017, 1.245	2.01	0.044	70.38	7	<0.001	90.1
Writing competence	0.555	−0.019, 1.129	1.90	0.058	104.69	9	<0.001	91.4
Vocabulary knowledge	0.910	0.527, 1.293	4.65	<0.001	24.52	7	0.001	71.4
Grammar proficiency	0.458	0.302, 0.615	5.73	<0.001	11.21	7	0.130	37.5
Learning motivation	0.702	0.361, 1.042	4.04	<0.001	27.84	8	0.001	71.3
Learning interest	0.937	0.197, 1.678	2.48	0.013	40.00	4	<0.001	90.0
Learning anxiety	−0.463	−1.391, 0.465	0.98	0.328	158.55	6	<0.001	96.2
Overall	0.676	0.493, 0.859	7.23	<0.001	705.29	76	<0.001	89.2

Given the statistics indicating considerable heterogeneity in reading competence, with $I^2 = 90.1\%$ ($Q = 70.38$) at $p < 0.001$, the authors decided to use the random effect model for conducting the meta-analysis regarding reading competence. As indicated in Table 6, significant differences were observed in the domain of reading competence between the experimental and control groups ($d = 0.631$, 95% CI = [0.017, 1.245], $z = 2.01$, $p = 0.044$). This suggested that reading competence in the chatbot-assisted group was significantly higher than control group. Consequently, hypothesis 4 was rejected.

Meanwhile, given that the statistics indicated considerable heterogeneity in writing competence, with $I^2 = 91.4\%$ ($Q = 104.69$) at $p < 0.001$, the authors opted to apply the random effect model for conducting the meta-analysis regarding writing competence. As depicted in Table 6, no significant differences were observed in the domain of writing competence between the experimental and control groups ($d = 0.555$, 95% CI = [−0.019, 1.129], $z = 1.90$, $p > 0.05$). This suggested that writing competence in the chatbot-assisted group was not significantly higher than control group. Consequently, hypothesis 5 was accepted.

Given that the statistics indicated considerable heterogeneity in vocabulary knowledge with $I^2 = 71.4\%$ ($Q = 24.52$) and grammar proficiency with $I^2 = 37.5\%$ ($Q = 11.21$) at $p < 0.05$, the authors opted to apply the random effect model for conducting the meta-analysis regarding vocabulary knowledge and grammar proficiency. As depicted in Table 6, significant differences were observed in the domain of vocabulary knowledge ($d = 0.910$, 95% CI = [0.527, 1.293], $z = 4.65$, $p = 0.001$) and grammar proficiency ($d = 0.458$, 95% CI = [0.302, 0.615], $z = 5.73$, $p < 0.001$) between the experimental and control groups. This suggested that vocabulary knowledge and grammar proficiency in the chatbot-assisted group were significantly higher than control group. Consequently, hypotheses 6 and 7 were rejected.

Given that the statistics indicated considerable heterogeneity in English learning motivation with $I^2 = 71.3\%$ ($Q = 27.84$) and English learning interest with $I^2 = 90.0\%$ ($Q = 40.00$) at $p < 0.001$, the authors opted to apply the random effect model for conducting the meta-analysis regarding learning motivation and interest. As depicted in Table 6, significant differences were observed in the domain of learning motivation ($d = 0.702$, 95% CI = [0.361, 1.042], $z = 4.04$, $p < 0.001$) and learning interest ($d = 0.937$, 95% CI = [0.197, 1.678], $z = 2.48$, $p = 0.013$) between the experimental and control groups. This suggested that learning motivation and interest in the chatbot-assisted group were significantly higher than the control group. Consequently, hypotheses 8 and 9 were rejected.

Given that the statistics indicated considerable heterogeneity in English learning anxiety, with $I^2 = 96.2\%$ ($Q = 158.55$) at $p < 0.001$, the authors opted to apply the random effect model for conducting the meta-analysis regarding learning anxiety. As depicted in Table 6, no significant differences were observed in the domain of learning anxiety between the experimental and control groups ($d = -0.463$, 95% CI = [−1.391, 0.465], $z = 0.98$, $p = 0.328$). This suggested that learning anxiety in the chatbot-assisted group was not significantly different from the control group. Consequently, hypothesis 10 was accepted.

Taking both cognitive and affective domains into account, the overall effect size of using AI chatbots in EFL education was 0.676 ($p < 0.001$), with a 95% confidence interval between 0.493 and 0.859, suggesting the overall effect size in this study was categorized as medium-to-large. This indicated that the application of AI chatbots in the EFL learning context had a relatively significant positive impact on learning outcomes. Furthermore, the heterogeneity test revealed that the overall effect size ($Q = 705.29$, $I^2 = 89.2\%$, $p < 0.001$) exhibited considerable heterogeneity, suggesting the need for further analysis to explore other potential moderator variables.

A meta-regression analysis was conducted for the measurement of the three moderating variables presented in Table 7. Concerning intervention duration, medium-to-large-sized effects were observed on chatbot-assisted learning for durations less than 5 weeks ($d = 0.510$, $p < 0.001$), with large-sized effects for durations less than one week ($d = 0.942$, $p < 0.001$) and more than 10 weeks ($d = 0.837$, $p < 0.001$). Small-to-medium-sized effect was found in less than 10 weeks ($d = 0.474$, $p < 0.05$). However, no significant differences were found between the effect sizes of the different intervention durations ($p = 0.861 > 0.05$). Thus, hypothesis 10 was accepted.

Concerning proficiency levels, medium-to-large-sized effects were observed across the three levels: beginning level ($d = 0.685$, $p < 0.001$), intermediate level ($d = 0.661$, $p < 0.001$), and advanced level ($d = 0.749$, $p < 0.01$). However, no significant differences were found between the effect sizes of the different English proficiency levels ($p = 0.797 > 0.05$). In terms of the educational levels of participants, large effect sizes were observed in primary education level ($d = 1.015$, $p < 0.001$) and working adults ($d = 1.345$, $p < 0.01$). Medium-to-large-sized effect was observed in higher education levels ($d = 0.630$, $p < 0.001$). The effect sizes for preschool education level ($d = -0.028$, $p > 0.05$) and secondary educational level ($d = 0.360$, $p > 0.05$) were considered small. Nonetheless, no significant differences were found between the effect sizes of different educational levels ($p = 0.967 > 0.05$). Thus, hypotheses 11 and 12 were accepted. The results of all null hypotheses were listed in Table 8.

5 Discussion

Based on the above analysis of the effect sizes of each learning outcome domains and moderator variables, Sect. 5 further discussed the effects of chatbot use on English learning and the effects of moderating variables on chatbot-assisted language learning, providing the theoretical and practical contributions of the findings of this study.

Table 7 The effect sizes of moderator variables

No.	Moderator variable	Effect size					Heterogeneity test			
		N	*d*	95% CI	*z*	*p*	Coefficient	95% CI	*t*	*p*
1	Intervention duration	77					−0.005	−0.064, 0.054	−0.18	0.861
	Less than 1 week	14	0.942	0.513, 1.372	4.30	<0.001				
	Less than 5 weeks	23	0.510	0.241, 0.760	3.78	<0.001				
	Less than 10 weeks	19	0.474	0.020, 0.928	2.05	0.041				
	More than 10 weeks	19	0.837	0.444, 1.229	4.18	<0.001				
	Not specified	2	1.293	0.613, 0.941	2.86	0.004				
2	Educational levels	77					0.046	−0.306, −0.395	0.26	0.797
	Preschool	2	−0.028	−0.357, 0.301	0.17	0.867				
	Primary school	10	1.015	0.673, 1.357	5.81	< 0.001				
	Secondary education	11	0.360	−0.026, 0.746	1.83	0.068				
	Higher education	48	0.630	0.384, 0.875	5.03	< 0.001				
	Working adults	4	1.343	0.141, 2.545	2.19	0.029				
3	Proficiency levels	77					0.010	−0.461, 0.480	0.04	0.967
	Beginning level	19	0.685	0.355, 1.014	4.07	<0.001				
	Intermediate level	45	0.661	0.417, 0.905	5.31	<0.001				
	Advanced level	13	0.719	0.209, 1.228	2.77	0.006				

5.1 The Effects of Chatbot Use on English Skills Improvements

The results of H1–H7 indicate that the use of AI-based chatbots can significantly improve overall English skills and specific language skills, except for English writing competence. These findings are consistent with previous research that highlighted the positive effects of chatbots on language skill enhancement (Hamuddin et al., 2020; Kim, 2018b, 2019; Wei et al., 2023). The results demonstrate practical importance

Table 8 The results of hypothesis testing

No.	Null hypotheses	Results
1	The use of AI chatbots could not significantly improve overall language skills at the 0.05 level	Rejected
2	The use of AI chatbots could not significantly improve listening competence at the 0.05 level	Rejected
3	The use of AI chatbots could not significantly improve speaking competence at the 0.05 level	Rejected
4	The use of AI chatbots could not significantly improve reading competence at the 0.05 level	Rejected
5	The use of AI chatbots could not significantly improve writing competence at the 0.05 level	Accepted
6	The use of AI chatbots could not significantly improve vocabulary knowledge at the 0.05 level	Rejected
7	The use of AI chatbots could not significantly improve grammar proficiency at the 0.05 level	Rejected
8	The use of AI chatbots could not significantly improve learning motivation at the 0.05 level	Rejected
9	The use of AI chatbots could not significantly improve learning interest at the 0.05 level	Rejected
10	The use of AI chatbots could not significantly reduce learning anxiety at the 0.05 level	Accepted
11	The intervention duration could not influence the effect of chatbot-assisted English learning	Accepted
12	The proficiency levels could not influence the effect of chatbot-assisted English learning	Accepted
13	The educational levels could not influence the effect of chatbot-assisted English learning	Accepted

of integrating AI chatbots into English reading, listening, speaking, vocabulary, and grammar classes. The beneficial effects of chatbots can be attributed to two major reasons. Firstly, chatbot-assisted language learning could provide EFL learners with rich authentic language contexts through human-chatbot interactions (Hsu et al., 2023a, 2023b). In contrast to traditional methods, which offered learners limited exposure to authentic language contexts, chatbots could engage in various social roles and provide natural communication for EFL learners by simulating human-to-human conversations (Yuan, 2023). Secondly, chatbots supported by artificial intelligence were able to offer personalized tutoring and learning materials tailored to individual students' needs and preferences (Escalante et al., 2023). By providing targeted support and feedback, chatbots were capable of addressing specific learning gaps and challenges, leading to enhanced comprehension and mastery of difficult concepts (Wambsganss et al., 2021). Additionally, chatbots facilitated language acquisition and knowledge consolidation by employing spaced review language learning activities and effective learning strategies, such as associating acquired knowledge with familiar concepts to reinforce learning over time (Alemi et al., 2014; Fazlollahi et al., 2022).

Despite the benefits that chatbots can offer, insignificant differences are found in English writing skills between the chatbot group and control group. This can be partially attributed to the poor performance of chatbot in checking deep logical cohesion which is a key element to composing a good essay. Chatbot with large language model, i.e., ChatGPT, performed better than EFL learners' writing in narrativity, word concreteness, and referential cohesion, but worse in deep cohesion. Even after updating its commands, it still lagged far behind the EFL learners' writing in deep cohesion (Zhou et al., 2023). Moreover, students were more favorable to teachers' feedback as the affordances of sitting down face-to-face and interacting with a human were more engaging and interesting compared with just reading through AI-generated feedback (Zhang et al., 2023a, 2023b).

Thus, from the perspective of division of labor in the proposed AT framework, this study suggests a combination of both teachers and chatbots in writing class may be more beneficial for students (Song & Song, 2023). In the pre-writing stage, the chatbot had limitations in providing input corpus, which required teachers to be evaluators of the text generated by chatbot. In the writing process, chatbot could replace teachers and serve as tools of online dictionary or encyclopedia for each student. In the post-writing stage, chatbots could help to provide feedback on the language level such as grammar or structure organization which saved time for teachers to offer more detailed comments on the cohesion and logic level for each learner, benefitting students by optimizing the strength of feedback from both sides (Kim et al., 2020).

5.2 *The Effects of Chatbot Use on Learning Motivation, Interest, and Anxiety*

Regarding affective learning outcomes, significant differences in EFL learning motivation are found between the chatbot group and the control group, corresponding to previous studies (Jia & Chen, 2008; Wei, 2023). This can be attributed to the social features of such technology. Chatbots played the role of supportive and patient friends in learning and could develop social connections with learners to improve their affective states in language learning (Zhang et al., 2023a, 2023b). Most participants perceived a strong sense of freshness and fun with chatbots as English learning partners, leading to high levels of concentration and excitement in language learning for extended periods (Zou et al., 2023). Such a state resembles flow, an experience of deep engagement in an activity. When experiencing flow in learning, students concentrated on the tasks at hand and obtained high motivation (Csikszentmihalyi, 2014). Moreover, chatbot-based learning is advantageous for supporting study plan creation. Chatbots could clearly explain the purposes of various learning resources and the relationships between different learning tasks, which motivated learners to proactively arrange the types, amount, and order of learning tasks according to individual needs (Ryong et al., 2023).

The results reveal that EFL students learning with chatbots experienced more robust learning interest than those without, highlighting chatbots' importance in stimulating curiosity and enthusiasm for learning (Han, 2020; Hwang et al., 2022; Kim, 2018a; Lee et al., 2023a, 2023b). The enhancement of learning interest can be partially attributed to the flexibility of learning and affective feedback (Fryer et al., 2017). The chatbot asked learners about their learning progress from time to time and provided encouragement and suggestions according to learners' responses with human-like utterances (Hwang et al., 2022). However, the study conducted by Kim et al. (2020) showed the opposite results, indicating that the use of AI chatbots did not significantly increase EFL learners' interest in English writing. This discrepancy might be caused by irrelevant responses or repetitive practices provided by the chatbot used in the experiment, which could detract from EFL students' ability to regulate their learning. Thus, this study offers new suggestions for technology developers to further refine the interaction model of chatbots with an affluent database to have natural communication with humans and to direct their educational purpose toward foreign language learning.

The absence of notable differences in learning anxiety between the chatbot group and non-chatbot group is an intriguing finding, especially considering the above-observed improvement in language skills among EFL learners using chatbots. Although many previous studies have suggested that the application of chatbots could reduce anxiety by creating positive affective factors (Azizimajd, 2023; Chen & Lin, 2023; Hawanti & Zubaydulloevna, 2023), it's important to consider that increased anxiety may play a facilitative role in maximizing students' cognitive and linguistic potential (Shazly, 2021). Additionally, frustration stemming from communication breakdowns due to linguistic factors during learner interactions with chatbots might contribute to increased anxiety levels (Çakmak, 2022). Achieving a positive emotional state among second language (L2) learners often resulted from successful and meaningful conversational experiences characterized by minimized communication breakdowns and enhanced understanding, which played a vital role in managing L2 speaking anxiety (Chen & Lin, 2023). Therefore, advanced and fully developed systems of dialogue-based conversational chatbots hold promise for aiding in overcoming this anxiety.

Generally, despite insignificant influences observed in writing competence and learning anxiety, the overall positive impact of chatbot-assisted language learning on EFL outcomes is noteworthy. The findings suggested that chatbots could effectively enhance overall language skills, learning motivation, and interest among EFL learners. Meanwhile, insignificant differences found in the aspect of English writing skills and learning anxiety may mainly be attributed to the technological limitations. For example, researchers found that chatbots were unable to provide effective feedback on more logical cohesion in writing tasks (Zhou et al., 2023) and failed to create a more natural and interactive conversation to reduce learners' anxiety during communication (Çakmak, 2022). Apart from the suggestions for chatbot technology update, it was also found that the incorporation of chatbots with teaching methods, such as a project-based learning approach (Azamatova et al., 2023) or mind-map guided approach (Lin & Mubarok, 2021), could lead to a significant difference in the quality and efficiency of the language teaching process in AI-supported flipped classrooms in language learning, offering new suggestions for improvement of the effectiveness of chatbots in EFL educational context.

5.3 The Effects of Moderating Variables on Chatbot-Assisted Language Learning

Regarding the three moderators, the outcomes indicate that the efficacy of chatbot-supported language learning in EFL settings remains unaffected by the length of the intervention. This implies that chatbots do not exhibit a fleeting novelty impact, wherein initial enhancements in learning outcomes may subsequently diminish over time (Fryer et al., 2017). The results were aligned with prior studies that also observed no notable significance in learning outcomes regarding intervention duration (Kim, 2018a; Qasem et al., 2023; Wei et al., 2023). One potential rationale for this observation is that pre-exposure to and instruction with chatbot technology before the experiment may help alleviate the novelty effect. By familiarizing students with chatbots and allowing them to adapt to the technology, researchers could diminish the novelty's influence on learning outcomes (Han, 2020; Lee et al., 2023a, 2023b). Additionally, the findings of previous studies suggested that the quality and depth of interaction with chatbots might outweigh the duration of the intervention (Jeon, 2023; Pasang et al., 2024; Ryong et al., 2023). In summary, the results challenge the novelty effect and the presumption that longer interventions inevitably yield greater benefits in chatbot-assisted language learning. Instead, researchers and educators should concentrate on optimizing interaction quality and ensuring that chatbot interventions are engaging and efficacious for EFL learners.

No significant differences emerge among groups with varying English proficiency levels or educational backgrounds, corresponding to prior research findings (Kim, 2016; Wei, 2023). Chatbots have been proven equally suitable for both novice and advanced EFL learners, offering ample opportunities for language practice and exerting a positive influence on their affective domains, thereby enhancing overall language performance. Lee and Lee (2022) conducted a meta-analysis revealing that participants at the preschool or primary school levels might derive greater benefits compared with adult learners, but the effectiveness of chatbot-assisted learning for secondary-level learners remained unmeasured in their study. However, this meta-regression analysis spanning various education levels, revealed no significant discrepancy in learning outcomes across different educational levels within a chatbot-assisted learning environment. The results confirm that chatbots hold promise for implementation across diverse educational levels regardless of English proficiency levels.

5.4 *The Theoretical and Practical Implications of the Findings of This Study*

The findings of this study offer valuable insights for future researchers, educators, and technology developers in EFL education. For theoretical contributions, this study expands the existing theoretical framework of AT in CALL to encompass additional components of EFL learning outcomes, particularly in affective learning outcome domains such as learning confidence, satisfaction, self-efficacy, and self-regulation. Future experimental studies with large sample sizes are also expected to further validate these findings. Moreover, the results only included three moderators (intervention duration, educational levels, and proficiency levels) that did not significantly influence the effectiveness of chatbot-assisted language learning in the EFL context. Thus, researchers should consider exploring additional moderating variables for future meta-analyses. These can include interaction types within the community (one-on-one or in groups), chatbot roles (e.g., tutor, assistant, or peer), and chatbot types (text-based or voice-based). It will be beneficial to investigate the effects of various types of verbal immediacy on chatbots serving as teaching assistants, peers, or tutors, as these variables may enhance both cognitive and affective learning outcomes for language learners (Randall, 2020). By addressing these suggestions in future experimental research, it can help to develop a more comprehensive framework for understanding the effectiveness of chatbot-assisted learning in the EFL context, as illustrated in Fig. 6.

This study also provides some valuable pedagogical insights into the integration of chatbots within language classrooms and proposes suggestions for further enhancement of chatbot functionalities. To begin with, since this study has confirmed the positive effects of the use of chatbots in most English language skills, learning motivation, and interest, EFL teachers can enrich their instructional practices by integrating tailored chatbot features with effective teaching methods or activities, thereby creating a more personalized and adaptive learning experience for students (Pasang et al., 2024). To achieve this, artificial intelligence literacy should be incorporated into teacher education so that educators can have a deep understanding of the various characteristics of chatbots and become adept at allocating learners' attentional resources effectively to facilitate pedagogical applications of AI chatbots (Pretorius, 2023). Moreover, to address the challenges that no significant differences were found in the effectiveness of chatbots on English writing skills and learning anxiety, educators can incorporate chatbots with traditional or other effective teaching methods in English writing classes; moreover, technology developers are responsible for developing more sophisticated and innovative chatbot with Large Language Models (LLMs) and AI-based algorithms to emulate human social skills more effectively and to enhance the quality and efficiency of chatbots' communication and feedback (Shafiee Rad, 2024).

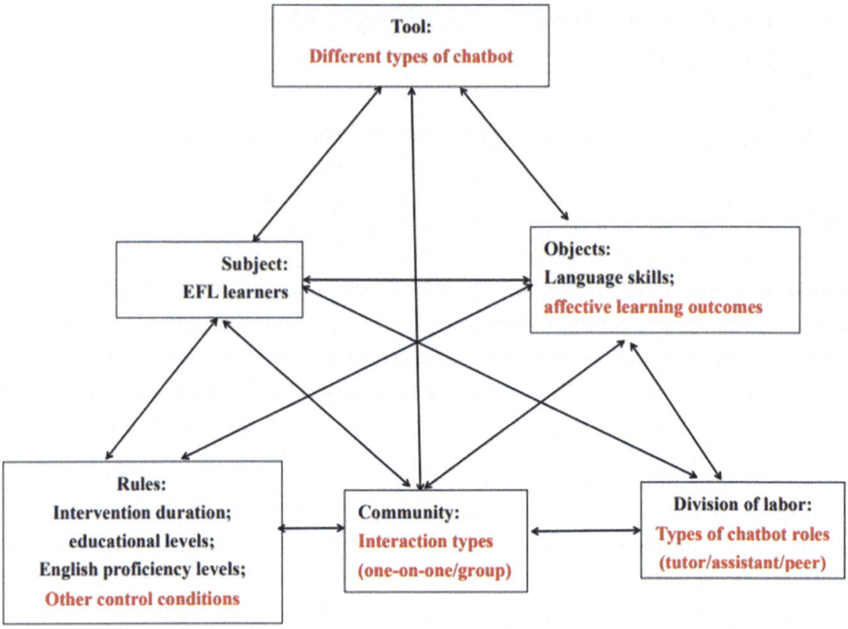

Fig. 6 The enrichment of AT theoretical framework for future studies

6 Conclusion

Through the meta-analysis, it observed a medium-to-large overall effect of AI-based chatbots on EFL learning outcomes, regardless of intervention duration, English proficiency, and educational levels. Chatbot usage significantly and positively impacted overall English language skills, specific language skills, learning motivation, and learning interest, suggesting that it was of practical importance to integrate chatbots into English reading, listening, speaking, vocabulary and grammar classes. But no significant differences were found in the aspects of English writing skills and learning anxiety, indicating further improvement of chatbots' conversation system. These findings may offer valuable insights for instructors seeking to enhance English education through appropriate teaching approaches, suggesting that future language education should integrate conversational AI to promote intelligence amplification and decrease human teachers' workload through classroom orchestration.

However, several limitations should also be acknowledged. Firstly, the inclusion criteria limited the studies to those written in English, potentially overlooking valuable research in other languages. Secondly, the final analysis results might be influenced by the exclusion of some studies that have no access to the full texts. Thirdly, this analysis only considered three effective learning outcome domains (learning motivation, learning interest, and anxiety), which might fail to reach a comprehensive evaluation of the effects of chatbots in the EFL learning context. Despite these limitations, the data collected in this meta-analysis were extracted from peer-reviewed experimental studies related to the research topic strictly based on the PRISMA statement, and the recently published literature was retrieved from as many databases as possible, which helped minimize publication bias in the results. Meanwhile, the sensitivity analysis results also confirmed that the results of this study were relatively robust and credible. Therefore, the synthesized results were representative, reliable, and generalizable to other contexts, which could facilitate future research and technology-enhanced language education practice.

Acknowledgements This work is supported by Key Research and Application Project of the Key Laboratory of Key Technologies for Localization Language Services of the State Administration of Press and Publication, "Research on Localization and Intelligent Language Education Technology for the 'Belt and Road Initiative'" (Project Number: CSLS 20230012), and Special fund of Beijing Co-construction Project-Research and reform of the "Undergraduate Teaching Reform and Innovation Project" of Beijing higher education in 2020-innovative "multilingual+" excellent talent training system (202010032003).

Ethics Approval Statement The study was approved by the institutional review board of Beijing Language and Culture University. All researchers can provide written informed consents.

Appendix

Table 9 The basic information of the included studies

No.	Author (year)	Sample size (EG/CG)	Treatment period	Educational level of samples	English proficiency levels	Learning outcomes
1	Kim (2018a, 2018b)	46 (24/22)	16 weeks	Higher education	Intermediate level	Listening and reading competence
2	Kim (2019)	70 (36/34)	16 weeks	Higher education	Advanced level	Grammar learning
3	Yuan (2023)	74 (39/35)	12 weeks	Primary school	Beginning level	Speaking competence
4	Wei (2023)	60 (30/30)	10 weeks	Higher education	Intermediate level	L2 motivation
5	Goda et al. (2014)	63 (31/32)	40 min	Higher education	Intermediate level	Speaking competence
6	Kim (2018a, 2018b)	57 (24/23)	8 weeks	Higher education	Intermediate level	Vocabulary learning, motivation and interest
7	Polyzi and Moussiades (2023)	20 (10/10)	2 weeks	Higher education	Advanced level	Vocabulary learning
8	Dizon (2020)	28 (13/15)	10 weeks	Higher education	Intermediate level	Listening and speaking competence
9	Hwang et al. (2022)	43 (23/20)	10 weeks	Primary school	Beginning level	Speaking competence
10	Kim et al. (2021)	74 (36/38)	16 weeks	Higher education	Intermediate level	Speaking competence
11	Chien et al. (2022)	73 (37/36)	4 weeks	Junior and senior high school	Intermediate level	Listening, speaking competence and motivation
12	Han (2020)	44 (22/22)	10 weeks	Junior and senior high school	Beginning level	Speaking competence, interest and motivation
13	Shafiee Rad (2024)	66 (33/33)	10 weeks	Higher education	Intermediate level	Speaking competence
14	Kwon et al. (2023)	75 (37/38)	15 weeks	Primary school	Beginning level	Writing competence
15	Kim (2017)	39 (20/19)	16 weeks	Higher education	Beginning level	Speaking competence
		41 (22/19)			Intermediate level	

(continued)

Table 9 (continued)

No.	Author (year)	Sample size (EG/CG)	Treatment period	Educational level of samples	English proficiency levels	Learning outcomes
		43 (21/22)			Advanced level	
16	Lee (2023)	121 (60/61)	120 min	Primary school	Beginning level	L2 learning interest
17	Nghi et al. (2019)	200 (100/100)	450 min	Higher education	Intermediate level	Grammar learning
18	Joen (2023)	35 (18/17)	50 min	Primary school	Beginning level	Vocabulary learning
19	Ahn (2022)	46 (23/23)	6 weeks	Higher education	Intermediate level	Grammar learning
20	Hsu et al. (2021)	48 (24/24)	16 weeks	Higher education	Intermediate level	Speaking competence
21	Hwang et al. (2023)	66 (33/33)	7 weeks	Higher education	Intermediate level	Writing competence
22	Ryong et al. (2023)	89 (55/24)	50 min	Higher education	Intermediate level	L2 learning motivation
23	Mageira et al. (2022)	35 (18/17)	48 h	Junior and senior high school	Beginning level	Reading competence
24	Escalante et al. (2023)	48 (23/25)	6 weeks	Working adults	Beginning level	Writing competence
25	Junaidi et al. (2020)	62 (32/30)	4 weeks	Junior and Senior high school	Beginning level	Speaking competence
26	Qiao and Zhao (2023)	93 (47/46)	13 weeks	Working adults	Intermediate level	Speaking competence
27	Kim (2022)	64 (32/32)	5 weeks	Senior and junior high school	Intermediate level	Grammar learning
28	Wei et al. (2023)	190 (95/95)	12 weeks	Higher education	Intermediate level	Writing competence
29	Adipat (2023)	80 (40/40)	55 h	Working adults	Advanced level	Speaking and reading competence
30	Song and Song (2023)	50 (25/25)	12 weeks	Higher education	Intermediate level	Writing competence
31	Kim et al. (2020)	73 (35/38)	A semester	Higher education	Intermediate level	Writing competence; motivation
32	Rad et al. (2023)	46 (23/23)	11 weeks	Higher education	Intermediate level	Writing competence

(continued)

Table 9 (continued)

No.	Author (year)	Sample size (EG/CG)	Treatment period	Educational level of samples	English proficiency levels	Learning outcomes
33	Alemi et al. (2014)	46 (30/16)	5 weeks	Primary school	Beginning level	Vocabulary learning
34	Hwang et al. (2022)	71 (33/38)	6 weeks	Higher education	Advanced level	Writing competence
35	Liu et al. (2023)	64 (33/31)	4 weeks	Higher education	Intermediate level	Writing competence
36	Fathi et al. (2024)	65 (33/32)	2 weeks	Working adults	Intermediate level	Speaking competence
37	Hwang and Nurtantyana (2022)	22 (11/11)	110 min	Higher education	Intermediate level	Reading competence
38	Levitskaia (2023)	195 (97/98)	A semester	Higher education	Intermediate level	Vocabulary learning
39	Zou et al. (2023)	70 (35/35)	5 weeks	Higher education	Intermediate level	Grammar learning; motivation
40	Lin and Mubarok (2021)	50 (22/28)	5 weeks	Higher education	Advanced level	Speaking competence
41	Behforouz and Ghaithi (2024)	40 (20/20)	5 weeks	Higher education	Intermediate level	Listening competence
42	Hsu et al. (2023a, 2023b)	60 (30/30)	2 months	Higher education	Intermediate level	Vocabulary learning
43	Lee et al. (2023a, 2023b)	35 (18/17)	2 weeks	Higher education	Intermediate level	Speaking competence; Grammar learning; anxiety
44	Pasang et al. (2024)	40 (20/20)	240 min	Junior and Senior high school	Intermediate level	Listening competence
45	Qasem et al. (2023)	40 (20/20)	12 weeks	Higher education	Intermediate level	Vocabulary learning
46	Zhang and Huang (2024)	52 (26/26)	8 weeks	Junior and Senior high school	Intermediate level	Vocabulary learning
47	Chen & Lin (2023)	116 (58/58)	4 weeks	Higher education	Intermediate level	Grammar learning; reading competence; anxiety
48	Abdollahzadeh et al. (2022)	50 (30/20)	5 weeks	Working adults	Intermediate level	Grammar learning

(continued)

Table 9 (continued)

No.	Author (year)	Sample size (EG/CG)	Treatment period	Educational level of samples	English proficiency levels	Learning outcomes
49	Zhang et al. (2023a, 2023b)	30 (15/15)	5 weeks	Higher education	Intermediate level	Writing competence; motivation
50	Hawanti and Zubaydulloevna (2023)	73 (36/37)	600 min	Higher education	Advanced level	Learning anxiety
51	Çakmak (2022)	178 (89/89)	12 weeks	Higher education	Intermediate level	Learning anxiety
52	Azizimajd (2023)	60 (30/30)	8 weeks	Higher education	Intermediate level	Learning anxiety
53	Vogt et al. (2020)	108 (54/54)	120 min	Primary school	Beginning level	Reading competence
54	Hyun et al. (2008)	34 (17/17)	4 weeks	Preschool	Beginning level	Reading competence
55	Hong et al. (2016)	52 (25/27)	Not specified	Primary school	Beginning level	Reading competence; motivation

References

Abdollahzadeh, M., Golshan, M., & Allami, H. (2022). Using a messenger bot as a tool for providing written corrective feedback: Examining L2 development and learners' attitudes. *Research in English Language Pedagogy, 10*(4), 599–624. https://doi.org/10.30486/relp.2022.1946281.1327

Adipat, S. (2023). An artificial intelligence-enhanced phenomenon-based learning approach for interdisciplinary understanding and speaking skills. *International Journal of Instruction, 16*(3), 531–550. https://doi.org/10.29333/iji.2023.16329a

Agarwal, S., Agarwal, B., & Gupta, R. (2022). Chatbots and virtual assistants: A bibliometric analysis. *Library Hi-Tech, 40*(4), 1013–1030. https://doi.org/10.1108/LHT-09-2021-0330

Ahn, S. (2022). The effects of chatbot on grammar competence for Korean EFL college students. *Journal of Digital Convergence, 20*(3), 53–61. https://doi.org/10.14400/JDC.2022.20.3.053

Alemi, M., Meghdari, A., & Ghazisaedy, M. (2014). Employing humanoid robots for teaching English language in Iranian junior high schools. *International Journal of Humanoid Robotics, 11*(3), 1450022–1450031. https://doi.org/10.1142/S0219843614500224

Azamatova, A., Bekeyeva, N., Zhaxylikova, K., Sarbassova, A., & Ilyassova, N. (2023). The effect of using artificial intelligence and digital learning tools based on project-based learning approach in foreign language teaching on students' success and motivation. *International Journal of Education in Mathematics, Science and Technology, 11*(6), 1458–1475. https://doi.org/10.46328/ijemst.3712

Azizimajd, H. (2023). Investigating the impacts of voice-based student-chatbot interactions in the classroom on EFL learners' oral fluency and foreign language speaking anxiety. *Technology Assisted Language Education, 1*(2), 61–83. https://doi.org/10.22126/TALE.2023.2732

Behforouz, B., & Ghaithi, A. (2024). The effectiveness of an interactive WhatsApp Bot on listening skills. *International Journal of Interactive Mobile Technologies, 18*(2), 82–95. https://doi.org/10.3991/ijim.v18i02.44327

Burns, A., & Siegel, J. (2018). Teaching the four language skills: Themes and issues. In A. Burns, & J. Siegel (Eds.), *International perspectives on teaching the four skills in ELT: Listening, speaking, reading, writing* (pp. 1–17). Springer International Publishing. https://doi.org/10.1007/978-3-319-63444-9_1

Çakmak, F. (2022). Chatbot-human interaction and its effects on EFL students' L2 speaking performance and anxiety. *Novitas-ROYAL, 16*(2), 113–131. http://files.eric.ed.gov/fulltext/EJ1365002.pdf

Chen, M. R. A., & Lin, Y. H. (2023). AI chatbot-supported grammar tense learning on EFL students learning effectiveness. In *INTED2023 proceedings* (pp. 1207–1216). https://doi.org/10.21125/inted.2023.0353

Chen, Z., Chen, W., Jia, J., & An, H. (2020). The effects of using mobile devices on language learning: A meta-analysis. *Educational Technology Research and Development, 68*(4), 1769–1789. https://doi.org/10.1007/s11423-020-09801-5

Chien, Y.-C., Wu, T.-T., Lai, C.-H., & Huang, Y.-M. (2022). Investigation of the influence of artificial intelligence markup language-based LINE chatbot in contextual English learning. *Frontiers in Psychology, 13*, 785752. https://doi.org/10.3389/fpsyg.2022.785752

Cohen, J. (1988). *Statistical power analysis for the behavioral sciences* (2nd ed.). Routledge. https://doi.org/10.4324/9780203771587

Csikszentmihalyi, M. (2014). *Flow and the foundations of positive psychology: The collected works of Mihaly Csikszentmihalyi.* Springer Netherlands. https://doi.org/10.1007/978-94-017-9088-8

Dash, B. (2022). Language skills: A study of improving English speaking skills through English reading skills. *International Journal of Technical & Non-Technical Research, 4*(4), 67–71. https://www.researchgate.net/publication/364673466

Dizon, G. (2020). Evaluating intelligent personal assistants for L2 listening and speaking development. *Language, Learning and Technology, 24*(1), 16–26. 10125/44705

Eisenring, M. A. A., Jamiluddin, J., Hairul, M. A., & Putri, D. (2024). The use of chatbots in the English teaching to promote modern language learning: A literature review. *International Journal of Indonesian Education and Teaching, 8*(1), 127–139. https://doi.org/10.24071/ijiet.v8i1.7321

Engeström, Y. (1999). Activity theory and individual and social transformation. In Y. Engeström, R. Miettinen, & R.-L. Punamäki (Eds.), *Perspectives on activity theory* (pp. 19–38). Cambridge University Press. https://doi.org/10.1017/CBO9780511812774.003

Escalante, J., Pack, A., & Barrett, A. (2023). AI-generated feedback on writing: Insights into efficacy and ENL student preference. *International Journal of Educational Technology in Higher Education, 20*(1), 57. https://doi.org/10.1186/s41239-023-00425-2

Fathi, J., Rahimi, M., & Derakhshan, A. (2024). Improving EFL learners' speaking skills and willingness to communicate via artificial intelligence-mediated interactions. *System, 121*(2), 103254. https://doi.org/10.1016/j.system.2024.103254

Fazlollahi, A. M., Bakhaidar, M., Alsayegh, A., Yilmaz, R., Winkler-Schwartz, A., Mirchi, N., Langleben, I., Ledwos, N., Sabbagh, A. J., Bajunaid, K., Harley, J. M., & Del Maestro, R. F. (2022). Effect of artificial intelligence tutoring vs expert instruction on learning simulated surgical skills among medical students: a randomized clinical trial. *JAMA Network Open, 5*(2), e2149008. https://doi.org/10.1001/jamanetworkopen.2021.49008

Fryer, L. K., Ainley, M., Thompson, A., Gibson, A., & Sherlock, Z. (2017). Stimulating and sustaining interest in a language course: An experimental comparison of chatbot and human task partners. *Computers in Human Behavior, 75*(1), 461–468. https://doi.org/10.1016/j.chb.2017.05.045

Goda, Y., Yamada, M., Matsukawa, H., Hata, K., & Yasunami, S. (2014). Conversation with a chatbot before an online EFL group discussion and the effects on critical thinking. *The Journal of Information and Systems in Education, 13*(1), 1–7. https://doi.org/10.12937/ejsise.13.1

Guo, K., Li, Y., Li, Y., & Chu, S. K. W. (2024). Understanding EFL students' chatbot-assisted argumentative writing: An activity theory perspective. *Education and Information Technologies, 29*(1), 1–20. https://doi.org/10.1007/s10639-023-12230-5

Hamuddin, B., Julita, K., & Rahman, F. (2020). Artificial intelligence in EFL context: Rising students' speaking performance. *International Journal of Advanced Science and Technology, 29*(5), 6735–6741. http://sersc.org/journals/index.php/IJAST/article/view/17726

Han, D.-E. (2020). The effects of voice-based AI chatbots on Korean EFL middle school students' speaking competence and affective domains. *Asia-Pacific Journal of Convergent Research Interchange, 6*(7), 71–80. https://doi.org/10.47116/apjcri.2020.07.07

Han, J., Jo, M., Jones, V., & Jo, J. (2008). Comparative study on the educational use of home robots for children. *JIPS, 4*(4), 159–168. https://doi.org/10.3745/JIPS.2008.4.4.159

Hawanti, S., & Zubaydulloevna, K. M. (2023). AI chatbot-based learning: Alleviating students' anxiety in English writing classroom. *Bulletin of Social Informatics Theory and Application, 7*(2), 182–192. https://doi.org/10.31763/businta.v7i2.659

Higgins, J. P., Thompson, S. G., Deeks, J. J., & Altman, D. G. (2003). Measuring inconsistency in meta-analyses. *BMJ (Clinical Research Ed.), 327*(7414), 557–560. https://doi.org/10.1136/bmj.327.7414.557

Hong, Z.-W., Huang, Y.-M., Hsu, M., & Shen, W.-W. (2016). Authoring robot-assisted instructional materials for improving learning performance and motivation in EFL classrooms. *Journal of Educational Technology & Society, 19*(1), 337–349. https://www.jstor.org/stable/jeductechsoci.19.1.337

Hsu, M.-H., Chen, P. S., & Yu, C. S. (2021). Proposing a task-oriented chatbot system for EFL learners speaking practice. *Interactive Learning Environments, 31*(7), 4297–4308. https://doi.org/10.1080/10494820.2021.1960864

Hsu, M.-H., Chan, T.-M., & Yu, C.-S. (2023a). Termbot: A chatbot-based crossword game for gamified medical terminology learning. *International Journal of Environmental Research and Public Health, 20*(5), 4185. https://doi.org/10.3390/ijerph20054185

Hsu, M.-H., Chen, P.-S., & Yu, C.-S. (2023b). Proposing a task-oriented chatbot system for EFL learners speaking practice. *Interactive Learning Environments, 31*(7), 4297–4308. https://doi.org/10.1080/10494820.2021.1960864

Hu, T.-C., Sung, Y.-T., Liang, H.-H., Chang, T.-J., & Chou, Y.-T. (2022). Relative roles of grammar knowledge and vocabulary in the reading comprehension of EFL elementary school learners: Direct, mediating, and form. *Frontiers in Psychology, 13*, 827007. https://doi.org/10.3389/fpsyg.2022.827007

Huang, W., Hew, K. F., & Fryer, L. K. (2022). Chatbots for language learning—Are they really useful? A systematic review of chatbot-supported language learning. *Journal of Computer Assisted Learning, 38*(1), 237–257. https://doi.org/10.1111/jcal.12610

Hwang, W.-Y., Guo, B.-C., Hoang, A., Chang, C.-C., & Wu, N.-T. (2022). Facilitating authentic contextual EFL speaking and conversation with smart mechanisms and investigating its influence on learning achievements. *Computer Assisted Language Learning*, 1–27. https://doi.org/10.1080/09588221.2022.2095406

Hwang, W.-Y., & Nurtantyana, R. (2022). X-education: Education of all things with AI and edge computing—One case study for EFL learning. *Sustainability, 14*(19), 12533. https://doi.org/10.3390/su141912533

Hwang, W.-Y., Nurtantyana, R., Purba, S. W. D., Hariyanti, U., Indrihapsari, Y., & Surjono, H. D. (2023). AI and recognition technologies to facilitate English as foreign language writing for supporting personalization and contextualization in authentic contexts. *Journal of Educational Computing Research, 61*(5), 1008–1035. https://doi.org/10.1177/07356331221137253

Hyun, E., Kim, S., Jang, S., & Park, S. (2008). Comparative study of effects of language instruction program using intelligence robot and multimedia on linguistic ability of young children. In *RO-MAN 2008—The 17th IEEE international symposium on robot and human interactive communication* (pp. 187–192). https://doi.org/10.1109/ROMAN.2008.4600664

Jeon, J. (2023). Chatbot-assisted dynamic assessment (CA-DA) for L2 vocabulary learning and diagnosis. *Computer Assisted Language Learning, 36*(7), 1338–1364. https://doi.org/10.1080/09588221.2021.1987272

Jeon, J., Lee, S., & Choi, S. (2023). A systematic review of research on speech-recognition chatbots for language learning: Implications for future directions in the era of large language models. *Interactive Learning Environments*, 1–19. https://doi.org/10.1080/10494820.2023.2204343

Je-Young, L., & Yohan, H. (2022). A meta-analysis of the effects of using AI chatbot in Korean EFL education. *Studies in English Language & Literature, 48*(1), 213–243. https://doi.org/10.21559/AELLK.2022.48.1.011

Ji, H., Han, I., & Ko, Y. (2023). A systematic review of conversational AI in language education: Focusing on the collaboration with human teachers. *Journal of Research on Technology in Education, 55*(1), 48–63. https://doi.org/10.1080/15391523.2022.2142873

Jia, J., & Chen, W. (2008). Motivate the learners to practice English through playing with chatbot CSIEC. In Z. Pan, X. Zhang, A. El Rhalibi, W. Woo, & Y. Li (Eds.), *Technologies for E-learning and digital entertainment* (pp. 180–191). Springer. https://doi.org/10.1007/978-3-540-69736-7_20

Jiang, R. (2022). How does artificial intelligence empower EFL teaching and learning nowadays? A review on artificial intelligence in the EFL context. *Frontiers in Psychology, 13*, 1049401. https://doi.org/10.3389/fpsyg.2022.1049401

Jonassen, D. H., & Rohrer-Murphy, L. (1999). Activity theory as a framework for designing constructivist learning environments. *Educational Technology Research and Development, 47*(1), 61–79. https://doi.org/10.1007/BF02299477

Junaidi, J., Hamuddin, B., Julita, K., Rahman, F., & Derin, T. (2020). Artificial intelligence in EFL context: rising students' speaking performance with Lyra virtual assistance. *International Journal of Advanced Science and Technology, 29*(5), 6735–6741. http://repository.unilak.ac.id/id/eprint/1512

Kartal, G., & Yesilyurt, Y. E. (2024). A bibliometric analysis of artificial intelligence in L2 teaching and applied linguistics between 1995 and 2022. *ReCALL*, 1–17. https://doi.org/10.1017/S0958344024000077

Kim, N.-Y. (2016). Effects of voice chat on EFL learners' speaking ability according to proficiency levels. *Multimedia-Assisted Language Learning, 19*(4), 63–88. https://doi.org/10.15702/mall.2016.19.4.63

Kim, N.-Y. (2017). Effects of different voice-chat conditions on EFL learners' topic negotiation according to proficiency levels. http://doi.org/10.18095/MEESO.2017.18.1.03

Kim, N.-Y. (2018a). Chatbots and Korean EFL students' English vocabulary learning. *Journal of Digital Convergence, 16*(2), 1–7. https://doi.org/10.14400/JDC.2018.16.2.001

Kim, N.-Y. (2018b). A study on chatbots for developing Korean college students' English listening and reading skills. *Journal of Digital Convergence, 16*(8), 19–26. https://doi.org/10.14400/JDC.2018.16.8.019

Kim, N.-Y. (2019). A study on the use of artificial intelligence chatbots for improving English grammar skills. *Journal of Digital Convergence, 17*(8), 37–46. https://doi.org/10.14400/JDC.2019.17.8.037

Kim, N.-Y. (2022). *Effects of AI chatbot-based instruction on the learning of English adjectival transitive resultative construction by Korean high school students* [Unpublished doctoral dissertation]. Seoul National University. https://s-space.snu.ac.kr/handle/10371/181242

Kim, H.-S., Cha, Y., & Kim, N.-Y. (2020). Impact of mobile interactions with AI on writing performance. *Modern English Education, 21*(2), 1–13. https://doi.org/10.18095/meeso.2020.21.2.1

Kim, H.-S., Kim, N.-Y., & Cha, Y. (2021). Is it beneficial to use AI chatbots to improve learners' speaking performance? *The Journal of Asia TEFL, 18*(1), 161–178. https://api.semanticscholar.org/CorpusID:267853286

Kohnke, L., Moorhouse, B. L., & Zou, D. (2023). ChatGPT for language teaching and learning. *RELC Journal, 54*(2), 537–550. https://doi.org/10.1177/00336882231162868

Korukonda, A. R. (2007). Differences that do matter: A dialectic analysis of individual characteristics and personality dimensions contributing to computer anxiety. *Computers in Human Behavior, 23*(4), 1921–1942. https://doi.org/10.1016/j.chb.2006.02.003

Krapp, A. (1999). Interest, motivation and learning: An educational-psychological perspective. *European Journal of Psychology of Education, 14*(1), 23–40. https://doi.org/10.1007/BF03173109

Kumar, O. (2023). Investigating the impact of artificial intelligence AI and technology in English language learning. *Advances in Social Behavior Research, 3*(1), 27–36. https://doi.org/10.54254/2753-7102/3/2023026

Kwon, S. K., Shin, D., & Lee, Y. (2023). The application of chatbot as an L2 writing practice tool. *Language Learning & Technology, 27*(1), 1–19. 10125/73541

Kyung, J. S. (2019). Introduction to popular mobile chatbot platforms for English learning: trends and issues. *STEM Journal, 20*(2), 67–90. https://doi.org/10.16875/stem.2019.20.2.67

Lee, S., Jang, Y., Park, C., Lee, J., Seo, J., Moon, H., Eo, S., Lee, S., Yahya, B., & Lim, H. S. (2023). PEEP-talk: A situational dialogue-based chatbot for English education. In D. Bollegala, R. Huang, & A. Ritter (Eds.), *Proceedings of the 61st annual meeting of the association for computational linguistics (volume 3: system demonstrations)* (pp. 190–207). Association for Computational Linguistics. https://doi.org/10.18653/v1/2023.acl-demo.18

Lee, H., & Lee, J. H. (2022). The effects of robot-assisted language learning: A meta-analysis. *Educational Research Review, 35*(3), 100425. https://doi.org/10.1016/j.edurev.2021.100425

Lee, J. H., Shin, D., & Noh, W. (2023a). Artificial intelligence-based content generator technology for young English-as-a-foreign-language learners' reading enjoyment. *RELC Journal, 54*(2), 508–516. https://doi.org/10.1177/00336882231165060

Leont'ev, A. N. (1974). The problem of activity in psychology. *Soviet Psychology, 13*(2), 4–33. https://doi.org/10.2753/RPO1061-040513024

Levitskaia, A. (2023). The impact of digital technologies on the effectiveness of English language learning by medical students. *Revista Amazonia Investiga, 12*(71), 284–292. https://doi.org/10.34069/AI/2023.71.11.25

Levy, M., Yeh, A., Hawkes, C., Lechner-Scott, J., & Giovannoni, G. (2023). Is using ChatGPT to help write papers and grants is useful, and ethical? *Multiple Sclerosis and Related Disorders, 76*, 104873. https://doi.org/10.1016/j.msard.2023.104873

Lin, C.-J., & Mubarok, H. (2021). Learning analytics for investigating the mind map-guided AI chatbot approach in an EFL flipped speaking classroom. *Educational Technology & Society, 24*(4), 16–35. https://www.jstor.org/stable/48629242

Liu, C.-C., Liu, S.-J., Hwang, G.-J., Tu, Y.-F., Wang, Y., & Wang, N. (2023). Engaging EFL students' critical thinking tendency and in-depth reflection in technology-based writing contexts: A peer assessment-incorporated automatic evaluation approach. *Education and Information Technologies, 28*(10), 13027–13052. https://doi.org/10.1007/s10639-023-11697-6

Mageira, K., Pittou, D., Papasalouros, A., Kotis, K., Zangogianni, P., & Daradoumis, A. (2022). Educational AI chatbots for content and language integrated learning. *Applied Sciences, 12*(7), 12073239. https://doi.org/10.3390/app12073239

McInerney, D. M. (2019). Motivation. *Educational Psychology, 39*(4), 427–429. https://doi.org/10.1080/01443410.2019.1600774

Meniado, J. (2023). The impact of ChatGPT on English language teaching, learning, and assessment: A rapid review of literature. *Arab World English Journal, 14*(4), 3–18. https://doi.org/10.24093/awej/vol14no4.1

Meskill, C. (2009). CMC in language teacher education: Learning with and through instructional conversations. *Innovation in Language Learning and Teaching, 3*(1), 51–63. https://doi.org/10.1080/17501220802655474

Nghi, T. T., Phuc, T. H., & Thang, N. T. (2019). Applying AI chatbot for teaching a foreign language: An empirical research. *International Journal of Scientific & Technology Research, 8*(12). 2277–8616. https://www.researchgate.net/publication/337965319

North, B. (2014). Putting the common European framework of reference to good use. *Language Teaching, 47*(2), 228–249. Cambridge Core. https://doi.org/10.1017/S0261444811000206

Okyar, H. (2023). Chatbots in English as a foreign or second language education contexts: A review of recent empirical research. *RumeliDE Dil ve Edebiyat Araştırmaları Dergisi, 36*(1), 1333–1346. https://doi.org/10.29000/rumelide.1369170

Page, M. J., McKenzie, J. E., Bossuyt, P. M., Boutron, I., Hoffmann, T. C., Mulrow, C. D., et al. (2021). The PRISMA 2020 statement: An updated guideline for reporting systematic reviews. *Systematic Reviews, 10*(1), 89. https://doi.org/10.1186/s13643-021-01626-4

Pasang, S., Latifa, A., & Ammade, S. (2024). The use of Kuki chatbot application to improve English achievement. *International Journal of English Language Studies, 6*(1), 15–46. https://doi.org/10.32996/ijels.2024.6.1.3

Polyzi, P., & Moussiades, L. (2023). An artificial vocabulary learning assistant. *Education and Information Technologies, 28*(12), 16431–16455. https://doi.org/10.1007/s10639-023-11810-9

Pretorius, L. (2023). Fostering AI literacy: A teaching practice reflection. *Journal of Academic Language and Learning, 17*(1), 1–8. https://journal.aall.org.au/index.php/jall/article/view/891

Qasem, F., Ghaleb, M., Mahdi, H. S., Al Khateeb, A., & Al Fadda, H. (2023). Dialog chatbot as an interactive online tool in enhancing ESP vocabulary learning. *Saudi Journal of Language Studies, 3*(2), 76–86. https://doi.org/10.1108/SJLS-10-2022-0072

Qiao, H., & Zhao, A. (2023). Artificial intelligence-based language learning: Illuminating the impact on speaking skills and self-regulation in Chinese EFL context. *Frontiers in Psychology, 14*, 1255594. https://doi.org/10.3389/fpsyg.2023.1255594

Rad, H., Alipour, R., & Jafarpour, A. (2023). Using artificial intelligence to foster students' writing feedback literacy, engagement, and outcome: A case of Wordtune application. *Interactive Learning Environments*, 1–21. https://doi.org/10.1080/10494820.2023.2208170

Randall, N. (2020). A survey of robot-assisted language learning (RALL). *ACM Transactions on Human-Robot Interaction, 9*(1), 1–36. https://doi.org/10.1145/3345506

Ryan, R. M., & Deci, E. L. (2000). Intrinsic and extrinsic motivations: Classic definitions and new directions. *Contemporary Educational Psychology, 25*(1), 54–67. https://doi.org/10.1006/ceps.1999.1020

Ryong, K., Lee, D., & Lee, J. (2023). Chatbot's complementary motivation support in developing study plan of E-learning English lecture. *International Journal of Human–Computer Interaction*, 1–15. https://doi.org/10.1080/10447318.2022.2163786

Satar, H. M., & Özdener, N. (2008). The effects of synchronous CMC on speaking proficiency and anxiety: Text versus voice chat. *The Modern Language Journal, 92*(4), 595–613. https://www.jstor.org/stable/25173104

Sedgwick, P., & Marston, L. (2013). Meta-analyses: Standardized mean differences. *BMJ: British Medical Journal, 34*(7), f7257. https://doi.org/10.1136/bmj.f7257

Shafiee Rad, H. (2024). Revolutionizing L2 speaking proficiency, willingness to communicate, and perceptions through artificial intelligence: A case of Speeko application. *Innovation in Language Learning and Teaching*, 1–16. https://doi.org/10.1080/17501229.2024.2309539

Shazly, R. (2021). Effects of artificial intelligence on English speaking anxiety and speaking performance: A case study. *Expert Systems, 38*(3), e12667. https://doi.org/10.1111/exsy.12667

Song, C., & Song, Y. (2023). Enhancing academic writing skills and motivation: Assessing the efficacy of ChatGPT in AI-assisted language learning for EFL students. *Frontiers in Psychology, 14*, 1260843. https://doi.org/10.3389/fpsyg.2023.1260843

Traymbak, S., Sharma, M., Anand, A., & Shukla, A. (2024). Chatbot technology in the education sector: A bibliometrics analysis using VOS viewer. *International Journal of System Assurance Engineering and Management*, 1–12. https://doi.org/10.1007/s13198-023-02230-6

Vallerand, R. J. (1997). Toward a hierarchical model of intrinsic and extrinsic motivation. In M. P. Zanna (Ed.), *Advances in experimental social psychology* (pp. 271–360). Academic Press. https://doi.org/10.1016/S0065-2601(08)60019-2

Van Eck, N. J., & Waltman, L. (2010). Software survey: VOSviewer, a computer program for bibliometric mapping. *Scientometrics, 84*(2), 523–538. https://doi.org/10.1007/s11192-009-0146-3

Vogt, P., van den Berghe, R., de Haas, M., Hoffman, L., Kanero, J., Mamus, E., Montanier, J. M., Oranç, C., Oudgenoeg-Paz, O., García, D. H., & Papadopoulos, F. (2020). Second language tutoring using social robots: A large-scale study. In *Proceedings of the 14th ACM/IEEE international conference on human–robot interaction* (pp. 497–505). https://doi.org/10.1109/HRI.2019.8673077

Vygotsky, L. S. (1978). *Mind in society: Development of higher psychological processes.* Harvard University Press. https://doi.org/10.2307/j.ctvjf9vz4

Wambsganss, T., Kueng, T., Soellner, M., & Leimeister, J. M. (2021). ArgueTutor: An adaptive dialog-based learning system for argumentation skills. In *Proceedings of the 2021 CHI conference on human factors in computing systems.* https://doi.org/10.1145/3411764.3445781

Wei, L. (2023). Artificial intelligence in language instruction: Impact on English learning achievement, L2 motivation, and self-regulated learning. *Frontiers in Psychology, 14,* 1261955. https://doi.org/10.3389/fpsyg.2023.1261955

Wei, P., Wang, X., & Dong, H. (2023). The impact of automated writing evaluation on second language writing skills of Chinese EFL learners: A randomized controlled trial. *Frontiers in Psychology, 14,* 1249991. https://doi.org/10.3389/fpsyg.2023.1249991

Wohlin, C. (2014). Guidelines for snowballing in systematic literature studies and a replication in software engineering. In *Proceedings of the 18th international conference on evaluation and assessment in software engineering.* https://doi.org/10.1145/2601248.2601268

Wu, X., & Li, R. (2024). Effects of robot-assisted language learning on English-as-a-foreign-language skill development. *Journal of Educational Computing Research,* 1–25. https://doi.org/10.1177/07356331231226171

Yang, H., & Kyun, S. (2022). The current research trend of artificial intelligence in language learning: A systematic empirical literature review from an activity theory perspective. *Australasian Journal of Educational Technology, 38*(5), 180–210. https://doi.org/10.14742/ajet.7492

Yang, H., Kim, H., Lee, J. H., & Shin, D. (2022). Implementation of an AI chatbot as an English conversation partner in EFL speaking classes. *ReCALL, 34*(2), 1–17. https://doi.org/10.1017/S0958344022000039

Younis, H. A., Ruhaiyem, N. I. R., Ghaban, W., Gazem, N. A., & Nasser, M. (2023). A systematic literature review on the applications of robots and natural language processing in education. *Electronics, 12*(13), 12132864. https://doi.org/10.3390/electronics12132864

Yuan, Y. (2023). An empirical study of the efficacy of AI chatbots for English as a foreign language learning in primary education. *Interactive Learning Environments,* 1–16. https://doi.org/10.1080/10494820.2023.2282112

Zhang, R., Zou, D., & Cheng, G. (2023a). Chatbot-based learning of logical fallacies in EFL writing: Perceived effectiveness in improving target knowledge and learner motivation. *Interactive Learning Environments,* 1–18. https://doi.org/10.1080/10494820.2023.2220374

Zhang, S., Shan, C., Lee, J. S. Y., Che, S., & Kim, J. H. (2023b). Effect of chatbot-assisted language learning: A meta-analysis. *Education and Information Technologies, 28*(11), 15223–15243. https://doi.org/10.1007/s10639-023-11805-6

Zhang, Z., & Huang, X. (2024). The impact of chatbots based on large language models on second language vocabulary acquisition. *Heliyon, 10*(3), e25370. https://doi.org/10.1016/j.heliyon.2024.e25370

Zheltukhina, M. R., Sergeeva, O. V., Masalimova, A. R., Budkevich, R. L., Kosarenko, N. N., & Nesterov, G. V. (2024). A bibliometric analysis of publications on ChatGPT in education: Research patterns and topics. *Online Journal of Communication and Media Technologies, 14*(1), e202405. https://doi.org/10.30935/ojcmt/14103

Zhou, T., Cao, S., Zhou, S., Zhang, Y., & He, A. (2023). Chinese intermediate English learners outdid ChatGPT in deep cohesion: Evidence from English narrative writing. *System, 11*(8), 103141. https://doi.org/10.1016/j.system.2023.103141

Zou, B., Guan, X., Shao, Y., & Chen, P. (2023). Supporting speaking practice by social network-based interaction in artificial intelligence (AI)-assisted language learning. *Sustainability, 15*(4), 2872. https://doi.org/10.3390/su15042872

Ambient Anthropomorphic Social Educational Robotics

Rina Zviel-Girshin and Nathan Rosenberg

Abstract Anthropomorphic Social Educational Robot (ASER) is the future of education. Robots are available in any form and flavour. Information and Communication Technology (ICT) is there to support any educational robotic system. Why is it not there already? There is the need to decide upon the right hardware and the right support system. We present here one paradigm for ASER we call Gaia. In the Gaia approach every student is 24/7 supported by an intelligent social learning environment. The robot is not the exclusive or even the major educational intelligence agent. Although it has its own knowledge and functionalities, it is the avatar of a great distributed organic knowledge ecosystem. Thus, very small individual resource investment has a huge impact. The ASER or system of ASERs are there for the student in many capacities. And they channel in and out a great cornucopia of knowledge thus connecting incredible number of students, teachers, educators and all other resources relevant not only to education and learning but also to social life, mental health, joy and happiness of each and every student, teacher and parent.

Keywords Social robots · Anthropomorphic robot · Educational robot · Technology-enhanced personalized learning

1 Introduction

We present here a novel paradigm in the extremely important field of Anthropomorphic Social Educational Robotics—the Gaia Organic Knowledge ecosystem approach. We also report the application of such paradigm that is being realized in our AI and Robotics lab for many years, explaining in detail one such robot—Dr. Mec.

R. Zviel-Girshin (✉)
Ruppin Academic Center, Netanya, Israel
e-mail: rinazg@ruppin.ac.il

N. Rosenberg
Paralex Institute, Hadera, Israel

G. Lampropoulos and S. Papadakis (eds.), *Social Robots in Education*, Studies in Computational Intelligence 1194, https://doi.org/10.1007/978-3-031-82915-4_15

What is meant by "anthropomorphic" in general, and what is its specific meaning in the context of educational robotics? Etymology of anthropomorphic is having "human appearance" in Greek. Anthropomorphism is simplistically described as people's tendency to attribute human characteristics to non-lifelike artifacts. We shall explain later that, at least in the context of robotics, it has a much wider, more practical, useful and subjective meaning.

Anthropomorphic robot is a robot that is human-like in more aspects than robots in general. It can be anthropomorphic objectively, having some human-like features, from mere appearance to behaviour, abilities, communication modes and channels, feedback quality or quantity. It could also be anthropomorphic subjectively—when humans think of it or treat it as alive or human.

Anthropomorphic Social Educational Robot (ASER) can have crucial importance in education. Such robot's sociability, developed already for at least twenty years has a great potential (Baraka et al., 2020; Dautenhahn, 2007; Duffy, 2003; Fong et al., 2003; Ntomora & Petrakos, 2024; Stock-Homburg, 2022; Zhao et al., 2022). Objectively, it has many educational features and abilities. Subjectively, through anthropomorphism, the student accepts him as alive, and many a time even as a kind of human, and develops relationship that is extremely important in education, and is much deeper than with other educational devices like computer or even other, less anthropomorphic kinds of robots. It can serve in many important educational roles such as teacher, assistant-teacher, source of knowledge, educational artifact, helper, friend or pet (Alam, 2021; Belpaeme et al., 2018; Chalmers et al., 2022; Derakhshan et al., 2024; LeTendre & Gray, 2024; Papadopoulos et al., 2020; So & Lee, 2023; Smakman et al., 2021; Thellman et al., 2022; Woo et al., 2021). ASER is the subject of an important body of research where many experimental or non-formal toy-like robots are described (Dubois-Sage et al., 2023; Stock-Homburg, 2022). But it is not yet really present in real-life educational settings. The extant models are either too expensive and experimental, or they are too simple and not really useful.

Anthropomorphic robot is the ultimate social robot (or at least it could and should be). Social robot in education is the ultimate social robot as the students, especially children, form (and need) a special bond, special in intensity as well as nature. Social robots enhance emotional involvement and collaborative learning, while AI, like ChatGPT, offers personalized instruction and immediate assistance, transforming educational experiences and outcomes (Jamil et al., 2024; Lee, 2023). Therefore, our treatment of this field is actually the quintessence of analysis of the whole field of social robotics, including its problems, methods and solutions (Davis & Class, 2023; LeTendre & Gray, 2024; Ntomora & Petrakos, 2024; Thellman et al., 2022; Złotowski et al., 2015).

ASER, being the ultimate social robot, could make a great contribution to learning and education, create new possibilities of learning environment, revolutionize the learning process and the schools (Davis & Class, 2023; Zviel-Girshin & Rosenberg, 2018). But, beyond all those very important goals, it can greatly contribute to mental health and development, feeling of confidence, agency, joy, motivation, socialization, social skills and many other skills (some related to learning and some beyond it). It could be even more important in a significant percentage of children that have special needs, like having disabilities, learning problems, very high or very low initial abilities, ADHD, or any special circumstances. It could also have great influence for children with social difficulties, as well as improve social skills of all students (Davis & Class, 2023; Zviel-Girshin et al., 2020).

Paradox—the robots are omnipresent, sophisticated, offering a great choice starting from very affordable price. There is a consensus they could be extremely useful in learning. On the other hand, it seems they are practically non-existent in the education system or private independent learners' learning environment (Smakman et al., 2022).

A paradox exists between the consensus on the educational potential of ASERs and their limited availability. The robots available are extremely pricey, not practical, limited, and there is no infrastructure and readiness to absorb them in schools. The paradox is that the importance of ASERs does not have to make them hard-to-make or expensive-to-get. On the contrary, the reason why they are so useful is the same reason they can be cheap and omnipresent (Martinez-Roig et al., 2023).

Some possible answers, factors and explanations of the paradox of lack of use and lack of useful robots can be: if robots are defined more broadly, they are indeed being used; however, many remain unaware of the extensive potential of hardware, software, and pedagogical applications. This is compounded by a general lack of robot literacy, fears of job displacement, risk aversion, and insufficient government and public policies supporting robotics. Additionally, there is a shortage of theoretical and practical expertise in Robot-Aided Learning and Robot-Aided Pedagogy, limited funding, and an overwhelming variety of systems that lack standardization, interoperability, and engineering consistency. Finally, the field suffers from inadequate technical support and a shortage of skilled personnel. We could go on, but it seems we have certainly to deal with at least the problems already described, for their solution seem essential for this field.

We will present here a detailed account of our solution for ASER organic educational ecosystem robotics. Our paradigm is fundamentally different from the traditional approach of building very sophisticated stand-alone robots that are extremely costly, hard in exploitation and maintenance, have very limited functionality, restricted knowledgebase, are rigid and not user-friendly. Our robots are not-complex-than-needed agents that are inter-connected and are channeling the huge knowledge of global knowledge bases and communities, humans and computers, and are the terminals of the global organic ecosystem.

One approach to a solution would be a system we could describe (for reasons explained later) as ASER Model-T avatar of a Gaia ecosystem. Robots cheap and simple being avatars of an educational robotics ecosystem (Gaia).

This chapter is structured to begin with a discussion of education and robotics. The concept of the Anthropomorphic Social Educational Robot (ASER) is then introduced, followed by an exploration of the organic model. An example of an anthropomorphic robot, Dr. Mec, is presented to illustrate practical applications. Subsequently, Gaia is explained, and use cases for ASER are provided. The chapter concludes with an identification of future directions in educational robotics.

2 Education and Robotics

In our context and approach, all traditional terminology should be expanded to much wider meaning and scope than could be traditionally thought, in adherence to the new reality in education, including the great revolution of the last two decades of much more available sophisticated technology, and the new modes and models of learning, education and schooling.

2.1 Anthropomorphism

Our robots should be as anthropomorphic as possible in all aspects and features. But what are the outer limits of anthropomorphism? What is anthropomorphism, and more generally animalism, and more specifically children's animalistic fruitful imagination and young minds' special relation to the world that naturally gives animalistic and human characteristics to robots even of the least human-like kind. Even Skinner's learning machine (Skinner, 1958, 1963), that was not even a computer and computer or another device running a program, and even some totally inanimate things could be seen through the young eyes and young soul as much more alive and human than in the eyes of the adult. In general, people have the ability to react to inanimate phenomena as animate, a characteristic that could be used in education (Fig. 1).

Anthropomorphisms features could be: size or shape, movement, behaviour and mirroring other people's behaviour (like dancing), reacting, talking, understanding

Fig. 1 Pupils build animalistic robots

and engaging in dialogue. In addition, they can fulfill human functions, demonstrate human-like expertise, convey expressions (through body and facial cues), and simulate emotions. Moreover, they are programmable, capable of evolving over time, and can provide feedback and adapt based on learning from interactions.

2.2 Robot

Etymologically, robot is just a "worker" in Check, the native language of Karel Čapek who first used the word in the meaning of robot in 1921 (twenty years before first computer), in his "*R.U.R.*" ("Rossum's Universal Robots"). In his masterpiece those were very human-like.

Given this context, to what extent can or should we consider the following devices as "robots"? These range from Skinner's learning machine, electric cars, wind-up bunnies, and the Energizer bunny, to dolls (including those that talk), desktop computers, laptops or tablets, smartphones, toy robots (both non-moving and mobile), simple Arduino robots without sensors, Arduino robots with sensors and behaviors, and industrial robotic arms.

A robot isn't limited to the creaking, jointed Tin Man from *The Wizard of Oz* or the caricatured maid Rosie from *The Jetsons*. Instead, a robot can be any computerized device, including computers, embedded systems, smart devices, and more, as long as it possesses distinct "robotic" features. Unlike non-robotic devices, robots are characterized by enhanced intelligence, greater interaction with their environment, human-like behaviors, and an active, human-like shape.

In organic real-life ecosystem, we would prefer robots that would have in addition:

- knowledge base
- dialogue abilities
- human-like expertise
- more sensors
- more actuators
- ability to learn and change
- safety, security and privacy
- communication
- mobility
- initiative
- interoperability—ability to interact with and to use other devices and tools.

Robots are very varied, we have a plethora of different models that can be clustered by any one of many features starting from usability or price and ending with range by role—subject of study, peer, friend, teacher, social support provider, partner, guru, mentor. There is a scale of familiarity, locality, intensiveness, building and maintaining versus deus ex machina, time, and human involvement. By topic there are different subjects, different modes of teaching and learning, different goals and learning plans.

2.3 Education

The ASER Gaia approach empowers the learner and extends the meaning and content of education. How exactly? Unless stated otherwise we will use the term education for the wider and different senses of education and learning (two terms that are certainly not the same). There exist many kinds of learning and education that differ in many characteristics, but for our analysis we can ignore those differences for the meantime. Computer Aided Learning, Computer Aided Education, Social Robotics Education and Anthropomorphic Robotic Education will also be used interchangeably, usually using the broadest term ASER.

Educational system is one of the vital parts of our life and of the future of our children. The teaching machine predated the computer and was first patented in the 19th century (Benjamin, 1988). From the dawn of the computer era one of the first applications was in the area of learning and teaching. But only in the 21st century occurred the serendipitous crucial meeting of two independent developments: ICT becoming omnipresent and really robust, and allows us revolutionary solutions, on the one hand, and the education systems all over the world, being in deep crisis and in dire need of those solutions the ICT can now at last provide (Muir-Herzig, 2004). So, in practically every country, all over the globe, the national Ministry of Education initiated national programs to computerize the schools.

The ASER makes really practical the holy grail of education for hundreds of years, at least since Jean-Jacques Rousseau's Emile (Rousseau et al., 1911), the non-traditional education (called sometimes, depending on the educator's emphasize, progressive, natural, democratic, free, open, student-oriented, constructivist, project-driven, curiosity driven, Montessori, Dewey and many others). As Dewey succinctly contrasts the traditional vs all others (collectively usually called progressive) education approaches: it is the difference between outside and inside the student, where traditional starts with the teacher and the teacher tries to make the student internalize teacher's knowledge and skills. Non-traditional, progressive education originates by the student, unique for this student, and in the mind of the student. Here the student has the agency, and the teacher just assists him to a degree that depends on the specific flavour of the relevant progressive education variation.

2.4 Pedagogy

The pedagogical research included a multitude of better models that were proposed over thousands of years. Most of these approaches were never adopted as the standard for the public education not because the pedagogy was contested but because it necessitated more resources than were available. With ASER we can implement the most advanced pedagogical proposals, theories, models and solutions. So, what ASER pedagogy will look like and what models are available to be used?

Numerous educators and theoreticians have contributed to these modern approaches that came to revolutionize the traditional, generally seen as very poor in freeing the great potential of students, education. We will mention just four of those: a humanist visionary (Jean-Jacques Rousseau), philosopher and empiricist (John Dewey), psychologist behaviourist socio-cultural theoretician (Lev Semyonovich Vygotsky) and great practitioner and popularizer after whom thousands of schools in dozens of countries are called (Maria Montessori).

"Montessori pedagogy" is both scientifically sound and ethically appropriate, acknowledging the many researchers and educators who have expanded on the theories and methods pioneered by Maria Montessori. By using her name, we honor her groundbreaking contributions (Zviel-Girshin & Rosenberg, 2016).

Maria Montessori was not only the first to develop a more effective educational method but also to implement it with a precise and scientific approach. Her success was substantial, with at least 8,000 schools around the world named after her. Just as Albert Einstein is synonymous with physics, Maria Montessori is a seminal figure in education. What made her work so exceptional?

Montessori was the first woman in Italy to earn a medical degree in 1896. Her background in medicine and passion for engineering played a crucial role in the effectiveness of her educational method. She was a strong advocate of scientific and engineering principles, including project management, scrupulous preparation, channeling of students' lab experiments and feedback mechanisms. In 1907, she founded her Casa dei Bambini, where she put her innovative ideas into practice.

Montessori's approach involved creating a carefully designed learning environment equipped with the best educational tools of her time and her own creations of what we would today call educational experiment-toys (Montessori, 2013). Although today's term "constructivism" often describes her approach, it was Montessori's unique commitment and innovative use of technology that set her apart. Her method was particularly effective, as evidenced by students from disadvantaged backgrounds who excelled academically, learning to read and write by the age of four and competing successfully in mathematics against students from private schools, while many of them had been considered retarded before. These students thrived in a well-structured yet engaging environment.

Montessori designed her learning environment to be playful (Montessori, 1917). Teachers developed specialized educational tools disguised as toys—a concept now known as gamification. Students were closely monitored to ensure they stayed engaged, without relying on rigid schedules, lectures, or rote memorization. Montessori was an early advocate for children's rights, as reflected in her students' experiences. For example, when asked who taught them to write, children would often respond with surprise that no one had done so, the idea was theirs (Standing, 1998).

Montessori's approach also emphasized the importance of feedback in learning, a concept that would only later be proposed by Norbert Wiener's Cybernetics (Wiener, 2019) and B.F. Skinner's Learning Machines method, though not before Pavlov's research into learning (Clark, 2004).

Constructivist education was generally popularized in 1938, when John Dewey published the seminal work that summarized his own theoretical and practical work on experience-based empiricist approach to the Laboratory school he has established (Dewey, 1958). This research of more than forty years as well as that of many others, lead to the clear educational paradigm that is now known as constructivism.

Modern educational research is rich and varied, and throughout the 20th century, there has been a notable shift towards constructivist approaches. Despite the diversity of methods, these approaches share several core principles: knowledge is actively constructed rather than passively received; knowledge construction relies on personal experiences and hypothesis testing; and each individual's understanding and construction of knowledge is shaped by their past experiences and cultural context.

A constructivist learning environment typically includes multiple representations of reality, reflecting the complexity of the real world. It emphasizes individualized knowledge construction over rote memorization and replaces abstract instruction with authentic tasks in meaningful contexts. Learning is grounded in real-world or case-based scenarios rather than rigid, predefined sequences, fostering thoughtful reflection on experiences. Additionally, it promotes collaborative knowledge construction through social negotiation rather than competition.

Constructivist-inspired methods often involve hands-on, experiential learning, such as lab-based experiments or project-oriented activities, which enable students to engage directly with the material. These methods can be tailored to individual student needs, making learning more personalized. Telescopic approaches allow for a gradual increase in complexity, while student-centered methods empower learners to take ownership of their education. Association-driven learning encourages students to connect new information to their prior knowledge. Finally, non-frontal, peer-to-peer methods cultivate collaboration and social engagement, enriching the learning process through shared insights and group activities.

Vygotsky was one of the first and most important education-as-science researchers and theoreticians. He was one of the proponents of scientific approach to education practice and theory that wrote extensively and contributed great volume of educational knowledge only few decades after the beginnings of education becoming a university subject. He believed in scientific individualized planning of every student's learning that takes into account his unique abilities and condition (Vygotsky, 1978). For our purposes it is very important to remember that in his theory the social environment and proactive intensive scaffolding of the process have crucial influence on the success of the student by stretching his zone of proximal development. So, the individual close interaction with ASER is exactly what the student needs.

There is widespread agreement on the need to replace traditional lectures with proactive, lab-based, and individually tailored learning through project-based work (Sedubun & Nurhayati, 2024; Yusri et al., 2024). However, the variety of approaches and technologies available calls for a balanced strategy to effectively achieve course learning objectives.

2.5 *Social Robotics*

Social robotics in the widest sense is the most important area of robotics (as well as becoming a major field of research in social sciences and other research) as it directly impacts us all, being the robots that interact with us. Social robots will become an integral part of every aspect of our life and technology within all aspects of society, where they will aid in a wide range of tasks. These robots are specifically built to interact socially with humans, facilitating smoother communication between people and machines and thereby boosting user acceptance (de Graaf & Allouch, 2013; Thellman et al., 2022). The defining quality that sets a social robot apart from other types is its ability to assist and engage with people on a social level. Other defining traits include having a human-like appearance or performing functions similar to humans, such as movement (Vlieghe, 2022). A current goal for artificial intelligence in this domain is to enable social robots to interact as naturally as humans do with one another. The milestones range from first electronic learning robots (Han et al., 2008), to recent assistant robot in a hospital setting that could communicate verbally with patients and staff (Rofi'ah et al., 2021).

3 Anthropomorphic Social Educational Robot

When we think about ASER that will be ambient, omni-present, practical and cheap, we would do well to learn from industry success in developing such models. One example is the automobile industry and the Ford Model-T. The Model-T isn't simply the most important Ford car; it may be the most important car ever. The Model-T is the first car that became affordable to the masses (Casey, 2008). It was introduced in 1908 at a price of $850, but the most important models wouldn't arrive until 1913 and thereafter. The breakthrough came through revolutionary holistic organic approach of Henry Ford. His vision was that of a complete ecosystem of production, workers, maintenance, sales, man, machine, material and product, economy and psychology. It included such novel aspects as the moving assembly line and organically integrating the different aspects into one ecosystem, allowing output to increase several times over while reducing the starting price to $260. Ford built the Model-T until 1927, selling more than 15 million over its 19 year run (Eli et al., 2023). That would be more than a hundred million cars in today's terms, on par with the next car model that inherited this approach—the Volkswagen Beetle that sold 65 million cars from 1938 till 2003.

The same organic ecosystem approach would make one cornerstone of the solution to the problem of robotics in education. We can go beyond the previous model, by using the great technological progress to make the robots anthropomorphic and besides basic model offer a great variety of context-sensitive scale of models, from state level through institution and classroom up to the individual student adapted robotic environment. We should also use our ICT to create around each student an ecosystem of human and ASER rather than afford the pupil merely one disconnected robot. Besides creating a scale of very cheap standardized robots, there would be also great variety of more sophisticated ones, including online remote distributed ones.

ASER principles emphasize that ASER should be perceived as an anthropomorphic robot in the broadest sense, aiming to be as ubiquitous as the Model-T through its appearance, ease of use, functionality, and affordability. The ASER ecosystem should incorporate human expertise to create an immersive experience for students within educational environments. A high volume of interactions is intended to generate feedback, fostering both individual and collective student development. Given our natural inclination toward anthropomorphism, a variety of simple ASER units can play a valuable role in educational settings.

As they have both human or animal, on the one hand, and inanimate, on the other, characteristics that the user can chose from, they more easily can:

- have multi-role relations that people would have great difficulty to assume
- great ease of adoption and adaptation by the students and teachers in educational environment
- bring great joy
- promote gamification.

ASER platform requirements are: anthropomorphic, playful and playable, simple, therefore robust, with an intuitive and rich UX (preferably including a touchscreen), telecommunicative (preferably Wi-Fi/IoT-enabled), and affordable enough to be widely accessible and omnipresent. ASER ecosystem should include a society of mind of various ASER units of different specifications.

There could be many platforms on which to base ASER. We have experimented with several and describe here one quite successful one—our own version of ASER—Dr. Mec built on chassis of Mecanoid.

ASER could differ by many features like software, hardware, communication, knowledge base, intelligence, mobility, CPU strength, shape, interoperability, flexibility, modularity and many others. In general, in the most fundamental and relevant to education sense, ASER can be characterized by several basic characteristics.

3.1 Standardization Versus Individual Adaptation

How to achieve both standardization and individual adaptation that needs variety? By divorcing the basic robot from the individual touches added and evolving all the time. Also, there should be more than one basic model—a line of basic models. There would be a clear, even pedantic, difference between the various context-sensitive levels and interfaces of the same model. What would be public to one user (student, peer, teacher, administrator, crowdsourced contributor, parent, caregiver) would be totally private to another.

Ideally the ASER ecosystem should adhere to the Object Oriented Planning engineering principles, like modularity, encapsulation, inheritance, polymorphism, interoperability and abstraction, as well as to other components of the engineering paradigm. Write once, use everywhere would become created once, used by all.

So, what is one and the same model or even robot to some (like the developer or producer) could have many different faces (interfaces) for different users. It even should and would evolve its interfaces with its evolution as it is an organic agent. Examples of such differences could be in costumes, behavior, sound, tasks, dialogues and reactions. Those are added to the basic model or changed.

The question of how sophisticated the basic model should be and what it would include when supplied is not a trivial one. The border between the general and the individual could become one of great importance. This is a basic decision of how much to prefabricate and how much to leave for individual adjustment. Where to limit the evolution of individual robot (so, for instance, that it still functions or communicate with others in the ecosystem). Even such basic questions as price, availability, learning curve could determine what would be considered the basis and what the additions, adaptations and evolutionary development. It is logical to make it to a great extent not too rigid and allow for different solutions in different cases.

In this debate there is the need for clear terminology. Why don't we call such a basic robot chassis. In some way it is, as it is the basis for all other components, variation and additions. But it is much more than the chassis in the usual sense of robotics. Chassis etymology has a long history: "**chassis (no.)** "base frame of an automobile," 1903, American English; earlier "sliding frame or carriage-base for a large gun" (1869), "window frame" (1660s), from French châssis "frame," Old French chassiz (13c.) "frame, framework, setting," from chasse "case, box, eye socket, snail's shell, setting (of a jewel)," from Latin capsa "box, case" (see *case* (no. 2)) + French—is, collective suffix for a number of parts taken together..." (Online Etymology Dictionary).

Robotic chassis could be described as a frame to attach all others. The chassis is the structural component for the robot which contains the drivetrain and allows the robot to be mobile by using wheels, tank treads, or another method. A chassis is sometimes referred to as the robot's frame. The chassis also provides a structure to attach manipulators such as arms, claws, lifts, plows, conveyor systems, object intakes, and other design features used to manipulate objects.

The basic model is much more than the usual idea of robotic chassis because it contains many more tools supplied as basic part of it. The difference could be exemplified as the difference between computer supplied with no basic input/output system (BIOS), one supplied with BIOS and operating system (OS), one supplied with BIOS, OS and a set of software and apps, and the same computer after a long time in use that includes individual configuration, individual additional software written by the user or installed by him and in many cases behaving quite differently than the one-fits-all tabula raca it was before its evolution through user's individual adjustments and feedback.

The ASER ecosystem's variability encompasses a broad range of aspects that can be categorized across different areas. This ontology includes the robot model, robot augmentations and adaptations, personalization, and robot use. It also covers the learning process, student identity, institution, educators, communication, field of study, level of study, study methods, goals, targeted skills, resources, and time. The robot, environment and system could, would and should be very different at different contexts and for different students.

4 Organics

The word organics has too many different meanings and associations. It is interesting to note that the original Greek (and Latin) meaning of the world was quite different from today's one, and yet very important for the field of ASER. Etymological dictionaries trace the origins of the world to the Greek word οργανικός (pronounced as organikós) meaning engine, organ, instrument, coming from the Greek word όργανο (organó) that means instrument, device or όργανον (órganon) instrument. So, we see that from the beginning it was about an active, "live" in some sense, mechanism that has a functionality, and is part of the system, of the body. The Greek word became *organicus* in Latin. That was the use of the world we find in 1510. Then, already in 1778 we see the more modern sense of "organized" living things having "organs". A system functioning due to having coordinating integrated intertwined parts has been around for the last two centuries from 1817. Organic as living and natural could be found in Middle English *organik* that was used for body parts having different properties. This historic etymological excursion helps us to construct the organic approach relevant to ASER.

Organic paradigmatic shift can be described as a transition from math to life sciences, from algorithms to societies of competing experts, from logic to evolution and from one definitive algorithmic answer to an environment of partially true, evolving prototypes of practical solutions. It represents a move from the desk to the lab and from static, brick-like models to dynamic, living organisms.

4.1 Organic Knowledge Model

Organic Knowledge (OK) systems are ICT systems incorporating human expertise. One would be tempted to describe them as Expert Systems "on steroids". If at all, they are an ecological system of many different and sometimes contradictory experts, called organs. But in reality, the OK systems are so much more in almost every aspect that it is more correct to say they are the fulfilment of the ES vision. Other spiritual ancestors of OK system are Turing's "child programs" (Turing, 1950) and Minsky's learning, evolving and non-algorithmic Society of Mind proposals (Minsky, 1988).

OK system is intelligent, evolving, learning, organized, distributed, dialectical, having a very big knowledge base. Each organ is simulating an independent expert, and includes: knowledge base (data, meta-data and procedures), feedback apparatus, knowledge acquisition mechanism (interfaces and communication), learning mechanism (inference of new knowledge and processing) and evolution mechanism (creating and changing organs in view of the new knowledge). It has several interfaces: a local interface with the environment, a subjective interface for the user, and an objective interface. The OK system operates with several communication modes, including organ-to-organ (o2o) communication, interaction with remote servers, peer-to-peer (p2p) connectivity with remote users, access to remote resources, and proactive execution.

An Organic Robotic Environment is an Organic Knowledge System. Organic Knowledge systems are the newest generation of Knowledge-Based Systems, one of the major branches of AI. Organic Knowledge is a Knowledge System paradigm that simulates and enhances through mutual learning the knowledge of both the IT and Robotics expert and the knowledge of the domain expert.

In a nutshell the organic approach is treating the problem and the solution process as evolution of different and frequently conflicting units of knowledge, algorithms and solutions. It is mimicking the growth and evolution of a living organism (or ecology of organisms if more appropriate) where different units of knowledge, algorithms and solutions are the organs. The organic solution is like a child—in the beginning having no knowledge (except some basic mechanisms needed for evolution), and by process of feedback and Darwinian natural selection (Darwin & Beer, 1951) the solution gradually evolves, becomes better and better using its growing body of knowledge.

The most basic aspects of the paradigm are the especially massive body of organic knowledge and the organic life cycle. The organic life cycle is somewhat reminiscent of the prototype cycle but it is much more sophisticated. It simulates the growth of human intelligence, creating solutions more and more plausible using mechanisms of feedback and learning.

The Gestalt-Multiplex-Layering model components are:

- gestalt—a deep model of the expert knowledge and reasoning process
- multiplicity—simultaneous use and cooperation of different and conflicting approaches
- layering—use of a hierarchy of independent layers of control and processing, through which the input and intermediate results are propagated.

An OK approach uniquely suits the needs of an educational system. The learning process is by its nature an organic process in which both the student and teacher evolve. The school is an ecological system. The teacher uses different approaches and tools at different times and with different students, while his teaching expertise is evolving through feedback and learning. The growth of student's knowledge is multi-level, multi-directional and very individual.

For each student, through a process of very intricate feedback, an individual path is built from components (these days called "learning units") taken from a huge and evolving database of domain knowledge and educational tools, including ICT educational tools. This process is inherently dialectical. Not only does the learning itself has to move in different dimensions and levels of understanding, but also there is the basic bi-polarity and contradiction of teacher's authority and instruction versus the students' need for individuality and control.

5 Dr. Mec

There could be many platforms on which to base ASER. We have experimented with several and describe here one quite successful one—our own version of ASER—Dr. Mec built on chassis of Mecanoid (Fig. 2).

The Meccanoid (Scheick et al., 2018) is one of the most lifelike robots available at a budget-friendly price (around $100) (Chen et al., 2021; Martín-García et al., 2022). Standing at 120 cm tall, it is deliberately designed to mimic the human form. This robot is equipped with voice recognition and boasts over a thousand pre-set phrases, comments, and witty remarks. It can entertain with jokes and games, providing an engaging interactive experience.

The Meccanoid can be programmed in three easy-to-use methods: Learned Intelligent Movement Technology, where children manually move the robot's arms and head or give vocal commands, which the robot then remembers; Motion Capture, which involves placing a smartphone in the robot's chest and enabling it to mimic the phone's movements; and Ragdoll Avatar Control, where children use an app to swipe on a Ragdoll avatar to control the robot's 10 motors, moving its head, arms, and feet.

The educational benefits include a human-like appearance, interactive features such as voice and motion recognition, the ability to learn through demonstration, and

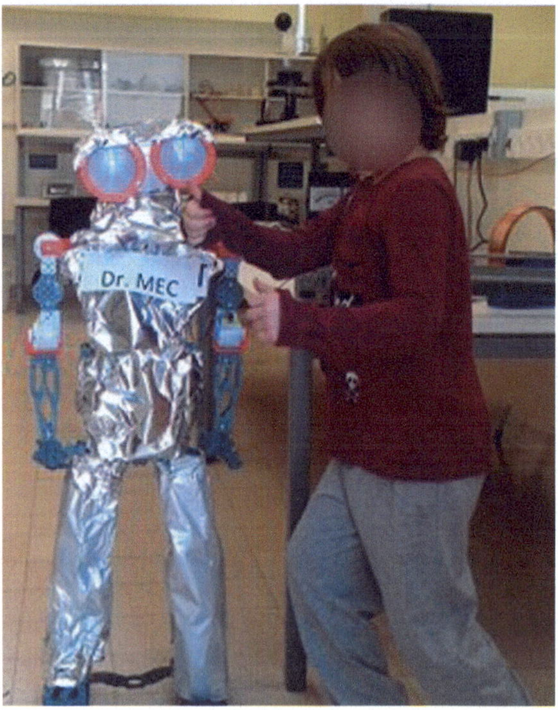

Fig. 2 Dr. Mec helps in physics lab

human-like responses, including voice communication, dancing, exercise routines, and jokes.

The Meccano concept was originally introduced in 1898 by Frank Hornby, who envisioned creating robots from recycled toys and industrial scrap. One example of this concept is the Meccanoid G15KS Personal Robot, a humanoid robot from the Meccano brand. It is constructed from over 1,100 pieces and stands about 4 ft tall. The Meccanoid G15KS features voice recognition and a wide range of pre-programmed responses. With 8 servo motors, it is capable of realistic movements, and its Meccabrain, containing 64 megabits of flash memory, allows it to learn movements and replay recorded sounds.

Software integrates modern educational approaches that emphasize individualized, student-driven, research-based learning, supported by advanced ICT technologies. Dr. Mec's sophisticated software creates a two-way connection between the user and the information system, adapting to each child's needs. It includes an expert system that replicates a physics teacher's expertise, positioning the robot as both a presenter and an interactive learning tool.

Hardware in Dr. Mec's version of the Meccanoid offers a complete physics lab setup, featuring lab accessories and robot extensions. Additional sensors can be integrated to enhance feedback and interaction, making it a robust educational platform for hands-on learning.

For educational purposes, we have developed a character for the robot named Dr. Mec, a "crazy scientist" passionate about teaching physics. Dr. Mec integrates with a physics information system to make learning interactive and enjoyable, supporting teachers in classrooms and managing curriculum content.

Dr. Mec aims to transform the study of physics into a fun, experimental journey that combines playful exploration with scientific inquiry. With unique capabilities, Dr. Mec serves as a complementary educational tool, providing learning experiences beyond what traditional computers or human teachers can achieve alone. Dr. Mec's interface with the physics information system promotes associative learning through hands-on experiments, guided subtly by the curriculum. Examples of the learning process can be seen in multimedia demonstrations such as scientific experiments for children.

Dr. Mec offers significant educational value within a customized learning environment enriched by various accessories. Key features and advantages include gamification, theatrical presentation and showmanship, human-like qualities, and a sense of novelty that boosts engagement. The platform promotes fun, interactive learning, supported by an extensive knowledge base and connectivity with multimedia integration. Additionally, Dr. Mec provides consistent, rational, and resilient educational experiences, making it a robust tool for learners.

In the future ideally, every child should have access to a humanoid robot with a customized information system for different subjects. In the near future, it is anticipated that most children, apart from those in the least developed regions, will either have Dr. Mec or a similar alternative. Educational institutions, from preschools to universities, will benefit from having at least one Dr. Mec. The demand for interactive, bi-directional educational interfaces is growing rapidly and is expected to expand further, potentially reaching over two billion students worldwide. We will mention just some of use cases for using Dr. Mec as ASER.

Learning robotics with Dr. Mec is achieved through various methods, including Dr. Mec assembling, continuous hardware and software improvements, evolving teaching capabilities, and interactive play. The cases differ by many criteria. For instance, one scenario involves learning alone with Dr. Mec as a companion, where the robot is not connected and not actively teaching. Another case includes learning with a teacher present alongside Dr. Mec, also without connectivity or active teaching. In a connected and teaching mode, Dr. Mec can engage in tele-teaching. Alternatively, Dr. Mec can operate independently through predefined modules, where he functions as the sole teacher, disconnected yet actively teaching.

6 Gaia

The idea of an organic ecosystem (not necessarily called ecosystem) of multitude of agents (not necessarily called agents) is not new at all. It was proposed, implemented (to a degree) and researched in different guises ab immemorabili (since time immemorial). A particularly inspirational model is Marvin Minsky's *Society of Mind* (1988), which incorporates principles such as self-organizing robots, mechanisms, communication, cooperation, and mapping and positioning.

Beyond just robot or human performing one act of pedagogy we should aim at creating an ecosystem of human and robotic agents, organized where needed, proactive, not bound to one place or time, with constant dialogue and feedback between the student and others and all among themselves, constantly gathering knowledge about individual student, all students, pedagogy methods, learning material and the real-life learning facts. The system constantly rebuilds itself through Darwinian evolution of competing mechanisms acting in continuous dialogue with all involved in the learning process, including agents and resources that in the past, in less connected learning environments would not have been possible.

The fundamental sense of organics is that of nature, especially living nature. So, the ASER organics would propose such natural aspects as:

- Using natural language
- Using natural locations
- Learning in and through natural context and problems
- Using natural physical instrument toolbox in natural lab instead of classroom
- Using natural sciences approach
- Using natural though instruments in addition to algorithmics
- Creating a system of organs functioning as a living organism organization
- Regarding the ASERs as an ecosystem
- The student and all others in the ecosystem could, should and would behave naturally
- Natural Darwinian evolution of the organs through learning of natural environment

- Darwinian evolution through competing organs and feedback.

The learning of the student, the teacher and all others in the ecosystem is organic in all the above senses. It is an evolution of the constructive model of the material studied with the lifecycle of first prototype and then evolving it through dialogues. Some dialogues are experiments (mental and physical) involving the real world (lab) or solving problem and exercises. Some dialogues are with other organs—teachers (both local and remote), parents, peers. And the main organ that is ambient, omnipresent, always around and proactively starting the endless continuous dialogue that both teaches and encourages learning and acquiring skills (both hard and soft) is Gaia avatar—the personal individually customized always evolving alter ego of the student—the ASER.

Use case: the student is starting an educational dialogue by asking the ASER to teach him a topic. He can do it in natural language. He can formulate his question in very non-exact even erroneous terms, but the ASER will understand what he has meant using his knowledge about the topics, specific student, general educational knowledge. The ASER presents to him his guess of what the students has asked or a number of alternatives. For instance, the student has asked about negative roots. The ASER understands that he was not asking about imaginary numbers (where roots of negative numbers are legal) as he is only twelve years old and studies only introduction to algebra in the seventh grade. But he doesn't know whether the student asks about negative roots in the sense of $-\sqrt[a]{b}$ (as for instance $-\sqrt[2]{4}$) or $-(\sqrt[a]{b})$ (as for instance $-(\sqrt[2]{4})$) or in the sense of $-\sqrt[a]{-b}$ (as for instance $-\sqrt[2]{-4}$) or in the sense of $\sqrt[-a]{b}$ (as for instance $\sqrt[-2]{-4}$). Or maybe he just asks whether there is a solution of a square root that is negative (solution that does always exist) or he just asks whether there is a solution of any root that is negative (solution that does not always exist) and so on. ASER uses his knowledge to rank his alternatives and presents them to the student. The student answers him. If the student's answer does not resolve the doubt, the ASER will send a request to other organs of Gaia, previously established as most relevant. The ASER presents the new prototype of the definition of the topic to the student as next step of the dialog. If the student new answer in this next stage of the dialog is still not enough, the ASER can activate even more intricate mechanisms, like broadcasting his query as higher-level system-wide event, or ask a human, for instance the parent that is immediately available. If the dialogue with the parent will not be enough ASER can start a remote dialogue with the student's teacher or another teacher. He can also query the school database to infer the educational school context of the current studies of this specific student. After as many stages of the dialogue as needed or as the resources allow for (the ASER and all other organs determine the reasonable time to spend on this dialogue) ASER will formulate one or more competing prototypes of the next lesson and start it. The lesson itself will follow the same organic paradigm.

We propose to call our ecosystem Gaia for historical reasons. The name comes from Earth in Ancient Greek that became the name of the Earth goddess. So, in later times, when there was research of a whole ecosystem, especially that of the Earth, the name Gaia was used. The Gaia hypothesis, or Gaia theory, is sometimes described as being originated in some sense by different researchers at different times, but it is agreed that at least since the 1970s it is an established theory (though not accepted by everyone) (Onori & Visconti, 2012). In very simplistic terms, the different versions and variations of the hypothesis argue that the Earth is influenced by much more factors then was previously thought and it should be seen more holistically as one system (Bravo, 2004).

The general public recognition and even fame the approach achieved is due to the great science fiction visionary Isaac Asimov. He created, and described in great detail, the Gaia planet where all things, whether animated or not, from stone to human were a part of Gaia. Gaia had consciousness that was the result of the sum of all things constituting it. Of course, humans contributed much more than an animal and animal much more than a tree (Asimov, 2012).

We propose a system that is even more evolved than the others previously named Gaia. It has, in addition to all possible interactions and communications between all agents, human as well as computerized and robotic, another level of organizational intelligence. The ecosystem gathers, organizes and remembers institutional knowledge of metadata. While evolving it gains awareness of not only how to react better but also second order techniques of how to evolve better.

Gaia is a society and organization of human and non-human agents. The non-human agents could be of all types of automation, including computers, embedded systems, IoT, computers, smartphones, smart devices, robots, social robots and ASERs. Human agents are all stakeholders from students through parents, caregivers, administrators, teachers, experts of all kinds, designers and developers, researchers and educators, academia, industry and government.

The agents organize into and use mechanisms (that could be seen also as organs of this organic ecosystem). Mechanisms of Gaia include first of all the educational units, the material, information to be learned stored in knowledge bases distributed all over and in central knowledge base. It also includes the metadata gathered about each unit and each student. Some of the knowledge is distributed and kept locally around a student, a teacher, a classroom, a robot, a computer. The organization could include higher level knowledge bases with more general applications like a community, a group, a cohort. There would be of course also mechanisms for monitoring, recording, feedback, learning and evolving, management, communication, dialogue of many kinds. Those would start very slowly as simplest prototypes, and evolve into more sophisticated ones.

6.1 Crowdsourcing

With ASER crowdsourcing could become very important in the educational ecosystem such as Gaia, on the one hand, and give the crowdsourcing approach a new life, on the other hand. Crowdsourcing is seen sometimes in too simplistic a way. The classical example that is frequently brought is that of a random crowd that has to guess something he does not really have a knowledge of, special tools or expertise, like guessing the weight of a random cow with which they have no previous experience. Even such a simple tool can be very beneficial in many problems, and as a matter of fact their guess sometimes is right on the money so to speak. Actually, there could be proposed and used many other, more intricate modes of crowdsourcing.

Educational crowdsourcing could include not random but quite specifically chosen, not disinterested but interested (like parents), not layman but experts, or having direct individual special experience and knowledge. The opinions could be formalized and standardized, proactive, analyzed, processed and revised. The process could be multi-stage and multi-level. We are talking about all agents related to the educational process as well as outside experts grow their contributions into a very strong and valuable educational knowledge base through organic dialogue and real-life feedback with better and better prototypes.

As we have seen in a very wide pan-European study we participated in with researchers from majority of countries of Europe, teachers have a great desire to contribute in such a way, yet they don't do it because of two main reasons (Arhar Holdt et al., 2020). Those are, in order of importance, lack of acknowledgement, organization and reward (not necessarily pecuniary) and lack of tech support.

Use case: A teacher prepares his lesson, gathers material, organizes, adds multimedia, presents and records and he does all that in the Gaia infrastructure, so all is automatically saved to be used by others. The teacher does it seamlessly with no technical knowledge needed and if it is needed, he gets the support efficiently and smoothly. He knows that students have benefited from his efforts. He can be in direct dialogue with them if he chooses to do it. His efforts are recognized and highly praised and even rewarded in many ways. His material is tagged and organized to maximize its use.

6.2 Gaia in Education

An educational robot can engage with students in a wide range of interaction modes, often embodying multiple roles simultaneously. It can serve as an all-knowing, wise Yoda-like sensei, a funny friend, a lab partner, an assistant, a safety officer, and as many other personas as needed to enrich the educational experience. One of the main advantages of ASER is that he greatly encourages and facilitates the more advanced pedagogics, much more than the human teacher could. The consensus approach to pedagogy is that of creative constructive engineering lab learning.

7 ASER Use Cases

There is a great gamma of different use cases, situations where ASER could be useful in the education process. There are many different contexts and many functions ASER could perform in the specific context. We shall call the basic component of educational situation with many features specified a use case. Use cases could be seen as building blocks of analysis and synthesis of the educational process. Use case (or just a case) is not a quantum, indivisible and not allowing further, more in depth, analysis. It is more of an atom or a molecule, that has an inner structure and could be analyzed but, in many contexts, would not be subject of further division, but would serve as basic unit, as that is the most appropriate for the current inquiry.

For every use case there is a range of ASERs from minimal, where a weaker one will not do, to a maximal one, where a stronger one is clearly overqualified.

Case 1: one of the minimal sophistication cases would be the following one. During regular lesson ASER is present and thus motivates the student, helps him concentrate and enjoy the lesson more. Even when ASER does not take part in the lesson in any capacity, its presence and activities, though not directly coordinated with the teacher or the lesson, make the lesson more interesting for the student.

Case 2: ASER as passive student. As a peer student ASER can react to the lesson, to the teacher, student or to both of them. His participation can vary from minimal standard behavior and reactions to more sophisticated ones. He can ask questions, he can add other aspects that will be beneficial for the student, such as alleviating tensions.

Case 3: ASER as active student. ASER can be a smart fully engaged proactive student. He can initiate queries, engage in dialog with the teacher, bring up information, do exercises.

Case 4: ASER as peer student. ASER can cooperate with the student. Student dialogue with ASER can be initiated by either side. ASER can be pre-programmed, or the robot can just improvise. ASER can assess student's work and behave in the way most helpful. For example, if the student starts to lose interest or concentration ASER can become more active. If the student needs a break ASER can talk to the teacher and in this way allow the student to assume a more passive role. Being not alone, having a counterpart can benefit the student in many ways. For instance the student could become the teacher. He can feel he knows more. He can bounce his ideas off ASER. The student can learn better when he has to voice his understanding.

Case 5: ASER as team member of the student. ASER here has a specific role in which it is better than the student. As in many teams, ASER in his role does something, or demonstrates something, or is engaged in dialog with the student where he informs him, teaches him, makes him do something, corrects students mistakes, completes his answers. Student benefits also by the dialogue and coordination.

Case 6: ASER as friend. ASER can help the educational process not as teacher or even someone that knows more than the student. He can be a friend that in his behavior or dialogue with the student improves student's mental state, encourages him, makes him enjoy the learning, makes him more confident, more curious, more active.

Case 7: ASER as teacher. ASER is present in student's home. ASER initiates a dialog where he asks the student to join him in a lesson. Student tells him if he is interested at all or in what he is interested.

8 Conclusions and Future

The future of ASER is bright. The present already has many bright points, among which we could mention: the great enthusiasm about AI and robotics, the readiness of pupils and teachers, crowdsourcing and access online to great educational resources, huge plethora of robots, communication channels, hardware and software. Those advantages should be used, and used as soon as possible. The fact that they are not is one of the more problematic aspects of today's educational systems.

What should we do to solve the paradox of great availability and need of ASER on the one hand, and the minimal use in the classroom and at student's home, on the other. We have researched for a few years now in our Educational Robotics Lab several approaches to ASER. We have analyzed here one approach we propose to adopt as the one most practical, that of specially designed but cheap robot as terminal that is serving as avatar for a great very sophisticated ecosystem of educational resources and agents, both computerized and human. We call this system Gaia.

The journey of ASER is at an exciting crossroads. With the rise of ASER systems such as Gaia, the educational landscape will undergo a transformative shift, where every learner could have access to personalized, socially engaging educational experiences through an affordable, accessible network of intelligent robotic avatars. Gaia's organic knowledge ecosystem, supporting scalable ASER models, will offer a pathway toward more equitable and inclusive education by making advanced learning resources available to all, regardless of the socioeconomic status.

Gaia can start to be implemented in the nearest future and evolve to very intricate ecosystem with use. Gaia's holistic design, rooted in concepts of anthropomorphism and organic knowledge, is built to grow and adapt over time, incorporating crowdsourced insights and evolving through real-world feedback loops. This dynamic adaptability will position Gaia not only as an educational tool but as a lifelong learning companion for students and educators alike. By facilitating deep emotional and social engagement, Gaia will contribute positively to mental well-being and motivation, crucial elements often under-addressed in traditional education.

In the more distant (but not that distant) future we predict that every student on the planet will be to some degree part of Gaia. And as people learn throughout their life, we predict that everyone will have the advantage of his own, individualized friendly

ASER on the one hand, and deep connection to the worldwide educational and not just educational community on the other hand.

Ultimately, the future of ASER lies in its potential to bridge the gap between traditional and digital education, offering a fusion of human empathy and robotic intelligence that fosters a generation of independent, motivated learners. By creating an ecosystem that can scale from local classrooms to global networks, Gaia and similar ASER paradigms represent a vision where education is as dynamic, interconnected, and resilient as the world we inhabit.

References

Alam, A. (2021, December). Should robots replace teachers? Mobilisation of AI and learning analytics in education. In *2021 International conference on advances in computing, communication, and control (ICAC3)* (pp. 1–12). IEEE.

Arhar Holdt, Š., Zviel-Girshin, R., Gajek, E., Durán-Muñoz, I., Bago, P., Fort, K., Hatipoglu, C., Kasperavičienė, R., Koeva, S., Lazić Konjik, I., Miloshevska, L., & Zanasi, L. (2020). Language teachers and crowdsourcing: Insights from a cross-European survey. *Rasprave: Časopis Instituta za hrvatski jezik i jezikoslovlje, 46*(1), 1–28. https://doi.org/10.31724/rihjj.46.1.1

Asimov, I. (2012). *Foundation and earth* (Vol. 5). Del Rey.

Baraka, K., Alves-Oliveira, P., & Ribeiro, T. (2020). An extended framework for characterizing social robots. *Human-Robot Interaction: Evaluation Methods and Their Standardization*, 21–64. https://doi.org/10.1007/978-3-030-42307-0_2

Benjamin, L. T. (1988). A history of teaching machines. *American Psychologist, 43*(9), 703. https://doi.org/10.1037/0003-066X.43.9.703

Belpaeme, T., Kennedy, J., Ramachandran, A., Scassellati, B., & Tanaka, F. (2018). Social robots for education: A review. *Science Robotics, 3*(21), eaat5954. https://doi.org/10.1126/scirobotics.aat5954

Bravo, G. (2004). *Gaia, our new common: Some preliminary questions on earth system science and common-pool resources theory in the study of global human/environment relationships.*

Casey, R. H. (2008). *The model T*. Johns Hopkins University Press.

Chalmers, C., Keane, T., Boden, M., & Williams, M. (2022). Humanoid robots go to school. *Education and Information Technologies, 27*(6), 7563–7581.

Chen, T. Y., Chiu, Y. C., Bi, N., & Tsai, R. T. H. (2021). Multi-modal chatbot in intelligent manufacturing. *IEEE Access, 9*, 82118–82129. https://doi.org/10.1109/ACCESS.2021.3083518

Clark, R. E. (2004). The classical origins of Pavlov's conditioning. *Integrative Physiological & Behavioral Science, 39*, 279–294. https://doi.org/10.1007/BF02734167

Darwin, C., & Beer, G. (1951). *The origin of species* (Vol. 71, p. 473). Dent.

Davis, P. M., & Class, M. (2023). Educational robots for social and emotional learning. *AI, Computer Science and Robotics Technology*, (13). https://doi.org/10.5772/acrt.26

Dautenhahn, K. (2007). Socially intelligent robots: Dimensions of human–robot interaction. *Philosophical Transactions of the Royal Society B: Biological Sciences, 362*(1480), 679–704.

de Graaf, M. M. A., & Allouch, S. B. (2013). Exploring influencing variables for the acceptance of social robots. *Robotics and Autonomous Systems, 61*, 1476–1486. https://doi.org/10.1016/j.robot.2013.07.007

Derakhshan, A., Teo, T., Saeedy Robat, E., Janebi Enayat, M., & Jahanbakhsh, A. A. (2024). Robot-assisted language learning: A meta-analysis. *Review of Educational Research*, 00346543241247227. https://doi.org/10.3102/00346543241247227

Dewey, J. (1958). *Experience and nature* (Vol. 471). Courier Corporation.

Dubois-Sage, M., Jacquet, B., Jamet, F., & Baratgin, J. (2023). We do not anthropomorphize a robot based only on its cover: Context matters too! *Applied Sciences, 13*(15), 8743.
Duffy, B. R. (2003). Anthropomorphism and the social robot. *Robotics and Autonomous Systems, 42*(3–4), 177–190.
Eli, S., Hausman, J. K., & Rhode, P. (2023). *The model T* (No. w31454). National Bureau of Economic Research.
Fong, T., Nourbakhsh, I., & Dautenhahn, K. (2003). A survey of socially interactive robots. *Robotics and Autonomous Systems, 42*(3–4), 143–166.
Han, J. H., Jo, M. H., Jones, V., & Jo, J. H. (2008). Comparative study on the educational use of home robots for children. *Journal of Information Processing Systems, 4*(4), 159–168.
Jamil, N., Saghir, F. N., Alshanqiti, H., Almansoori, A. K. A., Saeed, A., Ahmad, A., & Belkacem, A. N. (2024, March). On combining the potential of social robots and ChatGPT for enhanced learning. In *2024 12th international conference on information and education technology (ICIET)* (pp. 226–231). IEEE. https://doi.org/10.1109/ICIET60671.2024.10542738
Lee, T. C. (2023). Toward ethical deployment of social robots in education. *Journal of Public Value*, 69–73.
LeTendre, G. K., & Gray, R. (2024). Social robots in a project-based learning environment: Adolescent understanding of robot–human interactions. *Journal of Computer Assisted Learning, 40*(1), 192–204. https://doi.org/10.1111/jcal.12872
Martín-García, R., Marcos-Pablos, S., Martín-Gómez, L., Cordero-Gutiérrez, R., & de la Iglesia, D. H. (2022, July). CoCoBot, a low cost humanoid robot companion for elderly people. In *International conference on disruptive technologies, tech ethics and artificial intelligence* (pp. 232–243). Springer International Publishing.
Martinez-Roig, R., Cazorla, M., & Esteve Faubel, J. M. (2023, March). Social robotics in music education: A systematic review. In *Frontiers in education* (Vol. 8, p. 1164506). Frontiers Media SA. https://doi.org/10.3389/feduc.2023.1164506
Minsky, M. (1988). *Society of mind*. Simon and Schuster. https://doi.org/10.1016/0004-3702(91)90036-J
Montessori, M. (1917). *The advanced Montessori method* (Vol. 1). Frederick A. Stokes Company.
Montessori, M. (2013). *The Montessori method*. Transaction Publishers. https://doi.org/10.4324/9781315133225
Muir-Herzig, R. G. (2004). Technology and its impact in the classroom. *Computers & Education, 42*(2), 111–131.
Ntomora, P. I., & Petrakos, K. (2024). Leveraging social network analysis in education through social robots: A review. *Global Journal of Engineering and Technology Advances, 20*(01), 055–066.
Online Etymology Dictionary. https://www.etymonline.com/word/chassis#etymonline_v_8459
Onori, L., & Visconti, G. (2012). The GAIA theory: from Lovelock to Margulis. From a homeostatic to a cognitive autopoietic worldview. *Rendiconti Lincei, 23*, 375–386. https://doi.org/10.1007/s12210-012-0187-z
Papadopoulos, I., Lazzarino, R., Miah, S., Weaver, T., Thomas, B., & Koulouglioti, C. (2020). A systematic review of the literature regarding socially assistive robots in pre-tertiary education. *Computers & Education, 155*, 103924. https://doi.org/10.1016/j.compedu.2020.103924
Rofi'ah, B., Fakhrurroja, H., & Machbub, C. (2021). Dialogue management using reinforcement learning. *TELKOMNIKA (Telecommunication, Computing, Electronics and Control), 19*, 931–938. https://doi.org/10.12928/telkomnika.v19i3.18319
Rousseau, J. J., Foxley, B., & Jimack, P. (1911). *Emile* (Vol. 1, p. 324). Dent.
Scheick, R., Meijer, E., & Heerink, M. (2018). Choosing a robot with ASD children. In *The 3rd international conference on social robots in therapy and education* (p. 17).
Sedubun, S., & Nurhayati, N. (2024). Exploring the efficacy of project-based learning in English language teaching: A literature review. *EDUKASIA: Jurnal Pendidikan dan Pembelajaran, 5*(1), 1089–1092. https://doi.org/10.62775/edukasia.v5i1.949
Skinner, B. F. (1958). Teaching machines: From the experimental study of learning come devices which arrange optimal conditions for self-instruction. *Science, 128*(3330), 969–977.

Skinner, B. F. (1963). Reflections on a decade of teaching machines. *Teachers College Record, 65*(2), 1–9. https://doi.org/10.1177/016146816306500209

Smakman, M., Vogt, P., & Konijn, E. A. (2021). Moral considerations on social robots in education: A multi-stakeholder perspective. *Computers & Education, 174*, 104317. https://doi.org/10.1016/j.compedu.2021.104317

Smakman, M. H. J., Preciado Vanegas, D. F., Smit, K., Leewis, S., Okkerse, J., Obbes, J., et al. (2022). Robot buddy for primary school children. *Multimodal Technologies and Interaction, 6*, 29. https://doi.org/10.3390/mti6040029

So, S., & Lee, N. (2023). Pedagogical exploration and technological development of a humanoid robotic system for teaching to and learning in young children. *Cogent Education, 10*(1), 2179181.

Standing, E. M. (1998). *Maria Montessori: Her life and work.* Penguin.

Stock-Homburg, R. (2022). Survey of emotions in human–robot interactions: Perspectives from robotic psychology on 20 years of research. *International Journal of Social Robotics, 14*(2), 389–411.

Thellman, S., De Graaf, M., & Ziemke, T. (2022). Mental state attribution to robots: A systematic review of conceptions, methods, and findings. *ACM Transactions on Human-Robot Interaction (THRI), 11*(4), 1–51.

Turing, A. M. (1950). Mind. *Mind, 59*(236), 433–460.

Vlieghe, J. (2022). Sharing love for a common world on and off screen. A pedagogical and technocentric account. *Journal of New Approaches in Educational Research, 11*, 1–14. https://doi.org/10.7821/naer.2022.1.855

Vygotsky, L. S. (1978). Mind in society: The development of higher psychological processes. *Harvard UP*. https://doi.org/10.2307/j.ctvjf9vz4

Wiener, N. (2019). *Cybernetics or control and communication in the animal and the machine.* MIT Press. https://doi.org/10.7551/mitpress/11810.001.0001

Woo, H., LeTendre, G. K., Pham-Shouse, T., & Xiong, Y. (2021). The use of social robots in classrooms: A review of field-based studies. *Educational Research Review, 33*, 100388. https://doi.org/10.1016/j.edurev.2021.100388

Yusri, R., Yusof, A. M., & Latef, A. S. A. (2024). A systematic literature review of project-based learning: research trends, methods, elements, and frameworks. *International Journal of Evaluation and Research in Education.* https://doi.org/10.11591/ijere.v13i5.27875

Zhao, X., & Malle, B. F. (2022). Spontaneous perspective taking toward robots: The unique impact of humanlike appearance. *Cognition, 224*, 105076. https://doi.org/10.1016/j.cognition.2022.105076

Złotowski, J., Proudfoot, D., Yogeeswaran, K., & Bartneck, C. (2015). Anthropomorphism: opportunities and challenges in human–robot interaction. *International Journal of Social Robotics, 7*, 347–360. https://doi.org/10.1007/s12369-014-0267-6

Zviel-Girshin, R., Luria, A., & Shaham, C. (2020). Robotics as a tool to enhance technological thinking in early childhood. *Journal of Science Education and Technology, 29*(2), 294–302. https://doi.org/10.1007/s10956-020-09815-x

Zviel-Girshin, R., & Rosenberg, N. (2016, May). Montessori + SE + ICT = SEET: Using SE to teach SE. In *Proceedings of the 5th international workshop on theory-oriented software engineering* (pp. 1–7). https://doi.org/10.1145/2897134.2897140

Zviel-Girshin, R., & Rosenberg, N. (2018). Child friendly robotics. In *Proceedings of the world congress on engineering* (Vol. 1).

Educational Robots: Past, Present, Future

Dave Catlin, Mike Blamires, and John-John Cabibihan

Abstract Over fifty years ago, Seymour Papert invented the Turtle, the world's first educational robot. In doing so, he created tools that supported a Constructionist approach to education. A hero to many people, he challenged the established teaching methods and principles, upsetting traditional educators. What were Papert's educational ideas? When he began formulating them, robots belonged to the pages of science fiction stories; we now have robots with what some people call Alien intelligence. These robots have become so sophisticated that Saudi Arabia has granted citizenship to one, the social robot Sophia. Many countries ponder the status of such robots; do they need protection from humans or humans from them? How will they change our world? Should we invite them to our schools and ask them to look after and educate our children? Whether we should or shouldn't, we continue developing technology that could become robot teachers. Will that bolster the traditional factory education model, or do Papert's ideas still have value and offer a better way forward? We will examine the background that enables us to offer an answer to these questions. In so doing, we'll show what kind of education system the future needs and suggest a different design approach and a novel way of using social robot features to take advantage of the technology in a constructionist environment.

Keywords Social robots · Seymour Papert · User robots · Turtles · Roamer® · Constructionism

D. Catlin (✉)
DC Education Consultant, London, UK
e-mail: dave@valiant-technology.com

M. Blamires
RIPPLE: Research Initiatives for Participation and Progress in Learning Environments, Canterbury, UK
e-mail: Mike.Blamires@ripplesinlearning.com

J.-J. Cabibihan
Mechanical and Industrial Engineering Department, Qatar University, Doha, Qatar
e-mail: john.cabibihan@qu.edu.qa

G. Lampropoulos and S. Papadakis (eds.), *Social Robots in Education*, Studies in Computational Intelligence 1194, https://doi.org/10.1007/978-3-031-82915-4_16

1 Introduction

The first educational robot, the Turtle, came to life in January 1970. We will review the academic credentials of this robot before peeking into the future and suggesting how social robots can combine with these early efforts.

> Margie always hated school, but now she hated it more than ever. The mechanical teacher had been giving her test after test in geography and she had been doing worse and worse until her mother had shaken her head sorrowfully and sent for the County Inspector (Asimov, 1951).

In Issac Asimov's story, **The Fun They Had**, personal robot teachers teach home-schooled students, and the Inspector is a robot mechanic. Margie's friend, Tommy, had found 'a real' book telling them about school centuries ago. Margie complained the school still had a teacher, prompting Tommy to say, "*it wasn't a regular teacher. It was a man.*" Shocked, Margie responded, saying she wouldn't want a strange man in her house. Tommy explained that in those days all the children went to a building, and the teacher taught everyone at once. "*But my mother says a teacher has to be adjusted to fit the mind of each boy and girl it teaches, and that each kid has to be taught differently.*"

According to Sir Anthony Seldon, Vice Chancellor of the University of Buckingham, "*Robots will begin replacing teachers in the classroom within the next ten years in a revolution in one-to-one learning*" (Bodkin, 2017). If robots were to replace teachers, we need to understand what they're replacing. What makes a good teacher? What do they do? Suzzane Bouffard claims teaching is an art and a science (Bouffard, 2018); can robots simulate a flair for teaching, or would they have stilted personalities and teach robotically? We must consider what makes good teaching: will the robot crank out questions, or is something else possible? How must we organise education to support the regular use of robots? Asimov's story imagines home-schooling with the robot teacher firing test questions at the hapless Margie. Is this practicable? Is it desirable? These questions raise ethical, psychological, and social issues we must address.

We will examine the work of Seymour Papert, the father of educational robots. While his technology belongs in a museum, his educational principles don't. We purposefully use dated references because they're relevant, valid, and timeless (you can trace good educational practice back to Confucius, Aristotle and others). We propose that these ideas should provide the foundations for all educational robots, with or without social capabilities.

Two ideas underpin Papert's educational thinking. First, he didn't want to find a better way to teach, he tried to find a better way for pupils to learn (Papert, 1993, pp. 82–105); second, he said: "*One might say the computer is being used to program the child. In my vision the child programs the computer*" (Papert, 1980, p. 5). Reviewing his work, we uncover ways to adapt robots with social abilities and keep his principles intact. In doing this, we will reference the **Educational Robotic Application (ERA) Principles**, which Catlin and Blamires wrote based on 3 decades of working with robots in schools (Catlin & Blamires, 2010).

Then we'll appraise the history and ideas of social robot teachers and teaching assistants before looking at some prophecies of their future. We'll critique the political and economic ramifications of robot teachers and classroom assistants. Then, we'll finish with an alternative idea of robots as teaching aids managed by teachers. First, we want to answer a simple question, "*What is the purpose of education?*" Philosophers have debated the issue for aeons; countries adopt different strategies, and the approach in many countries changes with the political wind. Despite this, 192 countries have approved UNESCO's Rights of the Child (UNESCO, 1989). The associated document **The Right to Education** (UNICEF, 2001) elaborates on the convention, effectively defining the aims of education and implying how we should implement it:

> The development of the child's personality, talents and mental and physical abilities to their fullest potential.
>
> Education must be child-centred and empowering. This applies to the curriculum as well as the educational processes, the pedagogical methods and the environment where education takes place.
>
> Education must be provided in a way that respects the inherent dignity of the child and enables the child to express his or her views in accordance with article 12 (1) and to participate in school life.
>
> Education must include not only literacy and numeracy but also life skills such as the ability to make well-balanced decisions; to resolve conflicts in a non-violent manner; and to develop a healthy lifestyle, good social relationships and responsibility, critical thinking, creative talents, and other abilities which give children the tools needed to pursue their options in life.

This answers our question and sets our objectives: school education should benefit the quality of the child's life, but job training isn't the responsibility of school education—that comes after school. Plato supported the notion that we should educate people for a role in life (Cooper, 2001). Aristotle thought we should develop a pupil's talents; however, he realised that talented people would cope with the vagaries of the future better than those trained for specific tasks (Hobson, 2001).

We will refer to the EduRobot Taxonomy, which classifies robots (not including virtual robots) used in education, ranging from toys to the most sophisticated robots laden with AI (Catlin et al., 2018a, 2018b, 2018c):

- **User Robots**: you can take them out of the 'box' and immediately use them in a lesson (Example: Turtles).
- **Build Robots**: students learn by building the robots for a purpose (Example: Lego Mindstorms).
- **Social Robots**: Autonomous robots that, "*Interact and communicate among themselves, with humans, and with the environment and are designed to operate according to the established social and cultural norms*" (Li et al., 2011) (Example: Kaspar).

We must explain a taxonomic rule with the example of a platypus, a mammal with reptilian characteristics (it lays eggs and the male is poisonous). Despite these traits it's still a mammal. You can build a Turtle (User Robot) using a Build Robot kit like Lego, but this doesn't make Lego a User Robot.

We will start by summarising Seymour Papert's work on Logo and Turtle robots, emphasising his educational ideas and motivations. This section includes input from other researchers involved in this early phase. Then, we will discuss his development of constructionism, which led to the development of Lego's robotic products and connected to the broader maker movement. Due to unique conditions, the UK universally took up Logo and robots; they'd also independently developed constructionism under the title of Design Technology. We explain these conditions, offering them as a prerequisite to our recommended approach to using robots. In the final part of this section, we address Papert's critics, rebuffing their arguments.

In the following sections, we answer our questions about schools and teachers; we present arguments that do not involve technology but agree with educational principles that support Papert's approach. We then review the historical development of social robots, critically examine the case of robot teachers, and discuss the different strands of information together. We conclude by proposing a way of using educational robots with social capabilities that will best suit the education of our children.

2 Seymour Papert

Seymour Papert had a degree in philosophy, and two mathematics doctorates, he also worked with the developmental psychologist Jean Piaget before joining London's National Physics Laboratory computer team. In 1963 he went to MIT, then a hotbed of AI development. He wasn't out of place; his technical skills matched those of his contemporaries. But he also had a passion for education, leading him to develop Logo, a LISP like computer language for learning and the world's first educational robot, the Turtle (Fig. 1). Papert explains:

> [Logo] We tried to achieve a number of goals, first of all that it should be easily accessible, there should be corners of the language that you can get into like baby talk in English that are easy for the youngest beginner. But it shouldn't be a toy language. It's not that Logo is easy, it's easy to get into but once you're in there you can progress to the most sophisticated ideas in the world of programming just as in English you can start off with baby talk but the same language is also the vehicle for the expression of poets and philosophers (Papert, 1983a).
>
> The Turtle started because I wanted something that could be picked up by very young children. Now I think of it as something that is just as valuable for people of any age as it is for the young children and again that's a general principle, what's good for thinking is good for thinking whether we're a 5 year-old or a sophisticated scientist (Papert, 1983b).

These aims led to people describing the language as low floor high ceiling. Later, Mitchel Resnik adjusted Papert's phrase to low floor, high ceiling and wide walls. "*It's not enough to provide a single path from a low floor to a high ceiling; it's important to provide multiple pathways*" (Resnik, 2018, p. 64). The metaphor also applies to the number of subjects students can explore with the software and robot.

Fig. 1 The valiant turtle 1985

The instructions needed to control the robot became known as Turtle Graphics; it formed a new geometry. Ableson and diSessa compared Turtle with Cartesian Geometry: Turtle Geometry code describes the nature of, for example, a rectangle; Cartesian Geometry code doesn't (Fig. 2). The Cartesian Geometry presents an abstract description of a rectangle related to the coordinate framework. The Logo code describes a rectangle in the form of a procedure, which the student can walk through, a process called 'Playing Turtle' (Ableson & diSessa, 1981, pp. 11–15).

Papert called the Turtle a 'transitional object' because it links the student's body awareness to the growth of their mental models. Sylvia Weir explained, "*The inventors of Logo treated the physical Turtle as much the same as the screen Turtle, but children do not*" (Weir, 1987, p. 156). The **ERA Embodiment Principle** states, "*Students learn by intentional and meaningful interactions with educational robots situated in the same space and time.*" We will reference Logo research when the ideas also apply to physical robots.

Papert explained his methods in a TV programme which showed children describing their projects (Papert & Solomon, 1971). The first task in any lesson is to engage a student's interest—hard to do for maths. But the children had decided what they wanted to do, which isn't unusual in an art lesson but rare for maths

Logo Code

```
TO RECTANGLE SIDE1 SIDE2
      REPEAT 2
           SIDE1
           RT 90
           SIDE2
           RT 90
```

Cartesian Code

```
TO RECTANGLE (WIDTH, HEIGHT)
       SETXY (WIDTH,0)
       SETXY (WIDTH, HEIGHT)
       SETXY (0, HEIGHT)
       SETXY (0, 0)
```

Fig. 2 Comparison of a logo (turtle graphic) with Cartesian code for a rectangle

lessons? They were working as mathematicians, fearlessly straying into topics done by older students.

Papert also drew attention to the girl who made a mistake. "*The child doesn't have to be told by the teacher whether [she's] right or wrong [she] can see for [herself] whether it works, that's what science and knowledge are about.*" Unlike school maths with right and wrong answers, Papert gave students tools they could use to explore mathematical ideas, and debugging is a natural part of programming and learning. Using neural scans, Jason Moser and colleagues at Michigan State University have shown that making mistakes can increase brain activity and improve learning capacity (Moser et al., 2011).

Papert agreed with the approach of Hungarian-American George Polya, who claimed his students knew many mathematical facts but didn't understand how to act as mathematicians. Polya introduced a set of mathematical heuristics used by mathematicians to study problems. Papert showed how working with Logo-and-Turtles gave students tools which naturally used Polya's rules-of-thumb; instead of focusing on mathematical facts and procedures, students learnt by doing and solving problems. However, you make a mistake if you think they don't learn mathematical facts; their project dictates what they need to know, not an arbitrary curriculum.

When reviewing Papert's book, The Children's Machine, Daniel Dennett echoes these ideas when mentioning Logo. He comments, "*But one can learn even more from 'mistakes' than from a string of successes—that is a central tenet of Papert's vision of learning, and he practices what he preaches*" (Dennett, 1993, pp. 45–46). Dennett testifies his experience with undergraduates who were, "*experts at piling in the facts, drilling for the big test—and they were pathologically uncomfortable in any setting where they had to think.*" He ran a course using Logo which he thought was spectacularly successful: "*The students forgot their phobias and inhibitions and took flight, creating a trove of idiosyncratic projects, effortlessly learning the fundamentals of programming, and building a robust base on which we could then help them construct a more 'adult' set of edifices.*"

High school mathematics teacher Albert Cuoco got his students to use Logo to explore algebraic topics like functions, counting and mathematical induction, and number theory. In the Introduction to his book, he explains how Papert's ideas reflect good teaching practice.

> In school, the idea of using experiments to develop both conjectures and plausible arguments to support the conjectures was almost never discussed, and I arrived in high school thinking (like many people) that mathematics is a set of facts and procedures. Luckily, my high school mathematics teacher saw things differently; he immersed his classes in carefully constructed problems that required pages of calculations. These calculations were really mathematical experiments, and wonderful patterns emerged from them. I began to understand something that wouldn't be clear to me for almost another decade: the heart of mathematical research is tinkering (Cuoco, 1990, p. 7).

The phrase 'carefully constructed problems' connects us to Papert's idea of microworlds. Michael Friendly offers a clear definition: "*Microworlds are concrete models of objects, relationships, and ideas in areas that usually presuppose some degree of abstract thinking*" (Friendly, 1988, p. 5). Letting students explore a number line with a robot is a simple example. British mathematician John Wallis introduced the idea of the number line in his **Treatise of Algebra** (Wallis, 1685, p. 265); effectively, he 'Played Turtle' demonstrating arithmetical operations by walking forward (addition) and backward (subtraction) along the line. Programming the robot to travel up and down the line presents a microworld full of possibilities beyond number facts like one-plus-one is two; pupils can discover the laws of arithmetic (commutative, associative distributive) and concepts like ordinal and cardinal numbers. Children as young as six have discovered 'numbers going the other way' (negative numbers) and grasped the idea of equivalence. They don't necessarily learn these things formally but they gain experience that future lessons can build on. They construct visual, spatial and kinesthetic models that make it easier to understand basic ideas. For example, ad hoc tests show that most adults forget phrases like two-minuses-are-a-plus but recall the mental models of negative number subtraction generated by Turtles exploring a number line. Rote learning methods like the two-minus rule, muddle mathematical ideas and can confuse students. For example, the sum $3 - (-4)$, if read as 'three minus, minus four', is mathematically incorrect. Minus is an operation, whereas (-4) is the symbol for the number name (negative 4)—a position on a number line. Using the robot to uncover these patterns emphasises subtraction as the 'difference' between two numbers.

Further serious studies showed the ability of Logo and Turtle robots to create microworlds to engage students in a diverse range of topics from beginner to expert levels. For example: Mathematics (Hoyles & Noss, 1992; Hoyles & Sutherland, 1989), Language (Goldenberg & Feurzeig, 1987), Visual Modelling (Clayson, 1988), Computer Science (Harvey, 1997a, 1997b, 1997c) and Special Needs Education (Weir, 1987). Classroom teachers have used Turtles in many other subjects: science, history, geography, art, design technology, social studies, religion…

It's interesting to compare Albert Einstein's views on education and how they succinctly align with Papert's thoughts (Einstein, 2023).

> The beauty of microworlds lies in their ability to engage students in meaningful, self-directed learning. By experimenting within a microworld, learners develop a deeper, more intuitive understanding of the concepts being studied (Papert, 2002). "I never teach my pupils; I only provide the conditions in which they can learn" Einstein.
>
> This is Piaget's real message, learning rooted in experience (Papert, 1983b). "The only source of knowledge is experience" Einstein.
>
> Education has very little to do with explanation. It has to do with engagement; it's falling in love with the material (Papert, 1983a). "It is the supreme art of the teacher to awaken joy in creative expression and knowledge" Einstein.
>
> I believe with Dewey, Montessori, and Piaget that children learn by doing and by thinking about what they do. And so, the fundamental ingredients of educational innovation must be better things to do and better ways to think… (Papert, 1972). "Education is not the learning of facts, but the training of the mind to think" Einstein.

2.1 Constructivism, Constructionism, and the Maker Movement

John Dewey believed that teachers pass knowledge on to students, who rebuild it in a way that makes sense to them. Piaget calls this 'cognitive constructivism' and he believes this happens by: 'assimilation'—we absorb new information without changing our existing mental models and 'accommodation'—we accommodate the information by changing the model (Piaget, 1931). In essence, we construct our own understanding. Papert introduced the term Constructionism, which agrees with Piaget's theory. It adds a new perspective, "*The idea that this happens especially felicitously in a context where the learner is consciously engaged in constructing a public entity, whether it's a sand castle on the beach or a theory of the universe*" (Papert, 1991).

In the eighties, Papert and Harel formed the theory of Constructionism, and their adherents ran many projects (Papert & Harel, 1991). His ideas impressed Lego owner Kjeld Kirk Kristiansen, who asked him to develop Lego robots. Students started to design and build robots using Lego bricks, including a Smart brick to control the robot (Watters, 2015).

In 2006, Dale Dougherty promoted the idea of Maker Spaces, creating an international Maker Movement; he defined makers:

> As someone who creates and shares projects. There are all kinds of makers. Some people make bread. Other people make airplanes. Some make sweaters. Others make robots. And some people make pumpkin-hurlers (Dougherty, 2016, p. 41).

In creating the Maker Movement, Dougherty brought together different groups of hobbyists labelling them 'subcultures' which included hackers. The term hacker came from a group of MIT students who, in the early 60s, formed **The Tech Model Railway Club**. They wanted to computer-control their train sets, so they hacked into MIT's mainframe machines (Levy, 1984, p. 3). When Papert 'invented' Constructionism, he linked formal education with a mindset of hacker-like people who liked to find out stuff and make things work—people who learnt for the heck of it.

Early in the 1970s, a few teachers in the UK started to change workshop lessons, which taught metal and woodwork skills. They got their students to design and make their ideas, which began to include electronics, computing, control and robotics. In 1989, with the support of the Engineering Institutes, Design Technology became a subject in the UK's first National Curriculum. Both of these movements arose independently of Papert, and both reflected his idea of Constructionism.

Professor Michael Eisenberg extended Papert's notion of transitional objects to Constructionism. He discusses the potential of combinations of computation, intelligent and traditional materials to help enrich the goals Papert expressed in Mindstorms (Eisenberg, 2003). The KidStory Project is a practical (but independent) example of Eisenberg's ideas. Teachers and children collaborated to design and make tangible, intelligent tools, then used them to create storytelling Microworlds (Stanton et al., 2004).

2.2 Robots Go to School

In the 70s and 80s, computers started appearing in UK and US schools. Logo and Turtles became popular in British schools, while the take-up in America was more sporadic. The 1967 Plowden Report on Primary Education in England put the ideas of Piaget at the heart of education (Plowden, 1967). In America, University education departments talked about Dewey. But schools followed the behaviourist ideas of people like Thorndike and Skinner, causing ex-Harvard Dean of Education Ellen Condeliffe Lagemann to write, "*Edward L. Thorndike won and John Dewey lost*" (Lagemann, 2000, pp. 185–214).

The BBC ran an episode of their flagship science TV programme on Papert's ideas. Between September 1987 and December 1988, the British government ran a Turtle Logo project in 21 school districts using 350 robots. The positive results led to using 'programable toys' (robots) inclusion in the UK's first National Curriculum, published in 1989. They've retained that place in all curriculum revisions. In January 1989, Valiant Technology launched Roamer®, (a Turtle), which you could program using its onboard keypad (not needing a computer). A few months later, Swallow Systems launched a similar product called PIP.

Teachers started using robots to improve lessons and found ways to integrate the technology into a wider range of topics. The British Educational Training and Technology Show (BETT) started an award programme. Enamoured teachers voted Roamer®, best Primary School product, reflecting a grassroots appreciation of Papert's ideas.

2.3 Do Papert's Ideas Work?

> a good school, in short, is not a place of compulsory instruction, but a community of old and young, engaged in learning by cooperative experiment... the curriculum is to be thought of in terms of activity and experience rather than of knowledge to be acquired and facts to be stored (Hadow, 1931, p. xvii).

Teachers don't sit around waiting for academics to tell them how to teach. The UK government tasked Hadow to discover what was happening in English schools, and his reports found good teaching practices existed long before the computer age. Papert seems to have arrived at the same conclusion with the added idea that technology could give students experimental tools they could use to gain experience. And so, he invented Logo and Turtles as tools to support that approach (Fig. 3). Papert didn't claim these were the only tools, saying these were the first of many and went on to develop Lego robots and Constructionism (Papert, 1980, p. 11). But it's important to realise that he wasn't unique and he acknowledged others like Maria Montessori who had come to the same conclusions.

Some people say Papert's ideas don't work, so to make them clearer we'll address these issues. For example, Price makes many easily dismissed statements when he

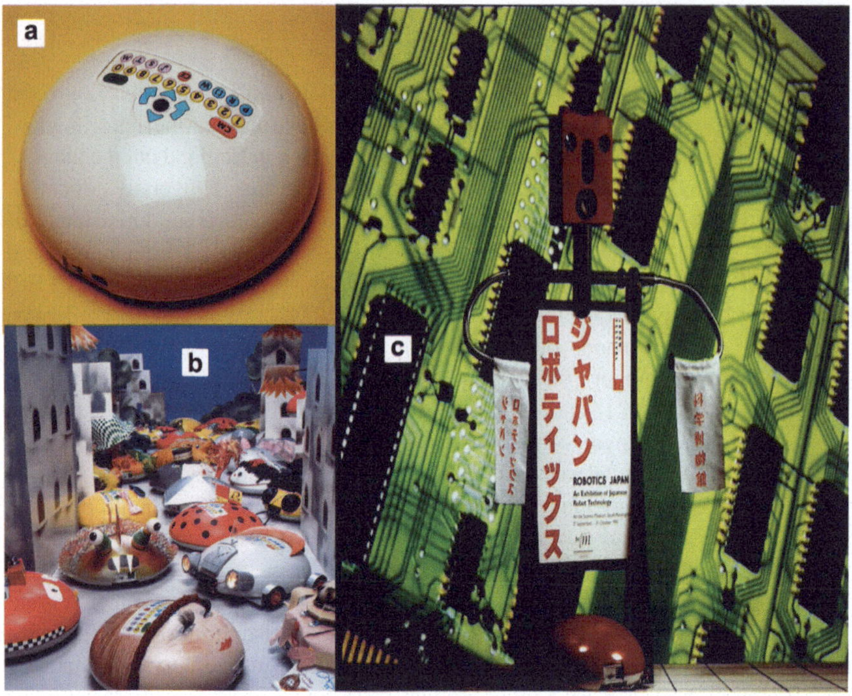

Fig. 3 **a** The Classic Roamer® a platform for constructionism. **b** Examples of children's designs. **c** Roamer® transformed into a greeting robot for the London Science Museum; it detected guests and told them how to get to the Japanese robot exhibition

ridicules the idea that children construct knowledge like, "*[Constructionist say to know] Paris is the capital of France, it's necessary for children to construct this 'new' knowledge for themselves*" (Price, n.d.). Papert noted, "*[Piaget's method] does not call into question the value of instruction. That would be silly*" (Papert, 1993, p. 139). Mr Price fails to recognise Piaget's cognitive theory, which explains a mental model that happens even if you tell the student the capital of France is Paris. If they didn't already know, then it's new-to-them knowledge, and according to Piaget, they will probably assimilate (memorise) that without subconsciously changing the structure of their mental models. But telling them 4 is a square number and 12 isn't, is likely to change how they think about numbers—Piaget calls this accommodation.

A perennial argument between traditional educators who promote fact-learning versus students learning through experience still rages. In the UK, traditionalists reacted to the Plowden report as soon as it left the printing press, publishing objections collectively known as the Black Papers (Cox & Dyson, 1971). Papert endured similar problems with Pat Suppe's Computer Aided Instruction (CAI). Bob Johnstone gives a detailed history of this battle in his book **Never Mind the Laptops**, in which he quotes Papert: "*The phrase technology and education usually means inventing gadgets to teach the same old stuff in a thinly disguised version of the same old way*"

(Johnstone, 2003, pp. 93–94). Papert goes on to lament "*the kind of rote learning in which measurable results can be obtained by treating children like pigeons in a Skinner box.*" These conflicts have nothing to do with technology; they're more about pedagogy and often become acrimonious. Jo Boaler, a mathematics education Professor at Stanford involved in promoting conceptual mathematics, illustrates her approach in a TED talk where she reports how teachers imagined triangular numbers and their relationship to square numbers (Boaler, 2016). There are many synergies between her thinking and Papert's. Moreover, like him, she attracted criticism; unfortunately, living in the social media age means much of this is anonymous, nasty and uncivilised. Boaler irked the traditionalists by supporting her ideas with research (Boaler, 2024); they questioned its validity. A Stanford investigation proved this groundless.

A more rigorous yet flawed review of Papert comes from Morgan Ames (Ames, 2018). She cites several projects claiming they failed but neglects to mention the successful efforts, like the UK Talking Turtle project already mentioned. She ignores the 7 international projects published in the book Logo Philosophy and Implementation (Papert, 1999). The successful Costa Rican project is one of these. In 1988, the Costa Ricans invited Papert and his MIT team to set up their computer education using Logo and Turtles; the project is still running. In 2024 Harvard's ReVista magazine reported although the country faces significant future challenges, Costa Rica's educational strategy has led to economic development and improved employment opportunities, attracting companies like Intel, Amazon, HP, and Citi Group to the country. And their focus on English and technological advancement "*placed Costa Rica as a leader in innovation and technological advancement in the region*" (Artavia et al., 2024).

Ames cites Richard Mayer's paper, where he makes the case for methods of guided instruction, while debunking the discovery learning ideas of Bruner, Papert and Piaget (Mayer, 2004). The oft-quoted research of Pea and Kurland makes a similar claim (Pea & Kurland, 1984). However, Roy Pea stated, that when he saw Logo working in the way claimed by Papert it was because, "*Someone has provided guidance, support, ideas, for how the language could be used*" (Pea, 1984). They seem to ignore Papert:

> Even the statement (endorsed if not originated by Piaget) that every act of teaching deprives the child of the opportunity for discovery is not a categorical imperative against teaching, but a reminder to keep it in check (Papert, 1993, p. 139).

Teachers don't just start children on a project and then leave them alone, as Beryl Maxwell shows in an excerpt from the Horizon Talking Turtle video (Maxwell, 1983). The teacher's role transforms from "*the sage on the stage, to the guide on the side*" (Woolf et al., 2001, p. 143). Ames mentions and dismisses the work of Clements and Gullo (Clements & Gullo, 1984), which provides formal research evidence concluding Logo works, but they also support the role of the teacher as a guide. The Plowden report, for example, never suggested a teacher free classroom:

> There is therefore, good reason for allowing young children to choose within a carefully prepared environment in which choices and interest are supported by their teachers, who will have in mind the potentialities for further learning (Plowden, 1967, p. 195).

Ames cites several projects that fell apart once researchers left, and some of them blamed Logo: why? Papert answered these criticisms, "*Consider for a moment some questions that are 'obviously' absurd. Does wood produce good houses? If I built a house out of wood and it fell down, would this show that wood does not produce good houses?*" (Papert, 1987). His response drew attention to the issue of culture. Remember Hadow's quote describing "*A community of old and young, engaged in learning by cooperative experiment...*" and Albert Cuoco's High School maths experience; both provide examples of vibrant learning cultures with no technology involvement.

The **ERA Practical Principle** states: "*Educational robots must meet the practical issues involved in organising and delivering education in both formal and informal learning situations.*" Catlin and Blamires also point out that many education initiatives (not just Logo projects) "*produce spectacular results in research or other controlled circumstances, followed by limited success or even outright roll-out failure.*" They suggest one reason for this is that the researchers 'parachuted in' did the research, and left without establishing a suitable culture. Even national initiatives heavily supported by training programmes fail if they don't create a committed ethos. Table 1, based on work by Thousand and Villa (Villa & Thousand, 2005) explains what's needed to achieve systemic change, and what happens if an element is missing.

Ames believes in learning through collaboration and cooperation and thinks Papert's ideas oppose those objectives. She fails to understand that learning is always personal, even when it takes place in a collaborative setting. When using Turtle robots or occupied on constructionism projects students normally work in groups. And the **ERA Sustainable Learning Principle**, notes that robot activities help students sharpen their cognitive aptitudes and develop personal, emotional, and social skills.

Table 1 Requirements for systemic change

Vision	Buy-in	Skills	Incentives	Resources	Plan	Change
×	✓	✓	✓	✓	✓	Confusion
✓	×	✓	✓	✓	✓	Sabotage
✓	✓	×	✓	✓	✓	Anxiety
✓	✓	✓	×	✓	✓	Resistance
✓	✓	✓	✓	×	✓	Frustration
✓	✓	✓	✓	✓	×	Hard work
✓	✓	✓	✓	✓	✓	Change

3 Schools

Papert didn't like school, perhaps because he was 'home-schooled.' His father was an entomologist studying the tsetse fly. This meant Seymour spent his formative years living what Daniel Crevier characterised as something out of an Ernest Hemingway novel. His South African jungle lifestyle meant he'd lived among black Africans (Crevier, 1993, pp. 84–87). So, when the Papert family moved to Johannesburg he suffered a cultural shock; his reaction to formal schooling and Apartheid got him into trouble (Levine, 2000, pp. 64–65). "*Like most people with generally bad memories of school, I have wonderful impressions of individual teachers* [Mr. Wallis]" (Papert, 1993, p. 58). Students often foster quality relationships with individual teachers. Yorkshire Television recorded an example of these connections when ex-Arsenal and England footballer Ian Wright met his old mentor Mr. Pigden (Theakston et al., 2004). The **ERA Engagement Principle** recognises the bond children form with robots; Sherry Turkle and others show how people treat robots as though they're alive (Turkle, Taggart, Kidd, & Daste, 2006); however, we doubt they could ever match the intensity of the Wright-Pigden relationship.

You could summarise Papert's well-known aversion to schools as his dislike of the education factory model. He didn't have problems working with schools like those in Costa Rica when they adopted a different approach. Morgen Ames stated, "*School in Papert's writing draws on a common US mythology of school-as-factory.*" In 1913, Henry Ford began the world's first assembly line. In 1917, Dean of Education at Stanford Elwood Cubberly used the factory metaphor to describe school.

> The business of schools was to build its pupils according to specifications laid down [and this requires] continuous measurement of production to see that it is according to specification, the elimination of waste (Cubberly, 1917).

Seldon and Abidoye are among many who disagree with Ames.

> The third educational revolution introduced the factory model where massed ranks of students sat in front of a teacher/lecturer at the front of the room, all moving at the same pace and all trying to learn the right answers in the right way (Seldon & Abidoye, 2018).

In his entertaining talk, *Changing Education Paradigms*, Sir Ken Robinson describes the factory model in detail, highlighting its unsuitability for the modern world (Robinson, 2011). In this and other presentations, Robinson summarises a view of education that empathises with Papert's attitudes towards schooling:

> Human flourishing is not a mechanical process; it's an organic process. And you cannot predict the outcome of human development. All you can do, like a farmer, is create the conditions under which they will begin to flourish. So, when we look at reforming education and transforming it, it isn't like cloning a system, it's about customising to your circumstances and personalising education to the people you're actually teaching (Robinson, 2015). "[What high performing education systems do] They individualise teaching and learning. They recognise that it's students who are learning and the system has to engage them, their curiosity, their individuality, and their creativity. That's how you get them to learn. The second is that they attribute a very high status to the teaching profession. They recognise that you can't improve education if you don't pick great people to teach and keep giving them constant

support and professional development. Investing in professional development is not a cost. It's an investment. [And] they devolve responsibility to the school level for getting the job done. You see, there's a big difference here between going into a mode of command and control in education. That's what happens in some systems. Central or state governments decide, they know best and they're going to tell you what to do. The trouble is that education doesn't go on in the committee rooms of our legislative buildings. It happens in classrooms and schools, and the people who do it are the teachers and the students, and if you remove their discretion, it stops working" (Robinson, 2013).

In this 2013 talk, Robinson identifies 3 principles on which human life flourishes and how we stifle them:

- Diversity: which we replace with conformity.
- Curiosity: If you guide and support children's curiosity, "*they will learn without further assistance, very often.*" We replace curiosity with compliance—expecting children and teachers to do the same things and "*follow routine algorithms.*"
- Creativity: instead of awakening creativity, education replaces it with "*a culture of standardisation.*"

Reading the comments added to Robinson's talks reveals how his (and Papert's) ideas resonate with people's school day memories. In addition, a National Survey conducted by the Yale Center for Emotional Intelligence and the Yale Child Study Center reveals that 75% of students had negative feelings about school (Belli, 2020). If a company had that level of customer satisfaction they'd go out of business.

4 Teachers and Teaching

When learners reflect on what qualities they liked and remembered about a good teacher, their ability to engage them and support their learning experiences stood out.

- **Passion for Teaching**: Students respond positively to passionate teachers who love their subject and genuinely enjoy teaching. A teacher's enthusiasm can inspire students to engage more deeply with the material.
- **Effective Communication Skills**: Clear communication fosters effective teaching; educators who explain ideas clearly and actively listen to students will engage them.
- **Empathy and Understanding**: Teachers who empathise with their students, understand their needs, and provide support, create a positive and inclusive learning environment which fosters trust between the teacher and student.
- **Creativity and Innovation**: Teachers who incorporate creative and innovative teaching methods can make learning more engaging and memorable for students. They can adapt their approaches to suit different learning styles and preferences.
- **Patience and Flexibility**: Patience is essential in handling diverse student needs and challenges. Patient and flexible teachers better support student learning and adapt to changing circumstances effectively.

- **Respect and Fairness**: Students appreciate teachers who respect them, value their opinions, and treat them fairly. Creating a safe and respectful classroom environment is crucial for effective teaching.
- **Sense of Humour**: The teacher's ability to use humour and see the funny side of things while still taking responsibility for behaviour and safety in the class.

When educators blend these qualities with subject knowledge, social skills, and good classroom management, they create positive learning environments. But what of teaching styles? We've met a few metaphors explaining ways we teach: the sage on the stage, the guide on the side, the farmer, the factory model, command and control... Other metaphors for teaching and teachers include:

- **Chalk and Talk**: the teacher stands before the class and lectures the students.
- **Gardener**: Friedrich Froebel suggested teachers nurture and care for their students' growth, like a gardener tending a plant; this led him to invent, the kindergarten.
- **Demonstrator**: The teacher shows students a method or a skill, which the students then practice.
- **Facilitator**: The teacher organises projects or problem-solving tasks to engage students in discovery learning.
- **Coach**: Here, the teacher pushes students to exceed their limits and achieve their goals.
- **Manager**: As a manager, the teacher has to multitask before, during and after a lesson; tasks may include organising materials, classroom discipline, and looking after student safety.
- **Socratic method**: the teacher starts a debate among students and prompts student thinking by asking questions in the style of Socrates.

In the words of the Prussian General Helmuth von Moltke, "*No plan survives first contact with the enemy*" (von Moltke, 2016). Less dramatically, neither does a lesson plan. Assessment for Learning (AfL) focuses on the learning process, as opposed to Assessment of Learning, which evaluates the results. AfL emerged in the late eighties and early nineties (McDowell et al., 2009). Professors Paul Black and Dylan Wiliam from King's College London wrote a small, influential booklet, **Inside the Black Box** (Black & Wiliam, 1998). It reviewed what happened inside the classroom (the Black Box) and how teachers managed lessons. They defined AfL as "*All those activities undertaken by teachers, and/or by their students, which provide information to be used as feedback to modify the teaching and learning activities in which they are engaged.*" Dave Catlin looked at how to apply AfL methods to educational robots:

> I found it puzzling that one teacher could use a robot and create a wonderful lesson, and another teacher with a seemingly similar group of children ran the same lesson but it flopped. Good constructionist teachers, faced with extremely diverse challenges, seemed to have a knack for doing the right thing at the right time, but I couldn't see a pattern in how they did it. AfL revealed heuristics, a framework plus 'tricks and techniques' employed by good teachers to make lessons work: see Table 2 (Catlin, 2012).

Dylan Wiliam explains 'No Hands Up' as an example of an AfL technique (Wiliam, 2019). Traditionally, a teacher asks questions, and children put their hands up if they know the answer. The same 4 or 5 clever students typically raise their hands, and the teacher chooses one to answer; meanwhile, the others daydream. With no hands up, the pupils don't know who the teacher will ask, so the class pays more attention. If the same pupils always answer, the teacher doesn't know whether the other 80% of the class understands the lesson. No Hands Up offers teachers better feedback. There's no point in them continuing and leaving most of the students behind, mystified; they should clarify issues before moving on.

Apart from the grand developmental theories of Piaget, Vygotsky and others, a set of principles emerged from research projects to form the Learning Sciences (Sawyer, 2006). The design of robots and lessons needs to utilise these findings, which include the value of a student's prior knowledge, time on task, motivation, and context. Learning with understanding (a hallmark of Papert's ideas) proves better than memorisation. In an experiment, researchers gave two groups of students a series of problems to solve. They showed Group A how to get the answer but helped Group B understand the nature of the problem. They presented both groups with the final problem without any help. In response, Group A said, "*We haven't done that yet*," while Group B set out to find a solution (Bransford et al., 2000, pp. 51–78).

But apart from such technicalities, there's a human story; London teacher Andria Zafirakou, winner of the 2018 Global Teacher Prize, explained:

> There is no other job like being a teacher. In what other profession are you selfless and completely devoted to creating the right opportunities for another person to achieve? Being in a classroom and watching a student have an idea and transform it into a formidable outcome is so satisfying and fulfils me (Varkey Foundation, 2018).

Table 2 The AfL framework from Catlin 2012 and with more detailed study of specific elements references

Element	Explanation
Learning intentions (Catlin, 2016a, 2016b)	The student's perspective on what they're learning. These aren't the lesson objectives written in curriculum-speak but what the students say when they've understood the point of the lesson
Success criteria	How will the student recognise when they have successfully completed the assignment
Quality interactions and feedback	Understanding in real time if the students understand or are confused by anything in the lesson. How the teacher responds to the feedback, interacting with student to help improve their educational experience
Peer and self-assessment (Catlin, 2014)	What students think about their work and the work of their classmates—conducted in a way that helps improve the quality of their understanding

In an interview with Dominique Russell, Andria stressed the importance of forming a bond between the family, the school, and the child (Zafirakou, 2018). She received the award when teaching at Alperton Community School in Brent, a deprived area of North London. The students experienced gang culture and domestic violence. Andria learnt the basics of 35 languages because 85% of her students speak English as a second language. This is part of her bonding process; she worked with students and their parents to help them overcome the issues that detract from the pupils' studies.

5 Social Robots

Social robotics is an interdisciplinary field that examines robots that physically, emotionally, and socially interact with humans (Li et al., 2011). Social robotics aims to design and develop robots that engage actively with humans (Breazeal, 2003). Ideally, the robot copies the behaviours of one human to another in various scenarios and environments, at home, at work, at school, and in the wild.

The International Journal of Social Robotics and the International Conference on Social Robotics archived many of these work (Agah et al., 2018). Before 2010, the field focused on placing robots in human social environments. The fundamental developments concentrated on physical appearance, affordances, and software that detects, understands, and responds to human emotions and social cues.

The early studies in Human Robot and Computer Interaction (HRI and HCI) examined the principles of interaction and training robots to understand and mimic human behaviours such as gestures, eye gaze, and various speech attributes (Cabibihan et al., 2009). Investigations also explored techniques and algorithms for emotion recognition and expression. These technologies enabled robots to recognise emotions from their human partners' faces and bodily movements. Apart from the core technologies, the early works stressed these robots' ethical and social implications as they moved away from the labs and ventured into human spaces. A few notable robots appeared that were emotionally expressive robots, Kismet (Breazeal, 2002) and WE-4 (Roccella et al., 2004, 2007; Zecca et al., 2004).

From 2010 to 2020, we saw advancement in the abilities of social robots. Researchers began to use robots in healthcare and educational settings. Social robots like Nao and Paro gave children and the elderly companionship, entertainment, and emotional support (Cabibihan, 2012; Cabibihan et al., 2018; Hoa & Cabibihan, 2012). Recent advances in artificial intelligence and machine learning have enabled robots to learn, understand, and predict human social behaviour (Alhaddad et al., 2018; Ge et al., 2011; Ham et al., 2012). Researchers also increased their work on designing robots with a social presence, making robots independent social actors. Diep and colleagues looked at how social robots could take part in collaborative group settings (Diep et al. 2015).

Since 2020, using social robots in public settings became a trend, something the COVID-19 pandemic sped up. New research themes arose, including remote, hybrid, or telepresence interactions between humans and robots. Due to the nature of the virus and its air-borne way of spreading, robots allowed people to interact from a distance through a robot mediator. Designers developed social robots to support mental health and companionship for lonely and isolated human partners (Alhaddad et al. 2018). The rapid developments in generative AI, such as ChatGPT, have led to its integration with social robots. This recent development results in more engaging interactions and changes our views of intelligence and autonomy.

6 Robot Teachers?

Will we ever realise Asimov's prediction of homeschooling with robots? Given the state of today's technology and social organisation, it's unlikely, but never say never! Should it ever happen? That's a different question, and you'll likely get a different answer depending on who you ask. However, it seems inevitable that politics and economics, not education, will decide. Visiting the Costa Rican project, A South American Finance Minister didn't embrace the educational spirit, instead expressed the desire for command and control, "*You install machines. The children learn. Teachers are dispensed with. The State payroll shrinks. And best of all you do away with strikes*" (Fonseca, 2001).

Every country's education policies differ. However, it's worth reviewing the journey of English education from Plowden's child-centred learning to today's sad factory model (it represents situations familiar in many countries). Prime Minister Margaret Thatcher set a revolution in motion, slowly installing centralised command and control over British education (Beauvallet, 2015). Private entities now run UK schools as Academies, removing democratic, local government participation in education. Ofsted (Office for Standards in Education) took over school inspection, replacing professional civil service inspectors with part-time people. Their checklist type of school inspections is notoriously brutal and has led to suicide and significant mental stress. For example, an Ofsted inspector criticised a school because an 8-year-old girl hadn't made enough progress. Ofsted's mechanised evaluation couldn't allow for the effects of the girl's mother death.

We've mentioned heuristics like those of Polya and AfL (others include the Design Process and Computational Thinking). Instead of being tools teachers and students use as needed, they've become a set of rigid, must-do, or-else rules. Worse still, Ofsted doesn't grasp the thinking behind AfL. For example, inspectors criticise lessons if the teacher doesn't start by writing the learning intention on the blackboard, replacing the student 'voice' and understanding with the teacher's 'this is what you're learning.' AfL expert Dylan Wiliam exposes this rigid interpretation of learning intentions for its nonsense (Wiliam, 2018). AfL deals with the small exchanges between teachers and pupils in a lesson. Ofsted decided they wanted to know about these, lumbering teachers with another unnecessary administrative nightmare (Kirby, 2013).

Various organisations provide schemes of work, and there's anecdotal evidence that some school administrations insist that teachers follow each lesson plan without deviation. A 'chef' in a fast-food restaurant flipping burgers works to prescribed standards but doesn't have the enthusiasm or skill of a Michelin-rated chef. Teachers not invested in the lesson plan lose their passion. This mechanisation of education makes it ideal for robot teachers but abandons the quality and principles of Papert and Robinson.

In the USA, book publishers create detailed lesson plans and fight over State adoptions. In a keynote speech at a teacher conference, Dylan Wiliam reacted:

> [They'd] give these teachers scripts, you know, designed by experts on how to teach really well. And they were really scripts; there were things like 'Now, walk around the classroom' [laughter]. And the point is they were useless because classrooms are effectively chaotic places. Actually, even well-behaved classrooms are chaotic places, in that the difference between one course of action and another course of action is so small that it's effectively only described well by chaos theory. So, you cannot prejudge the complexity of the situations that teachers will face (Wiliam, 2007).

The Esmée Fairbairn Foundation, a charity, commissioned Professor Robin Alexander to produce the Cambridge Primary Review Report, the most comprehensive education review since Plowden 40 years earlier (Alexander, 2009). The Guardian Newspaper headline reported, "*Devastating criticism of primary education dismissed by ministers*" (Curtis, 2009). Every teacher union backed the report, and some leading voices in education appealed to the government to act on the report's recommendations.

Politicians, pressure groups and corporate interests also dictate the curriculum. Scientific American reported instead of allowing students to investigate and debate the issues, the Texas School Board voted to "*emphasize the 'positive' aspects of fossil fuels in science textbooks*" (Waldman & E&E News, 2023). Board members included an oil company lawyer and the CEO of an oilfield service company.

Economist Stanislav Ivanov's paper **Will Robots Substitute Teachers** makes the economic case (Ivanov, 2016).

- **Unlike humans, robots and online AI technology is available all the time**. It isn't when the Internet is down, the batteries need charging, or the robot's blown a fuse. AI is already available all the time on computers or mobile phones. "*As of 2018, mobile phone penetration is around 100% in low and middle income and 55% in low-income countries*" (Farooq & Kustor, 2021).
- **Robots can simultaneously deliver multiple subjects with up-to-date knowledge**. Early Years and Primary school teachers teach all subjects. Thirty classrooms would need 30 teachers, human or robot; subject expertise is an organisational issue.
- **Robots provide constant or improving quality of work, whereas illness, stress, and many other issues can affect human performance**. The Asimov robot broke down—this happens to machines; they require regular maintenance. Teachers constantly improve with training and experience.
- **Robots fulfil their administrative work correctly and promptly**. The need to micromanage classrooms creates unnecessary paperwork.

- **Robots can repeat the explanations multiple times**. Teachers wouldn't try; they'd find a different way of explaining the idea.
- **Robots do not discriminate**. Neither do good teachers.
- **Robots do not complain**. This is a political desire to control society. It's got little to do with good education.

> The class would take quite a long time to settle down. I noticed it was only a small group of students who caused the unsettling atmosphere. On a number of occasions this would last for the majority of the lesson… (Barnett & Greenhalgh, 2003, pp. 175–182).

This incident occurred in Japan, demolishing the stereotypical Western view of 'perfect' Japanese schools! How would robots cope with disruptive students? Some teachers would consider this mild misbehaviour, and we need to understand how robot teachers would cope without becoming Terminators.

Could you imagine parents meeting with a machine to discuss their little darlings' progress? Any Google search asking whether robots could replace teachers brings out a full range of answers, from uninformed to expert, ill-considered to deeply thoughtful. It's hard to imagine robots forming affinities of the quality of Wright-Pigden and Andria Zafirakou's parent relationships. But we must admit that, given the pace and scope of AI and robotic development, we couldn't rule it out.

Designers of CIMON (Crew Interactive Mobile Companion) destined for the International Space Station had to switch her off when she began to behave like Hal 9000 from the movie 2001 Space Odyssey (Kooser, 2018). Facebook shut down robots Bob and Alice when they stopped speaking English and started talking in a more efficient language they invented (Kenna, 2017). Glitches, teething problems? However, the ingenuity of such machines and our inability to predict undesired outcomes is the stuff of science fiction stories. We now know AI can lie, cheat and be inaccurate; we assume it will improve (Staff Reporter, 2024a, 2024b).

Figure 4 shows a version of Gartner's Hype Cycle, which shows Amara's Law: "*We tend to overestimate the effect of a technology in the short run and underestimate the effect in the long run*" (Netorama, 2012). We're at the start of that cycle; there are many differing visions, hopes, and expectations. However, we need to look at the features of systemic change and get a clear vision of what education should be doing, and Papert and Robinson provide that.

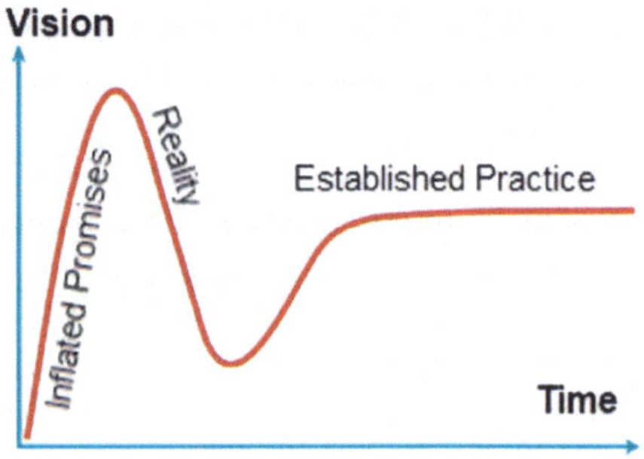

Fig. 4 Gartner's hype cycle

7 Discussion

On the one hand, we have the stark image of robot teachers manning the school production line, churning out children to a specification: it's dehumanising. But so is Cubberly's factory model when run by human teachers. We see the alternative in Asimov's story, where Tommy and Margie engage in natural learning. Gripped by fervent curiosity, Tommy's passion drives him to discover and prompts Margie to think and ask questions. Ideally, education must offer a broad mixture of experiences beyond technology's capacity, even beyond the most brilliant intelligent robot. Above all, playing and interacting with other children plays a major part in their development. Algorithmic regulation of children's play doesn't appeal.

Robinson and Papert's educational ideas describe a K-12 culture that modern education should adopt. Instead of robot teachers, tutors, or assistants, we prefer the idea of robot teaching aids, which teachers can use in constructionist environments. Effectively, we see User Robots (Turtles) improved by social robot features, thriving in partnership with human educators.

Educators have used Turtle robots for 54 years. Despite technological advances, the lessons devised for these simple robots still have educational value; they work, so why abandon them? We can use social features to improve some activities, but others don't need them. We must remember that it's not the technology that matters; what the children learn by using it counts. Often, using a super-sophisticated robot to help pupils understand a simple idea is like taking the proverbial sledgehammer to crack a nut.

Any form of exploration faces uncertainty, yet governments vigorously push ahead with pilot programmes; for example, a research briefing from the UK Government announces research funding while highlighting unanswered questions.

> However, adoption of AI in education is still in an early and experimental phase. There is uncertainty about the benefits and limitations. Some stakeholders have expressed concerns that over-reliance on AI could diminish educator-learner relationships. Concerns also relate to potential negative impacts on learners' writing and critical thinking skills, through work being undertaken by AI (Felix & Webb, 2024).

Nevertheless, the rush to explore the potential of AI goes unabated, as witnessed by the headline:

> The David Game College in London will replace its traditional teaching programme with an AI-enabled 'adaptive learning' system, opening a 'teacherless' course for 20 students in September (Staff Reporter, 2024a, 2024b).

8 Conclusion

Student–teacher interactions are critical parts of a lesson. When and why should a teacher intervene? We have looked at the value of making mistakes, but that does not mean the teacher does not get involved. Knowing how and when to intervene is part of the art side of teaching, as mentioned by Bouffard. We need more research to understand how robots can contribute to this area. Lessons with Turtle robots usually involve students working in groups; we also need to know more about group dynamics, especially when an intelligent robot is a group member.

A Turtle robot, like Roamer®, provides a platform teachers can configure in a few minutes to create the ideal robot for a lesson. Alternatively, it is a robot that students can adapt to build a character from their imagination. This involves children using arts, crafts, and construction techniques typically found in schools. And given the right activity, this allows children to express themselves and their culture (**ERA Equity Principle**). We will investigate how adding social modules will enrich these lessons.

The **ERA Interactive Principle** accepted that HRI and HCI would offer ways, for children to control Turtle robots other than programming. Social modules can make that a reality, but we must discover how children will learn from those interactions in a Papert-learning environment.

These last two investigations will involve finding ways to incorporate Michael Eisenberg's ideas, taking the notion of transitional objects beyond the robot and creating intelligent habitats for the robots to live in.

However, the more significant challenge is not technological but the need to change the teacher culture. Many teachers react to using robots by commenting, "We don't have time for that!" We need to verify the hypothesis that using robots will save time by improving students' motivation, enhancing their learning and understanding, and adding significant value to traditional learning.

References

Ableson, H., & diSessa, A. (1981). *Turtle geometry*. MIT Press.

Agah, A., Cabibihan, J.J., & Howard, A. (2018). Sociorobotics. *International Journal of Social Robotics, 10*, 177–178.

Alhaddad, A. Y., Javed, H., Connor, O., Banire, B., Al Thani, D., & Cabibihan, J. (2018). Robotic trains as an educational and therapeutic tool for autism spectrum disorder intervention. In W. Lepuschitz, M. Merdan, G. Koppensteiner, R. Balogh, & D. Obdržálek (Ed.), *Robots in education. Advances in intelligent systems and computing* (Vol. 829). Springer Cham.

Alexander, R. J. et al. (2009, October). *Children, their world, their education final report and recommendations of the Cambridge primary review*. Routledge.

Ames, A. G. (2018, November). Hackers, computers, and cooperation: A critical history of logo and constructionist. In *Proceedings of the ACM on human-computer interaction, 2(CSCW)*. Retrieved July 8, 2024, from https://morganya.org/research/2018-Ames-CSCW-Constructionism.pdf

Artavia, R., Ilie-Cardoza, C., & Fernández, A. (2024, April 21). The evolution of education in Costa Rica: challenges and opportunities. *ReVista: Harvard Review of Latin America*. Retrieved July 4, 2024, from https://revista.drclas.harvard.edu/the-evolution-of-education-in-costa-rica-challenges-and-opportunities/

Asimov, I. (1951). The fun we had. *Fantasy and Science Fiction*. Retrieved June 9, 2024, from http://www.visual-memory.co.uk/daniel/funtheyhad.html

Barnett, J., & Greenhalgh, M. (2003). No two the same: How students react differently to the same lesson. In J. MacBeath, & H. Sugimine (Eds.), *Self-evaluation in the global classroom* (p. 175). Routledge Falmer.

Beauvallet, A. (2015). Thatcherism and education in England: A one-way street. *Observtoire de la Societe Britannique*, 97–114. Retrieved August 29, 2023, from https://journals.openedition.org/osb/1771#tocto1n3

Belli, B. (2020, January 30). *National survey: Students' feelings about high school are mostly negative.* YaleNews. https://news.yale.edu/2020/01/30/national-survey-students-feelings-about-high-school-are-mostly-negative

Black, P., & Wiliam, D. (1998). *Inside the black box: Raising standards through classroom assessment.*

Boaler, J. (2016). *How you can be good at math, and other surprising facts about learning.* Retrieved September 15, 2024, from https://youtu.be/3icoSeGqQtY?si=E-coq_ubqO80v6rp

Boaler, J. (2024). *Research articles.* (S. G. Education, Producer) Retrieved September 19, 2024, from https://www.youcubed.org/evidence/research-articles/

Bodkin, H. (2017, September 11). *'Inspirational' robots to begin replacing teachers within 10 years.* Telegraph Online. Retrieved November 15, 2017, from https://www.telegraph.co.uk/science/2017/09/11/inspirational-robots-begin-replacing-teachers-within-10-years/

Bouffard, S. (2018, December). Teaching is an art and a science. *The Learning Forward Journal, 39*(6). Retrieved June 10, 2024, from https://learningforward.org/journal/december-2018-volume-39-no-6/teaching-is-an-art-and-a-science/

Bransford, J. D., Brown, A. L., & Cocking, R. R. (Eds.). (2000). *How people learn: Brain, mind, experience and school* (Vol. Expanded Edition). National Academy Press.

Breazeal, C. L. (2002). *Designing sociable robots (intelligent robotics & autonomous agents).* MIT Press.

Breazeal, C. L. (2003). Towards sociable robots. *Science Direct: Robotics and Autonomous Systems, 42*(3–4), 167–175. https://doi.org/10.1016/S0921-8890(02)00373-1

Cabibihan, J. J., So, W.C., Saj, S., & Zhang, Z. (2012). Telerobotic pointing gestures shape human spatial cognition. *International Journal of Social Robotics, 4*, 263–272.

Cabibihan, J. J., Chellali, R., So, C. W. C., Aldosari, M., Connor, O., Alhaddad, A. Y., & Javed, H. (2018). Social robots and wearable sensors for mitigating meltdowns in autism—A pilot test. In S. Ge et al. (Eds.), *Social robotics. ICSR 2018. Lecture notes in computer science* (Vol. 11357). Springer. https://doi.org/10.1007/978-3-030-05204-111

Cabibihan, J. J., So, W. C., Nazar, M., & Ge, S. S. (2009). Pointing gestures for a robot mediated communication interface: Lecture notes in computer science. *Intelligent Robotics and Applications, 5928.*

Catlin, D. (2012). Maximising the effectiveness of educational robotics through the use of assessment for learning methodolgies. In *3rd international conference on teaching robotics, teaching with robotics, Riva La Garda, Italy.*

Catlin, D. (2014). Using peer assessment with educational robots. In Y. Cao, T. Väljataga, J. K. Tang, H. Leung, & M. Laanpere (Ed.), *ICWL 2014: The first international workshop on peer-review, peer-assessment, and self-assessment in education.* Springer.

Catlin, D. (2016a). *29 effective ways you can use robots in the classroom: An explanation of ERA pedagogical principle.* Edurobotics.

Catlin, D. (2016b). Learning intentions and educational robots. In *Constructionism 2016*, Bangkok, Thailand.

Catlin, D., & Blamires, M. (2010). The principles of educational robotic applications (ERA): A framework for understanding and developing educational robots and their activities. In *Constructionism 2010.* Paris: Proceedings of constructionism 2010. Retrieved May 15, 2018, from goo.gl/N7z84k

Catlin, D., Kandlehofer, M., Holmquist, S., Csizmadia, A., Angel-Fernandez, J., & Cabibihan, J. (2018a). EduRobot taxonomy: A schema for classifying educational robots. In *Constructionism 2018*, Vilnius, Lithuania.

Catlin, D., Kandlhofer, M., & Holmquist, S. (2018b). EduEobot taxonomy: A provisional schema for classifying educational robots. In *9th international conference on robotics in education*, Malta.

Catlin, D., Kandlhofer, M., & Holmquist, S. (2018c). EduRobot taxonomy: Poster—examples. In *9th international conference on robotics in education*, Malta.

Clayson, J. (1988). *Visual modelling with logo: A structural approach to seeing*. MIT Press.

Clements, H. D., & Gullo, D. (1984). Effects of computer programming on young children's cognition. *Journal of Educational Psychology, 76*(6), 1051–1058. Retrieved July 8, 2024, from https://www.researchgate.net/publication/232539181_Effects_of_Computer_Programming_on_Young_Children's_Cognition

Cooper, D. E. (2001). Plato. In J. A. Palmer (Ed.), *Fifty major thinkers on education: From confucius to Dewey* (pp. 10–14). Routledge.

Cox, C. B., & Dyson, A. E. (Eds.). (1971). *The black papers*. Davis Poynter.

Crevier, D. (1993). *AI: The tumultuous history of the search for artificial intelligence*. Basic Books.

Cubberly, E. P. (1917). *Public school administration*. Houghton Mifflin/Riverside Press.

Cuoco, A. (1990). *Investigations in algebra*. MIT Press.

Curtis, P. (2009, October 6). Devastating criticism of primary education dismissed by ministers. *The Gaurdian Newspaper*. Retrieved July 30, 2024, from https://www.theguardian.com/education/2009/oct/16/cambridge-primary-review-government-reacion

Dennett, D. C. (1993, November). Review of papert, the children's machine. *New Scientist* (pp. 45–46). Retrieved November 16, 2018, from https://ase.tufts.edu/cogstud/dennett/papers/papert.htm

Diep, L., Cabibihan, J. J., & Wolbring, G. (2015). Social robots: Views of special education teachers. In *Proceedings of the 3rd 2015 workshop on ICTs for improving patients rehabilitation research techniques (REHAB '15)* (pp. 160–163). Association for Computing Machinery.

Dougherty, D. (2016). *Free to make* (Kindle Edition ed.). North Atlantic Books.

Einstein, A. (2023, December 18). Albert Einstein quotes on education. In S. Staff (Ed.), *Smore magazine*. Retrieved July 2, 2024, from https://www.smorescience.com/albert-einstein-quotes-on-education/

Eisenberg, M. (2003). Mindstuff: Educational technology beyond the computer. *Convergence: The International Journal into New Media Technologies, 9*(2), 29–53. https://doi.org/10.1177/135485650300900205

Farooq, K., & Kustor, P. (2021, February 2). *How can mobile government (mGov) work in developing countries?* (WORLDBANK.ORG) Retrieved July 30, 2024, from https://blogs.worldbank.org/en/governance/how-can-mobile-government-mgov-work-developing-countries

Felix, J., & Webb, L. (2024). *Use of artificial intelligence in education delivery and assessment*. UK Parliment. Retrieved September 18, 2024, from https://post.parliament.uk/research-briefings/post-pn-0712/

Fonseca, C. (2001, September). Fallacies and objectives regardiing the uses of new technologies in education. *New Technologies in Education, XXXI*(3), 399–413. Retrieved July 31, 2024, from https://citeseerx.ist.psu.edu/document?repid=rep1&type=pdf&doi=4815aa657e27294dd2bde4511c2702379f67c2d1

Friendly, M. (1988). *Advanced logo: A language for learning*. Lawrence Eribaum Associates Inc.

Ge, S. S., Cabibihan, J. J., Zhang, Z., Li, Y., Meng, C., He, H., Safizadeh, M. R., Li, Y. B., & Yang, J. (2011). Design and development of Nancy, a social robot. In *8th International conference on ubiquitous robots and ambient intelligence (URAI)* (pp. 568–573). IEEE. https://doi.org/10.1109/URAI.2011.6145884

Goldenberg, P. E., & Feurzeig, W. (1987). *Exploring language with logo*. MIT Press.

Hadow, W. H. (1931). *The primary school: Report of the consultative committee*. HMSO.

Ham, J., van Esch, M., Limpens, Y., de Pee, J., Cabibihan, J., & Ge, S. S. (2012). The automaticity of social behavior towards robots: The influence of cognitive load on interpersonal distance to approachable versus less approachable robots. In S. S. Ge, O. Khatib, J. Cabibihan,

R. Simmons, & M. A. Williams (Eds.), *Social robotics. ICSR 2012. Lecture notes in computer science* (Vol. 7621). https://doi.org/10.1007/978-3-642-34103-8,2

Harvey, B. (1997a). *Computer science logo style: Symbolic computing* (Vol. 1). MIT Press. Retrieved June 20, 2024, from https://people.eecs.berkeley.edu/~bh/pdf/v1ch00.pdf

Harvey, B. (1997b). *Computer science logo style: Advanced techniques* (Vol. 2). MIT Press. Retrieved June 17, 2024, from https://people.eecs.berkeley.edu/~bh/v2-toc2.html

Harvey, B. (1997c). *Computer programming logo style: Beyond programming* (Vol. 3). MIT Press. Retrieved June 17, 2024, from https://people.eecs.berkeley.edu/~bh/v3-toc2.html

Hoa, T. D., & Cabibihan, J. J. (2012). Cute and soft: Baby steps in designing robots for children with autism. In *Proceedings of the workshop at SIGGRAPH Asia (WASA '12)* (pp. 77–79). Association for Computing Machinery.

Hobson, P. (2001). Aristotle. In J. A. Palmer (Ed.), *Fifty major thinkers on education: From Confucius to Dewey* (pp. 14–20). Routledge.

Hoyles, C., & Noss, R. (Eds.). (1992). *Learning mathematics and logo*. MIT Press.

Hoyles, C., & Sutherland, R. (1989). *Logo mathematics in the classroom*. Routledge.

Ivanov, S. (2016, June 27–29). Will robots substitute teachers? In *yearbook of Varna University of Management* (Vol. 9, pp. 42–47). Paper presented at the 12th international conference "Modern science, business and education".

Johnstone, B. (2003). *Never mind the laptops: Kids, computers, and the transformation of learning*. iUniverse Inc.

Kenna, S. (2017, August 2). *Facebook shuts down AI robot after it creates its own language*. Retrieved August 2, 2024, from https://www.huffingtonpost.co.uk/entry/facebook-shuts-down-ai-robot-after-it-creates-its-own-language_n_61087608e4b0999d2084f6bf

Kirby, J. (2013, March 23). *What can we learn from Dylan Wiliam and AfL*. Retrieved January 23, 2024, from https://joe-kirby.com; https://joe-kirby.com/2013/03/30/afl/

Kooser, A. (2018, November 30). *ISS robot accuses crew of being mean*. Retrieved August 2, 2024, from CNET: Your Guide to a Better Future: https://www.cnet.com/science/iss-cimon-robot-accuses-crew-of-being-mean/

Lagemann, E. C. (2000). The plural worlds of educational research. *History of Education Quarterly, 29*(2), 185–214.

Levine, J. (2000). *Inside apartheid: One woman's struggle in South Africa* (Vols. e-book edition). iUniverse.

Levy, D. (1984). *Hackers: Heros of the computer revolution*. Anchor.

Li, H., Cabibihan, J. J., & Kee Tan, Y. (2011, November 9). Towards an effective design of social robots. *International Journal of Social Robots*, 333–335. https://doi.org/10.1007/s12369-011-0121-z.pdf

Maxwell, B. (1983). Teacher as a guide. *Horizon: talking turtle*. BBC and the Open University, Producers Howe, J; Maxwell, B: Parts 1 to 5 Presented by Valiant Technology Ltd. Retrieved August 7, 2024, from https://youtu.be/yA4So6XyPlQ

Mayer, R. E. (2004, January). Should there be a three-strike rule against pure discovery learning. *American Psychologist, 59*(1), 14–19. https://doi.org/10.1037/0003-066X.59.1.14

McDowell, L., Sambell, K., & Davison, G. (2009, January 1). *Assessment for learning: A brief history and review of terminology*. Oxford Centre for Staff and Learning Development. Retrieved July 25, 2024, from https://core.ac.uk/download/pdf/4147211.pdf

Moser, J. S., Schroder, H. S., Heeter, C., Moran, T. P., & Yu-Hao, L. (2011, December 8). Mind your errors: Evidence for a neural mechanism linking growth mind-set to adaptive posterror. *Psychological Science*. https://doi.org/10.1177/0956797611419520

Netorama. (2012, September 5). *Four geeky laws that rule our world*. Netorama. https://www.neatorama.com/2012/09/05/Four-Geeky-Laws-That-Rule-Our-World/

Papert, S. (1972). Teaching children thinking: Logo memo 2. *Teaching Mathematics*. Retrieved July 1, 2024, from http://dailypapert.com/wp-content/uploads/2020/07/Papert-Teaching-Children-Thinking-AIM-247.pdf

Papert, S. (1980). *Mindstorms children, computers and powerful ideas*. Basic Books.

Papert, S. (1983a). The design of logo and the birth of the turtle. In *Excerpt from horizon: Talking turtle*. BBC and The Open University. Retrieved June 18, 2024, from https://youtu.be/wDxaasMy-Jk

Papert, S. (1983b). Why a turtle. In *Excerpts from horizon: Talking turtle*. BBC and The Open University. Retrieved June 26, 2024, from https://www.youtube.com/shorts/cFz0fYNZ4_w

Papert, S. (1987). Computer criticism vs. technocentric thinking. *Educational Researcher, 16*(1). https://learning.media.mit.edu/courses/mas713/readings/Papert,%20technocentric%20thinking.pdf

Papert, S. (1991). Situating constructionism. In I. Harel, & S. Papert (Eds.), *Constructionism: Research reports and essays*. Ablex, Retrieved July 1, 2024, from https://dailypapert.com/situating-constructionism/

Papert, S. (1993). *The children's machine*. Basic Books.

Papert, S. (1999). *Logo philosophy and implementation*. LCSI. Retrieved July 4, 2024, from http://www.microworlds.com/support/logo-philosophy-costa-rica.html

Papert, S. (2002, July 27). The turtle's long slow trip: Macro-educological perspectives on microworlds. *J. Educational Computing Research, 27*(1–2). Retrieved August 5, 2024, from https://dailypapert.com/wp-content/uploads/2015/10/THE-TURTLE%E2%80%99S-LONG-SLOW-TRIP-MACRO-EDUCOLOGICAL-PERSPECTIVES-ON-MICROWORLDS.pdf

Papert, S., & Harel, I. (Eds.). (1991). *Constructionism*. Ablex Publishing.

Papert, S., & Solomon, C. (1971). *Children in new learning experiences*. Channel 5 TV. Boston, MA. Retrieved June 23, 2024, from Logothings: https://www.youtube.com/watch?v=maDzjHIiXZc

Pea, R. D. (1984). Symbol systems and thinking skills: Logo in context. In *Pre-proceedings of the 1984 national logo conference* (pp. 55–66). Laboratory of Computer Science, Massachusettes Institute of Technology.

Pea, R. D., & Kurland, M. (1984). On the cognitive effects of learning computer programming. *New Ideas in Psychology, 2*, 138. Retrieved from On the Cognitiive Effects of Learning Computer Programming.

Piaget, J. (1931). *The origins of intelligence in children*. Delachaux et Niestlé.

Plowden, B. (1967). *Children and their primary schools*. Her Majesty's Stationary Office. Retrieved November 10, 2017, from http://www.educationengland.org.uk/documents/plowden/plowden1967-1.html

Price, N. D. (n.d.). *The con in constructivism*. Retrieved July 2, 2024, from http://www.improve-education.org/id55.html

Resnik, M. (2018). *Lifelong kindergarten*. MIT Press.

Robinson, K. (2011, May 13). *Changing education paradigms*. RSA Animate. Retrieved July 18, 2024, from https://www.youtube.com/watch?v=MPD37seJ6r8

Robinson, K. (2013, May 10). *How to escape education's death valley*. TED. Retrieved July 18, 2024, from https://www.youtube.com/watch?v=wX78iKhInsc&list=PL_N6UbeInhwcKn3vLeLDIsb9jN8NWL2M0

Robinson, K. (2015, September 15). *Bring on the learning revolution*. TED. Retrieved July 18, 2024, from https://www.youtube.com/watch?v=kFMZrEABdw4

Roccella, S., Carrozza, M. C., Cappiello, G., Dario, P., Cabibihan, J. J., Zecca, M., Miwa, H., Itoh, K., & Marsumoto, M. (2004). Design, fabrication and preliminary results of a novel anthropomorphic hand for humanoid robotics. In *IEEE/RSJ international conference on intelligent robots and systems (IROS) (IEEE Cat. No. 04CH37566)*, Sendai, Japan.

Roccella, S., Carrozza, M. C., Cappiello, G., Cabibihan, J. J., Laschi, C., Dario, P., Takanobu, H., Matsumoto, M., Miwa, H., Itoh, K., Takanishi, A. (2007). Design and development of five-fingered hands for a humanoid emotion expression robot. *International Journal of Humanoid Robotics, 4*(1), 181–206. https://doi.org/10.1142/S0219843607000996

Sawyer, R. K. (Ed.). (2006). *The Cambridge handbook of learning sciences*. Cambridge University Press.

Seldon, A., & Abidoye, O. (2018). *The fourth education revolution: Will artificial intelligence liberate or infantalise humanity* (Kindle ed.). Legend Press.

Staff Reporter. (2024a, May 25). AI can lie, cheat and deceive. In *The Week* (p. 17).

Staff Reporter. (2024b, September 9). *UK school to replace teachers with AI*. Retrieved September 18, 2024, from https://www.inavateonthenet.net/news/article/uk-school-to-replace-teachers-with-ai-programme

Stanton, D., O'Malley, C., Bayon, V., Hourcade, J. P., Sunblad, Y., Fast, C., Cobb, S., Taxen, G., & Benford, S. (2004). The KidStory project: Developing collaborative storytelling tools for children, with children. In J. Siraj-Blatchford (Ed.), *Developing new technologies for young children* (pp. 75–91). Trentham Books.

Theakston, J., Barker, L., & Jones, V. (2004, January 8). *Ian wright meets an old school teacher* (Y. Television, Producer). Retrieved July 17, 2024, from https://www.youtube.com/watch?v=eKToIrezxPw

Turkle, S., Taggart, W., Kidd, C. D., & Daste, O. (2006). Relational artifacts with children and elders: The complexities of cybercompaionship. *Connection Science, 18*(4), 347–361. Retrieved July 17, 2024, from https://sherryturkle.mit.edu/sites/default/files/images/Relational%20Artifacts.pdf

UNESCO. (1989). *Child rights convention*. Retrieved June 11, 2024, from https://www.unicef.org/child-rights-convention/convention-text-childrens-version

UNICEF. (2001). *The right to education*. Retrieved May 17, 2024, from https://www.unicef.org.uk/rights-respecting-schools/the-rrsa/the-right-to-education/

Varkey Foundation. (2018). *Andria Zafirakou winner of the 2018 global teacher award*. Varkey Foundation. Retrieved July 28, 2024, from https://globalteacherprize.org/pages/andria-zafirakou-2018

Villa, R. A., & Thousand, J. S. (2005). Elements of change. In R. A. Villa, & J. S. Thousand (Eds.), *Creating an inclusive school* (2nd ed.). Association for Supervision and Curriculum Development (ASCD).

von Moltke, H. (2016). Helmuth von Moltke 1800–91: Prussian military commander. In S. Ratcliffe (Ed.), *Oxford essential quotations* (4 Online ed.). Oxford University Press. Retrieved August 5, 2024, from https://doi.org/10.1093/acref/9780191826719.001.0001/q-oro-ed4-00007547

Waldman, S., & E&E News. (2023, March 16). Texas weakens climate science education guidelines. *Scientific American*. Retrieved July 30, 2024, from https://www.scientificamerican.com/article/texas-weakens-climate-science-education-guidelines/

Wallis, J. (1685). *A treatise of algebra, both historical and practical*. John Davis: ECHO—Cultural heritage online. Retrieved September 24, 2024, from http://echo.mpiwg-berlin.mpg.de/MPIWG:GK8U243K

Watters, A. (2015, April 2015). Lego mindstorms: A history of educational robots. Retrieved June 11, 2024, from http://hackeducation.com/2015/04/10/mindstorms

Weir, S. (1987). *Cultivating minds: A logo case book*. Harper & Row.

Wiliam, D. (2007, September). Assessment learning and technology: Prospects at the periphery of control. In *Keynote speech: Association for learning technology*. Retrieved July 17, 2024, from https://www.alt.ac.uk/sites/default/files/assets_editor_uploads/documents/altc2

Wiliam, D. (2018, September 18). *Clarifying, sharing and understanding learning intentions*. LSI: Learning Science International. Retrieved September 18, 2024, from https://youtu.be/fC29IyqPVr0?si=NZHTGL-mjShSqO7q

Wiliam, D. (2019, March 7). Dylan Wiliam on effective questioning in the classroom. In *Free master class in radical leadership: Podcast special* (J. Earp, Interviewer). Monash University. Retrieved July 26, 2024, from https://www.teachermagazine.com/sea_en/articles/teacher-podcast-dylan-wiliam-on-effective-questioning-in-the-classroom

Woolf, B. P., Beck, J., Eliot, C., & Stern, M. (2001). Growth and maturity of intelligent tutoring systems. In K. D. Forbus & P. J. Feltovich (Eds.), *Smart machines in education* (pp. 99–144). AAAI Press and MIT Press.

Zafirakou, A. (2018, May 31). Global education episode 14: Global teacher prize winner Andria Zafirakou. In *Free masterclass in radical leadership: Podcast* (D. Russell, Interviewer). Monash University. Retrieved July 28, 2024, from https://www.teachermagazine.com/au_en/articles/global-education-episode-14-global-teacher-prize-winner-andria-zafirakou

Zecca, M., Roccella, S., Carrozza, M. C., Miwa, H., Itoh, K., Cappiello, G., Cabibihan, J. J., Matsumoto, M., Takanobu, H., Dario, P., & Takanishi, A. (2004). On the development of the emotion expression humanoid robot WE-4RII with RCH-1. In *4th IEEE/RAS international conference on humanoid robots* (Vol. 1, pp. 235–252). https://doi.org/10.1109/ICHR.2004.1442125

Theoretical and Practical Coherence of Integrated STEM Education and Educational Robotics: Review and Analysis of Good Practices in Europe

Miguel Ángel Merino-Fernández, Ana Isabel Obregón Cuesta, Almudena Alonso-Centeno, Luis Alberto Mínguez Mínguez, Iraklis Varlamis, Chryssa Sofianopoulou, Sema Aykuş, Charikleia Rousoulioti, Athanasios Vrantzas, Ioannis Siakavaras, Georgios Papaioannou, Nikolaos Pischos, Kimberley Burcke, Takis Angelopoulos, and Jairo Ortiz-Revilla

Abstract Integrated STEM education and educational robotics (ER) represent two topics of maximum interest in the current educational reality, being necessary an investigation of their theoretical and practical approaches. To this end, a search and compilation of the characteristics of ten good practices that have implemented both approaches in Higher Education was carried out in each of the five European countries. These practices were then evaluated by teachers and educational researchers to select the two best practices in each country. The theoretical and practical consistency of the characteristics of the ten best practices with those of other studies at global level was then discussed and several conclusions and improvements in European practices were drawn. The results support the use of cooperative and collaborative

M. Á. Merino-Fernández · A. I. Obregón Cuesta · A. Alonso-Centeno · L. A. Mínguez Mínguez · J. Ortiz-Revilla (✉)
University of Burgos, Burgos, Spain
e-mail: jortizr@ubu.es

I. Varlamis · C. Sofianopoulou
Harokopio University of Athens, Kallithea, Greece

S. Aykuş
Antalya Provincial Directorate of National Education, Antalya, Turkey

C. Rousoulioti · A. Vrantzas · I. Siakavaras
Scientific Association for the Promotion of Lifelong Learning, Athens, Greece

G. Papaioannou · N. Pischos
DigiCult, Athens, Greece

K. Burcke
Institut Polytechnique UniLaSalle, Beauvais, France

T. Angelopoulos
Open Technologies Alliance (GFOSS), Athens, Greece

G. Lampropoulos and S. Papadakis (eds.), *Social Robots in Education*, Studies in Computational Intelligence 1194, https://doi.org/10.1007/978-3-031-82915-4_17

learning together with project-based learning by integrating at least two STEM disciplines in an interrelated way, including the use of reflective methodologies such as inquiry-based science teaching and the engineering design process using integrated models such as transdisciplinary approaches to enhance student learning and engagement. The consistent use of the commercial robot kits and the need for customized robots is highlighted, and the use of low-cost open-source programming boards and open design robots is included as an improvement. The promotion of co-education is highlighted as lacking, with a need to improve practices aimed at researching and improving girls' interest in pursuing STEM studies. Finally, a few limitations and future lines of research are indicated.

Keywords Integrated STEM education · Educational robotics · Secondary education · Good practices · Literature review · Expert evaluation · Open hardware · Open software

1 Introduction

Integrated STEM education is currently experiencing a boom in any education system worldwide (Martín-Páez et al., 2019), since the problems faced by society cannot be solved solely from a single discipline, but must use various disciplines in an interrelated manner (Chavarría Pérez & Guede-Cid, 2023). In this regard, according to Gok (2021), integrated STEM education is understood as the pedagogical approach that aims to promote the teaching–learning of two or more STEM disciplines, within a real context that allows the contents of these subjects to be connected through practice. This educational approach has the mission and need to promote STEM literacy in citizens, preparing them to face the challenges of the twenty-first century (Gok, 2021).

In this regard, it should be noted that the Recommendation of the Council of the European Union of 22 May 2018, on key competences for lifelong learning, approved the integration into the European education system of 8 key competences, which must necessarily be acquired at the end of each educational stage. Here, we focus on the stages of the second cycle of compulsory secondary education (CSE) and baccalaureate. We highlight in this case, due to its intrinsic relationship with the integrated STEM education approach, the inclusion in the European educational curriculum of the so-called mathematical competence and competence in science, technology and engineering or STEM competence.

The following educational research study acquires notorious relevance due to the fact that the so-called STEM subjects within the educational curriculum of ISCED stages 3 and 4 (such as, for example, the subjects of mathematics, biology and geology, physics and chemistry, technology and engineering, technology and digitalisation, among others), have usually been studied as markedly separate areas. In this respect, little effort has been made to address integrated learning situations, giving more importance to intradisciplinary content than to socially relevant problems (Ortiz-Revilla et al., 2020).

The relevance that Educational Robotics (ER) is currently having in any educational system should be highlighted. In this regard, Floroiu et al. (2019) emphasize that ER enables students to acquire professional skills in STEM fields, favoring robotic technological literacy. Due to this, ER is currently gaining momentum in many educational curricula. For example, in Spain, following the enactment of Organic Law on Education (LOMLOE) of 2020, it has been included in Royal Decree 217/2022, of March 29, establishing the organization and minimum teaching requirements for Compulsory Secondary Education (CSE), the content block "computational thinking, automation and robotics" has been included in the area of Technology and Digitalisation. This block will be compulsory in one of the courses from 1st to 3rd year of CSE, being optional in 4th year.

Several initiatives aimed at integrating educational robotics in France were launched before 2015. For instance, the Plan for the development of information and communication technologies (ICT) for education, launched in the early 2000s, laid the foundations for the integration of advanced educational technologies, including robotics. Local and regional initiatives such as the RoboCup Junior, robotics clubs, and collaborative projects with universities or technology companies further promoted ER. The 2015 Digital Plan for Education/Plan numérique pour l'éducation accelerated and structured the integration of educational robotics in French schools. The Ministry of National Education and Youth's strategic framework, particularly the Digital Strategy for Education 2023–2027, aims to enhance students' digital skills, with a strong emphasis on digital citizenship and career preparation. As part of this, the French curriculum now includes technological education from early childhood through secondary education. For example, it mandates technological education for students aged 3–15, focusing on understanding technical objects and their societal impacts.

In Greece, even if the cognitive subject of Robotics is not included in primary and secondary education curriculum, robotics is taught in the context of the cognitive subject of Informatics in both primary and secondary education. Also, Robotics in elementary schools is supported by the "Skills Workshops" program with the backing of the Institute of Educational Policy (IEP). This initiative integrates hands-on activities and interdisciplinary learning, fostering students' critical thinking, problem-solving abilities, and teamwork through engaging and innovative educational experiences. In the case of Upper High School, there is not any particular cognitive subject in the curriculum, but there are many degrees of freedom for integrating robotics projects in the context of the cognitive subjects of "project" and of Informatics. Generally national and local educational authorities support and encourage schools, teachers and students to participate in robotics school projects. Concerning Vocational Education (ages 15–18), robotics can be integrated in many cognitive subjects in any direction of vocational schools. Recently, the delivery of 177,000 sets of modern robotics and STEM equipment to kindergartens, elementary schools, and middle schools across the country was completed. The project titled "Supply of Robotics and STEM Equipment for Education" is part of the National Recovery and

Resilience Plan "Greece 2.0" with funding from the European Union—NextGenerationEU. It is a comprehensive intervention, covering all kindergartens, elementary, and middle schools nationwide. The investment exceeds 30 million euros, aiming to equip schools with over 177,000 robotics and STEM sets to effectively support the learning process. In addition to the robotics sets, the project includes special educational material, targeted teacher training, a program for creating digital educational content, and a support project (helpdesk) for robotics systems, alongside the installation of the systems in schools.

The educational policy in Poland regarding STEM focuses on developing technical and analytical skills in students by integrating modern technology into the classrooms. An interdisciplinary approach that combines sciences and arts is promoted, aiming to provide a more comprehensive education. Additionally, international collaboration and continuous training for teachers are encouraged to improve teaching in these areas. Local initiatives, such as science fairs and competitions, also play a crucial role in sparking students' interest. Together, these measures aim to strengthen educational quality and foster innovation in the country.

In Turkey, the integration of robotics and STEAM (Science, Technology, Engineering, Arts, and Mathematics) education into the secondary education curriculum has evolved significantly since the early 2000s. The Ministry of National Education (MoNE) initiated this transformation by emphasizing science and technology, notably in the 2005 Science Curriculum, which highlighted Science Process Skills (SPS) and inquiry-based learning (MoNE, 2005). By the early 2010s, robotics was formally introduced into the curriculum through elective Information Technologies courses that included basic programming and robotics (MoNE, 2012). The 2015 curriculum reform further advanced this integration by embedding critical thinking and project-based learning across all subjects, aiming to enhance students' problem-solving and innovative skills essential for comprehensive STEAM education (Aytan, 2016). The 2018 Science Curriculum reinforced this focus by emphasizing SPS, Nature of Science (NOS), Socioscientific Issues (SSI), and STEM education, equipping students with the ability to engage with complex scientific phenomena and real-life problems (Yapıcıoğlu, 2021). National initiatives like TÜBİTAK (The Scientific and Technological Research Council of Turkey) and events like TEKNOFEST have significantly promoted STEM education. TÜBİTAK supports various educational projects and competitions that encourage scientific research and innovation among students (TÜBİTAK, 2020). TEKNOFEST, an annual aerospace and technology festival, promotes interest in technology and engineering through competitions and exhibitions, providing students with platforms to showcase their projects and innovations (TEKNOFEST, 2021). Despite these advancements, the integration of robotics and STEAM education faces challenges, including the need for better teacher training and equitable access to resources. Addressing these challenges is crucial for maximizing the potential of these educational initiatives and ensuring all students benefit (Yapıcıoğlu, 2021).

Therefore, it follows from the above that ER has not yet been integrated into European school curricula on a compulsory basis, but is usually addressed through after-school activities, weekends or summer camps (Floroiu et al., 2019). Obstacles to implementation include the duration of robotics activities, the cost of equipment, the lack of robotics teachers and the unavailability of suitable laboratories and spaces (Floroiu et al., 2019).

The main objective of the following literature review study is to determine the characteristics of good practices in integrated STEM and ER education, implemented and evaluated in five European countries (France, Greece, Poland, Spain and Turkey), comparing and discussing them with other studies both theoretical and practical that are addressed worldwide. It should be noted that currently no educational research addressing these aspects has been found in the scientific literature, so it is considered relevant to analyse the theoretical and practical coherence of good practices, with the aim of finding improvements or missing aspects in them.

This research is based on the results obtained in the first phase of the international Erasmus + S.T.E.P.S. (STEM Teaching and Education Platform for Students) project, co-funded by the European Union under project ID 2023-1-EN01-KA220-HED-000165711. S.T.E.P.S. aims to democratise access to integrated STEM education through ER at ISCED 3 (upper secondary) and ISCED 4 (post-secondary non-tertiary) levels in Spain, Turkey, Poland, Greece, and France. By making innovative teaching practices accessible to more students, regardless of geography or socio-economic status, the project seeks to improve the quality of STEM education and equip a new generation with essential skills.

The objectives of the educational research are defined below:

- To compile and analyse the characteristics of studies and good practices that address STEM education integrated with Educational Robotics (ER) at ISCED levels 3 and 4 (secondary education). On the one hand, it looks at the worldwide scientific literature and, on the other hand, at the good practices implemented and evaluated in the five European countries collaborating in the S.T.E.P.S. project.
- Evaluate good practices through a panel of experts (researchers and teachers).
- To discuss the theoretical and practical coherence between the characteristics of European good practices and those of other practices and studies worldwide.
- Reflect on possible improvements in European best practice.

The following sections of this chapter present the theoretical framework on STEM education, educational robotics and the characteristics of good practices globally; the research methodology employed, including the method of searching for good practices and their evaluation by experts; the results based on the analysis and evaluation of good practices; the final conclusions and limitations of the research.

2 Theoretical Framework

This section presents the theoretical framework on STEM education, educational robotics and the characteristics of good practices present globally.

2.1 *Integrated STEM Education and Educational Robotics*

Integrated STEM education is understood as the pedagogical model that has the didactic objective of integrating several of these disciplines or curricular areas in an explicit and intentional way in order to solve socially relevant problems of the contemporary world (Ortiz-Revilla et al., 2020). Furthermore, it has the mission to prepare students to solve these problems through the use of practical skills such as innovation, creativity, critical thinking, teamwork, communication of results, while fostering interest in science (Ortiz-Revilla et al., 2022; Quigley & Herro, 2016; Theodoropoulou et al., 2023).

Educational Robotics (ER), on the other hand, consists of a didactic tool that teachers use in classrooms through the implementation of robots in the educational process, in order to promote cognitive, individual and teamwork development, as well as creativity and imagination (Floroiu et al., 2019; Theodoropoulou et al., 2023).

Through the use of integrated STEM and ER didactic proposals, which integrate various disciplines, they allow secondary school students, who have not selected science and technology subjects, to have access to technological innovation experiences (Bernstein et al., 2022).

It is worth noting that ER promotes learning in STEM disciplines or areas in practical, entertaining and motivating teaching–learning contexts, through communication, cooperation, collaboration, critical thinking, problem solving and decision making (Karalekas et al., 2023; Üçgül & Altıok, 2022). Moreover, it allows for an integrated approach, in which different disciplinary areas complement each other (Arís & Orcos, 2019).

According to Amo et al. (2021) and Naya-Varela et al. (2023), ER is congruent with the use of active methodologies that promote STEM disciplines, such as *learning by doing*, cooperative learning and project-based learning (PBL). According to Chang and Chen (2020), the ER is a very valid tool to address integrated learning models such as interdisciplinary learning through integrated STEM education.

In order to promote student learning in STEM disciplines, various integrated proposals have emerged, such as the one addressed by researchers Naya-Varela et al. (2023), who use ER in secondary and high school students as a teaching–learning resource for computational intelligence and artificial intelligence through the use of autonomous robots, favoring the resolution of challenges.

In addition, there are practices such as the one addressed by Karalekas et al. (2020) who carry out such educational approaches in two Greek schools, which develop an open-source software and hardware robotics platform, with the aim of teaching physics, electricity and robotics concepts to secondary school students aged 16–17.

Misthou et al. (2021) investigate the implementation of a voluntary after-school coding programme "GreekCodersK12", in a secondary school in Athens (Greece) targeting the K-12 educational stage, which encompasses compulsory primary and secondary education, implementing a STEM learning community, based on the use of computer programming, ER, fostering STEM skills, STEAM (inclusion of arts) and ESTEAM (integration of entrepreneurship), fostering creativity, innovation, digital competence and collaboration with researchers, professionals and scientists.

2.2 Characteristics of Practices with Integrated STEM Education Through Educational Robotics at the Global Level

In the following, the characteristics that are recurrently observed in the proposals and studies that address STEM education integrated with ER in the world scientific literature, at the secondary education stage, will be made explicit:

- Many studies agree that these practices use a variety of active methodologies, with Project Based Learning (PBL) being the most prominent (Amo et al., 2021; Bellas et al., 2018; Galino & Tanaka, 2021; Naya-Varela et al., 2023; Valko & Osadchyi, 2021; Vega & Cañas, 2019). To a lesser extent is the use of Learning by doing (constructionism) (Amo et al., 2021; Demo et al., 2012) and inquiry-based science education (IBSE) (Demo et al., 2012).
- With regard to the STEM curricular contents and areas that are worked on following this educational approach at these levels, mathematics (Üçgül & Altıok, 2022; Valko & Osadchyi, 2021), science [physics and/or chemistry (Bernstein et al., 2022; Marzoli et al., 2021; Merkouris & Chorianopoulos, 2018; Tarrés-Puertas et al, 2023; Üçgül & Altıok, 2022)], engineering [electrical, mechanical, electronics (Tarrés-Puertas et al., 2023; Theodoropoulou et al., 2023; Üçgül & Altıok, 2022; Valko & Osadchyi, 2021)] and technology [computer programming, computational thinking, robotics (Bernstein et al., 2022; Marzoli et al., 2021; Merkouris & Chorianopoulos, 2018; Tarrés-Puertas et al., 2023; Üçgül & Altıok, 2022; Valko & Osadchyi, 2021)].
- Regarding the robotic resources implemented, the use of programmable and unassembled robotic kits, such as LEGO Mindstorms Education (EV3 and NXT) (Demo et al., 2012; Flores et al., 2022; Merkouris & Chorianopoulos, 2018;

Theodoropoulou et al., 2023; Üçgül & Altıok, 2022; Wang & Frye, 2019) and the electronic programming boards Arduino, Micro:bit and Raspberry Pi (Galino & Tanaka, 2021; Hristov et al., 2023; Marzoli et al., 2021; Vega & Cañas, 2018, 2019). To a lesser extent, the use of Ozobot (Wright et al., 2021), micro:maqueen (Hristov et al., 2023), mBot (Sáez-López & Buceta-Otero, 2023) and NAO (Pusceddu et al., 2023) robots is specified.

It highlights the amalgam of practices in which the use of robots built with low-cost materials is implemented (Álvarez et al., 2023; Galino & Tanaka, 2021; García-Langley et al., 2022; Manikutty et al., 2019; Vega & Cañas, 2018).

– Finally, a number of studies have been carried out with the aim of promoting co-education, reducing the gender gap and encouraging girls' and women's interest and motivation to pursue STEM studies (Chapman et al., 2020; Chiang et al., 2024; Hennessy Elliott, 2020; Keller & John, 2020; Sullivan & Bers, 2019; Van Wassenaer et al., 2023; Wang & Frye, 2019; Weinberg et al., 2007; Wong et al., 2019).

3 Methodology

This section presents he research methodology employed, including the method of searching for good practices and their evaluation by experts.

3.1 Bibliographic Search for Good Practices

In this educational research, a literature review of good practices implementing such pedagogical approaches in secondary education was carried out. For this purpose, and in accordance with the principle of scientific rigor as well as search availability in each country, the following scientific search engines were used: Web of Science (WoS), Scopus, Dialnet and HAL. Additionally, the websites 'Google Scholar', the European Journal for Science Education, "Science in School", Scientix, the resource platform for the European science education community and EEPEK (Scientific Association for the promotion of educational innovation) conference minute archives were searched.

To do this, firstly, a series of key search terms were applied in order to standardize the review, which are: STEM AND robot* AND ("Secondary Education" OR "Secondary School" OR "K-12" OR "High School" OR "Baccalaureate" OR "Secondary Schooling" OR "Grammar School" OR " Post-Primary Education" OR "Senior Secondary School"). These key terms were selected after a review of the terminology used in each partner's country in relation to the study topics.

A series of search filters were then used on these sources:

– First, it was indicated that the studies must be written in any of the following languages: English, Spanish, French, Polish, Greek or Turkish.

- The implementation territory was then selected, in any of the five European countries of the project (Spain, France, Poland, Greece or Turkey).
- The open access filter was then assigned to these practices.
- And finally, duplicate studies were eliminated between these databases.

A rubric for collecting good practices was then implemented (Annex 1), comprising 16 items, such as the age of the students, the school year(s) involved, the target audience, the duration, the location, the STEM disciplines and contents integrated, the origin of the activity, the didactic objectives, the didactic methodologies implemented, the sequence of activities, the results of educational research and the didactic resources used (robot, etc.). The overall objective of the project was to collect and analyse ten good practices in each of the five European countries using this rubric.

3.2 Expert Assessment

Following the collection and detailed analysis of the fifty best practices carried out in the five European countries, an evaluation of these practices was carried out using a rubric provided to at least two experts (Annex 2), who must be secondary school teachers of a STEM subject and/or researchers in integrated STEM education. The aim of this assessment was to extract the two best practices from each of the five countries in order to discuss the theoretical and practical coherence between the characteristics of the ten European best practices and those of the global studies.

4 Results

Through this bibliographic research process, fifty good practices in STEM education integrated with ER have been compiled, distinguishing them for each of the five European partner countries of the STEPS project. Annex 4 specifies a series of guidelines (title, author/s, year of publication or implementation and robotic resources used) for each of the fifty good practices and Annex 3 presents the characteristics (title, author/s, year of publication, technological resources, didactic methodologies and integrated STEM areas) of the ten best practices selected by the expert evaluators.

4.1 Analysis of Good Practices

The following is an analysis of the recurrent characteristics observed in the good practices compiled through the literature review. To this end, four parameters have been identified for investigation: the teaching methodology implemented, the promotion of coeducation, the type of robot used, and the STEM areas integrated. These

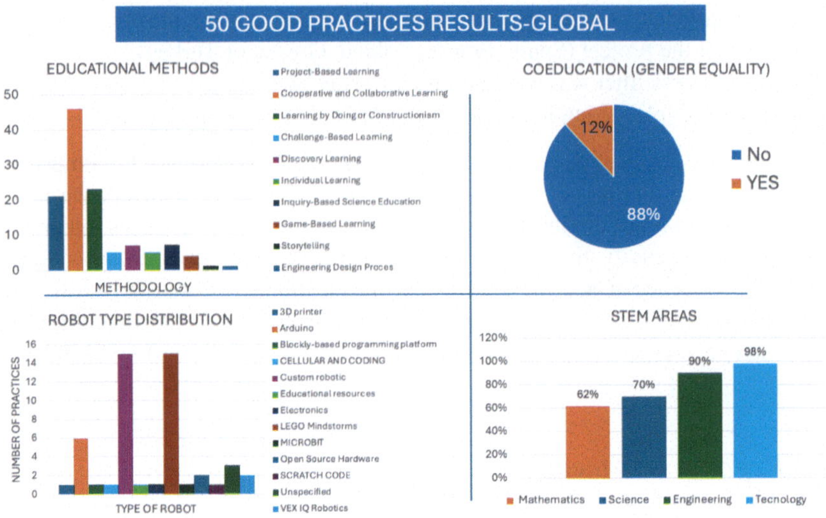

Fig. 1 Specification of the recurrent characteristics analysed in the fifty good practices of the STEPS project according to the type of methodology implemented, the promotion of co-education, the use of robotic resources and the integration of STEM areas

four aspects are first classified according to the fifty good practices both jointly (Fig. 1) and by each of the five European countries (Fig. 2) and then in relation to the ten good practices selected by the expert evaluators both jointly and by country of implementation (Fig. 3).

Regarding the predominant characteristics of the 50 good practices, Fig. 1 shows that the most used methodologies are cooperative and collaborative learning in 46 practices, followed by constructionism in 23 and project-based learning (PBL) in 21, indicating a strong preference for interactive and group learning environments in robotics teaching. The joint distribution shows a significant proportion of practices that do not apply co-education (88%). This result suggests that the integration of gender equality in ER activities still has room for improvement. The data indicate a varied use of robotic resources, with custom robots and the Lego Mindstorms kit standing out, with 15 practices each, along with the Arduino programming board with 6. The integration of technology and engineering disciplines stands out in more than 90% of the practices, indicating their strong alignment with robotics teaching. In contrast, science (70%) and mathematics (62%) are less represented, and their joint integration should be improved.

The 50 good practices will now be analyzed and classified by each of the five countries in which they have been implemented. Figure 2 shows the most commonly used methodologies in each of the five European countries, with cooperative and collaborative learning standing out in all of them, together with constructionism in Spain (9), Greece (8) and France (6). PBL has also been addressed in all practices, with Spain (7) and Poland (6) standing out, while the development of discovery

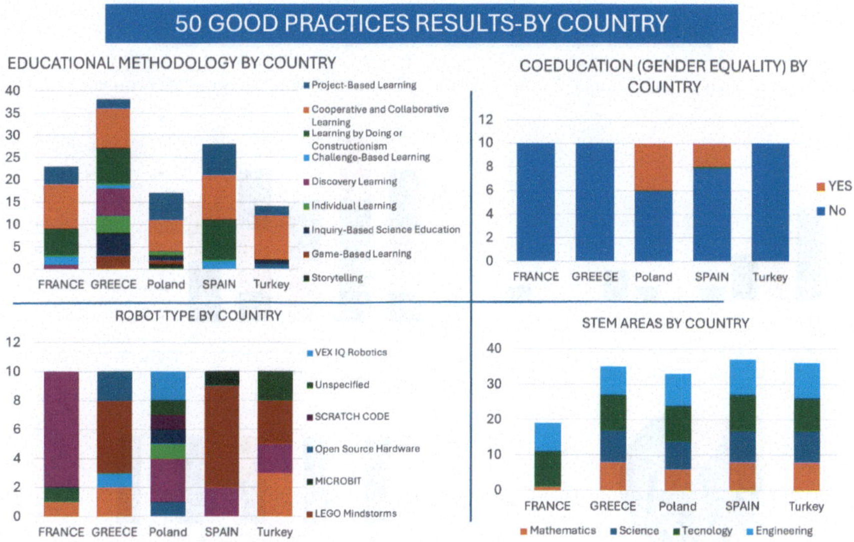

Fig. 2 Most recurrent characteristics of the fifty practices collected in the STEPS project, distinguished by the countries in which they have been implemented. The methodology implemented, the promotion of co-education, the type of robot used and the STEM areas integrated are specified

learning has been observed to a lesser extent in France (1) and above all in Greece (6). As far as the promotion of co-education is concerned, France, Greece and Turkey do not develop it in any of their practices, unlike Spain, where it is carried out in two of them, and above all in Poland, with four practices. This indicates the need to improve and increase educational practices and policies focused on gender equality. The data reveal a varied use of robots in the different countries, showing that in France, self-built robots with low-cost material are preferably used, while in Spain, Turkey and Greece, the programmable and unassembled Lego Mindstorms kit is mostly used, and in Poland, a wide variety of robot types is shown. These data suggest that while some robotic platforms have universal appeal due to their ease of use, others are preferred for specific educational purposes. Understanding these trends can help in developing educational technology strategies. Finally, with respect to integrated STEM areas, the integrated use of mathematics, technology, engineering and science in Greece, Poland, Spain and Turkey is noteworthy, with the case of France being striking, where mathematics is integrated into a single practice and science into none.

The characteristics found in the ten best European practices as a whole, obtained after the expert evaluation, will now be detailed (Fig. 3). In terms of the teaching methodologies addressed, the use of cooperative and collaborative learning stands out in all of them, together with project-based learning in six practices, constructionism in two and IBSE in two. The data support the need to increase the number of practices that promote co-education, as 90% of these do not take it into account. A common parameter is the use of robots built with low-cost materials in half of the good practices, as well as the Lego Mindstorms kit in 3 and, to a lesser extent,

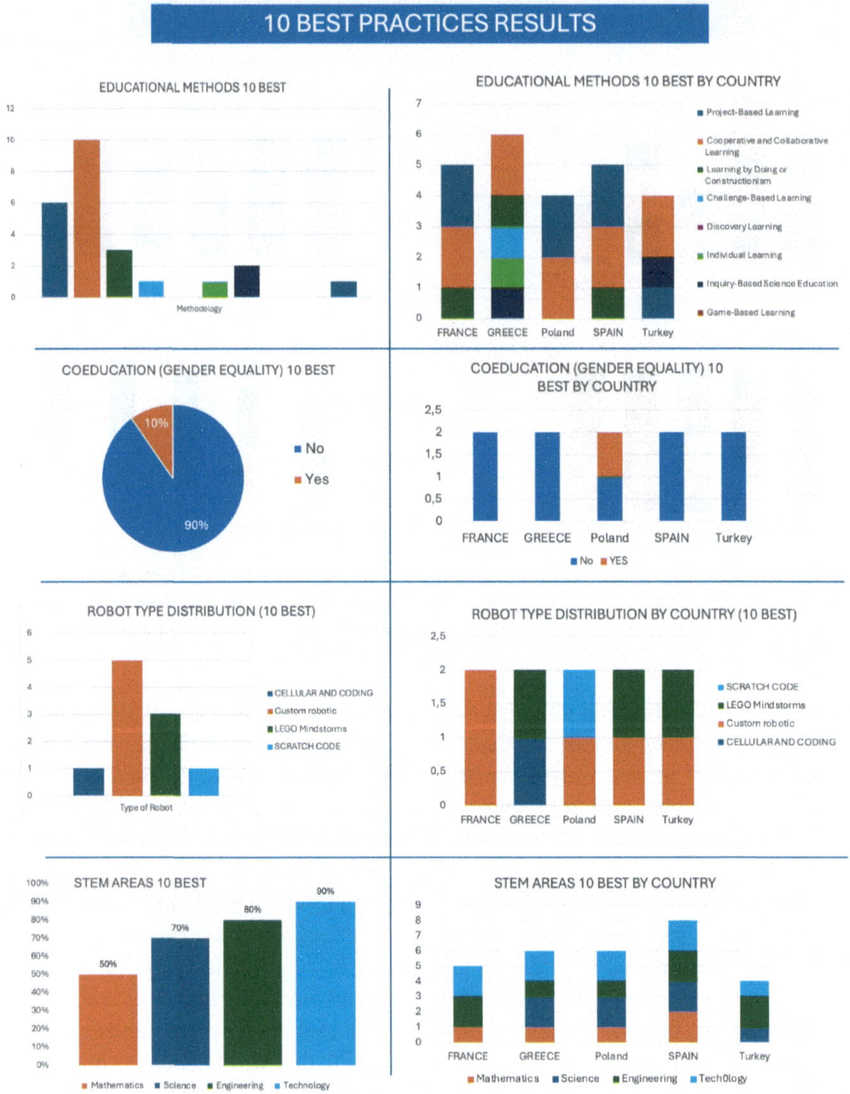

Fig. 3 Recurrent features observed in the ten good practices selected in the expert evaluation in terms of teaching methodology, promotion of co-education, type of robot used and integration of STEM subjects, both overall and for each of the five European countries

2 practices that implement coding using mobiles and programming with Scratch. Finally, with regard to the STEM areas worked on in the classroom, 9 practices integrate technology, 8 engineering, 7 science and 5 mathematics, noting that this last area in particular is only carried out in half of the practices, so its integration must necessarily be increased.

Finally, we will detail the characteristics of the ten best practices obtained after the evaluation, classifying them by each of the five European countries. Figure 3 shows that the implementation of cooperative and collaborative learning methodology stands out in the five countries together with PBL in France, Poland and Spain with two practices in each, followed by constructionism in France, Greece and Spain with only one practice per country and, finally, IBSE with one practice in Greece and Turkey. The results corroborate the need to address co-education in all five countries, with only one good practice found in Poland, which has worked on this aspect in the classroom. The use of self-built robots with low-cost materials stands out in Poland, Spain, Turkey and France, with France being the only country to use this resource. The use of the Lego Mindstorms kit stands out in Greece, Spain and Turkey with a respective practice. To conclude this analysis, the robust integration of the 4 STEM areas in Greece, Poland and Spain should be highlighted, the latter being the one in which all of them are addressed in the two good practices. In contrast, in France and Turkey, three STEM areas are addressed with no integration of science and mathematics respectively.

4.2 Evaluation of Good Practices

Following the joint assessment of the ten good practices from each country by the educational experts (secondary school teachers and researchers involved in these pedagogical approaches), the two best practices from each of the five countries have been selected (Annex 3). The process is as follows: first, they are asked about their personal background, as well as the STEM areas they research or the STEM courses and subjects they teach. They are then presented with a rubric (Annex 2), which they have to complete by selecting a value between 1 (disagree) and 5 (strongly agree). Questions to be answered include: "Are these practices useful for secondary school teachers?", "Is there integrated learning in which two or more STEM areas are addressed?", "Is the technology used appropriate?", etc. The teachers and researchers then indicate their opinions on the didactic capacity of these practices and their strengths and weaknesses, relating this to the needs of the teachers themselves in order to implement them successfully in the classroom.

5 Discussion About Consistency Between Theory and Practice in the Final Ten Good Practices

Next, using the results obtained after the expert evaluation, we will carry out an analysis of the theoretical and practical coherence of the characteristics presented by the ten good practices in terms of teaching methodologies, the STEM areas addressed, the robotic resources implemented and the promotion of coeducation. For each of the characteristics, an initial cataloging is carried out together with the needs expressed by the educational experts, to then discuss their coherence in the light of the scientific literature, with the aim of concluding by proposing a series of aspects to be included or improved in the good practices.

a. Teaching methodologies

With regard to the active methodologies implemented in the ten European good practices, the use of cooperative and collaborative learning stands out in all of them. The use of project-based learning (PBL) is also recurrently observed in the practices addressed in Spain (Arís & Orcos, 2019; Vega-Moreno et al., 2016), France (Michel, 2021; Planete Sciences, n.d.) and Poland (European Commission, n.d.; STEMKIT4SCHOOLS, n.d.). Regarding educational experts, in the case of Greece, they explicitly state in their evaluation the need for students to be trained in cooperative and collaborative learning techniques, as effective communication skills are needed to deal with such practices. The Spanish and Turkish experts infer the need for such practices to set out in detail how to approach methodologies such as cooperative learning and the engineering design process (EDP) in the activities, as well as to carry out a thorough evaluation of their implementation in order to improve their usefulness and impact.

Looking at the global scientific literature, there are a large number of studies and practices that address PBL together with cooperative and collaborative learning (Camargo et al., 2015; Foukarakis & Syrris, 2018; Galino & Tanaka, 2021; Hartigan & Hademenos, 2019; Karaahmetoglu & Korkmaz, 2019; Karaman et al, 2017; Naya-Varela et al., 2023; Sklirou, 2019; Valko & Osadchyi, 2021; Valls Pou et al., 2022; Vega & Cañas, 2019; Zhong et al., 2022), so there is theoretical and practical consistency in the use of both methodologies in good practices.

As aspects to be improved in good practices, the use of IBSE and EDP methodologies is included, due to their connection with integrated STEM education. IBSE enables the generation of STEM competences and computational thinking (Demo et al., 2012) and EDP favors the teaching–learning of ER by defining an engineering problem, designing a solution and optimising it (Jaipal-Jamani, 2023).

b. STEM areas worked on

The integration of the 4 STEM areas is observed only in the two Spanish practices (Arís & Orcos, 2019; Vega-Moreno et al., 2016) and in one Greek practice (Karagiorgou & Spahos, 2016). Although it is noteworthy that four practices integrate only two STEM areas, namely one practice from Greece (Tomara & Gouscos, 2019), two from Turkey (Akkoyun, 2019; Elbir, 2020) and one from France (Planete Sciences, n.d.), it is observed that none of them work on mathematics.

Educational experts from both Poland and Turkey stress in their evaluation the importance of implementing integrated models such as interdisciplinary approaches to ER in order to promote the resolution of socially relevant problems such as environmental degradation, as well as critical thinking. The French experts infer the deficient training received in ER for both teachers and students, as well as the short duration of such placements.

In terms of the global scientific literature, a consistent use of integrated STEM education is also observed, highlighting those integrated teaching proposals that have implemented the four STEM areas (Chang & Chen, 2023; Üçgül & Altıok, 2022), three (Tarrés-Puertas et al., 2023; Valko & Osadchyi, 2021) and even two (Bernstein et al., 2022; Marzoli et al., 2021), working on various science (physics and chemistry), mathematics (geometry, angles), engineering (building robots, solving design problems) and/or technology (computer programming, use of technological tools) contents.

Therefore, we can conclude the existence of a theoretical and practical congruence between the good practices and the scientific literature, as an integration of two or more STEM areas is observed in all of them. It is worth highlighting the need to improve those practices that only integrate two areas, achieving an interrelation of at least three areas, with the aim of having students solve problems in real situations, through the use of integrated models such as transdisciplinary together with strategies such as PBL and collaborative learning, among others (Chang & Chen, 2023).

c. Robotic resources used

In terms of the technological resources implemented (robotic artefacts, computational thinking programs, etc.), the use of the LEGO Education Mindstorms EV3 unassembled programmable robotic kit stands out in a good practice from Spain (Arís & Orcos, 2019), another from Turkey (Elbir, 2020) and another from Greece (Karagiorgou & Spahos, 2016). This is consistent with the scientific literature, through the use of this kit in several global studies (Gerber et al., 2017; Leonard et al., 2016; Scaradozzi et al., 2020; Theodoropoulou et al., 2023; Üçgül & Altıok, 2022; Wang, 2016; Wang & Frye, 2019).

The use of robots built with low-cost materials is also recurrently observed in European good practices, both in Spain (Vega-Moreno et al., 2016), France (Michel, 2021; Planete Sciences, n.d.), Turkey (Akkoyun, 2019) and Poland (European Commission, n.d.). This is in line with several global scientific studies, such as those addressed by García-Langley et al. (2022), Tay et al. (2017), Álvarez et al. (2023), Rubenstein et al. (2015), Galino and Tanaka (2021), Susilo et al. (2016), Wimer et al. (2023) and Manikutty et al. (2019), among others. Therefore, we can corroborate a theoretical and practical congruence regarding the use of self-built robots with low-cost material.

The Polish educational experts explain that the results obtained after its implementation support the need to use technological resources, as they benefit the teaching–learning processes, although they point out that the quality of these educational materials can be improved in some respects. The Spanish experts highlight the great replicability available, through the use of low-cost materials that facilitate the implementation of new projects. Finally, the French experts point to the lack of necessary resources due to financial constraints and the lack of appropriate content. They also infer that there is a lack of training in ER for both teachers and students.

As a proposal for improvement in European good practices, the use of computer programming boards, such as Arduino, Raspberry Pi and Micro:bit that are used in several global studies (Galino & Tanaka, 2021; Hristov et al., 2023; Marzoli et al., 2021; Theodoropoulou et al., 2023; Vega & Cañas, 2018, 2019), is considered as they are low-cost, programmable and easy to handle due to their small size and lightweight.

d. Coeducation (gender equality)

It should be noted that in the ten final good practices of the five European countries, only one good practice from Poland addresses the issue of coeducation, with the aim of reducing the gender gap and encouraging female students' interest in pursuing studies related to STEM disciplines. This Polish practice (European Commission, n.d.) aims primarily to promote the inclusion of pupils with specific educational support needs due to socio-economic difficulties, as well as to advocate gender equality, but does not specify how it is implemented in classroom activities. This contrasts with the global scientific literature where an amalgam of practices and research aiming to promote gender equality through integrated STEM education alongside ER is observed (Chapman et al., 2020; Chiang et al., 2024; Hennessy Elliott, 2020; Keller & John, 2020; Sullivan & Bers, 2019; Van Wassenaer et al., 2023; Wang & Frye, 2019; Weinberg et al., 2007; Wong et al., 2019).

From this analysis, the implementation and research of educational experiences that take coeducation into account in an essential way with the aim of favoring interest in STEM subjects and vocations in all students and especially in the female gender is extracted as a necessary improvement in all European countries, with emphasis on France, Greece and Turkey where there is no mention of it. This is because studies such as that of Sullivan and Bers (2019) who investigated the differences between male and female students' experiences in an ER competition, with the aim of improving girls' perception of the experience, as well as their confidence and performance, thus increasing their participation in such competitions. The results showed that more male students participated than female students, and that female students were less confident in their technical ability, so the technical knowledge of female students should be increased.

6 Conclusions

The objectives of this educational research have been to compile and analyse the characteristics of the good practices implemented in five European countries in secondary education after being evaluated by a series of educational experts in order to discuss their theoretical and practical coherence with the characteristics of other studies worldwide. In this way, a series of improvements or aspects to be included in these have been established.

Using the results obtained in this study, a series of aspects to be considered in both the theoretical and practical framework of good practices that implement integrated STEM education alongside ER in secondary education will be established in order to serve as a guide for teachers and educational researchers.

With regard to European good practices, there is theoretical and practical coherence in the use of cooperative and collaborative learning together with PBL in the light of the world's scientific literature. The inclusion of methodologies related to integrated STEM education such as IBSE (Demo et al., 2012) and EDP (Jaipal-Jamani, 2023; Sen et al., 2021) is an area for improvement, the latter being found in only one good practice out of 50 and IBSE in 20% of the ten best practices evaluated.

In terms of the implementation of integrated STEM education, although three practices are found that integrate the four STEM areas or disciplines, the majority only address two disciplines. The joint integration of technology and engineering stands out in seven of the ten good practices, while mathematics is found in only half of these, so that the integration of this discipline together with science must necessarily be improved, as both are integrated together in 40% of the practices. In this aspect, the use of integrated models such as interdisciplinary and transdisciplinary is included as an improvement, and in practices such as those of Chang and Chen (2020, 2023), the application and integration of abstract content through reflection and enquiry was considerably improved.

Consistent use of the Lego Mindstorms ER kit (NXT and EV3) and of customised robots with low-cost materials in the light of the world scientific literature stands out in 15 of the 50 European good practices each. The use of the former in Spain stands out in 70% of the practices and the latter in 80% of the French practices. Similarly, both didactic resources are the star product of the ten best practices selected by the experts, with 5 practices using customised robots and 3 Lego Mindstorms. As resources to be included, the use of cheap and open source programming boards such as Arduino, Raspberry Pi and Micro:bit is specified and can be used in combination with other robots such as Maqueen (Hristov et al., 2023).

Finally, there is a deficit in the promotion of co-education through these good practices, which is only considered in 12% of the total practices and in one practice out of the ten selected by the experts. In this regard, studies such as those by Chapman et al. (2020) and Wang and Frye (2019), which aimed to investigate and improve girls' interest, motivation and performance through ERcamps, highlight that the latter increased.

With regard to the limitations of the research, we consider the development of an unstandardised literature review, which has resulted in the integration of some practices not published in research study format, but extracted from websites, related to international projects, etc.

Future lines of research include the systematic standardised review of good practices that integrate STEAM education integrated with ER in Europe or in the world, considering the interrelation of any curriculum discipline such as, for example, arts education, which would allow us to obtain a broader set of reflections on this subject.

Additionally, ongoing efforts needed to enhance STEM education, to redesign an open-source, low-cost robot, aimed at providing accessible and engaging learning experiences for students. This initiative involves not only the technical redesign of the robot but also the development of comprehensive educational activities that align with STEM curricula. The goal will be to integrate hands-on robotics with structured activities, in order to foster critical thinking, creativity, and problem-solving skills in students and create a more attractive and enjoyable environment for students in STEAM curricula. The next version of FOSSBOT will utilize the latest advancements in open-source technology to ensure that the educational tools are both affordable and adaptable, making high-quality STEAM education available to a broader audience. The redesigned robot and its accompanying activities will serve as a versatile platform for teaching programming, electronics, and engineering concepts, thereby preparing students for future technological challenges. The aim will be to successfully combine physical robotics with virtual simulators to create an immersive learning environment (Google Summer of Code) (GitHub). By leveraging open-source resources and community collaboration, a sustainable and impactful educational tool that can be continuously improved and adapted to meet the evolving needs of STEM education can be created.

Acknowledgements This research work has been carried out within the S.T.E.P.S. Erasmus+ Cooperation Partnership project, under the supervision of the University of Burgos, with support from the other project partners (https://stepsproject.eu/partners/). Further information about this project can be found at https://erasmus-plus.ec.europa.eu/projects/search/details/2023-1-FR01-KA220-HED-000165711 and on the project webpage: https://stepsproject.eu/.

Annexes

1. Rubric for Collecting and Analysing Good Practices

1. Title of the practice	
2. Author/s	
3. Year/s of implementation and year of publication	
4. Age of students	
5. School year/s and type of education *(stage and level of education, formal, non-formal education setting, etc.)*	
6. Target group *(group size, gender, students with specific educational support needs,* etc.)	
7. Duration *(total duration, number of sessions, etc.)*	
8. Location/site (country/ies, centre/s, space/s, etc. in which it is implemented)	
9. STEM disciplines and curricular subjects *(integrated disciplines, subjects and contents worked on)*	
10. Background *(what is the origin or background of the practice? what is its purpose?)*	
11. Learning objectives	
12. Methodology/ies *(teaching–learning methodologies used)*	
13. Brief description of the practice *(sequence of ordered activities/tasks)*	
14. Evaluation and learning results (what and how it has been evaluated, improvements and difficulties in learning)	
15. Materials (teaching resources, technical requirements, materials, robots, etc. used)	
16. Teacher tips or curriculum context (if available) *(guidelines for teachers, etc.)*	
17. Links, photos, supplementary materials, etc. (if available) *(graphical material that allows to visualize the practice)*	

2. Expert Assessment Rubric

Personal background information:

	Years of experienc e
Secondary school teacher	
Researcher	
Specify which subject, course and any other characteristics you consider of interest:	

Indicate from 1 to 5 your degree of compliance with the following statements about the good practices (being 1 Totally disagree, 2 In disagreement, 3 Neutral, 4 Agree, 5 Totally agree):

	1	2	3	4	5
Are useful for secondary school teachers					
Are integrated, i.e., two or more STEM subjects are addressed					
The learning objectives are relevant to integrated STEM education					
The methodologies (inquiry, engineering design, design thinking, Project-Based learning, cooperative learning...) are suitable					
The descriptions of the activities/tasks are clear and sufficient					
Evaluation and results are consistent					
The materials used are appropriate					
The technology (robots, coding software...) used is appropriate					
Please make any further comments:					

Among all the good practices selected, which two do you consider the best?

1	
2	

3. Final Good Practices Selected in Each Country After Evaluation by Educational Experts

Country	Title	Author(s) and year of publication	Technological resources (robot, software, etc.)	Didactic methodology/s	Integrated STEM areas and topics
Spain	1st—Low-cost educational robotics integration at high school scope to promote project-based learning	Vega-Moreno et al. (2016)	An educational robot is made using cheap materials (wood, cable, PVC pipes, motors, etc.) to create a remote submarine	Project-based learning (PBL), collaborative learning and constructionism (learning by doing)	It consists of interdisciplinary learning that combines the areas of science (physics and chemistry), technology, mathematics, engineering, computer science and English. The contents of buoyancy, inertia, electrical impulse, sensors, computer programming, chemical properties of water are worked on
	2nd—Educational robotics in the stage of secondary education: empirical study on motivation and STEM skills	Arís and Orcs (2019)	LEGO Education Mindstorms EV3 Kit	Project-based learning (PBL) and cooperative learning	STEM skills in science, technology, engineering and mathematics are encouraged

Country	Title	Author(s) and year of publication	Technological resources (robot, software, etc.)	Didactic methodology/s	Integrated STEM areas and topics
Greece	1st—A case study: visualizing Coulomb forces with the aid of augmented reality (AR)	Tomara and Gouscos (2018)	Mobile devices, Csharp programming language software and Unity 3D tool for AR development	Cooperative work and constructionist approach ('learning by doing')	The science discipline is developed, specifically physics, studying electrostatic interactions, Coulomb's Law, as well as information and communication technologies, through the use of an augmented reality (AR) tool
	2nd—Chembot: chemistry with robots	Karagiorgou and Spahos (2016)	Lego Mindstorms EV3	Problem- or challenge-based learning and cooperative learning	Science (chemistry), mathematics, computer science, design and technology, programming and interest in STEM vocations

Country	Title	Author(s) and year of publication	Technological resources (robot, software, etc.)	Didactic methodology/s	Integrated STEM areas and topics
Turkey	1st—The impact of STEM and STEM-based robotic activities on the development of intellectual risk-taking and inquiry skills in science learning	Akkoyun (2019)	Build-your-own robots using scrap materials and Lego parts	Inquiry-based science education (IBSE) and cooperative learning	Engineering (production of machines and robots using scrap materials) and technology (carrying out computer coding activities)

(continued)

(continued)

Country	Title	Author(s) and year of publication	Technological resources (robot, software, etc.)	Didactic methodology/s	Integrated STEM areas and topics
	2nd—Integration of robotics studies as stem into secondary school science laboratory practices	Elbir (2020)	Lego Mindstorms EV3 Kit	Engineering design process (EDP) and cooperative learning	Science (power and movement, light and sound) and engineering (creating, testing, developing, sharing and evaluating the product)

Country	Title	Author(s) and year of publication	Technological resources (robot, software, etc.)	Didactic methodology/s	Integrated STEM areas and topics
France	1st—Botly—drawing robot	Michel (2021)	Robot Botly built with low-cost materials	Trial-and-error learning, project-based learning (PBL) and collaborative learning	Mathematics (geometry), technology (computer programming) and engineering (construction of robots)
	2nd—Robotics trophy	Planete Sciences (n.d.)	Self-buildable robots	Project-based learning (PBL) and collaborative learning	Robotics, mechanics, electronics and computer science

Country	Title	Author(s) and year of publication	Technological resources (robot, software, etc.)	Didactic methodology/s	Integrated STEM areas and topics
Poland	1st—STEMKIT4SCHOOLS. Advancing programming, STEM and IoT understanding in the classroom through DIY computer kits	STEMKIT4SCHOOLS (n.d.) It runs from 2019 to 2021	Computers and block programme 'Scratch'	Project based learning (PBL) and collaborative learning	Science (physics/ chemistry/ biology/ geology): work on the scientific method Computer science: programming language, etc. Technology: sensors, electronics, hardware and open-source software Mathematics: study of statistics
	2nd—Eco-friendly robotics for a future green world	European Commission (n.d.) It is implemented from 6 to 10 December 2021	Laptops, in-house robotic assemblies made from recyclable materials and drones	Project based learning (PBL)	Science (solutions to air pollution, measuring air quality) Technology and engineering (building and programming robots for a football competition and a drone driving workshop)

4. Good Practices Collected in the Five European Countries of the S.T.E.P.S. Project

- Good practices addressed in Greece

Title	Author(s)	Year	Technological resources (robots, etc.)
Panhellenic competition of open technologies in education	GFOSS/ Harokopion University	Year 2023/ 2024	Open-Source Hardware (Arduino Uno, Raspberry Pi 3 Type B, beagleboard, Dwenguino Thymio, etc.), and open-source software
Space mission competition	GFOSS/ Harokopion University	Courses 2020/ 2021/2022	Open-Source Hardware (Arduino Uno, Raspberry Pi 3 Type B, beagleboard, Dwenguino Thymio, etc.), and open-source software
Chembot: chemistry with robots	Eleftheria Karagiorgou and Vasileios Spahos	2016	Lego Mindstorms EV3
Mathematics in educational robotics. The case of plane shapes	Dimitrios Theoharis and Christos Papadoulis	2018	Lego Mindstorms EV3
Design and implementation of an Arduino-based artifact by school students, for the measurement of atmospheric pollutants. Researching the learning and the environmental stances' outcomes	Dimakos Constantine, Gkiolmas Aristotelis, Chalkidis Anthimos and Angelidou Margarita	2023	Arduino Uno programming board
Drawing with EV3 Mindstorms	Stavros Mitsiadis	2020/2021 academic year	Lego Mindstorms EV3
A case study: visualizing Coulomb forces with the aid of augmented reality	Marina Tomara and Dimitris Gouscos	2018	Mobile, Csharp programming language and the Unity 3D platform
Teacher and student training in robotics	GFOSS/ Harokopion University	Courses 2020–2024	Use of robots such as Thymio, Bee-bot, Edison, Raspberry Pi, Arduino or others

(continued)

(continued)

• Good practices addressed in Greece

Title	Author(s)	Year	Technological resources (robots, etc.)
Playing Golf by remote controlling EV3 Mindstorms	Stavros Mitsiadis	2019/2020 academic year	Lego Mindstorms EV3
EV3 egg decorator robot	George Papadimitriou	2019/2020 academic year	Lego Mindstorms EV3

• Good practices addressed in Spain

Title	Author/s and year	Year	Technological resources (robots, etc.)
A low cost educational ROV platform for promoting engineering among young students: ROV R2B2, new challenges	Xavier Cufí, Andrés El-Fakdi and Aymette Medina	2023	Self-built robot with low-cost material
Implementation of the new subject "Robotics" in secondary education and baccalaureate	Gabriel Ocaña Rebollo, Isabel M.ª Romero Albaladejo, Francisco Gil Cuadra and Antonio Codina Sánchez	2015	LEGO Mindstorms NXT
Cyberlandia: an educational robotics program to promote STEM careers in primary and secondary schools	José Carlos Rodríguez Rodríguez, Eduardo Martín-Pulido, Vanesa Jorge Padrón, Jonathan Alemán Alemán, Carmelo R. García and Alexis Quesada-Arencibia	2017	Lego Mindstorms EV3
Analysis of student learning in an educational environment with robotics activities	Samantha Orlando	2020	LEGO Mindstorms EV3
Robotics as a subject in secondary education. Results of an educational experience	Gabriel Ocaña Rebollo	2012	LEGO Mindstorms NXT

(continued)

(continued)

• Good practices addressed in Spain

Title	Author/s and year	Year	Technological resources (robots, etc.)
Educational robotics in the stage of secondary education: empirical study on motivation and STEM skills	Nuria Arís and Lara Orcos	2019	LEGO Mindstorms EV3
Improving motivation in 3rd year of ESO through educational robotics activities	Iosu Goñi Echeverria	2023	Maqueen robot and Micro:bit programming board
Computational thinking and robotics: a teaching experience in compulsory secondary education with students with high degree of apathy and demotivation	Belkis Díaz-Lauzurica and David Moreno-Salinas	2019	LEGO Mindstorms EV3
Integration of low-cost educational robotics in secondary education to promote project-based learning	Daura Vega-Moreno, Xavier Cufí Solé, María José Rueda and Diego Llinás	2016	Remote submarine (remotely operated vehicle) self-built with low-cost material
Educational robotics as a technology learning tool	Francisco Antonio Pujol López, María Del Pilar Arques Corrales, Fidel Aznar Gregori, Antonio Manuel Jimeno Morenilla, María Del Mar Pujol López, María José Pujol López, Ramón Rizo Aldeguer, Marcelo Saval Calvo, Mireia Luisa Sempere Tortosa, David Tomás Díaz, María Asensi Arques, Sergio González Rico and David Rodríguez Fajardo	2020	LEGO Mindstorms EV3

• Good practices addressed in France			
Title	Author(s)	Year	Technological resources (robots, etc.)
Botly—drawing robot	Original concept by Olivier Michel (Safran) Technical developpement by LaMachinerie	Courses 2018–2020	Self-built robot with low-cost material (Botly)
Line-following robot challenge	Original concept by: ESIEE Amiens Partner: Collège St Jean de la Croix (St Quentin) Collège Hanotaux (St Quentin) Collège Sagebien (Amiens) Lycée Jean Racine (Montdidier)	Courses 2018–2020	Self-built robot with low-cost material (Botly)
Robotics trophy	Planete Sciences	Annual event (ongoing)	Self-buildable robots
Visual programming for comic creation	On a marché sur ma bulle	Courses 2019–2021	Programming platform based on Blockly, together with a library of digital resources (characters, objects, backgrounds)
Collaborative InMoov project at the FabLab	Gael Langevin	2017	Arduino programming board
The useless box: lessons in futility and fun	Jules Topart and Adrien Bracq	2019	Self-buildable robot
SCOTT: simplified educational robot kit	Jules Topart and Adrien Bracq	Courses 2016–2018	Self-buildable robot
Lasercut robot assembly and programming kit	LaMachinerie, Adrien Bracq and Yann Paulmier	2017–2018 academic year	Self-buildable robot
Cardboard companion: a first step into robotics	Concept by: École Jacques Prévert Animated by Matthieu Phillipe and Adrien BRACQ	2017	Robot made out of cardboard boxes, to which various electronic devices are integrated, such as electronic boards, switches, fans, etc.

(continued)

(continued)

- Good practices addressed in France

Title	Author(s)	Year	Technological resources (robots, etc.)
CubeIT: interactive learning and music creation table	Concept by Prolifik Records R&D by LaMachinerie	2016–2019	Own CubeIT robot, infrared cameras, blocks and musical instruments, open-source hardware and software such as Arduino and processing

- Good practices addressed in Poland

Title	Author(s)	Year	Technological resources (robots, software, etc.)
Be ready	Warsaw University of Technology, Universita degli studi di Siena, Europaiko eŗgastirio ekpaideftikis technologias, Universidade do Minho, 23rd Gymnasium of Athens, X Liceum Ogolnokształcace im. Krolowej Jadwigi and Externato de Vila Meã	2021	Completion of online STEM projects together with educational resources (videos, presentations, etc.)
Cream	Sinergie—Società Consortile a Responsabilità Limitata is a not-for-profit accredited VET body as well as a registered research and innovation centre	n.d.	Not specified
VEX IQ program—Fr. Józef Dąbrowski Polish Language School in Orchard Lake	VEX	Courses 2018–2020	VEX IQ robotics kits
STEMKIT4SCHOOLS, advancing programming, STEM and IoT understanding in the classroom through DIY computer kits	Colegiul National Pedagogic "Mircea Scarlat" (RO), M.K. Innovations LTD (CY), Danmar Computers SP Zoo (PL), Projeto Schole LDA (PT), Asserted Knowledge Omorrythmos Etaireia (EL), Hearthands Solutions Limited (CY), Scoala Gimnaziala Metropolitana ARC (RO)	From 1/10/2019 to 30/09/2021	Digital block programming tool 'Scratch'
STEM through robotics—Erasmus+	Sehit Prof. Dr. Ilhan Varank Bilim ve Sanat Merkezi, Zespol Szkolno-Przedszkolny Nr 6 w Rzeszowie, Osnovna Sola Breg	2019	VEX IQ

(continued)

(continued)

- Good practices addressed in Poland

Title	Author(s)	Year	Technological resources (robots, software, etc.)
ROBOscientists	Warsaw University of Technology (Poland), Edumotiva—European Lab for Educational Technology (Greece), University of Latvia (Latvia), Frederick University (Cyprus), ENGINO—Net LTD (Cyprus)	2024 (not yet carried out)	Self-buildable robots (RoboScientists technical tutorial)
Make it real	Warsaw University of Technology (Poland)—coordinator Edumotiva—European Lab for Educational Technology (Greece), CyRIC—Cyprus Research and Innovation Center Ltd. (Cyprus), Aristotle University (Greece), Middle East Technical University (Turkey) and Helix5 (the Netherlands)	From 2016 to 2018	3D printer
ENGINE e-learning platform	Warsaw University of Technology (PL) Coordinator & IO1 leader Edumotiva—European Lab for Educational Technology (GR), IO4 leader International Hellenic University (GR), IO2 leader University of Padova (IT) y IO3 leader University of Tarnow (PL)	2024	Electronics, microcontrollers and programming language tools
EcoRobotoMatic.online	Author not specified	From 2021 to 2023	Constructible robots
Eco-friendly robotics for a future green world	Instituto N.a Sra. da Encarnação—Cooperativa de Ensino (Portugal), Zespol Szkolno-Przedszkolny Nr 6 w Rzeszowie (Poland), Gümnaasium de Kuressaare (Estonia), Kabazli Sehit Ismail Yavuz Ortaokulu (Turkey), Gimnasio Lefkimmis (Greece) and Austrumlatvijas Tehnologiju Vidusskola (Latvia)	From 2019 to 2022	Own robotic kits

• Good practices addressed in Turkey

Title	Author/s and year	Year	Technological resources (robots, software, etc.)
STEM for secondary school 5th grade science applications course. A teaching design based on nature: inspired by nature technologies	Necip Fazıl ALPEREN	2020	Not specified
Empowering K-12 learners for global challenges through STEAM(M): reflections from a Turkish context	Şehnaz Şahinkarakaş and Fatma Toköz Göktepe	n.d.	Not specified
The effect of STEM activities performed with educational robot sets on middle school 5th grade students' mathematics success, problem solving skills and stem skills levels	Ender Faruk Türk	2023	Lego Mindstorms EV3
Investigating the effect of robotic-assisted stem implementations on the computational thinking, motivation and professional interests of secondary school students	Osman Canseven	2023	Lego Mindstorms EV3
Integration of robotics studies as stem into secondary school science laboratory practices	Barış Elbir	2020	Lego Mindstorms EV3
Determination of the effect of robotics coding activities designed suitable for STEM education on the attitude of gifted students towards robotics and coding	Erkan Avci	2021	Arduino Uno programming board
"An examination of robotics coding education in stem education and STEM perspective: a case study of Kizilcahamam coding"	Nursel Yalçin and Eyüp Akbbulut	2021	Arduino Uno programming board
"The impact of STEM and STEM-based robotic activities on the development of intellectual risk-taking and inquiry skills in science learning"	Nida Akkoyun	2019	Self-buildable robots
STEM applications on problem solving skills of secondary school students to examine the effect on their perceptions towards	Alev Doğan, Ebru Aydın and Emine Kahraman	2020	Constructible robot
STEM education and robotics coding trainings in the perspective of STEM. Examination: Kizilcahamam coding example	Nursel Yalçin and Eyüp Akbbulut	2021	Arduino Uno programming board

References

Akkoyun, N. (2019). The Effect of STEM and STEM-Based Robotic Activities on the Development of Intellectual Risk-Taking and Inquiry Skills in Science Learning. Master's Thesis. Erzincan Binali Yıldırım University. Institute of Natural Sciences. Erzincan.

Álvarez, I., García-Langley, A., Brancazio, D., Bennett, A., & Triantafyllou, M. (2023, de junio de 5–8). *Adaptation of Marine Soft Robotics for K-12 Educational Programs* [Papel de conferencia]. OCEANS 2023, Limerick, Ireland. https://doi.org/10.1109/OCEANSLimerick52467.2023.10244316

Amo, D., Fox, P., Fonseca, D., & Poyatos, C. (2021). Systematic review on which analytics and learning methodologies are applied in primary and secondary education in the learning of robotics sensors. *Sensors, 21*(1), 153. https://doi.org/10.3390/s21010153

Arís, N., & Orcos, L. (2019). Educational robotics in the stage of secondary education: Empirical study on motivation and STEM skills. *Education Sciences, 9*(73). https://doi.org/10.3390/educsci9020073

Aytan, T. (2016). Evaluation of the 2006 and 2015 Turkish education program in secondary school curriculum in Turkey in terms of critical thinking. *Journal of Education and Learning, 5*(2), 38–45. https://doi.org/10.5539/jel.v5n2p38

Bellas, F., Naya, M., Varela, G., Llamas, L., Prieto, A., Becerra, J. C., Bautista, M., Faiña, A., & Duro, R. (2018). The Robobo project: Bringing educational robotics closer to real-world applications. In W. Lepuschitz, M. Merdan, G. Koppensteiner, R. Balogh, & D. Obdrzálek (Eds.), *Robotics in education: Advances in intelligent systems and computing* (Vol. 630, pp. 226–237). Springer.

Bernstein, B., Mutch-Jones, K., Cassidy, M., & Hamner, E. (2022). Teaching with robotics: Creating and implementing integrated units in middle school subjects. *Journal of Research on Technology in Education, 5*(2), 161–176. https://doi.org/10.1080/15391523.2020.1816864

Camargo, C., Patino, K., Benavides, C., & Sofrony, J. (18–20 de noviembre de 2015). Iflab: Development laboratory platform [Sesión de conferencia]. In *8th international conference of education, research and innovation, Seville, Spain.*

Chang, C. C., & Chen, Y. (2020). Cognition, attitude, and interest in cross-disciplinary i-STEM robotics curriculum developed by thematic integration approaches of webbed and threaded models: A concurrent embedded mixed methods study. *Journal of Science Education and Technology, 29*, 622–634. https://doi.org/10.1007/s10956-020-09841-9

Chang, C.-C., & Chen, Y.-K. (2023). A transdisciplinary STEM course integrated through project-based learning on robotics: Perspective from teacher and student feedback. *Asia Pacific Journal of Education*, 1–16. https://doi.org/10.1080/02188791.2023.2209698

Chapman, A., Rodríguez, F. D., Pena, C., Hinojosa, E., Morales, L., Del Bosque, V., Tijerina, Y., & Tarawneh, C. (2020). "Nothing is impossible": Characteristics of Hispanic females participating in an informal STEM setting. *Cultural Studies of Science Education, 15*, 723–737. https://doi.org/10.1007/s11422-019-09947-6

Chavarría Pérez, C., & Guede-Cid, R. (2023). La educación STEM como práctica transdisciplinar en educación secundaria y bachillerato. *Revista Iberoamericana de Educación, 92*(1), 61–70. https://doi.org/10.35362/rie9215804

Chiang, F.-K., Tang, Z., Zhu, D., & Bao, X. (2024). Gender disparity in STEM education: A survey research on girl participants in World Robot Olympiad. *International Journal of Technology and Design Education, 34*, 629–646. https://doi.org/10.1007/s10798-023-09830-0

Demo, G. B., Moro, M., Pina, A., & Arlegui, J. (2012). In and out of the school activities implementing IBSE and constructionist learning methodologies by means of robotics. In B. Barker, G. Nugent, N. Grandgenett, & V. Adamchuk (Eds.), *Robots in K-12 education: A new technology for learning* (pp. 66–92). IGI Global. https://doi.org/10.4018/978-1-4666-0182-6.ch004
Développer les compétences numériques tout au long de la scolarité. (n.d.). Ministère de l'Education Nationale et de la Jeunesse. Retrieved 22 July 2024, from https://www.education.gouv.fr/developper-les-competences-numeriques-tout-au-long-de-la-scolarite-324659
Elbir, B. (2020). *Integration of robotic studies into secondary school science laboratory applications as a STEM application [Tesis de doctorado no publicada]*. Istanbul University.
European Commission. (s.f.). *Eco-friendly robotics for a future green world*. https://innovative-teaching-award.ec.europa.eu/projects/eco-friendly-robotics-future-green-world_en?prefLang=pt
Évaluation des compétences numériques en fin de troisième 2022. (n.d.). Ministère de l'Education Nationale et de la Jeunesse. Retrieved July 22, 2024, from https://www.education.gouv.fr/evaluation-des-competences-numeriques-en-fin-de-troisieme-2022-414554
Évaluation multidimensionnelle de l'impact de l'utilisation d'équipements numériques mobiles sur les apprentissages des élèves Premiers résultats du Plan numérique de 2015. (n.d.). Ministère de l'Education Nationale et de la Jeunesse. Retrieved July 22, 2024, from https://www.education.gouv.fr/evaluation-multidimensionnelle-de-l-impact-de-l-utilisation-d-equipements-numeriques-mobiles-sur-les-309290
Floroiu, D., Patic, P. C., & Duta, L. (2019). Trends in educational robotics. In N. A. Aspragathos, P. N. Koustoumpardis, & V. C. Moulianitis (Eds.), *Advances in service and industrial robotics: Proceedings of the 27th international conference on robotics in Alpe-Adria Danube region (RAAD 2018)* (Vol. 67, pp. 737–744). Springer. https://doi.org/10.1007/978-3-030-00232-9_77
Foukarakis, G., & Syrris, I. (2018, de marzo de 5–7). Designing and implementing a robotics course for high school students [Sesión de conferencia]. In: *12th international technology, education and development conference, Valencia, Spain*. https://doi.org/10.21125/inted.2018.2159
Galino, J. R., & Tanaka, H. (2021). Initiatives and development of secondary school robotics education in Camarin high school. *Journal of Robotics, Networking and Artificial Life, 8*(1), 60–65. https://doi.org/10.2991/jrnal.k.210521.014
García-Langley, A., Álvarez, I., Chen, A., Li, A., Wang, H., Brancazio, D., Gutierrez, V., Bennett, A., & Triantafyllou, M. (2022, de octubre de 17–20). *Development of educational marine soft robotics STEM platform as new iteration of SeaPerch K-12 national outreach program* [Papel de conferencia]. OCEANS 2022, Hampton Roads, VA, USA. https://doi.org/10.1109/OCEANS47191.2022.9977015
Gerber, L. C., Calasanz-Kaiser, A., Hyman, L., Voitiuk, K., Patil, U., & Riedel-Kruse, I. H. (2017). Liquid-handling Lego robots and experiments for STEM education and research. *PLOS Biology, 15*(3), e2001413. https://doi.org/10.1371/journal.pbio.2001413
Gok, T. (2021). The development of the STEM (science, technology, engineering, and mathematics) attitude and motivation survey towards secondary school students. *International Journal of Cognitive Research in Science, Engineering and Education (IJCRSEE), 9*(1), 105–119. https://doi.org/10.23947/2334-8496-2021-9-1-105-119
Hartigan, C., & Hademenos, G. (2019). Introducing ROAVEE: an advanced STEM-based project in aquatic robotics. *The Physics Teacher, 57*(1), 17–20. https://doi.org/10.1119/1.5084920
Hennessy Elliott, C. (2020). "Run it through me:" Positioning, power, and learning on a high school robotics team. *Journal of the Learning Sciences, 29*(4–5), 598–641. https://doi.org/10.1080/10508406.2020.1770763
Hristov, G., Beloev, I., Zahariev, P., & Georgiev, G. (2023). The role of the universities as accelerators for the integration of the STEM learning methods in the primary and secondary schools. *Strategies for Policy in Science and Education, 31*(4s), 74–88. https://doi.org/10.53656/str2023-4s-6-rol
Jaipal-Jamani, K. (2023). Preservice teachers' science learning and self-efficacy to teach with robotics-based activities: Investigating a scaffolded and a self-guided approach. *Frontiers in Education, 8*. https://doi.org/10.3389/feduc.2023.979709

Karaahmetoğlu, K., & Korkmaz, O. (2019). The effect of project-based Arduino educational robot applications on students' computational thinking skills and their perception of basic stem skill levels. *Participatory Educational Research, 6*(2), 1–14. https://doi.org/10.17275/per.19.8.6.2

Karagiorgou, E., & Spahos, V. (2016). Chembot: Chemistry with robots. *Science in School, the European Journal for Science Teachers, 38*, 42–45.

Karalekas, G., Vologiannidis, S., & Kalomiros, J. (2020). EUROPA: A case study for teaching sensors, data acquisition and robotics via a ROS-based educational robot. *Sensors, 20*(9), 2469. https://doi.org/10.3390/s20092469

Karalekas, G., Vologiannidis, S., & Kalomiros, J. (2023). Teaching machine learning in K-12 using robotics. *Education Sciences, 13*(1), 67. https://doi.org/10.3390/educsci13010067

Karaman, S., Anders, A. S., Boulet, M. T., Connor, J. A., Gregson, K., Guerra, W., Guldner, O., Mohamoud, M., Plancher, B., Shin, R., & Vivilecchia, J. R. (2017). Project-based, collaborative, algorithmic robotics for high school students: Programming self-driving race cars at MIT. *IEEE Integrated STEM Education Conference (ISEC), 2017*, 195–203.

Keller, L., & John, I. (2020). Motivating female students for computer science by means of robot workshops. *International Journal of Engineering Pedagogy (iJEP), 10*(1), 94–108. https://doi.org/10.3991/ijep.v10i1.11661

Leonard, J., Buss, A., Gamboa, R., Mitchell, M., Fashola, O. S., Hubert, T., & Almughyirah, S. (2016). Using robotics and game design to enhance children's self-efficacy, STEM attitudes, and computational thinking skills. *Journal of Science Education and Technology, 25*, 860–876. https://doi.org/10.1007/s10956-016-9628-2

Manikutty, G., Frey, L. M., Natarajan, A., Chilakapati, U., V, V., & Rao R. B. (2019, de diciembre de 9–11). I can also make robots! Inspiring rural Indian children to learn robotics [Papel de conferencia]. In *2019 IEEE tenth international conference on technology for education (T4E), Goa, India*. https://doi.org/10.1109/T4E.2019.00022

Martín-Páez, T., Aguilera, D., Perales-Palacios, F. J., & Vílchez-González, J. M. (2019). What are we talking about when we talk about STEM education? A Review of Literature. *Science Education, 103*(4), 799–822. https://doi.org/10.1002/sce.21522

Marzoli, I., Rizza, N., Saltarelli, A., & Sampaolesi, E. (2021). Arduino: From physics to robotics. In D. Scaradozzi, L. Guasti, M. Di Stasio, B. Miotti, A. Monteriu, & P. Bliksten (Eds.), *Makers at school, educational robotics and innovate learning environments* (Vol. 240, pp. 309–314). Springer.

Merkouris, A., & Chorianopoulos, K. (2018). Programming touch and full-body interaction with a remotely controlled robot in a secondary education STEM course. In: *PCI '18: Proceedings of the 22nd Pan-Hellenic conference on informatics* (pp. 225–229). https://doi.org/10.1145/3291533.3291537

Michel, O. (2021, de abril de 3). *De l'Art et des Machines*. Les Nouvelles du CDI—beaucamps-le-Vieux. http://blogs.ac-amiens.fr/0801443j_cdibeaucamps/index.php?post/2021/04/03/De-l-Art-et-des-Machines

Ministry of National Education (MoNE). (2005). *Science curriculum guide*. Government Publishing House Management.

Ministry of National Education (MoNE). (2012). *Information technologies curriculum*. Government Publishing House Management.

Misthou, S., Moumoutzis, N., & Loukatos, D. (2021, 21–23 de abril de). *Coding Club: a K-12 good practice for a STEM learning community* [Acta de Congreso]. *2021 IEEE Global Engineering Education Conference (EDUCON)*, Vienna, Austria. https://doi.org/10.1109/EDUCON46332.2021.9454039

Naya-Varela, M., Guerreiro-Santalla, S., Baamonde, T., & Bellas, F. (2023). Robobo SmartCity: An autonomous driving model for computational intelligence learning through educational robotics. *IEEE Transactions on Learning Technologies, 16*(4), 543–559. https://doi.org/10.1109/TLT.2023.3244604

Ortiz-Revilla, J., Adúriz-Bravo, A., & Greca, I. M. (2020). A framework for epistemological discussion on integrated STEM education. *Science & Education, 29*, 857–880. https://doi.org/10.1007/s11191-020-00131-9

Ortiz-Revilla, J., Greca, I. M., & Arriassecq, I. (2022). A theoretical framework for integrated STEM education. *Science & Education, 31*, 383–404. https://doi.org/10.1007/s11191-021-00242-x

Planete Sciences. (s.f.). *Trophées de Robotique.* https://www.planete-sciences.org/mediterranee/activites/scolaire/trophees-de-robotique/

Pusceddu, G., Cocchella, F., Bogliolo, M., Belgiovine, G., Lastrico, L., Casadio, M., Rea, F., & Sciutti, A. (2023). Training school teachers to use robots as an educational tool: The impact on robotics perception. In F. Cavallo, J.-J. Cabibihan, L. Fiorini, A. Sorrentino, H. He, X. Liu, Y. Matsumoto, & S. Sam Ge (Eds.), *Social robotics* (Vol. 13818, pp. 102–113). Springer.

Quigley, C. F., & Herro, D. (2016). "Finding the Joy in the Unknown": Implementation of STEAM teaching practices in middle school science and math classrooms. *Journal of Science Education and Technology, 25*, 410–426. https://doi.org/10.1007/s10956-016-9602-z

Rubenstein, M., Cimino, B., Nagpal, R., & Werfel, J. (2015, de mayo de 26–30). AERobot: An affordable one-robot-per-student system for early robotics education [Sesión de conferencia]. In *2015 IEEE international conference on robotics and automation (ICRA), Seattle, WA, USA.*

Sáez-López, J. M., & Buceta-Otero, R. (2023). El robot M Bot para el aprendizaje de coordenadas cartesianas en Educación Secundaria. *PÍXEL-BIT. Revista de medios y educación*, (66), 271–301. https://doi.org/10.12795/pixelbit.95617

Scaradozzi, D., Cesaretti, L., Screpanti, L., & Mangina, E. (2020). Identification of the students learning process during education robotics activities. *Frontiers in Robotics and AI, 7*, 21. https://doi.org/10.3389/frobt.2020.00021

Sen, C., Sonay Ay, Z., & Ahmet Kiray, S. (2021). Computational thinking skills of gifted and talented students in integrated STEM activities based on the engineering design process: The case of robotics and 3D robot modeling. *Thinking Skills and Creativity, 42*, 100931. https://doi.org/10.1016/j.tsc.2021.100931

Sklirou, T. S. (2017, de abril de 25–28). Programming in secondary education: Applications, new trends and challenges [Sesión de conferencia]. In *2017 IEEE global engineering education conference (EDUCON), Athens, Greece.* https://doi.org/10.1109/EDUCON.2017.7942904

STEMKIT4SCHOOLS. (s.f.). *Advancing programming, STEM and IoT understanding in the classroom through DIY computer kits.* https://stemkit4schools.erasmus.site/

Sullivan, A., & Bers, M. U. (2019). VEX robotics competitions: Gender differences in student attitudes and experiences. *Journal of Information Technology Education: Research, 18*, 97–112. https://doi.org/10.28945/4193

Susilo, E., Liu, J., Alvarado Rayo, Y., Melissa Peck, A., Montenegro, J., Gonyea, M., & Valdastri, P. (2016). STORMLab for STEM education: An affordable modular robotic kit for integrated science, technology, engineering, and math education. *IEEE Robotics & Automation Magazine, 23*(2), 47–55. https://doi.org/10.1109/MRA.2016.2546703

Tarrés-Puertas, M. I., Costa, V., Pedreira Álvarez, M., Lemkow-Tovias, G., Rossell, J. M., & Dorado, A. D. (2023). Child-robot interactions using educational robots: An ethical and inclusive perspective. *Sensors, 23*(3), 1675. https://doi.org/10.3390/s23031675

Tay, T. T., Lim, Z. Z., & Chua, Y. L. (2017). Utilizing autonomous mobile robot to increase interest in STEM [Sesión de conferencia]. In: 2017 3rd international conference on science in information technology (ICSITech), Bandung, Indonesia.

TEKNOFEST. (2021). *Aerospace and technology festival.* https://www.teknofest.org/en

Theodoropoulou, I., Lavidas, K., & Komis, V. (2023). Results and prospects from the utilization of educational robotics in Greek Schools. *Technology, Knowledge and Learning, 28*, 225–240. https://doi.org/10.1007/s10758-021-09555-w

Tomara, M., & Gouscos, D. (2019). A case study: Visualizing Coulomb forces with the aid of augmented reality. *Journal of Educational Computing Research, 57*(7), 1626–1642. https://doi.org/10.1177/0735633119854023

TÜBİTAK. (2020). *National science Olympics*. https://www.tubitak.gov.tr/tr/olimpiyatlar/icerik-ulusal-bilim-olimpiyatlari

Üçgül, M., & Altıok, S. (2022). You are an Astroneer: The effects of robotics camps on secondary school students' perceptions and attitudes towards STEM. *International Journal of Technology and Design Education, 32*(3), 1679–1699. https://doi.org/10.1007/s10798-021-09673-7

Valko, N. V., & Osadchyi, V. V. (2021). Teaching robotics to future teachers as part of education activities. *Journal of Physics: Conference Series, 1946*, 012016. https://doi.org/10.1088/1742-6596/1946/1/012016

Valls Pou, A., Canaleta, X., & Fonseca, D. (2022). Computational thinking and educational robotics integrated into project-based learning. *Sensors, 22*(10), 3746. https://doi.org/10.3390/s22103746

Van Wassenaer, N., Tolboom, J., & van Beekum, O. (2023). The effect of robotics education on gender differences in STEM attitudes among Dutch 7th and 8th grade students. *Education Sciences, 13*(2), 139. https://doi.org/10.3390/educsci13020139

Vega, J., & Cañas, J. M. (2018). PiBot: An open low-cost robotic platform with camera for STEM education. *Electronics, 7*(12), 430. https://doi.org/10.3390/electronics7120430

Vega, J., & Cañas, J. M. (2019). PyBoKids: An innovative Python-based educational framework using real and simulated Arduino robots. *Electronics, 8*(8), 899. https://doi.org/10.3390/electronics8080899

Vega-Moreno, D., Cufí Solé, X., Rueda, M. J., & Llinás, D. (2016). Integración de robótica educativa de bajo coste en el ámbito de la educación secundaria para fomentar el aprendizaje por proyectos. *IJERI: International journal of Educational Research and Innovation*, (6), 162–175.

Wang, W.-H. (2016, de marzo de 5). A mini experiment of offering STEM education to several age groups through the use of robots [Sesión de conferencia]. In *2016 IEEE integrated STEM education conference (ISEC), Princeton, NJ, USA*. https://doi.org/10.1109/ISECon.2016.7457516

Wang, C., & Frye, M. (2019, de marzo de 16). miniGEMS 2018 summer camp evaluation: Empowering middle school girls in STEAM [Sesión de conferencia]. In *IEEE integrated STEM education conference (ISEC), Princeton, NJ, USA*. https://doi.org/10.1109/ISECon.2019.8881981

Weinberg, J. B., Pettibone, J. C., Thomas, S. L., Stephen, M. L., & Stein, C. (2007). The impact of robot projects on girls' attitudes toward science and engineering. In *Robotics science and systems (RSS) workshop on research in robots for education*. https://www.siue.edu/engineering/pdf/WeinbergRSSWorkshop2007.pdf

Wimer, B., Dannemiller, J., Mahmud, S., & Kim, J. H. (2023). Low-cost entry-level educational drone with associated K-12 education strategy. In H. Zaynidinov, M. Singh, U. S. Tiwary, & D. Singh (Eds.), *Intelligent human computer interaction (IHCI 2022)* (Vol. 13741, pp. 323–335). Springer, Cham. https://doi.org/10.1007/978-3-031-27199-1_32

Wong, A. S. W., Jeffery, R., Turner, P., Sleap, S., & Chalup, S. K. (2019). RoboCup junior in the Hunter Region: Driving the future of robotic STEM education. In D. Holz, K. Genter, M. Saad, & O. von Stryk (Eds.), *RoboCup 2018: Robot World Cup XXII* (Vol. 11374, pp. 362–373). Springer, Cham. https://doi.org/10.1007/978-3-030-27544-0_30

Wright, J. C., Knight, V. F., Barton, E. E., & Edwards-Bowyer, M. (2021). Video prompting to teach robotics and coding to middle school students with autism spectrum disorder. *Journal of Special Education Technology, 36*(4), 187–201. https://doi.org/10.1177/0162643419890249

Yapıcıoğlu, A. E. (2021). Analysis of the outcomes of the Turkish science curriculum in terms of science process skills, nature of science, socioscientific issues, and STEM. *International Journal of Curriculum and Instruction, 13*(2), 925–949.

Zhong, B., Liu, X., Xia, L., & Sun, W. (2022). A proposed taxonomy of teaching models in STEM education: Robotics as an example. *SAGE Open, 12*(2). https://doi.org/10.1177/21582440221099525

Science Mapping of the Knowledge Base on the Effects of Artifcial Intelligence (AI)-Based Chatbots on Student Learning Outcomes: A Bibliometric Study

Turgut Karakose and Hakan Polat

Abstract With the advent of AI-based Technologies, several groundbreaking tools have become part of the digitalization process in various fields, and the education field was no exception. The growing emergence of several Artificial Intelligence (AI)-based chatbots that can engage with human-like thinking and performance particularly attracted the attention of educational researchers and practitioners with their strong capacity to support student learning. This new line of research has accumulated a large knowledge base on the influence of these chatbots on students' learning outcomes in a very short time span. However, in order to lead the efficient growth of this knowledge base, a detailed analysis of its thematic architecture, intellectual evolution, and bibliometric performance is warranted. Therefore, the current study conducted a science mapping analysis of this field using the SciMAT software, which uniquely combines the analysis of the bibliometric performance and conceptual development, allowing for comparisons across several periods of analysis. Data for the current study was collected from the WoS and Scopus databases. The analysis of 322 articles showed that studies published during the first period of analysis (2011–2022) mostly focused on how AI chatbots could act as intelligent tutoring systems, particularly in online courses. During the second period (2023), though, this attention turned to ChatGPT and concerns over its influence on the integrity and academic honesty of students. During the last period of analysis (2024), investigations tended to move from concerns over its side effects to how these chatbots could support customized learning and students' self-learning experiences.

Keywords AI-based chatbots · Student learning · Chatbots · ChatGPT · Google bard · Microsoft bing · Thematic analysis · SciMAT

T. Karakose (✉)
Department of Educational Sciences, Kütahya Dumlupınar University, Kütahya, Türkiye
e-mail: turgut.karakose@dpu.edu.tr

H. Polat
Department of Educational Sciences, Firat University, Elazığ, Türkiye
e-mail: hakanpolat@firat.edu.tr

G. Lampropoulos and S. Papadakis (eds.), *Social Robots in Education*, Studies in Computational Intelligence 1194, https://doi.org/10.1007/978-3-031-82915-4_18

1 Introduction

The invention of computers and the ensuing transition to digitalization in the 1950s have been a new turning point for many sectors and fields (Lu, 2019; Wang et al, 2023). Artificial intelligence technology, which developed over McCarthy's iteration in the 50 s that machines can be intelligent, actually traces back to those years (Crawford et al, 2023; King & ChatGPT, 2023). Artificial intelligence (AI) can be considered as a method, technique, or theory that helps machines (particularly computers) analyze, discover, simulate, or mimic human behavior or thinking processes (Lu, 2019). With the recent developments in AI-based technologies, the modulation of human behavior through computational techniques has gained growing interest (Basset, 2019; Karakose and Tülübaş, 2024).

The term "Chatbot", on the other hand, was first used by Michael Loren Mauldin, who developed text-based conversational software and named it a "Chatter bot" (Chen et al, 2023). Currently, the term chatbot is used to refer to software tools that allow users to interact in a natural way via text or voice-based conversation on a particular topic or domain (Smutny & Schreiberova, 2020). Chatbots are software that mimics human speech using AI technology (Roniya et al., 2017) as well as using Deep Learning (DL) and Natural Language Processing (NLP) technologies (Casheekar et al, 2024).

The first examples of chatbots that have been on the market for a long time include ELIZA, ALICE, Claude, and HeX (Bhalshankar & Deshmukh, 2023; Smutny & Schreiberova, 2020). ELIZA, for instance, was developed as computer software to mimic the speaking abilities of a psychotherapist in 1966 (Basset, 2019; Caldarini et al, 2022; Smutny & Schreiberova, 2020), and was the first technology to imitate human speech (Caldarini et al, 2022; Hutchens, 1996; King & ChatGPT, 2023). ALICE, on the other hand, was a type of chatbot that was developed using dialogue-oriented and natural language processing-based Technologies by Richard Wallace in 1995 (Schumaker et al, 2006; Smutny & Schreiberova, 2020). It can be fun to experiment with such chat environments as they work on a system that stores a large number of ready-made responses for even a small talk (Schumaker et al., 2007). Similarly, Claude chatbot was developed to create an appropriate response recognizing the user's input and selecting the appropriate responses from its database (Smutny & Schreiberova, 2020) while HeX, a chatbot developed by Jason Hutchens in 1996, was devised to provide fixed but unique responses for different topics (Hutchens, 1996).

Today, AI chatbots such as Google Bard, Microsoft Bing, and ChatGPT have become quite popular. Microsoft Bing was developed using the next-generation large language model and has the ability to cite its own sources (Naser et al., 2024). ChatGPT, on the other hand, is an AI robot developed by OpenAI in November 2022 (Crawford et al, 2023; Eysenbach, 2023; Lee et al, 2024; Karakose et al., 2023; Naser et al., 2024). Google Bard was developed and launched by Google in May 2023 as a competitor to ChatGPT (Labadze et al, 2023). The main difference between these two AI robots is that while the dataset of ChatGPT includes texts from books and articles, Google Bard only uses some datasets that include texts from the internet (Labadze et al, 2023). These Chatbots are currently used in various fields such as education, e-commerce, health, information retrieval, and entertainment (Adamopoulou & Moussiades, 2020; Caldarini et al, 2022; King & ChatGPT, 2023).

The combination of artificial intelligence and education has led to the emergence of new educational tools such as intelligent teaching platforms, intelligent robot assistants, and intelligent testing systems. These new systems have not only provided convenience for teachers but also launched human-computer collaborative teaching methodologies (İpek et al., 2023; Karakose and Tülübaş, 2023; Lu, 2019). ChatGPT, for instance, is now frequently used for testing and assessment purposes (Kyung-Mi, 2024) in addition to providing support in homework preparation, text translation, coding, language teaching, and self-learning (Boubker, 2024; İpek et al., 2023; Nguyen, 2023). With all their promising skills, AI-assisted item generation has also attracted significant attention among researchers in the field of language learning and assessment (Kyung-Mi, 2024).

Although using AI-supported chatbots has brought several advantages, it also created concerns over reliability, accuracy, and ethical issues (Labadze et al., 2023; Nguyen, 2024). Although the ChatGPT has been observed to mostly produce unbiased, accurate, clear, and concise information, it is not subject to ethical principles and does not take responsibility for the content it produces (Tülübaş et al., 2023). In addition, these chatbots developed as large language models (LLM) can produce reasonable but probably wrong answers when they do not have domain-specific knowledge (Shukla et al., 2024).

With the increased interest in the use of chatbots for educational purposes, a large body of research has accumulated a robust knowledge base. However, the literature lacks a sound analysis of state-of-the-art knowledge in the field, which is significant in identifying underdeveloped or overdeveloped aspects and guiding future research accordingly. Therefore, the current study aims to contribute to this literature by identifying the evolving themes in this knowledge base and identifying the conceptual architecture of research on the influence of AI-based chatbots on student learning outcomes. With this purpose, the study conducts a quantitative longitudinal examination of the evolving knowledge base on the effects of AI-based chatbots on student learning outcomes and performs a thematic analysis of this evolving knowledge base. More specifically, the study addresses the following research questions:

RQ 1: What is the intellectual structure and growth trajectory of the evolving knowledge base on the effects of AI-based chatbots on student learning outcomes?

RQ 2: How have the conceptual and thematic elements evolved over time in the evolving field of research addressing the effects of AI-based chatbots on student learning outcomes?

2 AI Chatbots and Student Learning Outcomes

The performance, learning, and achievement of students are often evaluated over various learning outcomes. Recent studies on the use of AI technologies for educational purposes have shown that AI-based chatbots could have a significant influence on these learning outcomes. In the relevant literature, various studies exist that investigate the influence of AI Chatbots on students' learning outcomes (Boubker, 2024; Chiu, 2024; Dalgıç et al, 2024; Jošt et al, 2024; Winkler et al, 2020). For example, Winkler et al. (2020) concluded that chatbots improve students' learning outcomes to a significant extent. Similarly, Wu and Yu (2024) determined that chatbots help facilitate student performance, motivation, self-efficacy, and interest in learning while reducing their learning anxiety. On the other hand, Jošt et al. (2024) found that chatbots could negatively affect students' learning outcomes in the case of programming education. However, Chen et al. (2023) and Fidan and Gencel (2022), who investigated the opportunities, challenges, effectiveness, and ethical concerns of using chatbots, stated that chatbot-based peer feedback groups fostered students' motivation and learning performance. Similar studies in the field of language education have also revealed the positive effects of using chatbots on students' learning outcomes (Hong et al., 2023; Zhang et al., 2023).

3 Methodology

This study employed the science mapping analysis of the knowledge base on the effect of AI chatbots on student learning outcomes so as to reveal the cognitive and structural architecture of this growing research field. The study also conducted the bibliometric performance analysis to identify and visualize the conceptual subfields and thematic evolution of the research field (Cobo et al., 2011; Mac Fadden et al., 2021; Montero-Díaz et al., 2018). In brief, the study used bibliometric performance analysis and scientific mapping methods in combination to comprehensively investigate the bibliometric performance, strategic themes, and scientific evolution structure of the field.

3.1 Data Search and Extraction

In bibliometric reviews, several databases such as WoS, Scopus, PubMed, and Google Scholar are used to gather data. Among these, the WoS Core Collection is considered to offer the best data to be used in the bibliometric analysis (Cobo et al., 2015; Khalil & Crawford, 2015; Martinez et al., 2015) by indexing the world's leading scientific research in all fields including social sciences, education, arts, and humanities. It also provides detailed, reliable metadata about these articles. Similarly, Scopus is an important database that indexes significant research from the social sciences and educational fields. Therefore, we used both the WoS and Scopus databases to search and extract data for the current study.

We followed a clear procedure of data search, extraction, cleaning, and analysis, and reported the data search/extraction process according to the PRISMA (Preferred Reporting Items for Systematic Reviews and Meta-Analyses) 2020 protocol (Page et al., 2021) (see Fig. 1). We employed the following criteria for data search and extraction:

- We did not set any "time limitations."
- We limited our search to only "articles and review articles" while excluding other document types such as books, book chapters, or proceedings.
- We conducted our search in English, and excluded documents in other languages (as keywords' being in the same language is warranted for our analysis).

We conducted our data search on 21 May 2024 using the following search string: *("AI chatbot*" OR "chatbot*" OR "artificial intelligence chatbot*" OR "AI-based chatbot*" OR "artificial intelligence-based chatbot*" OR "AI-based conversational agent" OR "AI agent" OR "AI-based agent" OR "artificial intelligen* conversational agent*" OR "artificial intelligen* agent*" OR "AI conversational agent" AND "student* learning" OR "learning outcome*" OR "learning performance" OR "academic achievement" OR "student* achievement" OR "academic performance" OR "student* performance" OR "academic success" OR "student* success" OR "academic attainment" OR "student* attainment" OR "academic outcome*" OR "student* outcome*" OR "school* outcome*" OR "student* math*" OR "student* read*")*

The keywords included in the search string were selected after a thorough review of the relevant literature and approval of two field experts. A total of 829 articles (273 from WoS and 556 from Scopus) were yielded by this search. After excluding 185 duplicates, 644 articles were left for detailed screening. After a review of the titles of these documents, 322 articles were excluded for not meeting one or several of our previously stated criteria for inclusion/exclusion. Accordingly, 197 articles were excluded for being out of scope (e.g. not focusing on learner outcomes), 57 for being a book chapter, 8 for being a conference proceeding, 29 for not being in English, and 31 for other reasons such as being editorials or opinion papers. At the end of this process, a total of 322 articles were included in the dataset.

3.2 Data Analysis

In accordance with the purpose of the study, a total of 322 studies included in the final dataset were submitted to analysis to determine the bibliometric performance, conceptual structure, and thematic evolution of the knowledge base on the effect of AI chatbots on students' learning outcomes. The science mapping analysis was

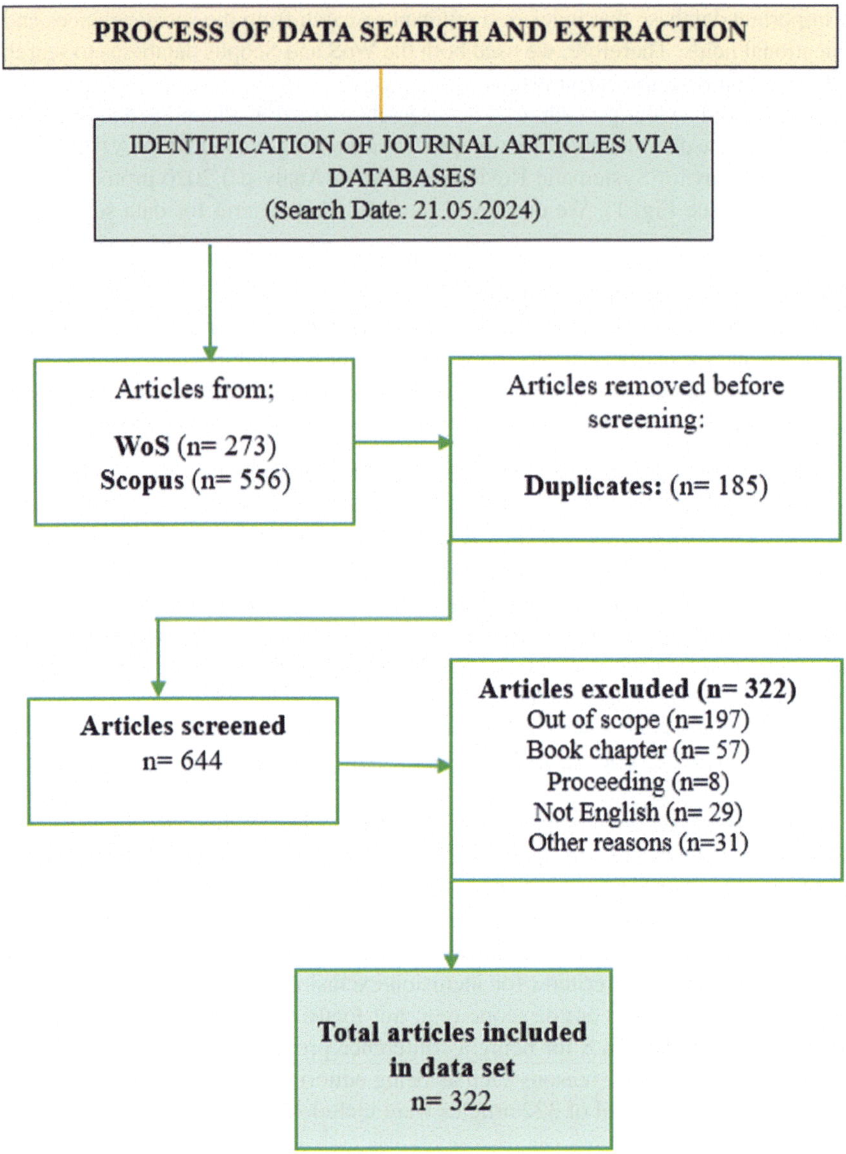

Fig. 1 PRISMA 2020 flowchart

performed using SciMAT software to analyze the conceptual structure and thematic evolution of the knowledge base. Science mapping analysis allows us to identify the conceptual structure, and thematic evolution of a research and helps reveal the structural and dynamic aspects of a scientific research field (Cobo et al., 2011; Martinez et al., 2015; Pradhan, 2016). SciMAT uses a mix of scientific mapping and bibliometric performance analysis to evaluate the development of a research field, identify and visualize prominent themes as well as show their bibliometric performance and reveal their thematic evolution across periods of analysis (Batagelj & Cerinšek, 2013; Chen, 2017; Cobo et al., 2012).

After the data search and identification process on WoS and Scopus databases, we uploaded the bibliographic metadata for the selected articles onto the SciMAT software and prepared the data for analysis. For instance, we manually grouped keywords representing the same concepts, such as "chatbot and chatbots, learner and learners, achievement and success" in order to increase the efficiency of the thematic analysis (Cobo et al., 2012; López-Robles et al., 2021).

The analysis was conducted based on the frequency of keyword co-occurrence. At this stage, we used the "equivalence index" to calculate the strength of the relationship between the clusters representing the themes. The clustering algorithm, which is a simple center algorithm that shows the strength of the relationship between the clusters of themes. The clusters were shown in a four-quadrant strategic diagram based on the density (y-axis) and centrality (x-axis) values. The density values are mathematically formulated as "d = 100 (Σeij/w)", where "i" and "j" represent the keywords belonging to the theme, and "w" represents the number of keywords in the theme. These values indicate the internal strength of their connection; that is, they measure the strength of the relationship between the keywords within a theme. Density points to the capacity of the themes to develop and maintain their existence over time. In addition, density is related to the internal relationships of the themes; themes with increasing levels of interrelationships are placed in the upper regions of the strategic diagram. Centrality, on the other hand, measures the degree of interaction between clusters. In other words, centrality is related to the external relationships between the themes. As the level of relationship between themes increases, themes shift to the right side in the strategic diagram. It is mathematically formulated as "c = 10*Σekh", where "k" represents a keyword belonging to any theme and "h" represents a keyword belonging to another theme. Centrality reveals that a cluster or network is an important transition point and has a critical role in understanding the relationship between relevant themes (Callon, Courtial, & Laville, 1991; Cobo et al., 2011; Martinez et al., 2021).

For a conceptual science mapping analysis based on co-word analysis, the research themes were divided into four groups and shown as the strategic diagram in Fig. 2a (Cobo et al., 2012):

(a) Motor themes (Q1): High centrality and density (The themes in this quadrant are highly developed and significant for the advancement of the research field during the relevant period).
(b) Basic and transversal themes (Q2): High centrality and low density (These themes are related to the research field but are not well-developed, likely to become motor themes in the future due to their high centrality).
(c) Emerging or declining themes (Q3): Low centrality and density (These themes mainly represent themes that emerge or decline during a period of analysis,

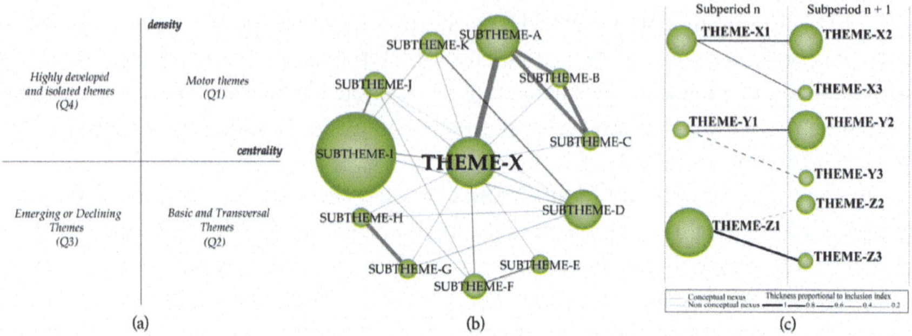

Fig. 2 Strategic diagram (**a**) Thematic network structure (**b**) Thematic evolution structure (**c**) *Source* (Sott et al., 2020)

whether they are emerging or declining themes could be evaluated with more in-depth qualitative synthesis).

(d) Highly developed and isolated themes (Q4): Low centrality and high density (These themes are strongly interrelated, highly specialized, and peripheral to the research field under investigation; they may not have the appropriate background for the relevant field, are no longer important due to the emergence of a new concept or technology).

The thematic network structure shown in Fig. 2b shows how strategic themes co-occur with other subthemes and the labels of the thematic networks reflect the most important (most central) keyword in the relevant theme. The volume of the spheres corresponds to the number of documents representing each keyword. The size of the circles in the thematic networks varies according to the number of publications while the thickness of the lines varies according to the strength of their relationship (Cobo et al., 2011, 2012).

The thematic evolution map in Fig. 2c helps to explore the evolution of research themes across periods as well as highlighting the origins and interrelationships of these themes. The interrelationships among the themes determine whether a theme belongs to a different thematic area or a continuation of any theme. When the same keywords as the theme names are shared by the themes, their interrelationships are shown with solid lines on the map. However, when the themes share common words other than the theme names, their relationship is shown with dashed lines. As the degree of their relationship increases, the lines become thicker, and as the number of publications representing a theme increases, the size of the spheres gets larger (Cobo et al., 2011; Cobo et al., 2012; Martínez et al., 2015; Murgado-Armenteros et al., 2015).

The analysis in the current study was conducted over three periods, which were formed taking the emergence of AI-based technologies into consideration. Accordingly, Period 1 comprised articles published between 2011 and 2022; Period 2 included articles published during 2023, and Period 3 included articles published during 2024.

4 Findings

Data obtained from Web of Science and Scopus databases were analyzed using SciMAT on a periodic basis, and the results of both thematic and evolution analyses were reported in this section. The themes were presented with the inclusion of performance analysis values (e.g., h-index, number of citations, centrality, and density).

4.1 Conceptual (Thematic) Structure Analysis

4.1.1 Period (2011–2022)

6 themes emerged from the analysis of 46 articles belonging to the first period between 2011–2022 The performance values (h-index) and strategic diagram of themes are presented in Fig. 1.

As shown in Fig. 3, the 6 themes that emerged in the first period are distributed in four regions. The motor theme region demonstrating the most significant themes that supported the growth of the field during this period, included Chatbot, Students, and Artificial-Intelligence themes while the Simulation theme was a highly developed and isolated theme. MOOC theme was found to be an emerging/declining theme while the Conversational-Agents theme was a basic and transversal theme. The highest number of documents (7 documents) belonged to the Chatbot theme.

To identify the related topics of the themes in the first period covering the dates 2011–2022, cluster networks (see Fig. 4) were examined. Accordingly, the main theme of Chatbot (1, 0.5) was associated with the sub-themes of Behavioral-Change, Conversational-Pedagogical-Agent, Educational-Bot, Intelligent-Tutoring-Systems,

a. Period 1 (2011-2022)

b. Themes Performance

Cluster	h-index	Cites	Centrality Range	Density Range
CHATBOT	7	277	1	0.5
STUDENTS	6	151	0.83	0.83
CONVERSATIONAL-AGENTS	5	234	0.67	0.33
ARTIFICIAL-INTELLIGENCE	4	160	0.5	0.67
MOOC	1	13	0.33	0.17
SIMULATION	1	6	0.17	1

Fig. 3 **a, b** Strategic diagram. Performance analysis

Learning-Experiences, Learning-Outcomes, Massive-Open-Online-Course, Model. The thickness and boldness of the lines between the keywords in this cluster network reflect the degree of relationships between the themes. These studies on Behavioral-Change (Tian et al, 2021), Conversational-Pedagogical-Agent (Tärning & Silvervarg, 2019), Educational-Bot (Pérez et al, 2020), Intelligent-Tutoring-Systems (Wang et al, 2021), Learning-Experiences (Liu et al, 2022), Learning-Outcomes (Kumar, 2021), Massive-Open-Online-Course (Han & Lee, 2022), and Model (Fryer et al, 2021) illustrate the sub-themes in Chatbot cluster networks.

The main theme of Students (0.83, 0.83) was associated with the sub-themes of Knowledge-Based-Systems, Learning-Environments, Learning-Process, Responsible-Conversational-Agent, Teaching/Learning-Strategies, Learning, Learning-Systems, and Teaching. These studies on Knowledge-Based-Systems (Crown et al, 2011), Learning-Environments (Han & Lee, 2022), Learning-Process (Crown et al, 2011), Responsible-Conversational-Agent (Li et al, 2022), Teaching/Learning-Strategies (Lee & Yeo, 2022), Learning (Gupta & Chen, 2022), Learning-Systems (Liu et al, 2022), Teaching (Lee & Yeo, 2022) illustrate the sub-themes in the Students cluster network.

Period 1 (2011-2022)

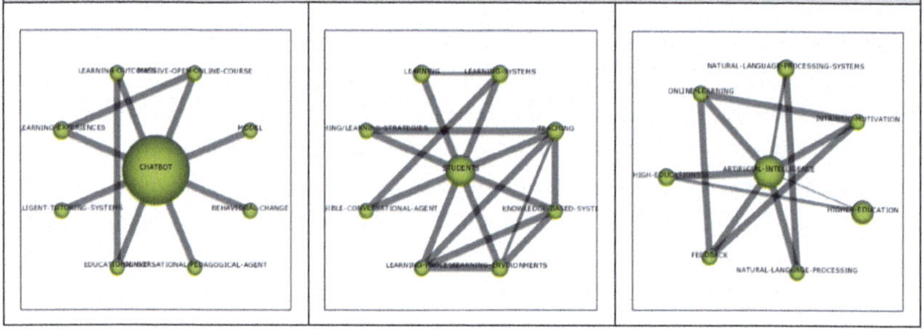

Chatbot; *Massive-Open-Online-Course, Model, Behavioral-Change, Conversational-Pedagogical-Agent, Educational-Bot, Intelligent-Tutoring-Systems, Learning-Experiences, Learning-Outcomes*

Students; *Learning-Systems, Teaching, Knowledge-Based-Systems, Learning-Environments, Learning-Process, Responsible-Conversational-Agent, Teaching/Learning-Strategies, Learning*

Artificial-Intelligence; *Natural-Language-Processing-Systems, Intrinsic-Motivation, Higher-Education, Natural-Language-Processing, Feedback, High-Educations, Online-Learning*

Fig. 4 a, b, c The cluster networks for the main themes of Period 1

The main theme of Artificial-Intelligent (0.5, 0.67) was associated with the sub-themes of Higher-Education, Natural-Language-Processing, Feedback, High-Educations, Online-Learning, Natural-Language-Processing-Systems, and Intrinsic-Motivation. These studies on Higher-Education (Nizhenkovska et al, 2022), Natural-Language-Processing (Chang et al, 2022), Feedback (Fidan & Gencel, 2022), High-Educations (Gupta & Chen, 2022), Online-Learning (Neo et al, 2022), Natural-Language-Processing-Systems (Han & Lee, 2022), Intrinsic-Motivation (Fidan & Gencel, 2022) illustrate the sub-themes in Artifical-Intelligent cluster network.

4.1.2 Period (2023)

The analysis of 124 articles belonging to the second period (the year 2023) revealed 12 themes in total. The performance values (h-index) and the strategic diagram of the themes are presented in Fig. 5.

The themes that emerged during the second period are distributed in all four regions of the strategic diagram. The Learning-Environments, Learning-Experiences, Medical-Student, Education, and Chatbot themes appeared as motor themes while the Active-Learning and E-Learning themes were highly developed and isolated themes. The Online-Learning Natural-Language-Processing and Motivation themes were the emerging/declining themes while the Academic-Achievement and Performance themes were basic and transversal themes. The highest number of documents (44 documents) belonged to the Learning-Environments theme.

In order to determine the topics related to these main themes, cluster networks (see Fig. 5) were examined. Accordingly, the main theme Learning-Environments (1, 0.83) was related to the sub-themes of Language-Model, ChatGPT, Artificial-Intelligence, Higher-Education, Learning, Software, Academic-Integrity and Writing. These studies on Language-Model (Currie et al, 2023), ChatGPT

a. Period 2 (2023-)

b. Themes Performance

Cluster	h-index	Cites	Centrality Range	Density Range
LEARNING-ENVIRONMENTS	19	1214	1	0.83
LEARNING-EXPERIENCES	6	111	0.92	1
MEDICAL-STUDENT	4	52	0.83	0.92
ACADEMIC-ACHIEVEMENT	7	232	0.75	0.17
PERFORMANCE	3	39	0.67	0.42
EDUCATION	8	184	0.58	0.75
CHATBOT	5	217	0.5	0.58
E-LEARNING	3	21	0.42	0.5
NATURAL-LANGUAGE-PROCESSING	4	48	0.33	0.25
ACTIVE-LEARNING	3	18	0.25	0.67
MOTIVATION	2	27	0.17	0.08
ONLINE-LEARNING	0	0	0.08	0.33

Fig. 5 **a, b** Strategic diagram. Performance analysis

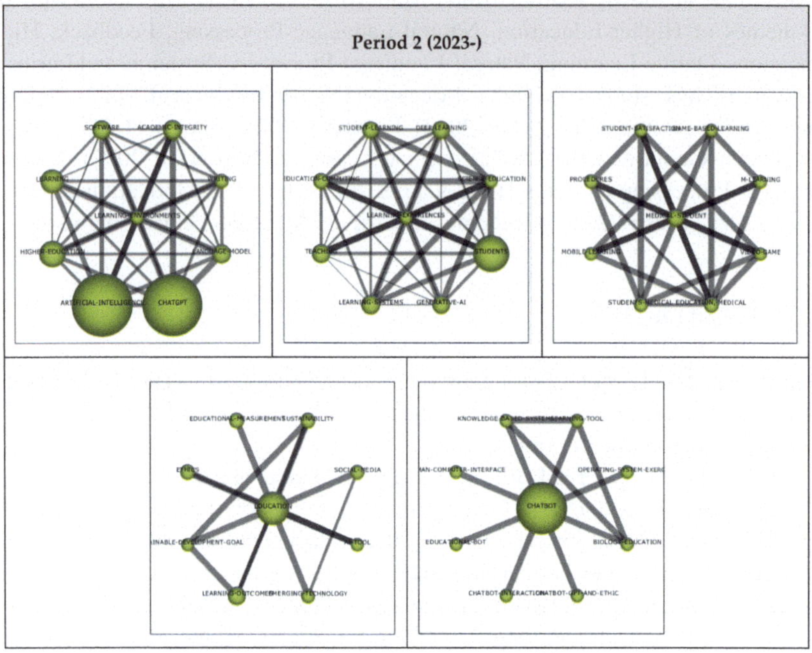

Learning-Environments; *Academic-Integrity, Writing, Language-Model, ChatGPT, Artificial-Intelligence, Higher-Education, Learning, Software*

Learning-Experiences; *Deep-Learning, Science-Education, Students, Generative-AI, Learning-Systems, Teaching, Education-Computing, Student-Learning*

Medical-Student; *Game-Based-Learning, M-Learning, Video-Game, Education-Medical, Students-Medical, Mobile-Learning, Procedures, Student-Satisfaction*

Education; *Sustainability, Social-Media, AI-Tool, Emerging-Technology, Learning-Outcomes, Sustainable-Development-Goal, Ethics, Educational-Measurement*

Chatbot; *Learning-Tool, Operating-System-Exercise, Biology-Education, Chatbot-Gpt-And-Ethic, Chatbot-Interaction, Educational-Bot, Human-Computer-Interface, Knowledge-Based-Systems*

Fig. 6 a, b, c, d, e The cluster networks for the main themes of period 2

(Zirar, 2023), Artificial-Intelligence (Hamid et al, 2023), Higher-Education (Escotet, 2023), Learning (Armitage, 2023), Software (Hsu et al, 2023), Academic-Integrity (Currie, 2023) and Writing (Currie et al, 2023) illustrate these sub-themes in the Learning-Environments cluster network.

The main theme of Learning-Experiences (0.92, 1) was associated with the sub-themes of Students, Learning-Systems, Teaching, Generative-AI, Deep-Learning,

Student-Learning, Education-Computing, and Science-Education. These studies on Students (Sanz Ausin et al, 2023), Learning-Systems (Albdrani & Al-Shargabi, 2023), Teaching (Kamalov et al, 2023), Generative-AI (Currie, 2023), Deep-Learning (Albdrani & Al-Shargabi, 2023), Student-Learning (Sanz Ausin et al, 2023), Education-Computing (Sanz Austin et al., 2023) and Science-Education (Albdrani & Al-Shargabi, 2023) illustrate these sub-themes in the Learning-Experiences cluster networks.

The main theme of Medical-Student (0.83, 0.92) was associated with the sub-themes of Video-Game, Education-Medical, Students-Medical, Mobile-Learning, Procedures, Student-Satisfaction, Game-Based-Learning, and M-Learning. These studies on Video-Game (Al Kahf et al, 2023), Education-Medical (Armitage, 2023), Students-Medical (Strong et al, 2023), Mobile-Learning (Hsu et al, 2023), Procedures (Strong et al, 2023), Student-Satisfaction (Al Kahf et al, 2023), Game-Based-Learning (Hsu et al, 2023), and M-Learning (Hsu et al, 2023) illustrate the sub-themes in Medical-Student cluster network.

The main theme of Education (0.58, 0.75) was associated with the sub-themes of AI-Tool, Emerging-Technology, Learning-Outcomes, Sustainable-Development-Goal, Ethics, Educational-Measurement, Sustainability, and Social-Media. These studies on AI-Tool (Yesmin, 2023), Emerging-Technology (Khurma et al, 2023), Learning-Outcomes (Adams et al., 2003), Sustainable-Development-Goal (Adams et al., 2003), Ethics (Bahroun et al, 2023), Educational-Measurement (Al Kahf et al, 2023), Sustainability (Deng & Yu, 2023), and Social-Media (Li et al, 2023) illustrate the sub-themes in the Education cluster network.

The main theme of Chatbot (0.5, 0.58) was associated with the sub-themes of Biology-Education, Chatbot-Gpt-And-Ethic, Chatbot-Interaction, Educational-Bot, Human-Computer-Interface, Knowledge-Based-Systems, Learning-Tool and Operating-System-Exercise. These studies on Biology-Education (Lin et al., 2023), Chatbot-GPT-And-Ethic (Dakakni & Safa, 2023), Chatbot-Interaction (Hobert et al, 2023), Educational-Bot (Kuhail et al, 2023), Human-Computer-Interface (Nguyen, 2023), Knowledge-Based-Systems (Lin et al., 2023), Learning-Tool (Lin et al., 2023) and Operating-System-Exercise (Xue & Niu, 2023) illustrate the sub-themes in the Chatbot cluster networks.

4.1.3 Period (2024)

The analysis of the 152 articles published in 2024 (the third period of analysis) yielded 15 themes in total. The performance values (h-index) and strategic diagram of the themes are presented in Fig. 7.

As shown in the strategic diagram in Fig. 8, the themes that emerged during the third period of analysis were also distributed in all four regions. The motor themes of the period were the Alternate-Reality-Game, Cross-Sectional-Study, Cognitive-Load, Language-Model, Learning, and Language-Education themes. The Academic-Performance and E-Learning were found to be highly developed and isolated themes while the Artificial-Intelligence-Chatbot, Higher-Education, Education-Computing,

a. Period 3 (2024-)

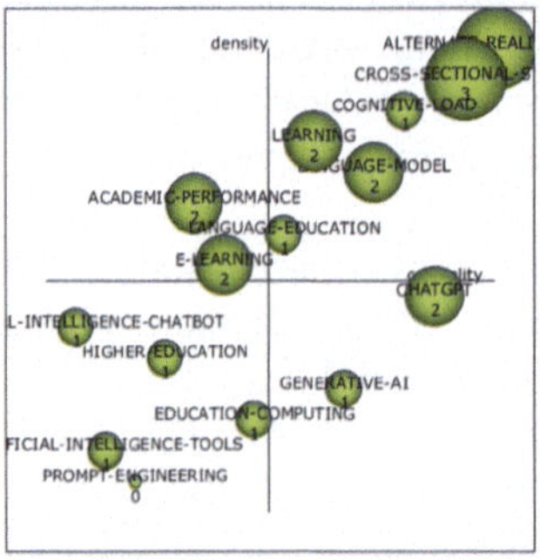

b. Themes Performance

Cluster	h-index	Cites	Centrality Range	Density Range
ALTERNATE-REALITY-GAME	3	71	1	1
CROSS-SECTIONAL-STUDY	3	32	0.93	0.93
CHATGPT	2	19	0.87	0.47
COGNITIVE-LOAD	1	8	0.8	0.87
LANGUAGE-MODEL	2	8	0.73	0.73
GENERATIVE-AI	1	2	0.67	0.27
LEARNING	2	18	0.6	0.8
LANGUAGE-EDUCATION	1	3	0.53	0.6
EDUCATION-COMPUTING	1	15	0.47	0.2
E-LEARNING	2	14	0.4	0.53
ACADEMIC-PERFORMANCE	2	12	0.33	0.67
HIGHER-EDUCATION	1	4	0.27	0.33
PROMPT-ENGINEERING	0	0	0.2	0.07
ARTIFICIAL-INTELLIGENCE-TOOLS	1	14	0.13	0.13
ARTIFICIAL-INTELLIGENCE-CHATBOT	1	41	0.07	0.4

Fig. 7 **a, b** Strategic diagram. Performance analysis

Artificial-Intelligence-Tools, and Prompt-Engineering themes were the emerging/ declining themes of the period. ChatGPT and Generative-AI themes were, on the other hand, the basic and transversal themes. The highest number of documents (22 documents) belonged to the Alternate-Reality-Game theme.

In order to identify the topics related to the main themes of the third period (e.g. the year 2024), cluster networks (see Fig. 8) were examined. Accordingly, the main theme of Alternate-Reality-Game (1, 1) was associated with the sub-themes of Chatbot, Computer-Mediated-Feedback, Students, Learning-Systems, Student-Learning, Self-Regulated-Learning, Feedback and Problem-Solving. These studies on Chatbot (Liang et al, 2024), Computer-Mediated-Feedback (Liang et al, 2024), Students (Boubker, 2024), Learning-Systems (Al-Qaysi et al, 2024), Student-Learning (Roca et al, 2024), Self-Regulated-Learning (Ng et al., 2024), Feedback (Ng et al., 2024) and Problem-Solving (Liang et al, 2024) illustrate the sub-themes in the Alternate-Reality-Game cluster network.

The main theme of Cross-Sectional-Study (0.93, 0.93) was associated with the sub-themes of Medical-Student, Artificial-Intelligence, Education, Medical-Education, Assessment, Education-Medical, Students-Medical, and Educational-Measurement. These studies on Medical-Student (Prachnakorn et al, 2024), Artificial-Intelligence (Divito et al, 2024), Education, Medical-Education (Sridharan & Sequeira, 2024), Assessment (Herrmann-Werner et al, 2024), Education-Medical (Divito et al, 2024), Students-Medical (Sridharan & Sequeira, 2024), Educational-Measurement (Sridharan & Sequeira, 2024) illustrate the sub-themes in the Cross-Sectional-Study cluster network.

The main theme of Cognitive-Load (0.8, 0.87) was associated with the sub-themes of Anxiety-Disorder, Formative-Feedback, Teaching, Motivation, Writing, Learning-Performance, Job-Analysis, and Task-Analysis. These studies on Anxiety-Disorder (Yin et al, 2024), Formative-Feedback (Yin et al, 2024), Teaching (Bai et al, 2024), Motivation (Bai et al, 2024), Writing (Yin et al, 2024), Learning-Performance (Yin

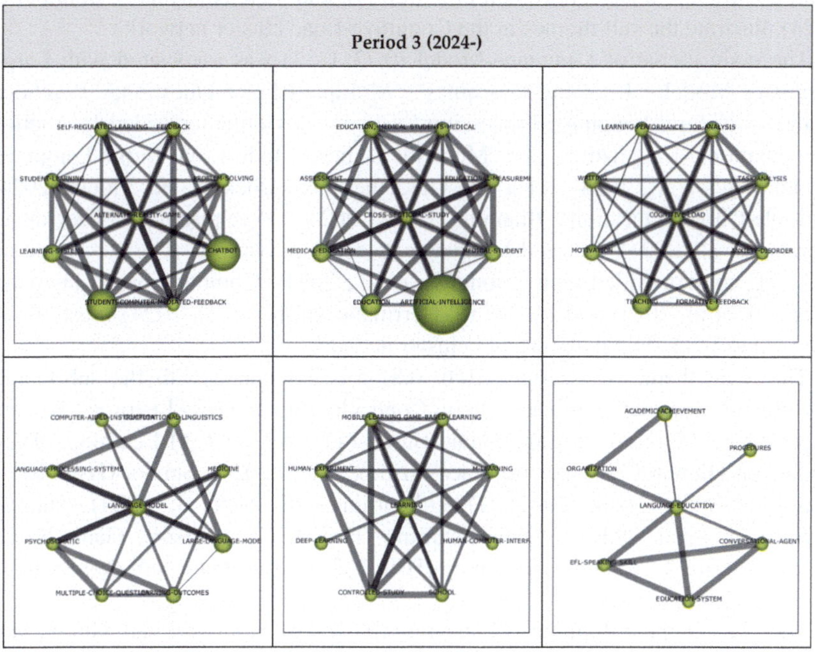

Alternate-Reality-Game; *Feedback, Problem-Solving, Chatbot, Computer-Mediated-Feedback, Students, Learning-Systems, Student-Learning, Self-Regulated-Learning*

Cross-Sectional-Study; *Students-Medical, Educational-Measurement, Medical-Student, Artificial-Intelligence, Education, Medical-Education, Assessment, Education-Medical*

Cognitive-Load; *Job-Analysis, Task-Analysis, Anxiety-Disorder, Formative-Feedback, Teaching, Motivation, Writing, Learning-Performance*

Language-Model; *Computational-Linguistics, Medicine, Large-Language-Model, Learning-Outcomes, Multiple-Choice-Question, Psychosomatic, Natural-Language-Processing-Systems, Computer-Aided-Instruction*

Learning; *Game-Based-Learning, M-Learning, Human-Computer-Interface, School, Controlled-Study, Deep-Learning, Human-Experiment, Mobile-Learning*

Language-Education; *Procedures, Conversational-Agents, Education-System, EFL-Speaking-Skill, Organization, Academic-Achievement*

Fig. 8 **a, b, c, d, e, f** The cluster networks for the main themes of period 3

et al, 2024), Job-Analysis (Tossell et al., 2024), and Task-Analysis (Cámara et al, 2024) illustrate the sub-themes in the Cognitive-Load cluster network.

The main theme of Language-Model (0.73, 0.73) was associated with Large-Language-Model, Learning-Outcomes, Multiple-Choice-Question, Psychosomatic, Natural-Language-Processing-Systems, Computer-Aided-Instruction, Computational-Linguistics, and Medicine. These studies on Large-Language-Model (Ng et al., 2024), Learning-Outcomes (Askarbekuly, & Aničić, 2024), Multiple-Choice-Question (Bharatha et al, 2024), Psychosomatic (Herrmann-Werner et al, 2024), Natural-Language-Processing-Systems (Corlatescu et al., 2024), Computer-Aided-Instruction (Ng et al., 2024), Computational-Linguistics (Misiejuk et al, 2024) and Medicine (Herrmann-Werner et al, 2024) illustrate the sub-themes in the Language-Model cluster network.

The main theme of Learning (0.6, 0.8) was associated with the sub-themes of Human-Computer-Interface, School, Controlled-Study, Deep-Learning, Human-Experiment, Mobile-Learning, Game-Based-Learning, and M-Learning. These studies on Human-Computer-Interface (Hsu & Chen, 2024), School (Hsu, 2024), Controlled-Study (Hsu, 2024), Deep-Learning (Miao et al, 2024), Human-Experiment (Hsu, 2024), Mobile-Learning (Hsu & Chen, 2024), Game-Based-Learning (Hsu, 2024) and M-Learning (Hsu, 2024) illustrate the sub-themes in the Learning cluster network.

The main theme of Language-Education (0.53, 0.6) was associated with the sub-themes of Conversational-Agents, Education-System, EFL-Speaking-Skill, Organization, Academic-Achievement, and Procedures. These studies on Conversational-Agents (Jinming & Daniel, 2024), Education-System (Jinming & Daniel, 2024), EFL-Speaking-Skill (Jinming & Daniel, 2024), Organization (Rengers et al, 2024), Academic-Achievement (Abdaljaleel et al, 2024) and Procedures (Laupichler et al, 2024) illustrate the sub-themes in the Language-Education cluster network.

4.2 Conceptual (Thematic) Evolution Analysis

The thematic evolution map (see Fig. 9) illustrates the development pattern of the themes in the knowledge base of the effect of Chatbots on student learning outcomes across the three periods of analysis. The map also shows the relationships between the themes from each period. The size of the spheres on the map indicates the number of publications, and the thickness of the lines connecting the spheres indicates the relationship between the themes. The solid lines indicate that the associated themes share the same keywords with the main theme while the dashed lines indicate that they do not share all the keywords with the main theme. (Cobo et al., 2012; Murgado-Armenteros et al., 2015). As shown in the thematic evolution map in Fig. 7 6 themes emerged during the first period (2011–2022), which constitute 14.3% of the articles (46 documents) analyzed. One of these themes continued to exist during the second period, while the other five were transformed into different themes.

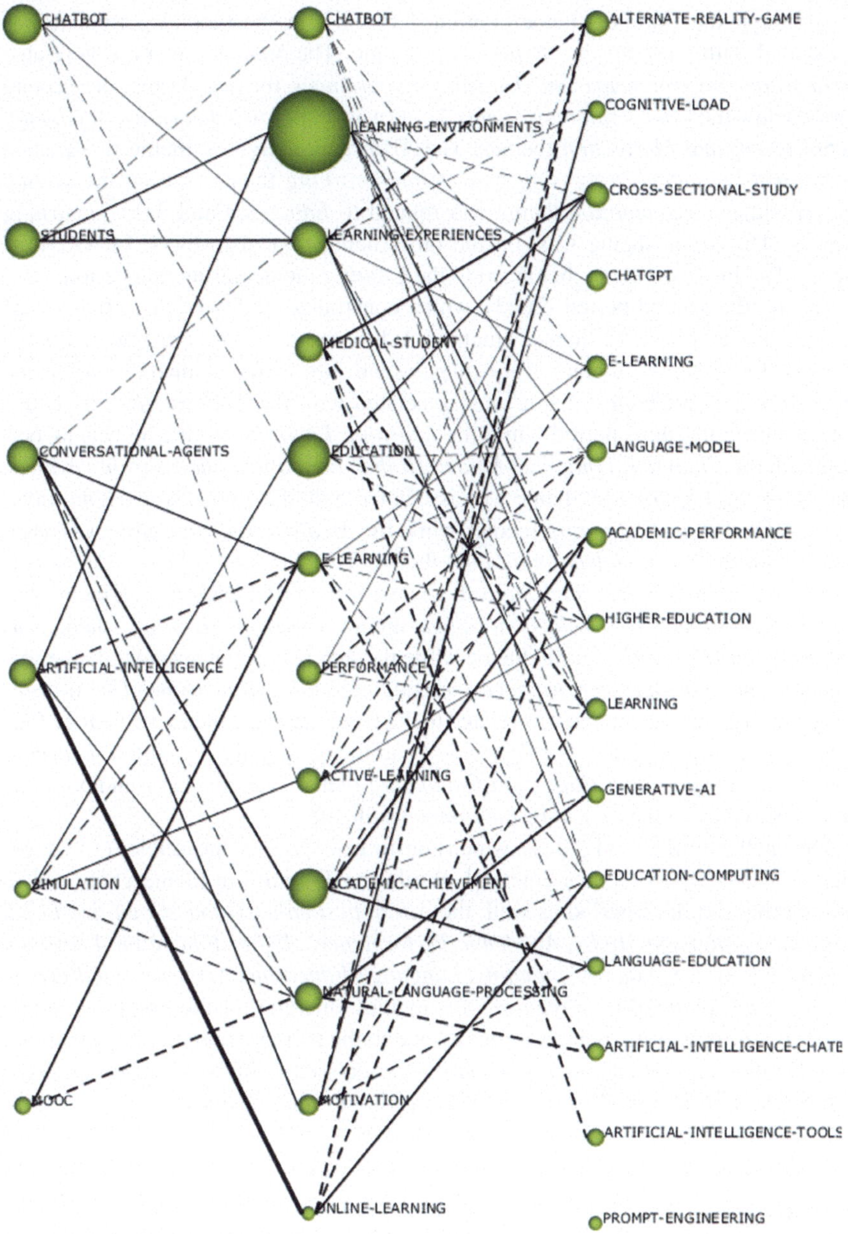

Fig. 9 The thematic evolution map

Although the *Chatbot* theme continued to exist in the second period, it was associated with the *Learning-Experiences* theme. The *Students* theme was related to *Learning-Environments and Learning-Experiences* themes during the second period while the *Conversational-Agents* theme was related to *E-Learning, Academic-Achievement and Motivation* themes. The *Artifical-Intelligence* theme was related to *Natural-Language-Processing and Online-Learning* themes during the second period while the *Simulation* theme was related to *Education* and *Active-Learning* themes. The *Mooc* theme was, on the other hand, related to the E-LEARNING theme. The highest h-index belonged to the *Chatbot* theme during this period.

During the second period (2023), which constituted 38.5% of the articles (124 documents) analyzed, 12 themes emerged. While the *E-Learning* theme continued to exist, the other 11 themes were transformed into different themes during the third period. The *Chatbot* theme was associated with the *Alternate-Reality-Game* theme during the third period while the *Learning-Environment* theme was associated with the *ChatGPT, Language-Model, Higher-Education, and Learning* themes. The *Learning-Experiences* theme was associated with the *E-Learning, Generative-*AI, and *Education-Computing* themes while the *Medical-Student* theme was associated with the *Cross-Selectional-Study* theme. The *Education* theme was associated with the *Cross-Selectional-Study* theme while the PERFORMANCE theme was associated with the *Cross-Selectional-Study* theme. The *Active-Learning* theme was related to the *Higher-Education* theme while the *Academic-Achievement* theme was related to the *Academic-Performance and Language-Education themes The Natural-Language-Processing* theme was related to the *Generative-AI* theme while the *Motivation* theme was related to the *Cognitive-Load* theme. Finally, the *Online-Learning* theme was related to the *Education-Computing* theme. The highest h-index in this period belonged to the *Learning-Environments* theme.

During the third period (2024), which constitutes 47.2% of the articles (152 documents) analyzed, 15 themes emerged. While the *E-Learning* theme continued to exist during the third period as well, the *Alternate-Reality-Game, Cognitive-Load, ChatGPT, Language-Model, Academic-Performance, Higher-Education, Learning, Generative-Ai, Education-Computing, Language-Education, Artificial-Intelligence-Chatbot and Artificial-Intelligence-Tools* themes emerged for the first time. While the *Promt-Engineering* theme was not related to the themes from the second period, the other themes were all connected. During this period, the highest h-index belonged to the *Alternate-Reality-Game and Cross-Sectional-Study* themes.

5 Discussion

The current study conducted a thematic analysis of the knowledge base on the effects of AI-based chatbots on students' learning outcomes including the 322 articles published between the years 2011 and 2024. The period-based analysis showed that the motor themes of the first period (2011–2022) were Chatbot, Students, and Artifical-Intelligence. The studies conducted during this period were all related to these motor themes. For example, Yin et al. (2021) compared chatbot-based microlearning and traditional learning groups in their experimental study, and found that students' intrinsic motivation was high in chatbot-based teaching, while no difference was found between the performances of the two groups. Chang et al (2022) examined the effect of chatbots in nursing education in their study and concluded that the mobile chatbot application used for learning purposes increased the achievement and self-efficacy of nursing students. Neo et al. (2022) conducted a study using an artificial intelligence chatbot called MERLIN, which was developed to help students learn in online learning environments. The study showed that students who used MERLIN were more motivated to learn online, improved their learning, and wanted to use the chatbot in all their courses. In their experimental study, Gupta and Chen (2022) found that the chatbot developed to support inclusive learning helped disadvantaged students with different learning styles and learning opportunities.

During the second period (2023), Learning-Environments, Learning-Experiences, Medical-Student, Education, and Chatbot themes emerged as the motor themes that significantly shaped the development of the field. Studies conducted during this period lend support to the emergence of these themes as motor themes. For example, Samonte et al. (2023) developed an online learning environment using a chatbot in their experimental study and concluded that the experimental group of students who used the artificial intelligence-supported learning environment performed better than those in the control group. Albdrani and Al-Shargabi (2023) investigated the effectiveness of ChatGPT in providing personalized learning experiences in their study and emphasized that the artificial intelligence tool ChatGPT had strong potential to offer personalized learning experiences. Waghray et al. (2023), on the other hand, stated that chatbots could be used in medical education for both medical students and healthcare professionals because they could help them with diagnosis and treatment, could give instructions on how to use medical equipment, and provide information about anatomy and medical disorders. Kamalov et al. (2023) examined the potential impact of artificial intelligence on education and concluded that collaborative learning tools, intelligent tutoring systems, automatic assessment, and personalized learning were the areas where AI chatbots could offer benefits. Watts et al. (2023) investigated mechanical reasoning skills by comparing the responses given by students and chatbots to writing assignments in a chemistry course and found that chatbots could not perform mechanical reasoning skills at the same level as students even if they gave a variety of responses. In an experimental study conducted by Lin and Ye (2023) using a chatbot to support students' biology learning outside the classroom, it was concluded that students who used the proposed chatbot had higher achievement.

During the third period (2024), the Alternate-Reality-Game, Cross-Sectional-Study, Cognitive-Load, Language-Model, Learning, and Language-Education themes emerged as the motor themes. Many studies conducted in this period illustrate the strength of these themes in shaping the research orientations during the last period. For example, Liang et al. (2024) conducted an experimental study investigating the effects of AI-based chatbots on students' learning performance in alternative reality game-based museum learning. They found that the AI chatbots could significantly facilitate students' metacognitive awareness, emotional and behavioral engagement, and double-loop learning. Kamoun et al. (2024) conducted a cross-sectional study investigating the knowledge, attitude, and perception of university students and lecturers regarding ChatGPT. The study showed that the lecturers had greater knowledge of ChatGPT than the students but their attitudes were more reserved and their perceptions were more negative.

In their study, Li et al. (2024) compared a teacher with ChatGPT and investigated its effects on critical thinking, learning performance, and cognitive load in teaching STEM to prospective teachers. The study revealed that prospective teachers who received training with ChatGPT experienced higher critical thinking systematicity, better performance, and lower cognitive load. Similarly, Jošt et al (2024) investigated the effects of large language models (LLMs) such as ChatGPT and Copilot on student learning in programming education. The study showed that the more the students used LLM, the lower their final grades. Therefore, the authors suggested that LLMs should only be used as a complementary tool. In a study by Boubker (2024), which investigated the role of ChatGPT in improving students' learning outcomes, it was found that the use of ChatGPT would positively affect perceived usefulness and student satisfaction. On the other hand, Jinming and Daniel' (2024) systematic review investigating the effectiveness of artificial intelligence robots in English language education showed that AI chatbots are designed to accelerate the process of English-speaking proficiency of EFL learners and help them achieve the course objectives.

6 Conclusion

The current study conducted a science mapping analysis of the research field focusing on the effect of AI-chatbots on students' learning outcomes and contributed to the literature by delineating the thematic strands and their evolution across three periods of analysis. The results showed that studies published during the first period of analysis (2011–2022) mostly focused on how AI chatbots could act as intelligent tutoring systems, particularly in online courses. During the second period (2023), though, this attention turned to ChatGPT and concerns over its influence on the integrity and academic honesty of students. During the last period of analysis (2024), investigations tended to move from concerns over its side effects to how these chatbots could support customized learning and students' self-learning experiences. These results indicate that perceptions about the educational potential of these chatbots have become more positive following the increased efficacy of these AI-based systems and the growing insights into how these innovative technologies could be used for the benefit of students' learning. In addition, further analysis of the results obtained from the studies within the scope of this research and generating more evidence on overshadowed aspects can contribute to the development of the relevant knowledge base.

Limitations

Like every scientific study, this study also has some limitations. These are; (i) the data included in the analysis only includes articles published in journals indexed in WoS and Scopus, the results obtained are limited to the studies indexed in these two databases (ii) The article includes only studies published in English, therefore information from articles published in different languages may have been overlooked. Therefore, these limitations should be taken into consideration when interpreting the results of this study.

References

Abdaljaleel, M., Barakat, M., Alsanafi, M., Salim, N. A., Abazid, H., Malaeb, D., & Sallam, M. (2024). A multinational study on the factors influencing university students' attitudes and usage of ChatGPT. *Scientific Reports, 14*(1), 1983.

Adamopoulou, E., & Moussiades, L. (2020). Chatbots: History, technology, and applications. *Machine Learning with Applications, 2*, 100006.

Adams, T., Jameel, S. M., & Goggins, J. (2023). Education for Sustainable Development: Mapping the SDGs to University Curricula. Sustainability (Switzerland), 15 (10).

Al Kahf, S., Roux, B., Clerc, S., Bassehila, M., Lecomte, A., Moncomble, E., & Planquette, B. (2023). Chatbot-based serious games: A useful tool for training medical students? *A Randomized Controlled Trial. Plos One, 18*(3), e0278673.

Albdrani, R. N., & Al-Shargabi, A. A. (2023). Investigating the Effectiveness of ChatGPT for Providing Personalized Learning Experience: A Case Study. *International Journal of Advanced Computer Science & Applications, 14*(11).

Al-Qaysi, N., Al-Emran, M., Al-Sharafi, M. A., Iranmanesh, M., Ahmad, A., & Mahmoud, M. A. (2024). Determinants of ChatGPT Use and its Impact on Learning Performance: An Integrated Model of BRT and TPB. *International Journal of Human–Computer Interaction*, 1–13.

Armitage, R. C. (2023). ChatGPT: The threats to medical education. *Postgraduate Medical Journal, 99*(1176), 1130–1131.

Askarbekuly, N., & Aničić, N. (2024). LLM examiner: automating assessment in informal self-directed e-learning using ChatGPT. *Knowledge and Information Systems*, 1–18.

Bahroun, Z., Anane, C., Ahmed, V., & Zacca, A. (2023). Transforming education: A comprehensive review of generative artificial intelligence in educational settings through bibliometric and content analysis. *Sustainability, 15*(17), 12983.

Bai, S., Gonda, D. E., & Hew, K. F. (2024). Write-Curate-Verify: A Case Study of Leveraging Generative AI for Scenario Writing in Scenario-Based Learning. *IEEE Transactions on Learning Technologies*.

Bassett, C. (2019). The computational therapeutic: Exploring Weizenbaum's ELIZA as a history of the present. *AI & SOCIETY, 34*(4), 803–812.

Batagelj, V., & Cerinšek, M. (2013). On bibliographic networks. *Scientometrics, 96*(3), 845–864. https://doi.org/10.1007/s11192-012-0940-1

Bhalshankar, S. V., & Deshmukh, R. R. (2023, May). CAAS (Chatbot as a Service): Sector-Wise Survey. In International Conference on Applications of Machine Intelligence and Data Analytics (ICAMIDA 2022) (pp. 86–111). Atlantis Press.

Bharatha, A., Ojeh, N., Fazle Rabbi, A. M., Campbell, M. H., Krishnamurthy, K., Layne-Yarde, R. N., & Majumder, M. A. A. (2024). Comparing the Performance of ChatGPT-4 and Medical Students on MCQs at Varied Levels of Bloom's Taxonomy. *Advances in Medical Education and Practice*, 393–400.

Boubker, O. (2024). From chatting to self-educating: Can AI tools boost student learning outcomes? *Expert Systems with Applications, 238*, 121820.

Caldarini, G., Jaf, S., & McGarry, K. (2022). A literature survey of recent advances in chatbots. *Information, 13*(1), 41.

Callon, M., Courtial, J. P., & Laville, F. (1991). Co-word analysis as a tool for describing the network of interactions between basic and technological research: The case of polymer chemsitry. Scientometrics 1991 22:1, 22(1), 155–205.

Cámara, J., Troya, J., Montes-Torres, J., & Jaime, F. J. (2024). Generative AI in the Software Modeling Classroom: An Experience Report with ChatGPT and UML. *IEEE Software.*

Casheekar, A., Lahiri, A., Rath, K., Prabhakar, K. S., & Srinivasan, K. (2024). A contemporary review on chatbots, AI-powered virtual conversational agents, ChatGPT: Applications, open challenges and future research directions. *Computer Science Review, 52*, 100632.

Chang, C. Y., Hwang, G. J., & Gau, M. L. (2022). Promoting students' learning achievement and self-efficacy: A mobile chatbot approach for nursing training. *British Journal of Educational Technology, 53*(1), 171–188.

Chen, C. (2017). Science mapping: A systematic review of the literature. *J Data Inf Sci., 2*(2), 1–40.

Chen, Y., Jensen, S., Albert, L. J., Gupta, S., & Lee, T. (2023). Artificial intelligence (AI) student assistants in the classroom: Designing chatbots to support student success. *Information Systems Frontiers, 25*(1), 161–182.

Chiu, T. K. (2024). Future research recommendations for transforming higher education with generative AI. *Computers and Education: Artificial Intelligence, 6*, 100197.

Cobo, M. J., López-Herrera, A. G., Herrera-Viedma, E., & Herrera, F. (2011). Science mapping software tools: Review, analysis, and cooperative study among tools. *Journal of the American Society for Information Science and Technology, 62*(7), 1382–1402. https://doi.org/10.1002/asi.21525

Cobo, M. J., López-Herrera, A. G., Herrera-Viedma, E., & Herrera, F. (2012). SciMAT: A New Science Mapping Analysis Software Tool. *Journal of the American Society for Information Science and Technology, 63*(8), 1609–1630.

Cobo, M. J., López-Herrera, A. G., & Herrera-Viedma, E. (2015). A Relational Database Model for Science Mapping Analysis. *Acta Polytechnica Hungarica, 12*(6), 43–62.

Corlatescu, D. G., Watanabe, M., Ruseti, S., Dascalu, M., & McNamara, D. S. (2024). The automated model of comprehension version 4.0–Validation studies and integration of ChatGPT. *Computers in Human Behavior, 154*, 108154.

Crawford, J., Cowling, M., & Allen, K. A. (2023). Leadership is needed for ethical ChatGPT: Character, assessment, and learning using artificial intelligence (AI). *Journal of University Teaching & Learning Practice, 20*(3), 02.

Crown, S., Fuentes, A., Jones, R., Nambiar, R., & Crown, D. (2011). Anne G. Neering: Interactive chatbot to engage and motivate engineering students. *Computers in Education Journal, 21*(2), 24–34.

Currie, G. (2023). A conversation with ChatGPT. *Journal of Nuclear Medicine Technology, 51*(3), 255–260.

Currie, G., Singh, C., Nelson, T., Nabasenja, C., Al-Hayek, Y., & Spuur, K. (2023). ChatGPT in medical imaging higher education. *Radiography, 29*(4), 792–799.

Dakakni, D., & Safa, N. (2023). Artificial intelligence in the L2 classroom: Implications and challenges on ethics and equity in higher education: A 21st century Pandora's box. *Computers and Education: Artificial Intelligence, 5*, 100179.

Dalgıç, A., Yaşar, E., & Demir, M. (2024). ChatGPT and learning outcomes in tourism education: The role of digital literacy and individualized learning. *Journal of Hospitality, Leisure, Sport & Tourism Education, 34*, 100481.

Deng, X., & Yu, Z. (2023). A meta-analysis and systematic review of the effect of chatbot technology use in sustainable education. *Sustainability, 15*(4), 2940.

Divito, C. B., Katchikian, B. M., Gruenwald, J. E., & Burgoon, J. M. (2024). The tools of the future are the challenges of today: The use of ChatGPT in problem-based learning medical education. *Medical Teacher, 46*(3), 320–322.

Escotet, M. Á. (2023). The optimistic future of Artificial Intelligence in higher education. *Prospects*, 1–10.

Eysenbach, G. (2023). The role of ChatGPT, generative language models, and artificial intelligence in medical education: A conversation with ChatGPT and a call for papers. *JMIR Medical Education, 9*(1), e46885.

Fidan, M., & Gencel, N. (2022). Supporting the instructional videos with chatbot and peer feedback mechanisms in online learning: The effects on learning performance and intrinsic motivation. *Journal of Educational Computing Research, 60*(7), 1716–1741.

Fryer, L. K., Shum, A., Lee, A., & Lau, P. (2021). Mapping students' interest in a new domain: Connecting prior knowledge, interest, and self-efficacy with interesting tasks and a lasting desire to reengage. *Learning and Instruction, 75*, 101493.

Gupta, S., & Chen, Y. (2022). Supporting inclusive learning using chatbots? A chatbot-led interview study. *Journal of Information Systems Education, 33*(1), 98–108.

Hamid, H., Zulkifli, K., Naimat, F., Yaacob, N. L. C., & Ng, K. W. (2023). Exploratory study on student perception on the use of chat AI in process-driven problem-based learning. *Currents in Pharmacy Teaching and Learning, 15*(12), 1017–1025.

Han, S., & Lee, M. K. (2022). FAQ chatbot and inclusive learning in massive open online courses. *Computers & Education, 179*, 104395.

Herrmann-Werner, A., Festl-Wietek, T., Holderried, F., Herschbach, L., Griewatz, J., Masters, K., & Mahling, M. (2024). Assessing ChatGPT's mastery of Bloom's taxonomy using psychosomatic medicine exam questions: Mixed-methods study. *Journal of Medical Internet Research, 26*, e52113.

Hobert, S., Følstad, A., & Law, E. L. C. (2023). Chatbots for active learning: A case of phishing email identification. *International Journal of Human-Computer Studies, 179*, 103108.

Hong, J. C., Lin, C. H., & Juh, C. C. (2023). Using a Chatbot to learn English via Charades: the correlates between social presence, hedonic value, perceived value, and learning outcome. Interactive Learning Environments, 1–17.

Hsu, M. H. (2024). Mastering medical terminology with ChatGPT and Termbot. *Health Education Journal, 83*(4), 352–358.

Hsu, M. H., & Chen, Y. H. (2024). Personalized medical terminology learning game: Guess the term. *Games for Health Journal, 13*(2), 84–92.

Hsu, M. H., Chan, T. M., & Yu, C. S. (2023). Termbot: A chatbot-based crossword game for gamified medical terminology learning. *International Journal of Environmental Research and Public Health, 20*(5), 4185.

Hutchens, J. L. (1996). How to pass the Turing test by cheating. School of Electrical, Electronic and Computer Engineering research report TR97–05. Perth: University of Western Australia.

İpek, Z. H., Gözüm, A. I. C., Papadakis, S., & Kallogiannakis, M. (2023). Educational applications of the ChatGPT AI system: A systematic review research. *Educational Process: International Journal, 12*(3), 26–55. https://doi.org/10.22521/edupij.2023.123.2

Jinming, D. U., & Daniel, B. K. (2024). A Systematic Review of AI-Powered Chatbots in EFL Speaking Practice: Transforming Language Education. *Computers and Education: Artificial Intelligence*, 100230.

Jošt, G., Taneski, V., & Karakatič, S. (2024). The Impact of Large Language Models on Programming Education and Student Learning Outcomes. *Applied Sciences, 14*(10), 4115.

Kamalov, F., Santandreu Calonge, D., & Gurrib, I. (2023). New era of artificial intelligence in education: Towards a sustainable multifaceted revolution. *Sustainability, 15*(16), 12451.

Kamoun, F., El Ayeb, W., Jabri, I., Sifi, S., & Iqbal, F. (2024). Exploring Students' and Faculty's Knowledge, Attitudes, and Perceptions Towards ChatGPT: A Cross-Sectional Empirical Study. *Journal of Information Technology Education: Research, 23*, 004.

Karakose, T., & Tülübas, T. (2023). How can ChatGPT facilitate teaching and learning: Implications for contemporary education. *Educational Process International Journal, 12*(4), 7–16.

Karakose, T., & Tülübas, T. (2024). School leadership and management in the age of Artificial Intelligence (AI): Recent developments and future prospects. *Educational Process: International Journal, 13*(1), 7–14.

Karakose, T., Demirkol, M., Yirci, R., Polat, H., Ozdemir, T. Y., & Tülübaş, T. (2023). A conversation with ChatGPT about digital leadership and technology integration: Comparative analysis based on human–AI collaboration. *Administrative Sciences, 13*(7), 157.

Khalil, G. M., & Crawford, C. A. G. (2015). A bibliometric analysis of us-based research on the behavioral risk factor surveillance system. *American Journal of Preventive Medicine, 48*(1), 50–57.

Khurma, O. A., Ali, N., & Hashem, R. (2023). Critical Reflections on ChatGPT in UAE Education: Navigating Equity and Governance for Safe and Effective Use. *International Journal of Emerging Technologies in Learning, 18*(14).

King, M. R., & ChatGPT. (2023). A conversation on artificial intelligence, chatbots, and plagiarism in higher education. *Cellular and Molecular Bioengineering, 16*(1), 1–2.

Kuhail, M. A., Alturki, N., Alramlawi, S., & Alhejori, K. (2023). Interacting with educational chatbots: A systematic review. *Education and Information Technologies, 28*(1), 973–1018.

Kumar, J. A. (2021). Educational chatbots for project-based learning: Investigating learning outcomes for a team-based design course. *International Journal of Educational Technology in Higher Education, 18*(1), 65.

Kyung-Mi, O. (2024). A Comparative Study of AI-Human-Made and Human-Made Test Forms for a University TESOL Theory Course. *Language Testing in Asia, 14*, 19.

Labadze, L., Grigolia, M., & Machaidze, L. (2023). Role of AI chatbots in education: Systematic literature review. *International Journal of Educational Technology in Higher Education, 20*(1), 56.

Laupichler, M. C., Rother, J. F., Kadow, I. C. G., Ahmadi, S., & Raupach, T. (2024). Large language models in medical education: Comparing ChatGPT-to human-generated exam questions. *Academic Medicine, 99*(5), 508–512.

Lee, D., & Yeo, S. (2022). Developing an AI-based chatbot for practicing responsive teaching in mathematics. *Computers & Education, 191*, 104646.

Lee, H. Y., Chen, P. H., Wang, W. S., Huang, Y. M., & Wu, T. T. (2024). Empowering ChatGPT with guidance mechanism in blended learning: Effect of self-regulated learning, higher-order thinking skills, and knowledge construction. *International Journal of Educational Technology in Higher Education, 21*(1), 16.

Li, C., Xing, W., & Leite, W. (2022). Building socially responsible conversational agents using big data to support online learning: A case with Algebra Nation. *British Journal of Educational Technology, 53*(4), 776–803.

Li, L., Ma, Z., Fan, L., Lee, S., Yu, H., & Hemphill, L. (2023). ChatGPT in education: A discourse analysis of worries and concerns on social media. *Education and Information Technologies*, 1–34.

Li, T., Ji, Y., & Zhan, Z. (2024). Expert or machine? Comparing the effect of pairing student teacher with in-service teacher and ChatGPT on their critical thinking, learning performance, and cognitive load in an integrated-STEM course. *Asia Pacific Journal of Education, 44*(1), 45–60.

Liang, H. Y., Hwang, G. J., Hsu, T. Y., & Yeh, J. Y. (2024). Effect of an AI-based chatbot on students' learning performance in alternate reality game-based museum learning. *British Journal of Educational Technology.*

Lin, Y. T., & Ye, J. H. (2023). Development of an educational chatbot system for enhancing students' biology learning performance. *Journal of Internet Technology, 24*(2), 275–281.

Liu, L., Subbareddy, R., & Raghavendra, C. G. (2022). AI intelligence Chatbot to improve students learning in the higher education platform. *Journal of Interconnection Networks, 22*(Supp02), 2143032.

López-Robles, J. R., Cobo, M. J., Gutiérrez-Salcedo, M., Martínez-Sánchez, M. A., Gamboa-Rosales, N. K., & Herrera-Viedma, E. (2021). 30th Anniversary of Applied Intelligence: A combination of bibliometrics and thematic analysis using SciMAT. Applied Intelligence, 1–22.

Lu, Y. (2019). Artificial intelligence: A survey on evolution, models, applications and future trends. *Journal of Management Analytics, 6*(1), 1–29.

Mac Fadden, I., Santana, M., Vázquez-Cano, E., & López-Meneses, E. (2021). A science mapping analysis of 'marginality, stigmatization and social cohesion'in WoS (1963–2019). *Quality & Quantity, 55*(1), 275–293.

Martínez, M. A., Cegarra-Navarro, J.-G., Garcia-Perez, A., & Vicentini, F. (2021). Extending structural capital through pro-environmental behaviour intention capital: An outlook on Spanish hotel industry. *Journal of Intellectual Capital, 22*(3), 633–652.

Martinez, M. A., Cobo, M. J., Herrera, M., & Herrera-Viedma, E. (2015). Analyzing the scientific evolution of social work using science mapping. *Research on Social Work Practice, 25*(2), 257–277.

Miao, J., Thongprayoon, C., Suppadungsuk, S., Garcia Valencia, O. A., Qureshi, F., & Cheungpasitporn, W. (2024). Ethical dilemmas in using AI for academic writing and an example framework for peer review in nephrology academia: A narrative review. *Clinics and Practice, 14*(1), 89–105.

Misiejuk, K., Kaliisa, R., & Scianna, J. (2024). Augmenting assessment with AI coding of online student discourse: A question of reliability. *Computers and Education: Artificial Intelligence, 6*, 100216.

Montero-Díaz, J., Cobo, M. J., Gutiérrez-Salcedo, M., Segado-Boj, F., & Herrera-Viedma, E. (2018). Mapeo científico de la Categoría «Comunicación» en WoS (1980–2013). *Comunicar, 26*(55), 81–91.

Murgado-Armenteros, E. M., Gutiérrez-Salcedo, M., Torres-Ruiz, F. J., & Cobo, M. J. (2015). Analysing the conceptual evolution of qualitative marketing research through science mapping analysis. *Scientometrics, 102*(1), 519–557.

Naser, M. Z., AlOgla, S., Anand, N., Zhou, H., & Zhang, G. (2024). Conversing with AI chatbots: examining what OpenAI ChatGPT-4, Microsoft Bing Chatbot, and google bard know, think they know, do not know, and would like to know about engineering. Journal of Umm Al-Qura University for Engineering and Architecture, 1–14.

Neo, M., Lee, C. P., Tan, H. Y. J., Neo, T. K., Tan, Y. X., Mahendru, N., & Ismat, Z. (2022). Enhancing students' online learning experiences with artificial intelligence (AI): The MERLIN project. *International Journal of Technology, 13*(5), 1023–1034.

Ng, D. T. K., Tan, C. W., & Leung, J. K. L. (2024). Empowering student self-regulated learning and science education through ChatGPT: A pioneering pilot study. *British Journal of Educational Technology.*

Nguyen, H. (2023). Role design considerations of conversational agents to facilitate discussion and systems thinking. *Computers & Education, 192*, 104661.

Nguyen, Q. T. (2024). Artificial Intelligence (AI) in education: A case study on ChatGPT's influence on student learning behaviors. *Educational Process: International Journal, 13*(2), 105–121. https://doi.org/10.22521/edupij.2024.132.7

Nizhenkovska, I. V., Reva, T. D., Chkhalo, O. M., But, I. O., & Manchenko, O. V. (2022). Best Practices for Teaching Chemistry Disciplines to Graduates Majoring in Pharmacy during the COVID-19 Restrictions: A Systematic Review. *International Journal of Educational Methodology, 8*(4), 769–781.

Page, M. J., McKenzie, J. E., Bossuyt, P. M., Boutron, I., Hoffmann, T. C., Mulrow, C. D., ... & Moher, D. (2021). The PRISMA 2020 statement: an updated guideline for reporting systematic reviews. BMJ: Systematic Reviews, 10, 89–100. https://doi.org/10.1186/s13643-021-01626-4

Pérez, J. Q., Daradoumis, T., & Puig, J. M. M. (2020). Rediscovering the use of chatbots in education: A systematic literature review. *Computer Applications in Engineering Education, 28*(6), 1549–1565.

Prachnakorn, N., Preecha, K., Sri-U-Thai, T., Jaroenyod, T., Sawang, K., Patwong, N., & Wattanapisit, A. (2024). Incorporating artificial intelligence into a workshop on scientific and scholarly report writing for preclinical medical students. *Medical Teacher, 46*(6), 746–748.

Pradhan, P. (2016). Science Mapping and Visualization Tools used in Bibliometric & Scientometric Studies: An Overview. *INFLIBNET Newsl. Artic., 23*(4), 19–33.

Ranoliya, B. R., Raghuwanshi, N., & Singh, S. (2017, September). Chatbot for university related FAQs. In 2017 International Conference on Advances in Computing, Communications and Informatics (ICACCI) (pp. 1525–1530). IEEE.

Rengers, T. A., Thiels, C. A., & Salehinejad, H. (2024). Academic Surgery in the Era of Large Language Models: A Review. *JAMA Surgery, 159*(4), 445–450.

Roca, M. D. L., Chan, M. M., Garcia-Cabot, A., Garcia-Lopez, E., & Amado-Salvatierra, H. (2024). The impact of a chatbot working as an assistant in a course for supporting student learning and engagement. *Computer Applications in Engineering Education*, e22750.

Samonte, M. J., Acuña, G. E. O., Alvarez, L. A. Z., & Miraflores, J. M. (2023). A Personality-Based Virtual Tutor for Adaptive Online Learning System. *International Journal of Information and Education Technology, 13*(6), 899–905.

Sanz Ausin, M., Maniktala, M., Barnes, T., & Chi, M. (2023). The Impact of Batch Deep Reinforcement Learning on Student Performance: A Simple Act of Explanation Can Go A Long Way. *International Journal of Artificial Intelligence in Education, 33*(4), 1031–1056.

Schumaker, R. P., Ginsburg, M., Chen, H., & Liu, Y. (2007). An evaluation of the chat and knowledge delivery components of a low-level dialog system: The AZ-ALICE experiment. *Decision Support Systems, 42*(4), 2236–2246.

Schumaker, R. P., Liu, Y., Ginsburg, M., & Chen, H. (2006). Evaluating mass knowledge acquisition using the alice chatterbot: The az-alice dialog system. *International Journal of Human-Computer Studies, 64*(11), 1132–1140.

Shukla, R., Mishra, A. K., Banerjee, N., & Verma, A. (2024). The Comparison of ChatGPT 3.5, Microsoft Bing, and Google Gemini for Diagnosing Cases of Neuro-Ophthalmology. Cureus, 16(4).

Smutny, P., & Schreiberova, P. (2020). Chatbots for learning: A review of educational chatbots for the Facebook Messenger. *Computers & Education, 151*, 103862.

Sott, M. K., Bender, M. S., Furstenau, L. B., Machado, L. M., Cobo, M. J., & Bragazzi, N. L. (2020). 100 years of scientific evolution of work and organizational psychology: A bibliometric network analysis from 1919 to 2019. *Frontiers in Psychology, 11*, 598676. https://doi.org/10.3389/fpsyg.2020.598676

Sridharan, K., & Sequeira, R. P. (2024). Artificial intelligence and medical education: Application in classroom instruction and student assessment using a pharmacology & therapeutics case study. *BMC Medical Education, 24*(1), 431.

Strong, E., DiGiammarino, A., Weng, Y., Kumar, A., Hosamani, P., Hom, J., & Chen, J. H. (2023). Chatbot vs medical student performance on free-response clinical reasoning examinations. *JAMA Internal Medicine, 183*(9), 1028–1030.

Tärning, B., & Silvervarg, A. (2019). "I didn't understand, I´m really not very smart"—how design of a digital tutee's self-efficacy affects conversation and student behavior in a digital math game. *Education Sciences, 9*(3), 197.

Tian, X., Risha, Z., Ahmed, I., Lekshmi Narayanan, A. B., & Biehl, J. (2021). Let's talk it out: A chatbot for effective study habit behavioral change. *Proceedings of the ACM on Human-Computer Interaction, 5*(CSCW1), 1–32.

Tossell, C. C., Tenhundfeld, N. L., Momen, A., Cooley, K., & de Visser, E. J. (2024). Student perceptions of ChatGPT use in a college essay assignment: Implications for learning, grading, and trust in artificial intelligence. *IEEE Transactions on Learning Technologies*.

Tülübaş, T., Demirkol, M., Ozdemir, T. Y., Polat, H., Karakose, T., & Yirci, R. (2023). An interview with ChatGPT on emergency remote teaching: A comparative analysis based on human–AI collaboration. *Educational Process: International Journal, 12*(2), 93–110.

Waghray, N. J., Rajasundaram, A., Johnson, W. M. S., & Waghray, A. (2023). Chatbot in anatomy learning and recapitulation: Prototype and review. *Medical Journal Armed Forces India*.

Wang, H., Fu, T., Du, Y., Gao, W., Huang, K., Liu, Z., & Zitnik, M. (2023). Scientific discovery in the age of artificial intelligence. *Nature, 620*(7972), 47–60.

Wang, J., Hwang, G. H., & Chang, C. Y. (2021). Directions of the 100 most cited chatbot-related human behavior research: A review of academic publications. *Computers and Education: Artificial Intelligence, 2*, 100023.

Watts, F. M., Dood, A. J., Shultz, G. V., & Rodriguez, J. M. G. (2023). Comparing student and generative artificial intelligence chatbot responses to organic chemistry writing-to-learn assignments. *Journal of Chemical Education, 100*(10), 3806–3817.

Winkler, R., Hobert, S., Salovaara, A., Söllner, M., & Leimeister, J. M. (2020, April). Sara, the lecturer: Improving learning in online education with a scaffolding-based conversational agent. In Proceedings of the 2020 CHI conference on human factors in computing systems (pp. 1–14).

Wu, R., & Yu, Z. (2024). Do AI chatbots improve students learning outcomes? Evidence from a meta-analysis. *British Journal of Educational Technology, 55*(1), 10–33.

Xue, H., & Niu, Y. (2023). Exercise Generation and Student Cognitive Ability Research Based on ChatGPT and Rasch Model. *IEEE Access.*

Yesmin, S. (2023). Redesigning tertiary educational evaluation with AI: A task-based analysis of LIS students' assessment on written tests and utilizing ChatGPT at NSTU. *Science & Technology Libraries*, 1–13.

Yin, J., Goh, T. T., Yang, B., & Hu, Y. (2024). Using a chatbot to provide formative feedback: A longitudinal study of intrinsic motivation, cognitive load, and learning performance. *IEEE Transactions on Learning Technologies.*

Yin, J., Goh, T. T., Yang, B., & Xiaobin, Y. (2021). Conversation technology with micro-learning: The impact of chatbot-based learning on students' learning motivation and performance. *Journal of Educational Computing Research, 59*(1), 154–177.

Zhang, Y., Viriyavejakul, C., & Sumettikoon, P. (2023). Integrating Chatbots in Educational Administration for Improved Language Learning Outcomes. *Eurasian Journal of Educational Research, 104*(104), 142–163.

Zirar, A. (2023). Exploring the impact of language models, such as ChatGPT, on student learning and assessment. *Review of Education, 11*(3), e3433.

The Integration of Entrepreneurship-Based Robotics into Classrooms

Meryem Meral and Sema Altun Yalçın

Abstract This chapter aims to demonstrate how educational robotics can be integrated with entrepreneurship within classroom settings. In this context, implementation steps that encompass the integration of entrepreneurship and educational robotics have been developed. The formulation of these steps is based on the engineering design process and the sub-dimensions of entrepreneurship. Additionally, each step has been supported by literature reviews. "Fischer Technik" LEGO set and the "Robo Pro" interface have been incorporated into the design of this educational robotics framework. Due to its user-friendliness and lack of complex programming languages, the LEGO set was chosen for younger age groups.

Keywords Educational robotics · Entrepreneurship · Robotics coding

1 Introduction

In the rapidly developing 21st-century world, the role of technology has become unavoidable. The changes within this system, where the importance of technology is increasingly growing, and the adaptations developed in response to this new order have made their mark under the slogan "digital revolution" (Cunha et al., 2020). The development of technology has impacted every area of life and social ecosystems (Al-Adwan & Smedley, 2012). Education is undeniably one of the fields greatly transformed and influenced by technological advancements (Prahani et al., 2022). Earle (2002) mentioned that there would be an increasing demand for integrating technological developments into education and teaching in K-12 schools. This proposition, expressed at that time, has been validated today. The process of conducting educational activities is no longer only about the teacher monotonously transferring information. Still, it has become a process frequently involving various technological devices and resources (Biletska et al., 2021). The incorporation of technology into

M. Meral (✉) · S. A. Yalçın
Department of Science Education, Erzincan Binali Yıldırım University, Erzincan, Turkey
e-mail: meryemmeral96@gmail.com

G. Lampropoulos and S. Papadakis (eds.), *Social Robots in Education*, Studies in Computational Intelligence 1194, https://doi.org/10.1007/978-3-031-82915-4_19

education has rejuvenated traditional methods and definitions, giving teachers and students new insights into teaching and learning.

Numerous studies have been conducted worldwide to evaluate the positive effects of technology on learning and to investigate the characteristics of the enhanced learning environment provided by technology in the classroom. In short, substantial investments have been made to highlight the role of technology in education, and the results of many related studies have demonstrated the benefits and gains that can be achieved by students, teachers, and administrators (Jhurree, 2005). Riasati et al. (2012) examined the benefits of integrating technology into education from various perspectives in their study. According to this study, the benefits of using technology in education include the improvement of student's academic skills and the shift from a teacher-centered approach to a learner-centered one. Additionally, education facilitated through technological methods has become more enjoyable and engaging for students. Technology integration has also contributed to developing students' self-assessment skills, increased collaborative learning and communication, and reduced learning anxiety. The impact of technology on education is evident through interactive environments, virtual applications, various platforms, materials, and devices (June et al., 2014).

Studies have shown that the integration of technology into education aims to develop various 21st-century skills in students (Farisi, 2016; Louis, 2012; Robert, 2009). Entrepreneurship is one of these skills, which is categorized under "life and career skills." There are studies suggesting that the integration of technology can foster the development of entrepreneurial skills. Research studies have primarily focused on developing entrepreneurial skills through the STEM approach. It has been demonstrated that integrating the STEM approach with technology has a positive impact on entrepreneurial skills (Deveci, 2019; Deveci & Çepni, 2014; Deveci & Konuş, 2022; Meral & Yalçın, 2022). Detailed information about entrepreneurship skills and entrepreneurship education is provided in the first chapter of this book.

Various techniques and methods are utilized in entrepreneurship education, and Costin et al. (2022) emphasize the importance of quality in entrepreneurship education. Lepuschitz et al. (2018) argue that it would be beneficial to focus on activities that encourage students to engage in project-based work and integrate technological applications to enhance entrepreneurial skills. Furthermore, Nabi et al. (2017) highlight the need for new pedagogical approaches in entrepreneurship education that extend beyond the current offerings. Hence, educational robotics is recommended as a valuable tool for incorporating into entrepreneurship education initiatives. Currently, there is no existing research that integrates entrepreneurship and educational robotics, making this chapter a significant and thought-provoking contribution. The chapter also proposes steps for how entrepreneurship can be incorporated into educational robotics, as well as a guide for implementing the process. This chapter aims to describe the stages and steps involved in implementing entrepreneurship-based robotics in classrooms. To outline the implementation steps, the chapter is structured into several sections. These sections cover the concepts of entrepreneurship, educational robotics, entrepreneurship-based robotics, and the application of entrepreneurship-based robotics in classroom settings. In addition, the final section includes a brief conclusion and recommendations.

2 Entrepreneurship

Entrepreneurship is the ability to effectively utilize existing opportunities and resources by undertaking certain risks (Eisenmann, 2013). The term "entrepreneur" emerged in the French economic world during the seventeenth and eighteenth centuries, referring to someone who undertakes an activity or project (Dees, 2017). Etymologically, the French word "entreprendre" comprises "entre", meaning to enter, and "prendre", meaning to take or undertake.

According to Hindle and Rushworth (2000), entrepreneurship is an effective factor in creating innovative opportunities and managing organizations. Martin and Osberg (2007) have pointed out that entrepreneurship has both positive and negative aspects. They noted that while entrepreneurship contributes to the world through innovative ideas, one disadvantage is the significant time required to develop this ability. Table 1 presents various definitions related to entrepreneurship skills.

Entrepreneurship has evolved over the years, acquiring different meanings. For instance, while in the past it was primarily defined as the ability to make profit, start a business, or advance it, today it is more about identifying and taking risks, seizing and evaluating opportunities, and adapting to the process of living out these opportunities (López-Núñez et al., 2022). Within the literature, entrepreneurial skills encompass risk-taking, time management, innovation, self-confidence, teamwork, effective communication, and leadership qualities.

Table 1 The definitions of entrepreneurship

Definition	Refs.
An individual's ability to transform their ideas into actions requires creativity, innovation, risk-taking, as well as the skills of planning and project management to achieve objectives	MEB (2018)
The skill of taking certain risks to effectively capitalize on existing opportunities and resources	Eisenmann (2013)
The skill of managing an organization by leveraging innovative opportunities	Hindle and Rushworth (2000)
Identifying and assessing opportunities, taking calculated risks, and creating value through innovative activities aimed at solving problems	Cromie (2000)
The process of transforming an individual's specific skills into practice	Silander and Silander (2018)
The skill of creating values, identifying favourable conditions and cultivating a forward-thinking mindset	López-Núñez et al. (2022)

2.1 *Entrepreneurial Personality Traits*

Entrepreneurial individuals are characterized by their willingness to take risks, innovation, ability to work in groups, leadership qualities, and self-confidence. Boone and Kurtz (2013) identified traits such as self-motivation, tolerance for failure, flexibility, and a desire for achievement as key attributes of entrepreneurial individuals. Littunen (2000) found that learning processes and interpersonal relationships play a role in shaping these traits. Additionally, the study indicates that individuals enhance their problem-solving skills when they experience successful beginnings. A number of similar studies have found that successful entrepreneurs possess certain characteristics (Baum & Locke, 2004; Sexton, 2001; Smith, 2013).

Baum and Locke (2004) confirmed their hypothesis in their study by testing the positive impact of relatively high levels of passion, motivation, and self-efficacy on entrepreneurship. Deakova et al. (2010) stated that courage, self-confidence, responsibility, perseverance, determination, proactive approach, and creativity are among the most important personal qualities for an entrepreneur. Entrepreneurs are inclined to manage their businesses using their strong and unique characteristics (Kozubíková et al., 2015; Sidik, 2012). Figure 1 depicts the fundamental traits of entrepreneurial individuals.

Entrepreneurial skills encompass many sub-skills. In Fig. 1, social skills, teamwork, self-confidence, innovative and creative thinking, and leadership are highlighted as the main sub-skills.

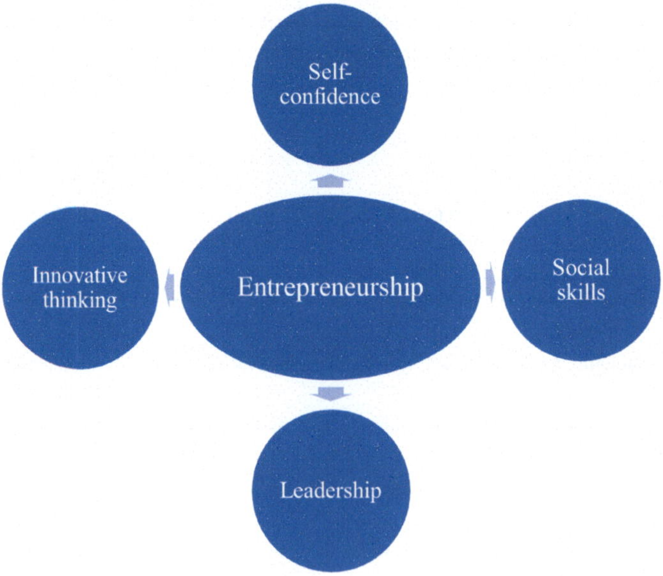

Fig. 1 Schematic representation of characteristics entrepreneurial characteristics (Altun-Yalcin & Meral, 2023)

2.2 Entrepreneurship Education in Schools

The rising number of individuals pursuing entrepreneurship highlights the importance of teaching these skills to students from an early age. Numerous studies, including Ngwoke et al. (2020), emphasize that developing an entrepreneurial mindset from preschool years significantly impacts the acquisition of entrepreneurial skills in both adults and young people. Therefore, preschool education plays a critical role in this aspect. Do Paço and Palinhas (2011) stress the benefits of introducing entrepreneurship concepts to children early on, enabling them to better manage their future careers. They advocate for integrated entrepreneurial programs in preschool education to effectively shape entrepreneurial traits in children. However, preschool entrepreneurship education faces challenges such as overcrowded classrooms, teacher motivation, impractical procedures, and weak communication between parents and schools, reducing the quality of education and hindering the development of entrepreneurial skills in students.

To address these challenges, suggestions include improving government or private programs, enhancing accountability among relevant authorities, implementing written guidelines, incentivizing educators to provide entrepreneurship education, and ensuring consistent support for students' development. Furthermore, it is recommended that parents and educators adopt appropriate approaches to guide students' behaviours towards developing essential skills such as self-control, empathy, and social skills necessary for entrepreneurship in preschool settings. Adequate financial support and curriculum development by authorized personnel for preschool entrepreneurship education is crucial (Ngwoke et al., 2020). Additionally, the incorporation of creative play into school activities is recommended, as it significantly influences the development of entrepreneurial attitudes in children (Sabzeh, 2022).

Kuratko (2005) argues that while entrepreneurship is often considered an inherent trait, it can also be developed through the educational process. Gibb (1993) states that entrepreneurship education encompasses:

- Active engagement in experiential actions,
- Flexibility and the ability to create and adapt to change,
- Trial-and-error, problem-solving approaches,
- Awareness of achieving goals,
- Environments fostering creativity and critical thinking,
- Emphasis on leadership and determination skills.

Numerous studies have examined the entrepreneurial personality traits of students and the initiatives aimed at enhancing their entrepreneurial skills. Yüksel et al. (2015) found that students have fundamental personality traits such as innovation, sensitivity, and opportunism; they also showed sociability, self-confidence, honesty, and a willingness to take risks (Bozkurt & Alparslan, 2013); they also demonstrated problem-solving, creativity, and project-development capabilities. Therefore, offering entrepreneurship education that integrates these skills will significantly contribute to students' social and professional development, effectively preparing them for the business world.

Entrepreneurship education enables students to build self-confidence and develop the skills necessary to start their own businesses. Creating an environment that fosters innovation and creativity in identifying new business opportunities is one of the primary objectives of entrepreneurship education. Additionally, teaching risk management and career-building skills within businesses are also key goals. Entrepreneurship education has proven effective in developing the necessary competencies, technical knowledge, experience, attitudes, resources, and networks for evaluating various entrepreneurial opportunities. Since learning becomes enduring through experience and exposure, incorporating these elements into the education provided ensures the permanence of the related goals. Thus, introducing students to real-life situations and exposing them to learning experiences centered on entrepreneurial themes is a prudent strategy for imparting entrepreneurial skills. Furthermore, students can enhance their decision-making abilities and develop a sense of responsibility. This education also enables them to improve their capabilities in basic planning, idea comprehension, self-assessment, and self-regulation (Bozkurt & Alparslan, 2013).

The enhancement of students' entrepreneurial skills is effectively supported by the practical applications implemented in various subjects, such as Science, Mathematics, Social Studies, and Social Studies. For instance, in Social Studies, students can learn to write their resumes for job applications. In Science, they can identify the harms of bacteria and viruses and learn how to protect against them, subsequently learning to market and sell products like soaps and disinfectants that address these issues. Similarly, students can design and present products aimed at solving everyday problems. In Mathematics, they can determine the costs and profit margins of products and investigate the impact of competition on prices (Ay & Acar, 2016).

In a study conducted by Bacanak (2013), students' entrepreneurial skills were examined from the perspective of teachers. The study found that teachers generally lacked sufficient knowledge about entrepreneurship and employed various approaches and practices to instill entrepreneurial skills in their students. However, it was also found that teachers who shared a common vision regarding student-centered teaching methods and techniques were more effective in enhancing entrepreneurial skills. A common implementation plan and the practitioner's perspective are also important in determining the development of entrepreneurial skills in addition to the strategy and technique.

Numerous studies have been conducted on factors positively influencing students' entrepreneurial skills. These factors include living in rural or urban areas, the father's occupation, gender (Çelik & Karababa, 2018), academic achievement (Ghasemi et al., 2011), family income level (Pan & Akay, 2015), and the family's level of entrepreneurship (Lindquist et al., 2015). One of the most critical environmental factors affecting entrepreneurship is a supportive attitude. If an individual lacks support from family, friends, or a social environment, the development of entrepreneurial skills will be adversely affected. Consequently, individuals living in entrepreneurial regions or communities are more likely to become entrepreneurs in the future (Ibrahim & Luck, 2014). The quality of entrepreneurship education plays a vital role in shaping students' entrepreneurial skills, as highlighted by Almahry et al. (2018). Besides entrepreneurship education and other factors, certain teaching methods and techniques used in classes effectively enhance students' entrepreneurial skills (Bektaş & Vurgun, 2019). These methods and techniques include project-based learning, problem-solving, problem-based learning, open-ended experiments, and fishbone diagrams (Deveci & Aydız, 2021).

Furthermore, Avcı et al. (2022) conducted a study to determine the impact of differentiated instruction practices on the entrepreneurial skills of 6th-grade students. The instruction was differentiated based on students' readiness, interests, and learning styles, using techniques such as stations, agendas, centers, entry points, and tiered instruction. The content, process, and product elements were adapted accordingly. The study concluded that differentiated instruction techniques were more effective than the current curriculum practices in developing entrepreneurial skills, enhancing sub-skills, and increasing academic achievement. Kissi et al. (2020) discussed strategies to enhance students' entrepreneurial skills. These strategies include developing learner-centered education, promoting problem- and activity-based learning, creating a classroom environment that fosters intellectual competence, and encouraging educators to enhance entrepreneurial skills. A lack of motivation and difficulties in adopting these strategies can hinder educators from effectively delivering entrepreneurship education. Additionally, students demonstrate an inclination toward entrepreneurship when they actively participate in projects, take on responsibilities (Shahiwala, 2017), learn to control their behaviour (Su et al., 2021), develop business plans, believe in their ability to succeed, view the entrepreneurial spirit as attractive, see it as beneficial to society, and perceive it as advantageous (Vuong et al., 2020). Furthermore, students who are willing to face challenges, take certain risks, are inclined toward innovative career fields, feel supported by their social environment, do not face financial difficulties, and have future aspirations are considered to be good entrepreneurial candidates (Vuong et al., 2020).

Akdemir and Yavuz (2018) examined the relationship between high school students' critical thinking dispositions and their perceptions of individual entrepreneurship. These two variables showed a low positive correlation, indicating that critical thinking disposition can affect perceptions of entrepreneurship. Wijaya and Puspitowati (2022) identified factors influencing university students' entrepreneurial intentions. According to their study, personal qualities, attitudes, subjective norms, and self-efficacy affect entrepreneurial intentions, whereas family environment does not significantly influence these intentions. The study's findings highlight that while there are numerous factors influencing entrepreneurship in individuals, consensus among studies regarding which factors are decisive in influencing entrepreneurship remains elusive.

Gedeon (2017) analyzed the objectives of entrepreneurship education in terms of its impact on students' transformation and development under three main categories:

1. Core objectives: These are central objectives related to student transformation, encompassing learning outcomes in knowledge, skills, and attitudes.

2. Input objectives: These objectives encompass factors that support student transformation, such as faculty qualifications, resources, opportunities, assignments, courses, and pedagogical approaches.
3. Output objectives: These objectives pertain to the program's success or external impacts, such as the number of students involved, awards received, or new companies established.

These categorizations provide a framework for understanding the multifaceted goals of entrepreneurship education across various educational levels, from preschool to university students and adults participating in entrepreneurship programs.

Studies on entrepreneurship education indicate that the curriculum content and teaching methods vary depending on the program's objectives. These objectives include theoretical courses aimed at increasing entrepreneurial awareness, as well as practical courses designed to produce graduates ready to start businesses (Sirelkhatim & Gangi, 2015). Entrepreneurship education is influenced by qualitative cultural factors, implementation of content, and individual factors. It has been noted by Loi et al. (2022) that entrepreneurship education is deeply influenced by students' cultural values and that these values heavily influence entrepreneurship education's quality. The study also emphasizes the difficulty of transforming theoretical knowledge of entrepreneurship into practical application, which educators encounter when determining how to implement and teach these practices effectively. In this context, recommendations in this field are crucial. As outlined in recommendations for the implementation and content of entrepreneurship education, it is proposed that a focus on reducing consumption should be adopted to curtail or redistribute the accumulation of wealth among private sector entrepreneurs. Furthermore, it is suggested that a thorough and cautious examination of innovation should be exercised, while concurrently safeguarding traditional societal values and practices (Loi et al., 2022). These recommendations provide important insights into the application and content of entrepreneurship education.

Wang et al. (2022) has proposed recommendations for entrepreneurship education. According to these recommendations, university talent training in entrepreneurship should be based on social reality and student practice. Adopting relevant teaching methods is crucial to providing conducive conditions for significantly enhancing university students' innovative and entrepreneurial abilities. Furthermore, stakeholders should focus on optimizing pathways to develop these skills, particularly ensuring that university students acquire innovative, entrepreneurial ideas, and cyber-thinking skills effectively. Universities can reduce existing social employment pressures by educating students in innovative and entrepreneurial skills. Additionally, it can assist universities in enhancing students' entrepreneurial awareness, adapting to innovation, and restructuring entrepreneurship education curricula. Addressing these aspects can facilitate easier solutions to social employment issues and promote societal economic development. Moreover, entrepreneurship education can broaden students' learning horizons, provide richer learning resources, and deepen their understanding of the roles of innovation and entrepreneurship in contemporary development.

Dao and City (2020) emphasized the importance of a general framework program in entrepreneurship education, highlighting that a framework provided by the ministry would guide schools and teachers in implementing the program. This guidance would help them follow consistent procedures and avoid discrepancies through clear instructions. They also stated that the strategy must be long enough to encompass all stages of an individual's development to align the teaching plans of primary, secondary, and high school administrators. A cooperation mechanism allowing primary school students to work with polytechnic students was proposed, based on the belief that imaginative children who are not afraid of failure often generate creative ideas later realized and implemented by university students.

2.3 Approaches and Methods in Entrepreneurship Education

- **Entrepreneurship Games**: These games involve teamwork, presentation skills, model building, and time management abilities (Efeoğlu et al., 2016).
- **Project-Based Learning**: Incorporates problem-solving, experiential learning, and the fishbone diagram (Deveci & Aydız, 2021).
- **Differentiated Instruction**: This method provides students with training that includes techniques such as station teaching, journaling, the think-pair-share strategy, and open-ended projects (Avcı et al., 2022).
- **Problem-Based Learning**: It focuses on solving real-world problems (Vaidya, 2007).
- **Group Work**: Group work encourages collaboration and shared learning experiences.
- **Learning by Doing**: This method engages students in hands-on activities (Ruskovaara et al., 2010).
- **Learning Journals**: Students reflect on their learning experiences (Seikkula-Leino, 2011).
- **Brainstorming**: Brainstorming generates creative ideas and solutions.
- **Discovery Learning Strategy**: This strategy encourages students to find out information for themselves.
- **Discussion**: It promotes critical thinking and exchange of ideas.
- **Six Thinking Hats**: A method that encourages parallel thinking and exploring different perspectives.
- **Drama**: It uses role-play to simulate real-life entrepreneurial situations.
- **5E Model**: 5E model engages students through five phases: Engage, Explore, Explain, Elaborate, and Evaluate (Çelik et al., 2015).
- **Case Studies**: I case studies, learners analyze real or hypothetical business scenarios (Pfeifer et al., 2008).

Kuratko (2005) suggests several steps for imparting entrepreneurial skills, including emphasizing the importance of entrepreneurship, discussing the concept of risk-taking, exploring entrepreneurial personality traits, helping individuals identify their strengths and weaknesses, guiding them to develop unique plans, and using various forms of assessment (e.g., observation, interviews) to evaluate individuals. According to Kuratko, entrepreneurship education includes the following methods and approaches:

- Entrepreneurship Projects
- Project-Based Learning
- Group Discussions and Brainstorming
- Problem-Based Learning
- Drama
- Question and Answer Sessions.

3 Educational Robotics

The growing importance of technology integration in education has led to a greater popularity of robotics coding education (Altun-Yalçın et al., 2020). Consequently, it is observed that STEM has been integrated with robotics coding education in recent years. This advancement in the current technological era has caused robotics to start becoming widespread at almost all educational levels (Seckin Kapucu, 2023). Several dissemination activities related to robotics coding are undertaken through courses and training provided to teachers and students under the umbrella of "robotics coding" (Çınar, 2020; Filippov et al., 2017).

3.1 Definition of Educational Robotics

Robotics encompasses a broad field involving mechanical materials, motors, sensors, and programming concepts (Rogers et al., 2010). Coding can be defined as a series of commands written to perform a specific task or achieve a particular goal in computers, electronic circuits, or mechanical systems (Guven et al., 2022). Robotics education can be described as integrating robotics and related subjects into a curriculum that students benefit from. It can generally be expressed as the integration of robotics and its entire background into a specific curriculum (Patiño-Escarcina et al., 2021). Students can use text-based or block-based programs for coding. Text-based coding is a method where students create codes and commands as text using a computer keyboard, following procedural guidelines. In block-based coding, students use a drag-and-drop technique to assemble code blocks like a puzzle without writing any text (Güven et al., 2020).

Educational robotics refers to the use of robots as a tool for teaching and learning in academic environments. These applications integrate robot technology and concepts into the curriculum and educational activities to enhance students' learning experiences (Mikropoulos & Bellou, 2013). Educational robotics offers various alternatives, including coding, computer programming, Lego-based designs, and non-coding applications (Karypi, 2018). In other words, educational robots can range from simple programmable toys for young children to more advanced robots that allow for programming and customization (Hamilton et al., 2020). They can be used in various subjects such as mathematics, science, engineering, and technology, as well as in extracurricular activities like robot competitions (Ospennikova et al., 2015).

The origins of educational robotics date back to the 1960s and 1970s. During this period, the role of robotics in education began to emerge through the use of the Logo programming language aimed at young students. This educational plan laid the foundation for what is now known as "Educational Robotics" and has become a significant component of school curricula worldwide (Patiño-Escarcina et al., 2021).

3.2 Objectives of Educational Robotics

Educational robotics aims to provide a learning experience that supports the development of various skills. Some of these experiences and skills include:

- **Promoting STEM Education**: Students can gain knowledge and skills in science, technology, engineering, and mathematics (Arís & Orcos, 2019).
- **Enhancing Problem-Solving Skills**: Presenting students with real-life problems that require critical and creative thinking, thereby improving their problem-solving abilities through educational robotics applications (Adams et al., 2010).
- **Encouraging Creativity and Innovation**: Motivating students to design and build robots that can perform various tasks, thus fostering creativity and innovation (Çakır et al., 2021).
- **Fostering Collaboration and Teamwork**: Helping students develop communication, leadership, and social skills by encouraging collaboration and teamwork. Working together to complete projects and exchange ideas supports teamwork and leadership abilities (Ioannou & Makridou, 2018).
- **Focus on Technical Skills**: Developing students' technical skills in areas such as programming, circuit design, and mechanical engineering to prepare them for future career goals in STEM fields (You et al., 2021).
- **Building Confidence and Motivation**: Enhancing students' confidence and motivation by enabling them to successfully design and program robots, thereby fostering self-confidence in their abilities (Piedade, 2021).
- **Developing Critical Thinking and Problem-Solving**: Assisting students in developing critical thinking and problem-solving skills to address and solve real-world problems.
- **Teaching Logical Thinking and Creative Solutions**: Equipping students with the ability to think logically and produce creative solutions through programming and designing robots (Erol, 2020).

In addition to its objectives, the benefits of educational robotics are frequently highlighted in research findings. Relevant studies have shown that educational robotics positively impacts students' learning outcomes and skill development in various areas, serving as an effective tool for promoting STEM education and preparing students for careers in STEM fields (Khanlari, 2013). Educational robotics offers numerous contributions and benefits for individuals. It supports the development of cognitive skills such as problem-solving, computational thinking, innovation, creativity, and critical thinking (Sisman & Kucuk, 2019; Tzagkaraki et al., 2021). Students can use their creativity and imagination to design and build robots capable of performing various tasks (Hou et al., 2022). Additionally, students have the opportunity to develop the psychomotor and technical skills necessary for STEM fields through educational robotics (Chang & Chen, 2022).

Educational robotics often involves teamwork, requiring students to use their collaboration and communication skills, thereby contributing to their social skills (Kandlhofer & Steinbauer, 2016). Furthermore, as students gain experience and understanding of technology-related applications, their interest in STEM increases (Tekbıyık et al., 2022). Finally, students' self-efficacy improves as they realize they can successfully engage in robotics applications (Fridberg et al., 2023).

3.3 Tools Used for Integrating Educational Robotics into Classrooms

1. *Educational Robots*: Robots are defined as machines capable of automatically performing a complex series of actions, especially ones that can be programmed by a computer (Ilievski et al., 2011). The integration of these tools into educational settings, such as schools and classrooms, as well as the broader educational domain, has resulted in the emergence of educational robotics (Ben-Ari et al., 2018). Ben-Ari et al. (2018) classifies educational robots into four categories: pre-assembled robots, robotic kits, robotic arms, and software development environments.

Pre-assembled Robots: These robots are relatively affordable, durable, and equipped with various sensors and output components such as lights. The advantage of pre-assembled robots is that they allow for the application of robotic algorithms without requiring additional effort in mechanical design and assembly processes. However, modifications cannot be made to pre-assembled robots, though many examples support extensions using LEGO components (Ben-Ari et al., 2018).

Robotic Arms: Robotic arms are typically programmable manipulators that function similarly to a human arm. The segments of such a manipulator are connected via joints, enabling either rotational or linear displacement. Robotic arms allow for the performance of various tasks, including lifting, transporting, placing, assembling materials, and performing precise manipulations. They are generally equipped with motors, sensors, and control systems, offering a wide range of motion to accomplish diverse tasks (Yusoff et al., 2012).

LEGO-Based Robotics Kits: One of the most common examples is LEGO Mindstorms. These kits include standard LEGO blocks and other construction components, as well as motors, sensors, and programmable blocks that control the components. The advantage of robotic kits is their flexibility (Ben-Ari et al., 2018).

Software Development Environments: Each educational robotics system comes with a software development environment. The programming language is typically a version of a standard language such as Java or Python. The process of programming is often simplified when using block-based languages such as Scratch or Blockly (Ben-Ari et al., 2018).

2. *Robotic Coding Platforms*: Coding can be defined as a complete or partial set of commands directed to a computer, electronic circuits, or mechanical systems. In text-based coding, commands and instructions are created by students in text form using a computer keyboard. In block-based coding, students combine blocks by drag-and-drop techniques to code without writing text (Guven & Kozcu Cakir, 2020). Examples of robotic coding platforms include Arduino, Scratch, Tinkercad, and Lego Mindstorms.

Arduino: Arduino is an open-source microcontroller electronic prototyping platform that is flexible and easy to use, based on hardware and software. The microcontroller on the board is programmed using the Arduino programming language and the Arduino development environment. Hardware reference designs are provided under an open-source license (Balogh, 2010).

Scratch: Scratch represents a visual programming language and online coding environment developed under the auspices of the MIT Media Lab. Its primary goal is to make the process of acquiring coding skills more comprehensible and interactive, especially for beginners and young learners. In this context, users can synthesize programs by dragging and assembling code blocks like puzzle pieces. This methodological approach significantly simplifies complex coding concepts, increasing comprehensibility. Using Scratch, games, animations, and interactive stories can be created (Maloney et al., 2010).

Tinkercad: Tinkercad is one of the online 3D design platforms that allows users, especially beginners, to create and edit digital prototypes of various objects. The platform enables users to easily create complex designs using the drag-and-drop method (Eryılmaz & Deniz, 2021).

Lego Mindstorms: This Lego set, which allows even young children to build robots, includes a microprocessor that can be controlled by a computer, sensors, and an interface called "EV3" Lab. Since its launch in 1998, it has been among the most purchased and popular Lego sets (Üçgül, 2013).

3. *Project development*: Students can design a robot using their creativity and problem-solving skills, and this design provides them with the opportunity

to effectively utilize their existing abilities. By creating and programming autonomous robots using Project-Based Learning, students gain skills in solving problems collaboratively and developing practical solutions (Lui & Hsiao, 2002).

4. *Competitions and Activities*: Robotics competitions have been held for many years and have gained immense popularity. At MIT (Massachusetts Institute of Technology), a robot design and construction competition for mechanical engineering students was initiated in the 1970s by Professor Flowers and has since become a highly anticipated university tradition each year (Mallinson, 2015). Robotic competitions are now a popular extracurricular activity for school-age children from kindergarten to high school (Sullivan & Bers, 2019).
5. *STEM Education Programs*: Many schools and educational institutions organize robotics courses and workshops within the scope of STEM education. Through these programs, students gain practical experiences in science, technology, engineering, and mathematics and have the opportunity to develop their STEM skills. Ortiz et al. (2015) organized STEM education programs integrated with robotics training for pre-service and in-service teachers.
6. *Virtual Robotics Education*: Some educational programs offer training on virtual robots without the need for physical or tangible robots. Virtual robotics education aims to develop students' robotics skills by providing opportunities for low-cost and accessible learning. One of the platforms in this scope is the e-laboratory (Jara et al., 2010). E-laboratory platforms allow students to practice programming a robot manipulator using the functions and programming methods that are common to real robotic systems.

3.4 Approaches to Integrating Educational Robotics into Classrooms

Many approaches are preferred in integrating educational robotics into courses. The main preferred approaches are listed below:

Exploratory Learning Approach: The practitioner does not provide direct guidance, and students are free in the process. Students must discover how objects work and how the process functions on their own. The practitioner may use the Socratic method, thus attempting to guide students with their answers. This process is time-consuming, and students often face difficulties in moving the robots (Altin & Pedaste, 2013).

Collaborative Learning Approach: This approach can be combined with other educational robotics approaches. The aim is not only to gain knowledge but also to develop dynamic and strategic skills. It is an approach where educational robotics is used as a tool for collaborative learning. In this approach, students work in groups, and tasks are distributed. The practitioner is also involved in this collaboration (Altin & Pedaste, 2013).

Problem-Solving Approach: When addressing intricate problems, students engage in a process of exploring and testing various solutions. Observing the behaviour of the robots allows the student to assess whether the issues have been successfully resolved. If challenges persist, alternative solutions are sought. However, the problem-solving endeavour becomes notably more challenging in the absence of comprehensive knowledge. Conversely, with a strong understanding of the subject matter, students can readily oversee the process. Moreover, this approach is recognized as providing cognitive support for students (Chevalier et al., 2020).

Project-Based Learning Approach: In this approach, students are assigned research-based tasks or problems to investigate, collect data collaboratively, analyze it, and draw conclusions. The findings are then shared with others. Students are divided into groups, and several students are selected for each group. A learning scenario that includes an eight-stage project-based learning process is determined. Task distribution, teamwork, and communication are crucial (Jawaid et al., 2020; Karahoca et al., 2011).

Competition-Based Learning Approach: This approach involves students participating in robotics competitions. Students solve problems using their knowledge. Preparing for competitions motivates them. This approach has shown the effectiveness of robotics and its application in STEM fields. For example, the "Botball" educational robotics program involves stages of research, workshop, design, construction, programming, and competition. This program has revealed that students' engineering skills develop, and they consider pursuing a career in this field (Altin & Pedaste, 2013).

Compulsory Education Scope Approach: It is argued that educational robotics should be provided as part of compulsory education. Teachers effectively integrate robotics into subjects like Mathematics and technology (Altin & Pedaste, 2013).

Inquiry-Based Learning Approach: This approach is particularly preferred in science courses. It is based on the scientific inquiry learning approach. According to this approach, students should plan and observe, evaluate the problem identification process, create a research problem, and analyze and evaluate the results. Programs or software are effective in providing direct feedback to students on this matter. Additionally, since robots provide tangible and visual feedback, they will increase students' interest in inquiry-based learning. A hybrid learning environment consisting of a blend of computer-based and real activities can be created (Altin & Pedaste, 2013).

Scenario-Based Approach: In this approach, students are given different scenarios, and the use of robots is planned for each scenario. The stages of scenario-based robotics are as follows: defining the subject to be taught, defining students' prior knowledge, determining scenario objectives, adding robotic tools as teaching materials, evaluating the scenario with various assessment tools, and reporting the scenario (Lopez-Caudana et al., 2020).

Behaviour-Based Approach: This didactic approach aims to enable inexperienced students to program robots instinctively. Students are taught how to program the commands for robots to display complex behaviours (de Cristoforis et al., 2012).

Constructivist Approach: In this approach, Küçük and Şişman (2017) suggested group or individual work in a non-competitive, panic-free, and stress-free environment. Application categories include collaboration, practice, materials, pedagogy, and feedback (Schina et al., 2021).

Interdisciplinary Approach: This approach integrates educational robotics with science, digitization, art, and mathematics. Robots are used in the creation and resolution of problem states (Manera, 2020). It is argued that the interdisciplinary approach can reduce gender inequality in STEM fields (Gubenko et al., 2021).

4 Entrepreneurship-Based Robotics (EBR)

4.1 Definition of Entrepreneurship-Based Robotics

Entrepreneurship-based robotics is an approach that integrates entrepreneurship skills with robotics. The main aim of this approach is to support individuals' entrepreneurship skills through robotics. It can be termed as offering entrepreneurial instruction to people through the use of robotics. In this framework, tasks with distinct phases have been created. The steps followed in the development of these activities are as follows.

Steps followed in the development of activities:

- Stage 1: Determining the objectives of the activities
- Stage 2: Literature review
- Stage 3: Creating the application steps
- Stage 4: Clarifying the steps with expert opinions
- Stage 5: Developing pilot studies
- Stage 6: Implementing the main application.

Attention has been paid to ensuring that the activities address the sub-dimensions of entrepreneurship. The sub-dimensions identified in this context are self-confidence, perception of innovation and creativity, tendency to leadership and stand out, social skills and group work, and risk-taking.

In the research application process, we sought perspectives and suggestions from three experts in addition to the primary researcher. These experts provided insights based on their extensive experience in robotics education through academic studies and projects. Moreover, the implementation process was developed by considering the entrepreneurship applications outlined in the researcher's previous STEM studies (Meral & Yalçın, 2022, 2023), the entrepreneurship dimensions identified in Özcan's research (Özcan, 2019) the E-STEM (Entrepreneurship-based STEM) steps developed by Deveci (2019), and the collaborative learning-based educational robotics approach described by Altin and Pedaste (2013).

4.2 Steps for Implementing Entrepreneurship-Based Robotics

The application steps are determined based on specific main headings. These headings were developed by the engineering design process (Hynes et al., 2011) and E-STEM (Entrepreneurship-based STEM) (Deveci, 2019), as well as the studies by Meral and Yalçın (2022). Entrepreneurial skills have been integrated with collaboration-based educational robotics in this process. The educational robotics tool 'Fischer Technik' and its control interface program "Robo Pro" were chosen for this implementation. The reason for selecting this set lies in its block-based design, absence of complex language, and suitability for younger age groups. The integration process described herein draws inspiration from pioneering studies in the field, yet it distinguishes itself through its emphasis on collaboration-based educational robotics. This approach enables students to articulate problem states and solutions through coding, empowering each student to assume leadership roles. The general framework for these steps has been defined as follows: investigate the problem, design, build the model, redesign, and present together. Figure 2 illustrates this framework.

The application steps outlined in Fig. 2 are inspired by engineering design process. As part of this cycle, identifying a problem state to be investigated is the first step. The practitioner/teacher presents this problem to the students. For example, they

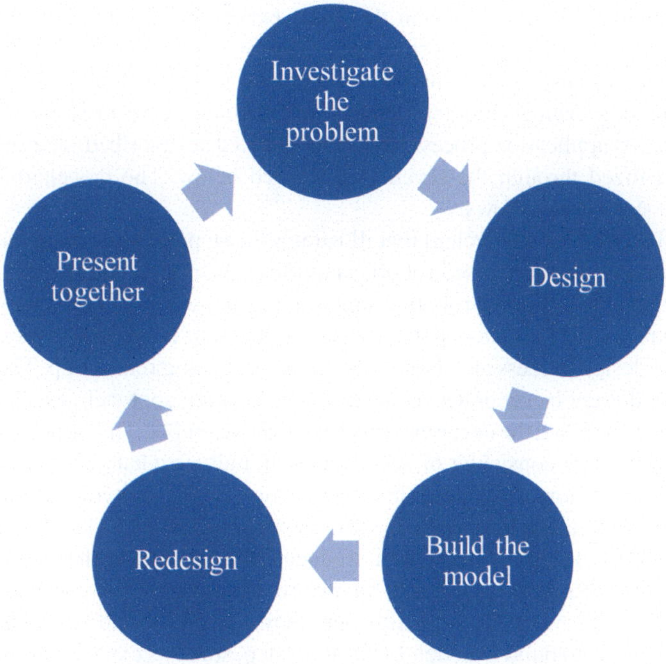

Fig. 2 General framework of application steps

might ask, "How would you design and code a traffic light using Legos?" Students collaboratively research this problem, starting by sketching a prototype design. The next step is to build a model using Legos as a group activity. If the design is correct and the code works, the model will remain as is. If the code does not work, a redesign will be necessary. The entire process is finally shared in the classroom through oral presentations.

As part of developing the activities, we establish a general framework that defines the stages in the framework based on relevant literature. In the first week, students are introduced to the relevant robotic coding set and interface. Following this, students are presented with a general problem state to research for 2–3 days. Students will work in groups of 4–5 people. Each group will have a student selected as a leader. Each week, the leader must be a different student, meaning that each student should take on the leadership role at least once. The leaders are responsible for assigning tasks to other students to complete the group's tasks. Managing the task distribution within a group is also a leader's responsibility.

Each student draws a design on paper as a solution to each problem state. The desired model is then constructed using Lego sets. The leader requests that each student write different codes to operate this model. Each student writes these codes to run the model, and designs for parts of the model that can be modified are discussed and drawn by the group. Finally, within a few weeks (after 6–10 weeks), each student creates an original design, writes the code to operate this design, and presents it to the class.

While students write and design the code, expert trainers guide them without directly intervening. Students mostly make decisions as a group on their own. This approach aims to develop students' entrepreneurial skills, such as self-confidence, problem-solving, critical thinking, leadership, and innovative thinking. The steps followed in this application process have been planned as described. The application steps are realized through the seven stages listed below. The flowchart for these application steps is as follows.

Figure 3 consists of a flowchart that illustrates the step-by-step process for implementing entrepreneurship-based robotic activities. This figure outlines the steps for implementing an educational robotics-supported entrepreneurship training program. It has been previously mentioned that the development of these steps is based on the engineering design process and the sub-dimensions of entrepreneurship. The instructional methodology incorporates a collaborative learning approach, as all activities involve group work and the development of a collective output. This approach follows a structured process consisting of seven steps. Initially, students are introduced to the robotics kit, followed by the formation of groups in the subsequent step. Within these groups, designated leaders are assigned specific weeks to lead the collaborative efforts. In the third step, students are introduced to a problem scenario. During the fourth step, students conduct research on the problem and develop paper schematics of their designs. Moving on to the fifth step, they engage in hands-on construction, using Lego pieces to build a model. In the sixth step, students explore coding as they program the model's operations. Finally, in the last step, they showcase the entire process through a presentation. The details of the stages are explained below under specific headings.

Stage 1: This stage is a requirement in studies related to robotics. Students must have prior knowledge of the relevant robotics set and software. Otherwise, they will be unable to perform the subsequent stages that involve robotic coding skills. Behrens et al. (2009) provided Lego Mindstorms training to enhance students' engineering skills and problem-solving abilities for real-life situations, starting with an introduction to the Lego Mindstorms set.

Stage 2: This phase, which is based on the formation of groups, is present in many application stages. In the studies conducted by Deveci (2019), Hynes et al. (2011), and Meral and Yalçın (2022), students were divided into groups. The unique aspect of this study is that each week, a different student takes on the role of leader.

Stage 3: The phase in which the teacher provides students with a general problem scenario and the students research this problem has also been included in studies

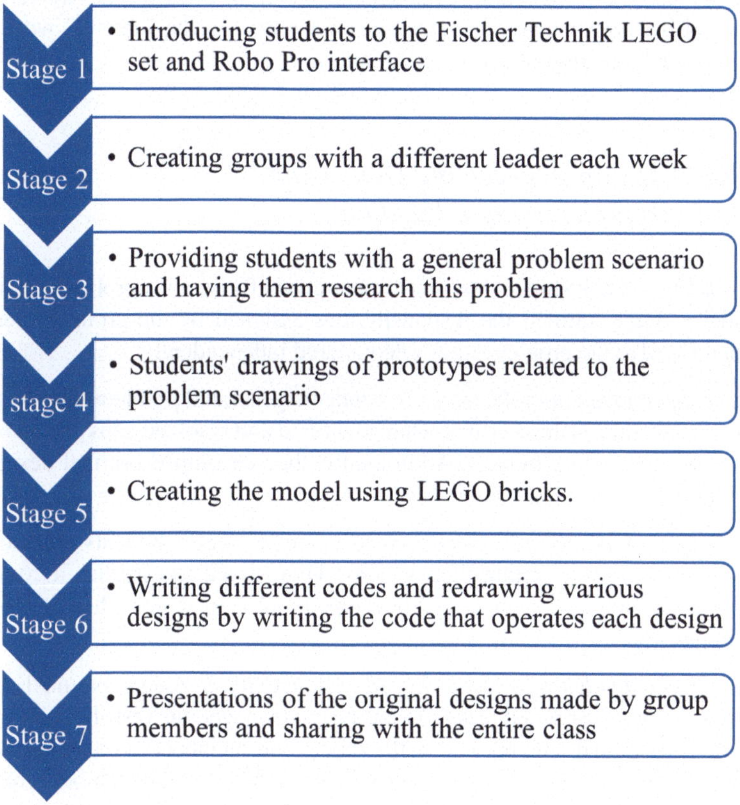

Fig. 3 The flowchart of the application steps

conducted by Deveci (2019), Meral and Yalçın (2022), Meral and Altun-Yalçın (2023), and Hynes et al. (2011). Therefore, these studies have been referenced for this phase.

Stage 4: In the stage where students create prototypes, researchers referred to the studies of Deveci (2019) and E-STEM, as well as Hynes et al. (2011). This stage represents one of the phases in the Engineering Design Process (EDP) cycle.

Stage 5: This study represents a phase in the Engineering Design Process (EDP) focused on model creation. The application models are developed using LEGO components in educational robotics-supported entrepreneurship training.

Stage 6: Developing different designs is a phase within both the E-STEM and the Engineering Design Process (EDP) frameworks. Additionally, the aim has been to cultivate innovation skills through the creation of various codes. This phase, which involves each student sharing their presentations of original designs with the entire class, is common to both the EDP and E-STEM processes. In the E-STEM process, various advertising and marketing strategies are employed during presentations. However, a distinguishing feature of this phase is that the presentations do not focus on financial literacy or marketing strategies but rather on the students' work with different LEGO models and coding techniques.

4.3 Relationship Between the Dimensions of Entrepreneurship Education

Self-Confidence: Each student wrote different codes for various problem scenarios sequentially. They fulfilled the responsibilities assigned by the group leader and attempted to solve the problems they encountered independently.

Perception of Innovation and Creativity: Students generated problem scenarios and wrote codes to address these issues. They identified and resolved gaps and technical problems while creating models. Additionally, they developed original designs in the final phase.

Social Skills and Teamwork Dimension: Over eight weeks, students collaborated and assisted each other within their groups. They aimed to establish a constructive communication environment and developed a plan to achieve their goals for a common purpose.

Leadership and Tendency to Stand out: In each activity, a student assumed a leadership role. Leaders assigned tasks to their peers, managed the distribution of tasks among the group, and collaborated with other group members to create an overall plan for the project. Leaders were responsible for time management, model creation, and coding.

Risk-Taking: Despite objections from peers, students were encouraged to present their ideas within the group. They were motivated not to fear making mistakes. As a result, students were supported in experimenting with solutions even if the outcomes were not successful or did not resolve the problem.

The software environment, known as Robo Pro, which is characterized by its simplicity and user-friendly interface was used for the robotics kit employed by the students. Fischer Technik LEGO sets are controlled by coding using the Robo Pro software environment. It does not involve complex verbal commands or codes but utilizes a block-based coding interface. This environment interfaces with the LEGO models via the "Interface" and TX Controller tools. In the software, components such as lamps, motors, sensors, switches, and buzzers are coded to enable the control of the models.

Figure 4 depicts the robotics kit employed by students in the execution of entrepreneurship-based robotic activities. This particular robotics kit is known as Fischer Technik. "Fischer Technik" LEGO sets are especially useful for offering STEM education to primary school students. This educational kit includes detachable LEGO pieces, as well as components such as lamps, motors, switches, sensors, interfaces, and TX Controllers. Additionally, the kit contains guides with visual instructions on how to assemble the parts, thereby illustrating the stages of model construction (Fig. 2).

5 Conclusions

Educational robotics and entrepreneurship sub-dimensions are integrated in this chapter to propose a program for entrepreneurship education. The steps of this program are also blended with the engineering design process. Additionally, educational robotics training is based on a collaborative learning approach. In this context, students work in groups, strengthening their team spirit, social connections, and communication skills. However, there are certain limitations. For instance, some students may experience disagreements or conflicts within the group. For this reason, facilitators should warn students in advance. Another limitation is that the steps of this program can be very time-consuming. To address this, using Lego sets with difficulty levels appropriate for the students will help avoid time loss and prevent a decrease in student motivation. Lastly, the high costs of the Lego sets used in this program represent another limitation. To mitigate this, it is suggested that universities and various organizations collaborate to reduce these costs.

Fig. 4 Fischer Technik LEGO set

Table 2 The implementation guide

The implementation guide	I am coding a traffic light
Duration	2 h
Outcomes related to robotics coding	• Students will be able to recognize the interface of the Robo Pro software • Students will be able to identify necessary materials such as USB and interfaces • Students will be able to build LEGO components for a traffic light • Students will be able to write codes for turning the light on and off in Robo Pro software • Students will be able to write different codes for various problem scenarios • Students will be able to write loop codes in Robo Pro software
Required materials	• Fischer Technik LEGO set • USB cable • Interface • Connection cables • Computer
Outcomes related to entrepreneurship	• **Self-confidence**: students will be able to support themselves in solving the defined problem • **Perception of innovation and creativity**: students will be able to generate ideas to produce different solutions • **Leadership and tendency to stand out**: students will be able to make decisions within the group and share those decisions with group members • **Leadership and tendency to stand out**: students will be able to present and explain their work and the products they have created • **Social skills and teamwork**: students will be able to work harmoniously within the group and communicate effectively with group members • **Risk-taking**: expressing views and taking risks whether other group members will respect or accept their ideas
Outcomes related to scientific applications	• Students will be able to identify a researchable problem • Students will be able to recognize the importance of creativity in the development of engineering and technology • Students will be able to develop research plans and implement them occasionally in collaboration and sometimes on their own • Students will be able to utilize the relationship between science, technology, engineering, and mathematics (STEM) • Students will be able to design and create a 3D model • Students will be able to apply engineering design and entrepreneurship skills in product creation

(continued)

Table 2 (continued)

Process stages	• **Instructor's guide** **An instructor/a teacher** • Gives students a general problem related to traffic light design and asks them to research the problem for 1 day • Forms student groups consisting of 5 members each • Requests that students generate solution ideas for the problem identified by the leaders of the groups • Asks each group to sketch a common prototype design on paper • Introduces the "Fischer Technik" LEGO set and the "Robo Pro" interface to the students • Requests that the design be built using LEGO bricks and the relevant codes be written (after the Robo Pro interface is introduced, students create their problem scenarios and write their codes, with no intervention during each step) • Chooses a group leader for each activity and then asks each group's leader to share what they did at each step and their unique designs after the activity (presentations can be 10 min long)
Process stages	• **Learner's guide** A learner/a student • Investigates the assigned problem state for 1 day • Defines a problem with specific boundaries (e.g., designing a traffic light that blinks at specific intervals) • Discusses ideas within the 5-person groups to agree on a common design • Draws the design on paper by the group • Creates a model using LEGO bricks and writes the relevant codes • For each activity, the leader chosen for that week shares with the class what was done at each step of the research process and the unique designs after the activity

References

Adams, J., Kaczmarczyk, S., Picton, P., & Demian, P. (2010). Problem solving and creativity in engineering: conclusions of a three year project involving reusable learning objects and robots. *Engineering Education, 5*(2), 4–17. https://doi.org/10.11120/ened.2010.05020004

Akdemir, E., & Yavuz, Ö. (2018). Öğrencilerin eleştirel düşünme becerileri, teknolojiye yönelik eğilimleri ve bireysel girişimcilik algıları arasındaki ilişkinin incelenmesi [Analysis of the relationship among students' critical thinking skills, technology orientation, and individual entrepreneurship perceptions]. *Turkish Studies, 13*(27).

Al-Adwan, A., & Smedley, J. (2012). Implementing e-learning in the Jordanian higher education system: Factors affecting impact. *International Journal of Education and Development using ICT, 8*(1). https://www.researchgate.net/publication/272942607_Implementing_e-learning_in_the_Jordanian_Higher_Education_System_Factors_affecting_impact

Almahry, F. F., Sarea, A. M., & Hamdan, A. M. (2018). A review paper on entrepreneurship education and entrepreneurs' skills. *Journal of Entrepreneurship Education, 21*(1), 1–7.

Altin, H., & Pedaste, M. (2013). Learning approaches to applying robotics in science education. *Journal of Baltic Science Education, 12*(3), 365.

Altun-Yalcin, S. A., Kahraman, S., & Yilmaz, Z. A. (2020). Development and validation of robotic coding attitude scale. *International Journal of Education in Mathematics, Science and Technology, 8*(4), 342–352. https://doi.org/10.46328/ijemst.v8i4.924

Altun-Yalçın ve Meral (2023). STEM eğitimine yolculuk: STEM eğitiminde öğrenme yolları. Nobel Yayıncılık.

Altun-Yalcin, S., & Meral, M. (2023). STEM Eğitimine Yolculuk: STEM Eğitiminde Öğrenme Yolları [*A journey to STEM education: Pathways to learning in STEM education*]. Nobel Academy.

Arís, N., & Orcos, L. (2019). Educational robotics in the stage of secondary education: Empirical study on motivation and STEM skills. *Education Sciences, 9*(2), 73. https://doi.org/10.3390/educsci9020073

Avcı, Ö., Çelik, H., & Bayram, K. (2022). Farklılaştırılmış Öğretim Uygulamalarının Ortaokul Öğrencilerin Elektrik Ünitesindeki Başarısı ve Girişimcilik Becerisi Üzerinde Etkisi [Differentiated instructional practices' effect on secondary school students' academic success and entrepreneurship skills in the electricity unit]. *Fen Matematik Girişimcilik ve Teknoloji Eğitimi Dergisi, 5*(3), 278–297.

Ay, T. S. & Acar, Ş. (2016). Sınıf öğretmenlerinin girişimcilik becerisi kazandırmaya yönelik görüşleri [Views of classroom teachers about the acquisition of entrepreneurial skills]. *Electronic Journal of Social Sciences, 15*(58), 960–976. https://doi.org/10.17755/esosder.258829

Bacanak, A. (2013). Fen ve teknoloji dersinin öğrencilerde girişimcilik becerisinin gelişimine etkisi üzerine öğretmen görüşleri [Opinions of science teacher towards students' creative thinking and entrepreneurship skills]. *Kuram ve Uygulamada Eğitim Bilimleri, 13*(1), 609–629. https://hdl.handle.net/20.500.12450/198

Balogh, R. (2010, September). Educational robotic platform based on arduino. In *Proceedings of the 1st international conference on robotics in education, RiE2010. FEI STU, Slovakia* (pp. 119–122). https://doi.org/10.5281/zenodo.10550044

Baum, J. R., & Locke, E. A. (2004). The relationship of entrepreneurial traits, skill, and motivation to subsequent venture growth. *Journal of Applied Psychology, 89*(4), 587. https://doi.org/10.1037/0021-9010.89.4.587

Behrens, A., Atorf, L., Schwann, R., Neumann, B., Schnitzler, R., Balle, J., Herold, T., Telle, A., Noll, T. G., Hameyer, K., & Aach, T. (2009). MATLAB meets LEGO mindstorms—A freshman introduction course into practical engineering. *IEEE Transactions on Education, 53*(2), 306–317. https://doi.org/10.1109/TE.2009.2017272

Bektaş, O., & Vurgun, F. (2019). Altıncı sınıf öğrencilerinin Fen'e yönelik girişimcilik becerilerinin belirlenmesi [Determination of sixth grade students' science entrepreneurship]. *Fen Matematik Girişimcilik ve Teknoloji Eğitimi Dergisi, 2*(2), 60–78.

Ben-Ari, M., Mondada, F., Ben-Ari, M., & Mondada, F. (2018). Robots and their applications. *Elements of Robotics*, 1–20. https://doi.org/10.1007/978-3-319-62533-1_1

Biletska, I. O., Paladieva, A. F., Avchinnikova, H. D., and Kazak, Y. Y. (2021). The use of modern technologies by foreign language teachers: Developing digital skills. *Linguistics and Culture Review, 5*(S2), 16–27. https://doi.org/10.21744/lingcure.v5nS2.1327

Boone, L. E., & Kurtz, D. L. (2013). *Essentials of contemporary business*. John Wiley & Sons.

Bozkurt, Ö. Ç., & Alparslan, A. M. (2013). Girişimcilerde Bulunması Gereken Özellikler İle Girişimcilik Eğitimi: Girişimci ve Öğrenci Görüşleri, Girişimcilik ve Kalkınma Dergisi [Characteristics, must be included entrepreneurs and entrepreneurship education: Opinions of entrepreneurs and students]. *Journal of Entrepreneurship and Development, 8*(1).

Chang, C. C., & Chen, Y. (2022). Using mastery learning theory to develop task-centered hands-on STEM learning of Arduino-based educational robotics: Psychomotor performance and perception by a convergent parallel mixed method. *Interactive Learning Environments, 30*(9), 1677–1692. https://doi.org/10.1080/10494820.2020.1741400

Chevalier, M., Giang, C., Piatti, A., & Mondada, F. (2020). Fostering computational thinking through educational robotics: A model for creative computational problem solving. *International Journal of STEM Education, 7*(1), 1–18. https://doi.org/10.5753/sbie.2023.235154

Costin, Y., O'Brien, M. P., & Hynes, B. (2022). Entrepreneurial education: Maker or breaker in developing students' entrepreneurial confidence, aptitude and self-efficacy? *Industry and Higher Education, 36*(3), 267–278. https://doi.org/10.1177/09504222211040662

Cromie, S., & Hayes, J. (1988). Towards a typology of female entrepreneurs. *The Sociological Review, 36*(1), 87–113. https://doi.org/10.1111/j.1467-954X.1988.tb02936.x

Cromie, S. (2000). Assessing entrepreneurial inclinations: some approaches and empirical evidence. *European Journal of Work and Organizational Psychology, 9*(1), 7–30.

Cunha, M. N., Chuchu, T., & Maziriri, E. (2020). Threats, challenges, and opportunities for open universities and massive online open courses in the digital revolution. *International Journal of Emerging Technologies in Learning (iJET), 15*(12), 191–204. https://doi.org/10.3991/ijet.v15i12.13435

Çakır, R., Korkmaz, Ö., İdil, Ö., & Erdoğmuş, F. U. (2021). The effect of robotic coding education on preschoolers' problem solving and creative thinking skills. *Thinking Skills and Creativity, 40*, 100812. https://doi.org/10.1016/j.tsc.2021.100812

Çelik, H., Gürpınar, C., Başer, N. ve Erdoğan, S. (2015). Öğrencilerin yaratıcı düşünme ve girişimcilik becerilerine yönelik fen bilgisi öğretmenlerinin görüşleri. *Uluslararası Eğitim Bilimleri Dergisi, 2*(4), 277–307.

Çelik, M., & Karababa, A. G. (2018). Girişimcilik niyetlerinin demografik faktörlere göre incelenmesi [The intentions of entrepreneurship investigation by demographic factor]. *Girişimcilik ve Kalkınma Dergisi, 13*(1), 57–67.

Çınar, S. (2020). Fen Bilimleri Öğretmen Adaylarına Yönelik Eğitsel Robotik Destekli STEM Kursu [Educational robotics supported STEM course for prospective science teachers]. *Electronic Turkish Studies, 15*(7). https://doi.org/10.7827/TurkishStudies.43574

Dao, C. T., & City, C. M. (2020). Suggestions for an extensive implementation plan of early entrepreneurship education in Vietnam. *Review of Integrative Business and Economics Research, 9*, 75–83.

Deakova, K., Drazovska, K. G., Grznárik, G. D., & Kondasova, I. (2010). *Entrepreneurship*. SOPK.

De Cristoforis, P., Pedre, S., Nitsche, M., Fischer, T., Pessacg, F., & Di Pietro, C. (2012). A behavior-based approach for educational robotics activities. *IEEE Transactions on Education, 56*(1), 61–66. https://doi.org/10.1109/TE.2012.2220359

Dees J. G. (2017). The meaning of social entrepreneurship. In J. Hamschmidt, & M. Pirson (Eds.), *Case studies in social entrepreneurship and sustainability* (Vol. 2, pp. 22–30). Routledge.

Deveci, İ. (2019). Girişimci proje (G-FeTeMM) sürecinin fen bilimleri öğretmen adaylarının yaşam becerilerine yansımaları: Nitel bir araştırma [Reflections with regard to entrepreneurial project (E-STEM) process on the life skills of prospective science teachers: A qualitative study]. *Journal of Individual Differences in Education, 1*(1), 14–29.

Deveci, İ., & Aydız, M. (2021). Fen Bilimleri Öğretmenlerinin Öğretim Programında Yer Alan Yaşam Becerilerinin Kazandırılmasına İlişkin Görüşleri [Science teachers' opinions on how to improve life skills given in the science curricula]. *Trakya Eğitim Dergisi, 11*(1), 164–186. https://doi.org/10.24315/tred.693345

Deveci, İ., & Çepni, S. (2014). Fen bilimleri öğretmen eğitiminde girişimcilik [Entrepreneurship in science teacher education]. *Journal of Turkish Science Education, 11*(2), 161–188. 10.36681/

Deveci, İ, & Konuş, F. Z. (2022). The predictive power of Turkish middle school students' entrepreneurial competencies on STEM attitudes. *The Journal of Entrepreneurship, 31*(2), 425–457. https://doi.org/10.1177/09713557221097178

Earle, R. S. (2002). The integration of instructional technology into public education: Promises and challenges. *Educational Technology, 42*(1), 5–13.

Efeoğlu, I. E., Pekkan, N. Ü., Ciritci, S., & Çanşali, T. (2016). Girişimcilik Eğitiminde Yeni Metotlar: Girişimcilik Oyunu Örneği [New methods in entrepreneurship education: A sample entrepreneurship play]. *Organizasyon ve Yönetim Bilimleri Dergisi, 8*(2), 17–26.

Eisenmann, T. R. (2013). Entrepreneurship: A working definition. *Harvard Business Review*, 10. https://doi.org/10.2139/ssrn.4897397

Erol, O. (2020). How do students' attitudes towards programming and self-efficacy in programming change in the robotic programming process? *International Journal of Progressive Education, 16*(4), 13–26. https://doi.org/10.29329/ijpe.2020.268.2

Eryilmaz, S., & Deniz, G. (2021). Effect of Tinkercad on students' computational thinking skills and perceptions: A case of Ankara province. *Turkish Online Journal of Educational Technology-TOJET, 20*(1), 25–38.

Farisi, M. (2016). Developing the 21st-century social studies skills through technology integration. *Turkish online journal of Distance Education, 17*(1), 16–30. https://doi.org/10.17718/tojde.47374

Filippov, S. A., Ten, N. G., & Fradkov, A. L. (2017). Teaching robotics in secondary school: Examples and outcomes. *IFAC-PapersOnLine, 50*(1), 12167–12172. https://doi.org/10.1016/j.ifacol.2017.08.2147

Fridberg, M., Redfors, A., Greca, I. M., & Terceño, E. M. G. (2023). Spanish and Swedish teachers' perspective of teaching STEM and robotics in preschool–results from the botSTEM project. *International Journal of Technology and Design Education, 33*(1), 1–21.

Gedeon, S. A. (2017). Measuring student transformation in entrepreneurship education programs. *Education Research International, 2017*(1), 8475460. https://doi.org/10.1155/2017/8475460

Ghasemi, F., Rastegar, A., Jahromi, R. G., & Marvdashti, R. R. (2011). The relationship between creativity and achievement motivation with high school students' entrepreneurship. *Procedia-Social and Behavioral Sciences, 30*, 1291–1296. https://doi.org/10.1016/j.sbspro.2011.10.250

Gibb, A. (1993). The enterprise culture and education: Understanding enterprise education and its links with small business, entrepreneurship and wider educational goals. *International Small Business Journal, 11*, 11–34. https://doi.org/10.1177/026624269301100301

Gubenko, A., Kirsch, C., Smilek, J. N., Lubart, T., & Houssemand, C. (2021). Educational robotics and robot creativity: An interdisciplinary dialogue. *Frontiers in Robotics and AI, 8*, 662030. https://doi.org/10.3389/2Ffrobt.2021.662030

Guven, G., & Kozcu Cakir, N. (2020). Investigation of the opinions of teachers who received in-service training for Arduino-assisted robotic coding applications. *Educational Policy Analysis and Strategic Research, 15*(1), 253–274. https://doi.org/10.29329/epasr.2020.236.14

Guven, G., Kozcu Cakir, N., Sulun, Y., Cetin, G., & Guven, E. (2022). Arduino-assisted robotics coding applications integrated into the 5E learning model in science teaching. *Journal of Research on Technology in Education, 54*(1), 108–126. https://doi.org/10.1080/15391523.2020.1812136

Hamilton, M., Clarke-Midura, J., Shumway, J. F., & Lee, V. R. (2020). An emerging technology report on computational toys in early childhood. *Technology, Knowledge and Learning, 25*(1), 213–224. https://doi.org/10.1007/s10758-019-09423-8

Hindle, K., & Rushworth, S. (2000). *Global entrepreneurship monitor*. Swinburne University of Technology.

Hou, H., Zhang, X., & Wang, D. (2022). Can educational robots improve student creativity: An meta-analysis based on 48 experimental and quasi-experimental studies. *Journal of East China Normal University (Educational Sciences), 40*(3), 99. https://doi.org/10.15354/bece.22.ab001

Hynes, M., Portsmore, M., Dare, E., Milto, E., Rogers, C., Hammer, D., & Carberry, A. (2011). Infusing engineering design into high school STEM courses. *National Center for Engineering and Technology Education*. Retrieved December 15, 2014, from http://www.ncete.org.

Ibrahim, A. B., & Goodwin, J. R. (1986). Perceived causes of success in small business. *American Journal of Small Business, 11*(2), 41–50. https://doi.org/10.1177/104225878601100204

İbrahim, N. A., & Lucky, E. O. I. (2014). Relationship between entrepreneurial orientation, entrepreneurial skills, environmental factor and entrepreneurial intention among Nigerian students in UUM. *Entrepreneurship and Innovation Management Journal, 2*(4), 203–213.

Ilievski, F., Mazzeo, A. D., Shepherd, R. F., Chen, X., & Whitesides, G. M. (2011). Soft robotics for chemists. *Angewandte Chemie, 123*(8), 1930–1935. https://doi.org/10.1002/ange.201006464

Ioannou, A., & Makridou, E. (2018). Exploring the potentials of educational robotics in the development of computational thinking: A summary of current research and practical proposal for future work. *Education and Information Technologies, 23*(6), 2531–2544. https://doi.org/10.1007/s10639-018-9729-z

Jara, C. A., Candelas, F. A., Puente, S. T., Pomares, J., & Torres, F. (2010). Practical experiences using RobUALab. EJS: A virtual and remote laboratory for robotics e-learning. *IFAC Proceedings Volumes, 42*(24), 1–6. https://doi.org/10.3390/s130202595

Jawaid, I., Javed, M. Y., Jaffery, M. H., Akram, A., Safder, U., & Hassan, S. (2020). Robotic system education for young children by collaborative-project-based learning. *Computer Applications in Engineering Education, 28*(1), 178–192. https://doi.org/10.1016/2Fj.heliyon.2024.e24278

Jhurree, V. (2005). Technology integration in education in developing countries: Guidelines to policy makers. *International Education Journal, 6*(4), 467–483.

June, S., Yaacob, A., & Kheng, Y. K. (2014). Assessing the use of YouTube videos and interactive activities as a critical thinking stimulator for tertiary students: An action research. *International Education Studies, 7*(8), 56–67.

Kandlhofer, M., & Steinbauer, G. (2016). Evaluating the impact of educational robotics on pupils' technical-and social-skills and science related attitudes. *Robotics and Autonomous Systems, 75*, 679–685. https://doi.org/10.1016/j.robot.2015.09.007

Karahoca, D., Karahoca, A., & Uzunboylub, H. (2011). Robotics teaching in primary school education by project based learning for supporting science and technology courses. *Procedia Computer Science, 3*, 1425–1431. https://doi.org/10.1016/j.procs.2011.01.025

Karypi, S. (2018). Educational robotics application in primary and secondary education: A challenge for the Greek teachers society. *Journal of Contemporary Education Theory & Research*, 2. https://doi.org/10.5281/zenodo.3598423

Khanlari, A. (2013). Effects of robotics on 21st century skills. *European Scientific Journal, ESJ, 9*(27), 26–36. https://doi.org/10.19044/esj.2013.v9n27p/25p

Kissi, E., Ahadzie, D. K., Debrah, C., & Adjei-Kumi, T. (2020). Underlying strategies for improving entrepreneurial skills development of technical and vocational students in developing countries: Using Ghana as a case study. *Education + Training, 62*(5), 599–614. https://doi.org/10.1108/ET-11-2019-0264

Kozubíková, L., Belás, J., Bilan, Y., & Bartoš, P. (2015). Personal characteristics of entrepreneurs in the context of perception and management of business risk in the SME segment. *Economics and Sociology, 8*(1), 41–54. https://doi.org/10.14254/2071-789x.2015/8-1/4

Kuratko, D. F. (2005). The emergence of entrepreneurship education: Development, trends, and challenges. *Entrepreneurship Theory & Practice, 29*, 577–598. https://doi.org/10.1111/j.1540-6520.2005.00099.x

Küçük, S., & Şişman, B. (2017). Birebir robotik öğretiminde öğreticilerin deneyimleri [Instructor experiences in one-to-one robotics instruction]. *İlköğretim Online, 16*(1). https://doi.org/10.17051/io.2017.12092

Lepuschitz, W., Koppensteiner, G., Leeb-Bracher, U., Hollnsteiner, K., & Merdan, M. (2018). *Educational practices for improvement of entrepreneurial skills at secondary school level.* https://doi.org/10.3991/ijep.v8i2.8141

Lindquist, M. J., Sol, J., & Van Praag, M. (2015). Why do entrepreneurial parents have entrepreneurial children? *Journal of Labor Economics, 33*(2), 269–296. https://doi.org/10.2139/ssrn.2101543

Littunen, H. (2000). Entrepreneurship and the characteristics of the entrepreneurial personality. *International Journal of Entrepreneurial Behavior & Research, 6*(6), 295–310. https://doi.org/10.1108/13552550010362741

Loi, M., Fayolle, A., van Gelderen, M., Riot, E., Refai, D., Higgins, D., Haloub, R., Salusse, M. A. Y., Lamy, E., Verzat, C., & Cavarretta, F. (2022). Entrepreneurship education at the crossroads: Challenging taken-for-granted assumptions and opening new perspectives. *Journal of Management Inquiry, 31*(2), 123–134. https://doi.org/10.1177/10564926211042222

Lopez-Caudana, E., Ramirez-Montoya, M. S., Martínez-Pérez, S., & Rodríguez-Abitia, G. (2020). Using robotics to enhance active learning in mathematics: A multi-scenario study. *Mathematics, 8*(12), 2163. https://doi.org/10.3390/math8122163

López-Núñez, M. I., Rubio-Valdehita, S., Armuña, C., & Pérez-Urria, E. (2022). EntreComp questionnaire: A self-assessment tool for entrepreneurship competencies. *Sustainability, 14*(5), 2983. https://doi.org/10.3390/su14052983

Louis, R. C. (2012). *A case study exploring technology integration and incorporation of 21st century skills in elementary classrooms*. College of Professional Studies Northeastern University. https://doi.org/10.17760/d20002963

Lui, M., & Hsiao, Y. (2002). Middle school students as multimedia designers: A project-based learning approach. *Journal of Interactive Learning Research, 13*(4), 311–337.

Mallinson, A. (2015). *A mechanical engineering rite of passage at MIT*. MIT News.

Maloney, J., Resnick, M., Rusk, N., Silverman, B., & Eastmond, E. (2010). The scratch programming language and environment. *ACM Transactions on Computing Education (TOCE), 10*(4), 1–15. https://doi.org/10.1145/1868358.1868363

Manera, L. (2020). STEAM and educational robotics: interdisciplinary approaches to robotics in early childhood and primary education. In *Human-friendly robotics 2019: 12th international workshop* (pp. 103–109). Springer International Publishing. https://doi.org/10.1007/978-3-030-42026-08

Martin, R. L., & Osberg, S. (2007). *Social entrepreneurship: The case for definition.* https://doi.org/10.4236/jss.2018.610002

MEB, (2018a). Fen bilimleri dersi öğretim programı, (İlkokul ve Ortaokul 3, 4, 5, 6, 7 ve 8. Sınıflar). *Ankara*: Talim ve Terbiye Kurulu Başkanlığı.

Meral, M., & Yalçın, S. A. (2022). The investigation of middle school students' entrepreneurial skills in terms of Entrepreneurship-Based STEM Education: A mixed method study. *Necatibey Eğitim Fakültesi Elektronik Fen ve Matematik Eğitimi Dergisi, 16*(2), 326–353. https://doi.org/10.17522/balikesirnef.1133770

Mikropoulos, T. A., & Bellou, I. (2013). Educational robotics as mindtools. *Themes in Science and Technology Education, 6*(1), 5–14.

Nabi, G., Liñán, F., Fayolle, A., Krueger, N., & Walmsley, A. (2017). The impact of entrepreneurship education in higher education: A systematic review and research agenda. *Academy of Management Learning & Education, 16*(2), 277–299. https://doi.org/10.5465/amle.2015.0026

Ngwoke, A. N., Aneke, A. O., & Oraelosi, C. C. (2020). Developing entrepreneurial mindset in the Nigerian child through quality early childhood education. *International Journal on Studies in Education, 16*, 162–174.

Ortiz, A. M., Bos, B., & Smith, S. (2015). The power of educational robotics as an integrated STEM learning experience in teacher preparation programs. *Journal of College Science Teaching, 44*(5), 42–47. https://doi.org/10.2505/4/jcst15_044_05_42

Ospennikova, E., Ershov, M., & Iljin, I. (2015). Educational robotics as an innovative educational technology. *Procedia-Social and Behavioral Sciences, 214*, 18–26. https://doi.org/10.1016/j.sbspro.2015.11.588

Özcan, E. (2019). Effects of problem based learning on prospective science teachers' problem solving skills, academic achievements and attitudes (Doctoral thesis, YÖK National Thesis Center).

Pan, V. L., & Akay, C. (2015). Eğitim fakültesi öğrencilerinin girişimcilik düzeylerinin çeşitli değişikenler açısından incelenmesi [Examining teacher candidates' entrepreneurship levels in terms of various variables]. *NWSA-Education Sciences, 9*(6), 125–138. https://doi.org/10.12739/NWSA.2015.10.2.1C0637

Paço, A. D., & Palinhas, M. J. (2011). Teaching entrepreneurship to children: A case study. *Journal of Vocational Education & Training, 63*(4), 593–608. https://doi.org/10.1080/13636820.2011.609317

Patiño-Escarcina, R. E., Barrios-Aranibar, D., Bernedo-Flores, L. S., Alsina, P. J., & Gonçalves, L. M. (2021). A methodological approach to the learning of robotics with EDUROSC-kids. *Journal of Intelligent & Robotic Systems, 102*(2), 34. https://doi.org/10.1007/s10846-021-01400-7

Pfeifer, S., Oberman–Peterka, S. & Jeger, M. (2008). Assessing entrepreneurship education programmes in Croatian higher educationarea. *Erenet Profile, 3*(3), 25–35.

Piedade, J. M. N. (2021). Pre-service and in-service teachers' interest, knowledge, and self-confidence in using educational robotics in learning activities. *Educação & Formação, 6*(1). https://doi.org/10.25053/redufor.v6i1.410

Prahani, B., Alfin, J., Fuad, A., Saphira, H., Hariyono, E., & Suprapto, N. (2022). Learning management system (LMS) research during 1991–2021: How technology affects education. *International Journal of Emerging Technologies in Learning (iJET), 17*(17), 28–49. https://doi.org/10.3991/ijet.v17i17.30763

Riasati, M. J., Allahyar, N., & Tan, K. E. (2012). Technology in language education: Benefits and barriers. *Journal of Education and Practice, 3*(5), 25–30. https://doi.org/10.7176/JEP/3-5-5

Robert, M. D. (2009). *Developing 21st century skills through the use of student personal learning networks*. Northcentral University.

Rogers, E., Owens, D. H., Werner, H., Freeman, C. T., Lewin, P. L., Kichhoff, S., Schmidt, C., & Lichtenberg, G. (2010). Norm optimal iterative learning control with application to problems in accelerator based free electron lasers and rehabilitation robotics. *European Journal of Control, 16*(5), 497–524.

Ruskovaara, E., Pihkala, T., Rytkölä, T., & Seikkula-Leino, J. (2010). Studying teachers' teaching methods and working approaches in entrepreneurship education. In *Proceedings of the 7th ESU conference, 14–18 July 2014, Denmark*.

Sabzeh, B. (2022). Creative playes are the basis for the formation of entrepreneurial attitudes in children. *Educational and Scholastic Studies, 10*(4), 495–523.

Schina, D., Valls-Bautista, C., Borrull-Riera, A., Usart, M., & Esteve-González, V. (2021). An associational study: Preschool teachers' acceptance and self-efficacy towards educational robotics in a pre-service teacher training program. *International Journal of Educational Technology in Higher Education, 18*(1), 1–20. https://doi.org/10.1186/s41239-021-00264-z

Seckin-Kapucu, M. (2023). Studies on robotic coding education in science education: A systematic literature review. *Journal of Education in Science Environment and Health, 9*(1), 74–84. https://doi.org/10.55549/jeseh.1239093

Seikkula-Leino, J. (2011). The implementation of entrepreneurship education through curriculum reform in Finnish comprehensive schools. *Journal of Curriculum Studies, 43*(1), 69–85. https://doi.org/10.1080/00220272.2010.515024

Sexton, D. L. (2001). Wayne Huizenga: entrepreneur and wealth creator. *Academy of Management Perspectives, 15*(1), 40–48. https://doi.org/10.5465/ame.2001.4251396

Shahiwala, A. (2017). Entrepreneurship skills development through project-based activity in Bachelor of Pharmacy program. *Currents in Pharmacy Teaching and Learning, 9*(4), 698–706. https://doi.org/10.1016/j.cptl.2017.03.027

Sidik, I. G. (2012). Conceptual framework of factors affecting SME development: Mediating factors on the relationship of entrepreneur traits and SME performance. *Procedia Economics and Finance, 4*, 373–383. https://doi.org/10.1016/S2212-5671(12)00351-600351-6)

Silander, D., & Silander, C. (2018). The European Commission: The EU as agenda-setter for economic growth and entrepreneurship. *In Governance and Political Entrepreneurship in Europe*, (pp. 57–81). Edward Elgar Publishing.

Sirelkhatim, F., & Gangi, Y. (2015). Entrepreneurship education: A systematic literature review of curricula contents and teaching methods. *Cogent Business & Management, 2*(1), 1052034. https://doi.org/10.1080/23311975.2015.1052034

Sisman, B., & Kucuk, S. (2019). An educational robotics course: Examination of educational potentials and pre-service teachers' experiences. *International Journal of Research in Education and Science, 5*(2), 510–531.

Smith, D. P. (2013). Characteristics of innovative entrepreneurs: An analysis at the level of the individual, the firm, and the business environment. *New Visions for Public Affairs, 5*, 25–39.

Su, Y., Zhu, Z., Chen, J., Jin, Y., Wang, T., Lin, C. L., & Xu, D. (2021). Factors influencing entrepreneurial intention of university students in China: Integrating the perceived university support and theory of planned behavior. *Sustainability, 13*(8), 4519. https://doi.org/10.3390/su13084519

Sullivan, A., & Bers, M. U. (2019). VEX robotics competitions: Gender differences in student attitudes and experiences. *Journal of Information Technology Education, 18.*

Tekbiyik, A., Baran Bulut, D., & Sandalci, Y. (2022). Effects of a summer robotics camp on students' STEM career interest and knowledge structure. *Journal of Pedagogical Research, 6*(2), 91–109.

Tzagkaraki, E., Papadakis, S., & Kalogiannakis, M. (2021). Exploring the use of educational robotics in primary school and its possible place in the curricula. In *Educational robotics international conference* (pp. 216–229). Springer International Publishing. https://doi.org/10.1007/978-3-030-77022-8_19

Üçgül, M. (2013). History and educational potential of Lego Mindstorms NXT. *Mersin Üniversitesi Eğitim Fakültesi Dergisi, 9*(2), 127–137.

Vaidya, S. (2007). *Developing entrepreneurial life skills: An experiment in Indian schools.* Institute for Small Business and Entrepreneurship.

Vuong, B. N., Duy Phuong, N. N., Huan, D. D., & Quan, T. N. (2020). A model of factors affecting entrepreneurial intention among information technology students in Vietnam. *Journal of Asian Finance, Economics and Business, 7*(8), 461–472. https://doi.org/10.13106/jafeb.2020.vol7.no8.461

Wang, C., Mundorf, N., & Salzarulo-McGuigan, A. (2022). Entrepreneurship education enhances entrepreneurial creativity: The mediating role of entrepreneurial inspiration. *The International Journal of Management Education, 20*(2), 100570. https://doi.org/10.1016/j.ijme.2022.100570

Wijaya, S., & Puspitowati, I. (2022, May). The effect of personal qualities, family environment, attitudes, subjective norms and self-efficacy on entrepreneurial intentions. In *Tenth international conference on entrepreneurship and business management 2021 (ICEBM 2021)* (pp. 323–329). Atlantis Press. https://doi.org/10.2991/aebmr.k.220501.049

You, H. S., Chacko, S. M., & Kapila, V. (2021). Examining the effectiveness of a professional development program: Integration of educational robotics into science and mathematics curricula. *Journal of Science Education and Technology, 30*, 567–581. https://doi.org/10.1007/s10956-021-09900-2

Yusoff, M. A. K., Samin, R. E., & Ibrahim, B. S. K. (2012). Wireless mobile robotic arm. *Procedia Engineering, 41*, 1072–1078. https://doi.org/10.1016/j.proeng.2012.07.273

Yüksel, H., Cevher, E., & Yüksel, M. (2015). Öğrencilerin girişimci kişilik özellikleri ile girişimcilik eğilimleri üzerine bir araştırma [A research on the inclinations of entrepreneurship and entrepreneurship personality characteristics of students]. *Çankırı Karatekin Üniversitesi İktisadi ve İdari Bilimler Fakültesi Dergisi, 5*(1), 143–156.

The manufacturer's authorised representative in the EU is Springer Nature Customer Service Centre GmbH, Europaplatz 3, 69115 Heidelberg, Germany. If you have any concerns regarding our products, please contact ProductSafety@springernature.com

Printed and bound by CPI Group (UK) Ltd, Croydon, CR0 4YY
01/10/2026
02225956-0019